Biomedical Ethics and the Law

Biomedical Ethics and the Law

Edited by

James M. Humber and
Robert F. Almeder
Georgia State University, Atlanta

PLENUM PRESS • NEW YORK AND LONDON

Library of Congress Cataloging in Publication Data

Main entry under title:

Biomedical ethics and the law.

Includes bibliographies and index.
1. Medical laws and legislation – Addresses, essays, lectures. 2. Medical ethics – Addresses, essays, lectures. I. Humber, James M. II. Almeder, Robert F. [DNLM: 1. Ethics, Medical – Collected works. 2. Jurisprudence – Collected works. W50 B615]
Law 179'.7 76-12495

ISBN 978-1-4684-2225-2 ISBN 978-1-4684-2223-8 (eBook)
DOI 10.1007/978-1-4684-2223-8

First Printing - July 1976
Second Printing - July 1977

Softcover reprint of the hardcover 1st edition 1976
A Division of Plenum Publishing Corporation
227 West 17th Street, New York, N.Y. 10011

PREFACE

In the past few years an increasing number of colleges and universities have added courses in biomedical ethics to their curricula. To some extent, these additions serve to satisfy student demands for "relevance." But it is also true that such changes reflect a deepening desire on the part of the academic community to deal effectively with a host of problems which must be solved if we are to have a health-care delivery system which is efficient, humane, and just. To a large degree, these problems are the unique result of both rapidly changing moral values and dramatic advances in biomedical technology.

The past decade has witnessed sudden and conspicuous controversy over the morality and legality of new practices relating to abortion, therapy for the mentally ill, experimentation using human subjects, forms of genetic intervention, suicide, and euthanasia. Malpractice suits abound and astronomical fees for malpractice insurance threaten the very possibility of medical and health-care practice. Without the backing of a clear moral consensus, the law is frequently forced into resolving these conflicts only to see the moral issues involved still hotly debated and the validity of existing law further questioned. In the case of abortion, for example, the laws have changed radically, and the widely publicized recent conviction of Dr. Edelin in Boston has done little to foster a moral consensus or even render the exact status of the law beyond reasonable question. To take another example, only recently have we seen the veil of secrecy lifted in the area of experimenting with human subjects. Indeed, as this is being written, well-known and respected agencies of our government are being publicly scrutinized and severely criticized for experimenting with drugs on human subjects without the informed consent of the subjects. And similar examples abound in areas involving forms of genetic intervention, therapy for the mentally-ill, and practices relating to euthanasia and proper care for the terminally ill.

As profound as the social and moral changes in this country have been in the recent past, they are outstripped by recent scientific and technological advances in the biomedical field. Deformed and mentally disabled children who once would have died at birth can now be kept alive. Should such beings be allowed to live? Like problems also arise at the other end of the life cycle. Should a person be kept alive even if he wishes a peaceful and "dignified" death? And when is a person dead anyway? With the development of the heart-lung machine and intravenous feeding, a person's body can be kept functioning indefinitely, long after his brain has ceased showing any activity. Does one die when his brain dies? If not, when does one die? If so, how does one determine when the brain is dead? And then there are the moral and legal problems arising from recent advances in genetics. Take cloning, for example. Should we reproduce our geniuses and attempt to breed a superior race? We do it with tomatoes and livestock, so why not with humans? Who is to make such decisions and what standards are to be employed? In a milieu of shifting moral values, the conceptual, moral, and legal problems generated by advances in scientific and medical technology are all the more perplexing and worthy of urgent attention.

The problems faced by medical practitioners, researchers, and geneticists raise questions having social, moral, legal, philosophical, and theological implications. In a very real sense, then, the field of biomedical ethics is essentially interdisciplinary; and in constructing this anthology we have done our best to represent the interdisciplinary character of these problems.

The book begins with an introduction by Professor Daniel Callahan, who discusses the area of bioethics as a discipline. The text itself is divided into five parts, each part being prefaced with a short introduction in which the theses of the various authors are briefly stated. Our division of the subject matter is not hard and fast, and the student will no doubt perceive that the problems dealt with in each part are interrelated.

It is impossible to put together an anthology of this sort without the aid and assistance of many. However, we are especially indebted to our graduate student, Dexter Christian, for compiling the bibliographies at the end of each chapter, and to Mr. Thomas Lanigan of Plenum Press for his insightful suggestions and encouragement. Special thanks also go to our wives, Lynn and Virginia, whose patience and understanding make our work considerably easier than it might otherwise be.

Georgia State University J.M.H.
R.F.A.

CONTRIBUTORS

Professor Henry Aiken
110 Deacon Haynes Road
Concord, Mass. 01742

David Bazelon, Chief Judge
United States Court of Appeals
Washington, D.C. 20001

Professor Thomas Beauchamp
Philosophy Department
Georgetown University
Washington, D.C. 20057

Dr. Henry K. Beecher
Harvard Medical School
10 Shattuck Street
Boston, Mass. 02115

Professor Richard Brandt
Philosophy Department
The University of Michigan
Ann Arbor, Mich. 48104

Professor Baruch Brody
Department of Philosophy
Rice University
Houston, Texas 77001

Dr. Daniel Callahan
Institute of Society,
Ethics, & Life Sciences
Hastings Center
623 Warburton Ave.
Hastings-on-Hudson, N.Y. 10706

Professor Alexander M. Capron
University of Pennsylvania
3400 Chestnut St.
Philadelphia, Pa. 19174

Professor Arthur J. Dyck
The Divinity School
Harvard University
Cambridge, Mass. 02138

Dr. Leon Eisenberg
Harvard Medical School
Department of Psychiatry
Children's Hospital Medical Center
300 Longwood Ave.
Boston, Mass. 02115

Rev. John C. Fletcher
Inter/Met Theological Education
1419 V. St., N.W.
Washington, D.C. 20009

Dr. Kurt Hirschhorn
Mount Sinai School of Medicine
The City University of New York
Fifth Ave. and 100th St.
New York, N.Y. 10029

Professor James Humber
Department of Philosophy
Georgia State University
University Plaza
Atlanta, Ga. 30303

Professor Hans Jonas
Department of Philosophy
New School for Social Research
66 West 12th St.
New York, N.Y. 10011

Dr. Leon R. Kass
The Joseph & Rose Kennedy Institute
Center for Bioethics
Georgetown University
Washington, D.C. 20007

Dr. Herbert Lubs
Department of Pediatrics
Univ. of Colorado Medical Center
4200 East Ninth Avenue
Denver, Col. 80220

Professor Barbara MacKinnon
Department of Philosophy
University Center, Room 538
University of San Francisco
San Francisco, Cal. 94117

Professor Ruth Macklin
Philosophy Department
Case Western Reserve University
Cleveland, Ohio 44106

Dr. Robert Morison
Box 277
Peterborough, N.H. 03458

Professor Nicholas Rescher
Department of Philosophy
University of Pittsburgh
Pittsburgh, Pa. 15213

Professor Herbert W. Richardson
St. Michael's College
University of Toronto
Toronto 5, Canada

Professor Jonas Robitscher
School of Law
Emory University
Atlanta, Ga. 30322

Dr. Michael Shimkin
Department of Community Medicine
School of Medicine
University of California at San Diego
LaJolla, Cal. 92037

Professor Tracy Sonneborn
Department of Zoology
Jordan Hall–224
Indiana University
Bloomington, Ind. 47401

Dr. Thomas Szasz
Department of Psychiatry
Upstate Medical Center
750 East Adams St.
Syracuse, N.Y. 13210

Professor Judith J. Thomson
Department of Philosophy
Massachusetts Institute of Technology
Cambridge, Mass. 02139

Professor Lawrence P. Ulrich
Philosophy Department
University of Dayton
Dayton, Ohio 45469

CONTENTS

Biomedical Ethics and the Law

BIOETHICS AS A DISCIPLINE

DANIEL CALLAHAN

One of the beguiling phrases I have picked up from reading scientific and medical journals is "anecdotal evidence." The careful researcher does not claim too much for evidence of that kind; he knows its scientific limitations. I must confess to the perversity of often finding evidence of that sort more suggestive than the solid, well-confirmed kind. It is surely far more interesting to collect. The topic of "bioethics as a discipline" invites a good deal of anecdotal evidence, and I intend to indulge fully in the pleasures of retailing my own experiences. For the sake of decency, I will maintain the confidentiality of my clinical records, shielding the names of the innocent and the guilty.

The topic also invites reflection on what I like to think of as "the politics of ethics." By that I mean the problem of taking the probings of professional philosophers and theologians and getting someone other than our lodge brothers and sisters to think that anything is being said at all. Or, in some richer sense of the word "politics," showing that serious ethical thinking has its place in the body politic of medicine and biology.

Finally, the topic invites at least some allusion to the concept of a "discipline," and particularly the place of disciplines in the academic world. The graduate school catalogue sense of the word denotes specific training, refined methodologies, distinctive approaches and commitments, a long apprenticeship, professional expertise. That is a flattering picture. My own connotation, having seen academic disciplines in action, is too often arrogance, insulation, neurosis and narrowness. Can bioethics as a discipline avoid that hazard?

I will begin with three anecdotes. The first was a criticism leveled at a book I had written on the ethics of abortion. It came from a passionate feminist, strongly pro-abortion, who on the whole liked the book and its conclusions. "But what *right* have you," she shouted at me, "to press all your heavy philosophical questions on people who aren't trained philosophers? What arrogance and cruelty! You should be trying to help women, not loading them down with a lot of hard intellectual problems which will just make them suffer all the more. You claim to be interested in ethics. Why don't you practice it!"

Reprinted from Daniel Callahan, "Bioethics as a Discipline," *Hastings Center Studies, 1,* No. 1 (1973), pp. 66–73, by permission of the author and the Institute of Society, Ethics and the Life Sciences, Hastings-on-Hudson, New York.

The second anecdote comes out of a seminar I once led on ethics and population control. The seminar members were professional sociologists and demographers. On one occasion, after some weeks of discussion, I was upbraided by a demographer for not making a precise enough distinction between "birthrates," "crude birthrates," and "fertility rates." "You philosophers," he gently chided me, "have just got to master the demographic literature and the technical distinctions if you're going to be any help to us." Chastened, I continued my lecture, which that day happened to be on the various philosophical distinctions between act- and rule-utilitarianism. After I had gone on a bit about that, my critic came back at me: "I'm afraid I'm finding this all very boring. You philosophers do nothing but make technical distinctions and split verbal hairs; and all that stuff about the different kinds of utilitarianism is too abstract to be of any use." As a footnote to that story, I might also note that the seminar participants diligently read all of the assigned readings on demography. But, with the exception of one young, unwashed research assistant, they hardly read any of the assigned reading in philosophical ethics; too tedious and irrelevant, they said. Even so, at the end of the seminar, one criticism I received was that I did not give enough time to ethics. "I wish we had been able to get more deeply into the problems," two or three people commented.

My third anecdote is this. Some colleagues and I had spent some months trying to convince a group of physicians that a good training in medicine did not necessarily qualify them to make good ethical decisions. They eventually conceded the point. But they conceded it with a twist of the knife. One day they came to us with a particularly agonizing case in hand, requiring a quick black-and-white decision. "What should we do?" they asked. "You're the philosopher, you tell us." The case was appalling, and I mumbled something about not being really qualified to tell them what to do. "But you've been telling us that, as physicians, we have no special qualifications for making ethical decisions. And now you're saying you don't either—even though you have a Ph.D. in philosophy and spend all of your time studying medical ethics. Just who *is* qualified to decide?"

I have painfully remembered each of these incidents because they well illustrate some of the fundamental problems of bioethics; and each of them illustrates more than one problem.

The first anecdote, involving my feminist critic, suggests three issues to me. The most obvious involves the hazards of pushing difficult questions on people who would prefer not to think of them; Socrates found out what that leads to. Another issue is the image of the ethicist as someone addicted to raising the philosophical ante, spinning out from one problem ten more, and then ten from each of those ten, the whole process pushing forward into every basic question about the meaning of life and existence; this can be called philosophical overkill. Still another issue is the exasperation ethicists arouse in those already committed to an answer to a specific moral problem, those for whom the intellectual probing is over and dedication to propagating the cause is the only item left on the agenda. The professional diffidence of the ethicist about seeing himself in the role of the advocate (I am less certain about the theologian) can only seem an evasion or a failure to take ethics seriously enough.

The second anecdote—my experience with the demographers—raises another series of issues. One is the very low tolerance level of most people, however, highly educated, for philosophical theories and niceties; their eyes quickly glaze over. "If I had wanted to be a philosopher," one biologist once told me, "I would have gone to graduate school in philosophy." This attitude, however, is quite apt to coexist with what certainly appears to be a *desire* on the part of these same critics to wrestle very hard with the issues. It's just that they don't find the characteristic way philosophers and theologians approach them very meaningful; which is a nice way of saying they really think it is mainly hot air. There is another way of putting this point. While everyone will agree in principle that there should be reciprocity in the interdisciplinary work of bioethics, it is utterly naive for the philosopher or theologian to think that many scientists and physicians will rush back to Plato, much less G.E. Moore.

The third anecdote pushes us into what is the underlying issue in all three of the stories. Just what is the role of the ethicist in trying to make a contribution to the ethical problems of medicine, biology, or population? I resisted, with utter panic, the idea of participating with the physicians in their actual decision. Who *me?* I much preferred the safety of the profound questions I pushed on them. But I also realized when faced with an actual case—and this is my excuse—that there was nothing whatever in my philosophical training which had prepared me to make a flat, clear-cut ethical decision at a given hour on a given afternoon. I had been duly trained in that splendid tradition of good scholarship and careful thinking which allows at least a couple thousand years to work through any problem. The *Review of Metaphysics* ran a contest some years ago, offering a prize to the person coming up with the best answer to the question: "Why can't philosophers make up their minds?" It is still a good question, and of course I haven't been able to make up my mind about the right answer. The propensity of physicians and lawyers to prefer the case-method, and for philosophers and theologians to offer no more than one concrete example every forty pages or so is just another side of the same coin.

Place of the Ethicist

Let me take up the question which I said was underlying my three anecdotes: what is the place of the ethicist in medicine and biology? Does he have anything to contribute and, if so, what? The answer at this point is that we just do not exactly know. But the fact that we may not know yet with any precision what the contribution can be, provides a splendid opportunity to shape the discipline in a way which might—just might—make an important difference in the long run.

Bioethics is not yet a full discipline. Most of its practitioners have wandered into the field from somewhere else, more or less inventing it as they go. Its vague and problematic status in philosophy and theology is matched by its even more shaky standing in the life sciences. The lack of general acceptance, disciplinary standards, criteria of excellence and clear pedagogical and evaluative norms provides, however, some unparalleled opportunities. It is a discipline not yet burdened by encrusted traditions and domineering figures. Its saving grace is that it is not yet a genuine discipline as that concept is usually understood in the academic and scientific communities. One has always to explain oneself, and that leaves room for creativity and constant redefinition; there are many advantages in being a moving target.

When we ask what the place of bioethics might be, we of course need to know just what the problems are in medicine and biology which raise ethical questions and need ethical answers. I will not retail the whole catalogue of issues here; suffice it to say that they begin with "A" (abortion and amniocentesis) and run all the way to "Z" (the moral significance of zygotes). One evident and first task for the ethicist is simply that of trying to point out and define which problems raise moral issues. A second and no less evident task is providing some systematic means of thinking about, and thinking through, the moral issues which have been discerned. A third, and by far the most difficult, task is that of helping scientists and physicians to make the right decisions; and that requires a willingness to accept the realities of most medical and much scientific life, that is, that at some discrete point in time all the talk has to end and a choice must be made, a choice which had best be right rather than wrong.

Not one of these tasks is easy, and one soon learns that all of the smart cracks one can direct at the dear innocent scientist who still thinks there is such a beast as a "value-free methodology" apply equally well to the methodologies of theology and philosophy. Feet will be wet before feet are even in the water. Moreover, one also soon discovers that well over half the in-fighting, and three-quarters of all punches to the groin, stem from arguments about whether such-and-such an issue actually poses any ethical dilemma; or whether what is *thought* to be the dilemma *is* actually the dilemma. In the battle of ethics, no less than in other forms of human warfare, the high ground goes to those who succeed in establishing their definitions of the issues-at-stake and their stipulations about what will count as a sound methodology; the rest is mainly a mopping-up operation.

I used above the phrase "the realities of life." Another one of these realities is that the ethical issues of medicine and biology rarely present themselves in a way nicely designed to fit the kinds of categories and processes of thought which philosophers and theologians traditionally feel secure about. They almost always start off on the wrong foot by coming encumbered with the technical jargon of some other discipline. And only in textbooks is one likely to encounter cases which present a clear occasion, say, for deciding on the validity of a deontological or utilitarian ethical solution. The issues come, that is, in a messy, jumbled form, cutting through many disciplines, gumming up all our clean theoretical engines, festooned with odd steamers and complicated knots.

The fact that this is the case immediately invites the temptation of what can be called "disciplinary reductionism." By that I mean a penchant for distilling out of an essentially complex ethical problem one transcendent issue which is promptly labeled *the* issue. Not coincidentally, this issue usually turns out to be a classic, familiar argument in philosophy or theology. By means of this kind of reductionism, the philosopher or theologian is thus enabled to do what he has been trained to do, deal with those classic disputes in a language and a way he is comfortable with—in a way which allows him to feel he is being a good "professional." The results of this tendency are doleful. It is one reason why most biologists and physicians find the contributions of the professional ethicist of only slight value. Their problems, very real to them in their language and their frame of reference, are promptly made unreal by being transmuted into someone else's language and reference system, in the process usually stripping the original case of all the complex facticity with which it actually presented itself. The whole business becomes positively pitiable when the philosopher or theologian, rebuffed or ignored because of this reductionism, can only respond by charging that his critics are obviously "not serious" about ethics, not interested in "real" ethical thinking.

I stress the problem of "disciplinary reductionism" out of a conviction that if a discipline of bioethics is to be created, it must be created in a way which does not allow this form of evading responsibility, of blaming the students for the faults of the teacher, of changing the nature of the problems to suit the methodologies of professional ethicists.

Toward this end, no subject would seem to me more worthy of investigation than what I will call the "ordinary language of moral thinking and discourse." Most people do not talk about their ethical problems in the language of philosophers. And I have yet to meet one professional ethicist who, when dealing with his own personal moral dilemmas, talks the language of his professional writings; he talks like everyone else, and presumably he is thinking through his own problems in banal everyday language like everyone else. Now of course it might be said that this misses the whole point of a serious professional discipline. Is it not like claiming that there must be nothing to theoretical physics simply because the physicist does not talk about the furniture in his house in terms of molecules and electrons? But the analogy does not work, for it is of the essence of moral decision-making to be couched in ordinary language and dealt with by ordinary, non-professional modes of thinking. The reason for this is apparent. An ethical decision will not be satisfactory to the person whose decision it is unless it is compatible with the way in which the person ordinarily thinks about himself and what he takes his life to be.

My point here is by no means to deny the validity of ethical theory, the value of technical ethical language, or the need for disciplined rigor. Nor am I prepared to grant a narrow scientist or physician the slothful luxury of demanding that the ethicist master his technical distinctions while he himself is dispensed from mastering the ethicist's. The ethicist can make no useful contribution *at all* unless he is capable, on occasion, of moving more deeply into the issues than others do, of giving them a coherence and clarity which they may lack in the formulations of ordinary language, and of bringing to them a nuanced methodology. I am only trying to make the point that unless, in the process of doing so, he can remain in contact with ordinary language and ordinary thought processes—constantly establishing the connections—both his theory and his pedagogy will fail. His theory will fail because it will not bear on psychological reality, and his pedagogy will fail because it will have nothing to do with the way in which the ethical problems actually present themselves. The physician who, in trying to deal with a hard ethical case, keeps in mind his own feelings and thoughts, hospital policy and public policy, the emotions of his patient and his patient's family, trying to be responsible and yet faced with multiple and often conflicting responsibilities—that physician will *not* be responsive (nor should he) to the ethicist who tells him that, in the end, the real and only issue is, say, utilitarianism.

If I may put the matter in the form of a paradox, the ethicist may be quite correct in his theoretical analysis—perhaps utilitarianism is, say, the largest philosophical issue at stake in many ethical dilemmas. Yet he will be quite clearly wrong if he does not recognize that the issue in particular cases—Mrs. Jones in Ward 5 at 4:10 in the afternoon—must and will involve far more than the status of utilitarian theory.

I want to add an ancillary point. I once knew a distinguished philosopher who said he had little interest in meeting and talking with other philosophers. After all, he could read their writings and get a far more rigorous, honed version of their views from that source than he could from conversation with them. Perhaps there is something to this when it is a question of one professional trying to understand the views of another professional. But I believe it an absolutely disastrous approach in the case of the ethicist trying to understand the moral thinking of the non-ethicist, particularly that of the scientist or physician. There is often no guarantee whatever that the ethical *language* used by the physician or scientist has any bearing whatever on the way in which he actually makes ethical decisions. Too often it is perfectly clear, in fact, that the actual principles which are coming into play are quite at variance with the expressed principles. I am not talking about hypocrisy here, nor about garden-variety logical inconsistency, but about the far more subtle phenomenon of someone who, because that is the only language he has been equipped with for public discourse, talks one kind of ethical terminology and yet acts in ways which are far more consistent with quite another kind of terminology, which no one has supplied him with. The existence of this phenomenon places a great premium on a good deal of oral exchange and probing, as well as a premium on a certain skepticism toward thinking that what non-ethicists say is actually what they necessarily mean and do.

The Meaning of "Rigorous"

In trying to create the discipline of bioethics, the underlying question raised by the foregoing remarks bears on what it should mean to be "rigorous" and "serious" about bioethics. I recently talked with a very good philosopher who had spent a year team-teaching a course in ethical problems of biology with a biology professor. He did not want to repeat the experience. He found it impossible, he said, to introduce any real philosophical and ethical rigor into the course, not only because he neither could nor would ever have the time to properly master biology but also because the very process of trying to talk across disciplinary lines seemed inherently inimical to rigorous, methodologically sound thinking; it was like mixing apples and avocados, with an inedible result. One can sympathize with his feelings, and it is common enough for ethicists to gather among themselves after some frustrating interdisciplinary session to mutter about the denseness and inanity of their scientific and medical colleagues.

There are two options open here. One is to continue the muttering, being quite certain that the muttering is being reciprocated back in the scientific lab. That is, one can stick to traditional notions of philosophical and theological rigor, in which case one will rarely if ever encounter it in the interdisciplinary work of bioethics. Or, more wisely, the thought may occur that it is definitions of "rigor" which need adaptation. Not the adaptation of expediency or passivity in the face of careless thinking, but rather a perception that the kind of rigor required for bioethics may be of a different sort than that normally required for the traditional philosophical or scientific disciplines.

This is to say no more than that the methodological rigor should be appropriate to the subject matter. I spoke above of three tasks for the bioethicist: definition of issues, methodological strategies, and procedures for decision-making. Each of these tasks requires a different kind of rigor. The first requires what I will paradoxically call the rigor of an unfettered imagination, an ability to see in, through and under the surface appearance of things, to envision alternatives, to get under the skin of people's ethical agonies or ethical insensitivities, to look at things from many perspectives simultaneously.

A different kind of rigor is needed for the development of methodological strategies. Here the traditional methodologies of philosophy and theology are indispensable; there are standards of rigor which can and should come into play, bearing on logic, consistency, careful analysis of terms, and the like. Yet at the same time they have to be adapted to the subject matter at hand, and that subject matter is not normally, in concrete ethical cases of medicine and biology, one which can be stuffed into a too-rigidly structured methodological mold.

I am not about to attempt here a full discourse on what should be the proper and specific methodology of bioethics. Some sketchy, general comments will have to do, mainly in the way of assertions. Traditionally, the methodology of ethics has concerned itself with ethical thinking; how to think straight about ethical problems. However, I believe that the province of the bioethicist can legitimately encompass a concern with three areas of ethical activity: thinking, feeling (attitudes), and behavior. The case for including feelings and behavior along with thinking rests on the assumptions (1) that in life both feelings and behavior shape thinking, often helping to explain why defective arguments are nonetheless, for all that, persuasive and pervasive; and (2) that it is legitimate for an ethicist to worry about what people do and not just what they think and say; a passion for the good is not inappropriate for ethicists.

If ethics was nothing other than seeing to it that no logical fallacies were committed in the process of ethical argumentation, it would hardly be worthy of anyone's attention. It is the premises of ethical arguments, the visions behind ethical systems, the feelings which fuel ethical (or nonethical) behavior, which make the real difference for human life. Verbal formulations and arguments are only the tip of the iceberg. An ethicist can restrict himself to that tip; he will be on safe enough professional grounds if he does so. But I see no reason why he can't dare more than that, out of a recognition that the source and importance of his field lie not in the academy but in private and public human life, where what people think, feel, and do make all the difference there is.

Even in individual decision-making, the purely rational part may quite legitimately play only a subordinate role in some cases. Personal or group experience, for one thing, may provide reasonable, though non-articulatable, principles for conduct. Traditions, for another, may provide still other principles, even though they lie buried well below the surface of consciousness.

To be sure, there will always be the critical task for ethics of asking that reasons and justifications be given for principles derived from experience and traditions. Nonetheless, in the nature of the case, these justifications may not be forthcoming, if only because they may be falsified if pressed into a verbally articulated form. Occasionally, too, we see situations in which behavior is clearly "good," but the reasons given for that behavior are weak or nonexistent. Only a maniac would want to subvert good behavior on the grounds of the defective arguments used to rationalize the behavior.

Criteria for Methodology

I will only offer one negative and one positive criterion for ethical methodology. The wrong methodology will be used if it is not a methodology which has been specifically developed for ethical problems of medicine and biology. This does not mean it cannot or should not bear many of the traits of general philosophical or theological methodology. But if it bears only those traits one can be assured that it will not deal adequately with specific issues which arise in the life sciences. My positive criterion for a good methodology is this: it must display the fact that bioethics is an interdisciplinary field in which the purely "ethical" dimensions neither can nor should be factored out without remainder from the legal, political, psychological and social dimensions. The critical question, for example, of who should make the ethical decisions in medicine and biology is falsified at the outset if too sharp a distinction is drawn between what, ethically, needs to be decided and who, politically, should be allowed to decide. It is surely important to ethical theory to make this kind of distinction; unfortunately, if pressed too doggedly it may well falsify the reality of the way decisions are and will continue to be made.

The problem of decision-making, which I include as the third task of the bioethicist, cannot be divorced from the methodological question. Actually it makes me realize that I have a second positive criterion to offer as a test of a good bioethical methodology. The methodology ought to be such that it enables those who employ it to reach reasonably specific, clear decisions in those instances which require them—in the case of what is to be done about Mrs. Jones by four o'clock tomorrow afternoon, after which she will either live or die depending upon the decision made. I have already suggested that philosophers are not very good at that sort of thing, and that their weakness in this respect is likely to be altogether vexing to the physician who neither has the right atmosphere nor the time to think through everything the philosopher usually argues *needs* to be thought through.

In proposing that a good methodology should make it possible to reach specific conclusions at specific times, I am proposing a utopian goal. The only kinds of ethical systems I know of which make that possible are those of an essentially deductive kind, with well-established primary and secondary principles and a long history of highly refined casuistical thinking. The Roman Catholic scholastic tradition and the Jewish *responsa* tradition are cases in point. Unfortunately, systems of that kind presuppose a whole variety of cultural conditions and shared world-views which simply do not exist in society at large.

In their absence, it has become absolutely urgent that the search for a philosophically viable normative ethic, which can presuppose some commonly shared principles, go forward with all haste. Short of finding that, I do not see how ethical methodologies can be developed which will include methods for reaching quick and viable solutions in specific cases. Instead, we are likely to get only what we now have, a lot of very broad and general thinking, full of vagrant insights, but on the whole of limited use to the practicing physician and scientist.

Much of what I have been saying presupposes that a distinction can be drawn between "ethics" understood broadly and ethics understood narrowly. In its narrow sense, to do "ethics" is to be good at doing what well-trained philosophers and theologians do: analyze concepts, clarify principles, see logical entailments, spot underlying assumptions, and build theoretical systems. There are better and worse ways of doing this kind of thing and that is why philosophers and theologians can spend much of their time arguing with each other. But even the better ways will, I think, not be good enough for the demands of bioethics. That requires understanding "ethics" in a very broad, well-nigh unmanageable sense of the term.

Impossible and Scandalous

My contention is that the discipline of bioethics should be so designed, and its practitioners so trained, that it will directly—at whatever cost to disciplinary elegance—serve those physicians and biologists whose position demands that they make the practical decisions. This requires, ideally, a number of ingredients as part of the training—which can only be life-long—of the bioethicist: sociological understanding of the medical and biological communities; psychological understanding of the kinds of needs felt by researchers and clinicians, patients and physicians, and the varieties of pressures to which they are subject; historical understanding of the sources of regnant value theories and common practices; requisite scientific training; awareness of and facility with the usual methods of ethical analysis as understood in the philosophical and theological communities—and no less a full awareness of the limitations of those methods when applied to actual cases; and, finally, personal exposure to the kinds of ethical problems which arise in medicine and biology.

This is an impossible list of demands, guaranteed in advance to scandalize all of those professionals in whose field one must trespass. The scientist will point out that you are not a trained scientist, the physician that you are lacking both a medical degree and the clinical experience of treating patients, the sociologist that your anecdotal evidence is worth noting, the philosopher that you are straying from solid philosophical work. Well, so what? That is what the discipline of bioethics requires.

One important test of the acceptance of bioethics as a discipline will be the extent to which it is called upon by scientists and physicians. This means that it should be developed inductively, working at least initially from the kinds of problems scientists and physicians believe they face and need assistance on. As often as not, they will be wrong about the real nature of the issues with which they have to wrestle. But no less often the person trained in philosophy and theology will be equally wrong in his understanding of the real issues. Only a continuing, probably tension-ridden dialectic will suffice to bridge the gap, a dialectic which can only be kept alive by a continued exposure to specific cases in all their human dimensions. Many of them will be very unpleasant cases, the kind which make one long for the security of writing elegant articles for professional journals on such manageable issues as recent distinctions between "rules" and "maxims."

FURTHER READINGS

FOR THE DISCIPLINE OF BIOETHICS

For lists of further sources see: *Bibliography of Society, Ethics and the Life Sciences* (1974), available from The Hastings Center, 623 Warburton Avenue, Hastings-on-Hudson, New York 10706 at a charge of $3.50; and *Ethical Issues in Health Services: A Report and Annotated Bibliography* Supplement I, prepared by James Carmody and published by the Department of Health, Education, and Welfare, Pub. No. HRA-74-3123, and available from the Bureau of Health Services Research, 5600 Fishers Lane, Rockville, Maryland 20852 at no charge.

Books

H. Tristram Engelhardt, Jr. and Stuart F. Spicker, eds., *Evaluation and Explanation in the Biomedical Sciences* (Dordrecht, Holland: Reidel Publishing Company, 1975), and also by this editor with Chester R. Burns, *The Humanities and Medicine,* a special issue of *Texas Reports on Biology and Medicine* Vol. 32, No. 1, (Spring, 1974); Willard Gaylin, and Robert Veatch, and Councilman Morgan, eds., *The Teaching of Medical Ethics* (Hastings-on-Hudson, New York: The Hastings Center, 1973); Samuel Gorovitz, *Teaching Medical Ethics: A Report on One Approach, Moral Problems in Medicine Project* (Case Western Reserve University, Cleveland, 1973); Robert M. Veatch, *Case Studies in Medical Ethics* (Cambridge, Mass.: Harvard University Press, forthcoming); Bernard Williams, *Utilitarianism: For and Against* (Cambridge University Press, 1973), and also by this author, *Morality: An Introduction to Ethics* (New York: Harper & Row, 1972).

Articles

Danner K. Clouser, "Humanities and the Medical School: A Sketched Rationale and Description," *British Journal of Medical Education,* Vol. 5, No. 3, (Sep., 1971), pp. 226–231, and also by this author the following list of articles: "Medical Ethics Courses: Some Realistic Expectations," *Journal of Medical Education,* Vol. 48 (April, 1973), pp. 373–74, and "Medical Ethics and Related Disciplines," in *The Teaching of Medical Ethics* eds. Veatch, Gaylin, and Morgan (Hastings Center, 1973), pp. 38–46, and "Medicine As Art: An Initial Exploration," *Texas Reports on Biology and Medicine* (Spring, 1974), pp. 267–74, and

"Philosophy and Medicine," *Proceedings of the Institute on Human Values in Medicine,* Society for Health and Human Values (Philadelphia, Pa., 1972), pp. 48–80, and "What Is Medical Ethics?" *Annals of Internal Medicine 80* (May, 1974), pp. 657–60; Willard Gaylin, "Teaching Medical Ethics: An Experimental Program," *Journal of Medical Education* (Oct., 1972), and also these articles by this author: "Science and Ethics, Part I," *Intellectual Digest* (Sep., 1973), "Science and Ethics, Part II," *Intellectual Digest* (Sep., 1973), "The Law and the Biological Revolution–A Symposium," *Columbia Journal of Law and Social Problems,* Vol. 10, No. 1 (Fall, 1973); Samuel Gorovitz, "Bioethics and Social Responsibility," *The Monist* (Jan., 1976); Robert M. Veatch, "Medical Ethics in a Revolutionary Age," in *Young Doctors, Health Care, and The American Future Second Staff Conference Proceedings* (1972), pp. 183–213, and "Medical Ethics: Professional or Universal?" Institute of Society, Ethics and Life Sciences Working Paper Series, No. 5 (Hastings-on-Hudson, New York, 1971), and "Medical Ethics: Professional or Universal?" Institute of Society, Ethics and Life Sciences Working Paper Series, No. 5 (Hastings-on-Hudson, New York, 1971), and "Medical Ethics: Professional or Universal," *Harvard Theological Review 65,* No. 4 (Oct., 1972), pp. 531–39, and "Medical Ethics: Updating the Hippocratic Oath," *Medical Opinion* (April, 1972), pp. 56–61, and "The Medical Model: Its Nature and Problems," *Hastings Center Studies 1,* No. 3 (1973), pp. 59–76, and "The Metaethical Foundation for an Ethic of the Life Sciences: Does Ethics Have an Empirical Basis?" *Hastings Center Studies 1,* No. 1 (1973), pp. 50–65, and "Models for Ethical Medicine in a Revolutionary Age," *Hastings Center Report 2,* No. 3 (June, 1972), pp. 5–7, and "Teaching Medical Ethics: An Experimental Program," *Journal of Medical Education 47* (Oct., 1972), pp. 779–785.

PART I: ABORTION

INTRODUCTION

In *Roe v. Wade* a pregnant single woman adopted the fictitious name Jane Roe, and brought a class action in federal court challenging the constitutionality of the Texas criminal abortion laws. Under the Texas statutes, abortions were prohibited except for the purpose of saving the mother's life. Roe claimed that she could not procure a safe abortion (i.e., an abortion performed by a licensed physician under clinical conditions) in Texas, because her life was not put in jeopardy by her pregnancy. She also claimed that she could not afford to travel to a state where safe abortions for persons such as herself were legal. Accordingly, she asked the court to declare the Texas criminal abortion statutes unconstitutional, and to enjoin the state from enforcing these laws.

In its decision, the Supreme Court refuses to rule on the difficult question as to when human life begins. Instead, the Court declares the Texas statutes unconstitutional because it holds that these laws violate the right to privacy guaranteed to all persons in the Fourteenth Amendment. But the right to privacy is not held to be an absolute or unqualified right. At some point, the Court says, the state's interest in health, medical standards, and prenatal life may supersede the mother's right to privacy. And at this "compelling" point, the state may choose to regulate abortion. Hence the court concludes:

(a) For the stage prior to approximately the end of the first trimester, the abortion decision and its effectuation must be left to the medical judgment of the pregnant woman's attending physician.

(b) For the stage subsequent to approximately the end of the first trimester, the State, in promoting its interest in the health of the mother, may, if it chooses, regulate the abortion procedure in ways that are reasonably related to maternal health.

(c) For the stage subsequent to viability the State, in promoting its interest in the potentiality of human life, may, if it chooses, regulate, and even proscribe, abortion except where necessary, in appropriate medical judgment, for the preservation of the life or health of the mother.

Given the Supreme Court's refusal to decide the question of when human life begins, Professor Brody's argument may be interpreted as a challenge to the ruling in *Roe v. Wade.* Brody asks if it is possible to consistently maintain: (1) that abortion is morally wrong because it is the taking of an innocent human life, and (2) that it is wrong (or at least inappropriate) for the state to have laws prohibiting abortion. He considers many of the arguments commonly offered in defense of the legalization of abortion, including the argument accepted by the Supreme Court (viz., the view that laws prohibiting abortion interfere with a woman's right to privacy by limiting her right to do what she wants with her own body). In the end, Brody concludes that there is no possibility of arguing for the legalization of abortion if (1) is true; and that as a consequence, the legal problem about abortion cannot be resolved independently of the status of the fetus problem.

If Brody is correct, one cannot argue for the legalization of abortion and consistently maintain that abortion is morally wrong because it is the killing of an innocent human being. But even if this fact is true, it does not follow that the Supreme Court was in error in not deciding the question of when human life begins. Both Professors Thomson and Brandt admit that abortion constitutes the destruction of human life; yet each claims that it is not morally wrong to perform such an act. In support of this position Professor Thomson argues that having a right to life does not guarantee either a right to be given the use of, or a right to be allowed continued use of, another person's body. And this holds true, she says, even if one needs the body of another in order to support life itself.

Professor Brandt takes a slightly different tack. Rather than seeking to justify all abortions, Brandt argues that it is not morally wrong to kill human fetuses in the first trimester of gestation. He contends that the right to life is not a right possessed by all human beings, and that consequently there are some kinds of humans whom it is not morally wrong to destroy. Using two separate arguments, Brandt concludes that (at the very least) fetuses under three months of age must be classed with those human beings whom it is not morally wrong to kill.

Professor Humber claims that it is contradictory to deny that human life begins at conception. He criticizes the arguments of Thomson, Brandt, and other pro-abortionists, and argues that the decision of *Roe v. Wade* gives legal sanction to immoral conduct. Still, he recognizes the emotive force of the pro-abortionists' stance, and sees present anti-abortionists' attempts to change the law as futile. As an alternative to this course of action, then, Humber proposes a new program for anti-abortion action—a program which he holds would, if successful, dissolve the problem of abortion by making such killings wholly unnecessary.

ROE v. WADE

410 U.S. 113, 93 S.CT. 705 (1973)

What follows are portions of the majority opinion by Justice Blackmun, together with the dissenting opinion of Justice White. The concurring opinions of Justices Douglas and Stewart have been omitted. Some footnotes have been dropped, with those remaining having been renumbered.

MR. JUSTICE BLACKMUN delivered the opinion of the Court.

The principal thrust of appellant's attack on the Texas statutes is that they improperly invade a right, said to be possessed by the pregnant woman, to choose to terminate her pregnancy. Appellant would discover this right in the concept of personal "liberty" embodied in the Fourteenth Amendment's Due Process Clause; or in personal, marital, familial, and sexual privacy said to be protected by the Bill of Rights or its penumbras . . . or among those rights reserved to the people by the Ninth Amendment. . . .

Three reasons have been advanced to explain historically the enactment of criminal abortion laws in the 19th century and to justify their continued existence.

It has been argued occasionally that these laws were the product of a Victorian social concern to discourage illicit sexual conduct. Texas, however, does not advance this justification in the present case, and it appears that no court or commentator has taken the argument seriously. The appellants and *amici* contend, moreover, that this is not a proper state purpose at all and suggest that, if it were, the Texas statutes are overbroad in protecting it since the law fails to distinguish between married and unwed mothers.

A second reason is concerned with abortion as a medical procedure. When most criminal abortion laws were first enacted, the procedure was a hazardous one for the woman.[1] This was particularly true prior to the development of antisepsis. Antiseptic techniques, of course, were based on discoveries by Lister, Pasteur, and others first announced in 1867, but were not generally accepted and employed until about the turn of the century. Abortion mortality was high. Even after 1900, and perhaps until as late as the development of antibiotics in the 1940s, standard modern techniques such as dilation and curettage were not nearly so safe as they are today. Thus, it has been argued that a State's real concern in enacting a criminal abortion law was to protect the pregnant woman, that is, to restrain her from submitting to a procedure that placed her life in serious jeopardy.

---

[1] See C. Haagensen & W. Lloyd, A Hundred Years of Medicine 19 (1943).

Modern medical techniques have altered this situation. Appellants and various *amici* refer to medical data indicating that abortion in early pregnancy, that is, prior to the end of the first trimester, although not without its risk, is now relatively safe. Mortality rates for women undergoing early abortions, where the procedure is legal, appear to be as low as or lower than the rates for normal childbirth.[2] Consequently, any interest of the State in protecting the woman from an inherently hazardous procedure, except when it would be equally dangerous for her to forgo it, has largely disappeared. Of course, important state interests in the areas of health and medical standards do remain. The State has a legitimate interest in seeing to it that abortion, like any other medical procedure, is performed under circumstances that insure maximum safety for the patient. This interest obviously extends at least to the performing physician and his staff, to the facilities involved, to the availability of after-care, and to adequate provision for any complication or emergency that might arise. The prevalence of high mortality rates at illegal "abortion mills" strengthens, rather than weakens, the State's interest in regulating the conditions under which abortions are performed. Moreover, the risk to the woman increases as her pregnancy continues. Thus, the State retains a definite interest in protecting the woman's own health and safety when an abortion is proposed at a late stage of pregnancy.

The third reason is the State's interest—some phrase it in terms of duty—in protecting prenatal life. Some of the argument for this justification rests on the theory that a new human life is present from the moment of conception.[3] The State's interest and general obligation to protect life then extends, it is argued, to prenatal life. Only when the life of the pregnant mother herself is at stake, balanced against the life she carries within her, should the interest of the embryo or fetus not prevail. Logically, of course, a legitimate state interest in this area need not stand or fall on acceptance of the belief that life begins at conception or at some other point prior to live birth. In assessing the State's interest, recognition may be given to the less rigid claim that as long as at least *potential* life is involved, the State may assert interests beyond the protection of the pregnant woman alone.

[2] Potts, Postconceptive Control of Fertility, 8 Int'l J. of G. & O. 957, 967 (1970) (England and Wales); Abortion Mortality, 20 Morbidity and Mortality 208, 209 (June 12, 1971) (U. S. Dept. of HEW, Public Health Service) (New York City); Tietze, United States: Therapeutic Abortions, 1963–1968, 59 Studies in Family Planning 5, 7 (1970); Tietze, Mortality with Contraception and Induced Abortion, 45 Studies in Family Planning 6 (1969) (Japan, Czechoslovakia, Hungary); Tietze & Lehfeldt, Legal Abortion in Eastern Europe, 175 J. A. M. A. 1149, 1152 (April 1961). Other sources are discussed in Lader 17–23.

[3] See Brief of *Amicus* National Right to Life Committee; R. Drinan, The Inviolability of the Right to Be Born, in Abortion and the Law 107 (D. Smith ed. 1967); Louisell, Abortion, The Practice of Medicine and the Due Process of Law, 16 U. C. L. A. L. Rev. 233 (1969); Noonan 1.

Parties challenging state abortion laws have sharply disputed in some courts the contention that a purpose of these laws, when enacted, was to protect prenatal life. Pointing to the absence of legislative history to support the contention, they claim that most state laws were designed solely to protect the woman. Because medical advances have lessened this concern, at least with respect to abortion in early pregnancy, they argue that with respect to such abortions the laws can no longer be justified by any state interest. There is some scholarly support for this view of original purpose. The few state courts called upon to interpret their laws in the late 19th and early 20th centuries did focus on the State's interest in protecting the woman's health rather than in preserving the embryo and fetus. Proponents of this view point out that in many States, including Texas, by statute or judicial interpretation, the pregnant woman herself could not be prosecuted for self-abortion or for cooperating in an abortion performed upon her by another. They claim that adoption of the "quickening" distinction through received common law and state statutes tacitly recognizes the greater health hazards inherent in late abortion and impliedly repudiates the theory that life begins at conception.

It is with these interests, and the weight to be attached to them, that this case is concerned.

The Constitution does not explicitly mention any right of privacy. In a line of decisions, however, going back perhaps as far as *Union Pacific R. Co.* v. *Botsford,* 141 U.S. 250,251 (1891), the Court has recognized that a right of personal privacy, does exist under the Constitution. In varying contexts, the Court or individual Justices have, indeed, found at least the roots of that right in the First Amendment . . . in the Fourth and Fifth Amendments . . . in the penumbras of the Bill of Rights . . . in the Ninth Amendment . . . or in the concept of liberty guaranteed by the first section of the Fourteenth Amendment. . . . These decisions make it clear that only personal rights that can be deemed "fundamental" or "implicit in the concept of ordered liberty," . . . are included in this guarantee of personal privacy. They also make it clear that the right has some extension to activities relating to marriage . . . procreation . . . contraception . . . family relationships . . . and child rearing and education. . . .

This right of privacy, whether it be founded in the Fourteenth Amendment's concept of personal liberty and restrictions upon state action, as we feel it is, or, as the District Court determined, in the Ninth Amendment's reservation of rights to the people, is broad enough to encompass a woman's decision whether or not to terminate her pregnancy. The detriment that the State would impose upon the pregnant woman by denying this choice altogether is apparent. Specific and direct harm medically diagnosable even in early pregnancy may be involved. Maternity, or additional offspring, may force upon the woman a distressful life and future. Psychological harm may be imminent. Mental and physical health may be taxed by child care. There is also the distress, for all concerned, associated with the unwanted child, and there is the problem of bringing a child into a family already unable, psychologically and otherwise, to care for it. In other cases, as in this one, the additional difficulties and continuing stigma of unwed motherhood may be involved. All these are factors the woman and her responsible physician necessarily will consider in consultation.

On the basis of elements such as these, appellant and some *amici* argue that the woman's right is absolute and that she is entitled to terminate her pregnancy at whatever time, in whatever way, and for whatever reason she alone chooses. With this we do not agree. Appellant's arguments that Texas either has no valid interest at all in regulating the abortion decision, or no interest strong enough to support any limitation upon the woman's sole determination, are unpersuasive. The Court's decisions recognizing a right of privacy also acknowledge that some state regulation in areas protected by that right is appropriate. As noted above, a State may properly assert important interests in safeguarding health, in maintaining medical standards, and in protecting potential life. At some point in pregnancy, these respective interests become sufficiently compelling to sustain regulation of the factors that govern the abortion decision. The privacy right involved, therefore, cannot be said to be absolute. In fact, it is not clear to us that the claim asserted by some *amici* that one has an unlimited right to do with one's body as one pleases bears a close relationship to the right of privacy previously articulated in the Court's decisions. The Court has refused to recognize an unlimited right of this kind in the past. *Jacobson* v. *Massachusetts,* 197 U.S. 11 (1905) (vaccination); *Buck* v. *Bell,* 274 U.S. 200 (1927) (sterilization).

We, therefore, conclude that the right of personal privacy includes the abortion decision, but that this right is not unqualified and must be considered against important state interests in regulation.

We note that those federal and state courts that have recently considered abortion law challenges have reached the same conclusion. A majority, in addition to the District Court in the present case, have held state laws unconstitutional, at least in part, because of vagueness or because of overbreadth and abridgment of rights. . . .

Although the results are divided, most of these courts have agreed that the right of privacy, however based, is broad enough to cover the abortion decision; that the right, nonetheless, is not absolute and is subject to some limitations; and that at some point the state interests as to protection of health, medical standards, and prenatal life, become dominant. We agree with this approach.

Where certain "fundamental rights" are involved, the Court has held that regulation limiting these rights may be justified only by a "compelling state interest," . . . and that legislative enactments must be narrowly drawn to express only the legitimate state interests at stake.

In the recent abortion cases . . . courts have recognized these principles. Those striking down state laws have generally scrutinized the State's interests in protecting health and potential life, and have concluded that neither interest justified broad limitations on the reasons for which a physician and his pregnant patient might decide that she should have an abortion in the early stages of pregnancy. Courts sustaining state laws have held that the State's determinations to protect health or prenatal life are dominant and constitutionally justifiable.

The District Court held that the appellee failed to meet his burden of demonstrating that the Texas statute's infringement upon Roe's rights was necessary to support a compelling state interest, and that, although the appellee presented "several compelling justifications for state presence in the area of abortions," the statutes outstripped these justifications and swept "far beyond any areas of compelling state interest." . . . Appellant and appellee both contest that holding. Appellant, as has been indicated, claims an absolute right that bars any state imposition of criminal penalties in the area. Appellee argues that the State's determination to recognize and protect prenatal life from and after conception constitutes a compelling state interest. As noted above, we do not agree fully with either formulation.

A. The appellee and certain *amici* argue that the fetus is a "person" within the language and meaning of the Fourteenth Amendment. In support of this, they outline at length and in detail the well-known facts of fetal development. If this suggestion of personhood is established, the appellant's case, of course, collapses, for the fetus' right to life would then be guaranteed specifically by the Amendment. The appellant conceded as much on reargument. On the other hand, the appellee conceded on reargument that no case could be cited that holds that a fetus is a person within the meaning of the Fourteenth Amendment.

The Constitution does not define "person" in so many words. Section 1 of the Fourteenth Amendment contains three references to "person." The first, in defining "citizens," speaks of "persons born or naturalized in the United States." The word also appears both in the Due Process Clause and in the Equal Protection Clause. "Person" is used in other places in the Constitution: in the listing of qualifications for Representatives and Senators, Art. I, §2, cl. 2, and §3, cl. 3; in the Apportionment Clause, Art. I, §2, cl. 3;[4] in the Migration and Importation provision, Art. I, §9, cl. 1; in the Emolument Clause, Art. II, §9, cl. 8; in the Electors provisions, Art. II, §1, cl. 2, and the superseded cl. 3; in the provision outlining qualifications for the office of President, Art. II, §1, cl. 5; in the Extradition provisions, Art. IV, §2, cl. 2, and the superseded Fugitive Slave Clause 3; and in the Fifth, Twelfth, and Twenty-second Amendments, as well as

[4] We are not aware that in the taking of any census under this clause, a fetus has ever been counted.

in §§2 and 3 of the Fourteenth Amendment. But in nearly all these instances, the use of the word is such that it has application only postnatally. None indicates, with any assurance, that it has any possible pre-natal application.[5]

All this, together with our observation, *supra,* that throughout the major portion of the 19th century prevailing legal abortion practices were far freer than they are today, persuades us that the word "person," as used in the Fourteenth Amendment, does not include the unborn.[6] This is in accord with the results reached in those few cases where the issue has been squarely presented. . . . Indeed, our decision in *United States* v. *Vuitch,* 402 U.S. 62, 91 S.Ct. 1294, 28 L.Ed.2d 601 (1971), inferentially is to the same effect, for we there would not have indulged in statutory interpretation favorable to abortion in specified circumstances if the necessary consequence was the termination of life entitled to Fourteenth Amendment protection.

This conclusion, however, does not of itself fully answer the contentions raised by Texas, and we pass on to other considerations.

B. The pregnant woman cannot be isolated in her privacy. She carries an embryo and, later, a fetus, if one accepts the medical definitions of the developing young in the human uterus. See Dorland's Illustrated Medical Dictionary 478–479, 547 (24th ed. 1965). The situation therefore is inherently different from marital intimacy, or bedroom possession of obscene material, or marriage, or procreation, or education, with which *Eisenstadt* and *Griswold, Stanley, Loving, Skinner,* and *Pierce* and *Meyer* were respectively concerned. As we have intimated above, it is reasonable and appropriate for a State to decide that at some point in time another interest, that of health of the mother or that

[5] When Texas urges that a fetus is entitled to Fourteenth Amendment protection as a person, it faces a dilemma. Neither in Texas nor in any other State are all abortions prohibited. Despite broad proscription, an exception always exists. The exception contained in art. 1196, for an abortion procured or attempted by medical advice for the purpose of saving the life of the mother, is typical. But if the fetus is a person who is not to be deprived of life without due process of law, and if the mother's condition is the sole determinant, does not the Texas exception appear to be out of line with the Amendment's command?

There are other inconsistencies between Fourteenth Amendment status and the typical abortion statute. It has already been pointed out . . . that in Texas the woman is not a principal or an accomplice with respect to an abortion upon her. If the fetus is a person, why is the woman not a principal or an accomplice? Further, the penalty for criminal abortion specified by Art. 1195 is significantly less than the maximum penalty for murder prescribed by Art. 1257 of the Texas Penal Code. If the fetus is a person, may the penalties by different?

[6] Cf. the Wisconsin abortion statute, defining "unborn child" to mean "a human being from the time of conception until it is born alive," Wis. Stat. §940.04(6) (1969), and the new Connecticut statute, Public Act No. 1, May 1972 Special Session, declaring it to be the public policy of the State and the legislative intent "to protect and preserve human life from the moment of conception."

of potential human life, becomes significantly involved. The woman's privacy is no longer sole and any right of privacy she possesses must be measured accordingly.

Texas urges that, apart from the Fourteenth Amendment, life begins at conception and is present throughout pregnancy, and that, therefore, the State has a compelling interest in protecting that life from and after conception. We need not resolve the difficult question of when life begins. When those trained in the respective disciplines of medicine, philosophy, and theology are unable to arrive at any consensus, the judiciary, at this point in the development of man's knowledge, is not in a position to speculate as to the answer.

It should be sufficient to note briefly the wide divergence of thinking on this most sensitive and difficult question. There has always been strong support for the view that life does not begin until live birth. This was the belief of the Stoics. It appears to be the predominant, though not the unanimous, attitude of the Jewish faith.[7] It may be taken to represent also the position of a large segment of the Protestant community, insofar as that can be ascertained; organized groups that have taken a formal position on the abortion issue have generally regarded abortion as a matter for the conscience of the individual and her family. As we have noted, the common law found greater significance in quickening. Physicians and their scientific colleagues have regarded that event with less interest and have tended to focus either upon conception, upon live birth, or upon the interim point at which the fetus becomes "viable," that is, potentially able to live outside the mother's womb, albeit with artificial aid.[8] Viability is usually placed at about seven months (28 weeks) but may occur earlier, even at 24 weeks.[9] The Aristotelian theory of "mediate animation," that held sway throughout the Middle Ages and the Renaissance in Europe, continued to be official Roman Catholic dogma until the 19th century, despite opposition to this "ensoulment" theory from those in the Church who would recognize the existence of life from the moment of conception.[10] The latter is now, of course, the official belief of the Catholic Church. As one brief *amicus* discloses, this is a view strongly held by many non-Catholics as well, and by many physicians. Substantial problems for precise definition of this view are posed, however, by new embryological data that purport to indicate that conception is a "process" over time, rather than an event, and by new medical

[7] Lader 97–99; D. Feldman, Birth Control in Jewish Law 251–294 (1968). For a stricter view, see I. Jakobovits, Jewish Views on Abortion, in Abortion and the Law 124 (D. Smith ed. 1967).

[8] L. Hellman & J. Pritchard, Williams Obstetrics 493 (14th ed. 1971); Dorland's Illustrated Medical Dictionary 1689 (24th ed. 1965).

[9] Hellman & Pritchard, *supra*, n. 59, at 493.

[10] For discussions of the development of the Roman Catholic position, see D. Callahan, Abortion: Law, Choice, and Morality 409–447 (1970); Noonan 1.

techniques such as menstrual extraction, the "morning-after" pill, implantation of embryos, artificial insemination, and even artifical wombs.[11]

In areas other than criminal abortion, the law has been reluctant to endorse any theory that life, as we recognize it, begins before live birth or to accord legal rights to the unborn except in narrowly defined situations and except when the rights are contingent upon live birth. For example, the traditional rule of tort law denied recovery for prenatal injuries even though the child was born alive.[12] That rule has been changed in almost every jurisdiction. In most States, recovery is said to be permitted only if the fetus was viable, or at least quick, when the injuries were sustained, though few courts have squarely so held. In a recent development, generally opposed by the commentators, some States permit the parents of a stillborn child to maintain an action for wrongful death because of prenatal injuries. Such an action, however, would appear to be one to vindicate the parents' interest and is thus consistent with the view that the fetus, at most, represents only the potentiality of life. Similarly, unborn children have been recognized as acquiring rights or interests by way of inheritance or other devolution of property, and have been represented by guardians *ad litem.*[13] Perfection of the interests involved, again, has generally been contingent upon live birth. In short, the unborn have never been recognized in the law as persons in the whole sense.

In view of all this, we do not agree that, by adopting one theory of life, Texas may override the rights of the pregnant woman that are at stake. We repeat, however, that the State does have an important and legitimate interest in preserving and protecting the health of the pregnant woman, whether she be a resident of the State or a nonresident who seeks medical consultation and treatment there, and that it has still *another* important and legitimate interest in protecting the potentiality of human life. These interests are separate and distinct. Each grows in substantiality as the woman approaches term and, at a point during pregnancy, each becomes "compelling."

With respect to the State's important and legitimate interest in the health of the mother, the "compelling" point, in the light of present medical knowledge,

[11] See Brodie, The New Biology and the Prenatal Child, 9 J. Family L. 391,397 (1970); Gorney, The New Biology and the Future of Man, 15 U. C. L. A. L. Rev. 273 (1968); Note, Criminal Law–Abortion–The "Morning-After Pill" and Other Pre-Implantation Birth-Control Methods and the Law, 46 Ore. L. Rev. 211 (1967); G. Taylor, The Biological Time Bomb 32 (1968); A. Rosenfeld, The Second Genesis 138–139 (1969); Smith, Through a Test Tube Darkly: Artificial Insemination and the Law, 67 Mich. L. Rev. 127 (1968); Note, Artificial Insemination and the Law, 1968 U. Ill. L. F. 203.

[12] W. Prosser, The Law of Torts 335–338 (4th ed. 1971); 2 F. Harper & F. James, The Law of Torts 1028–1031 (1956); Note, 63 Harv. L. Rev. 173 (1949).

[13] Louisell, Abortion, The Practice of Medicine and the Due Process of Law, 16 U. C. L. A. L. Rev. 233, 235–238 (1969); Note, 56 Iowa L. Rev. 994, 999–1000 (1971); Note, The Law and the Unborn Child, 46 Notre Dame Law. 349, 351–354 (1971).

is at approximately the end of the first trimester. This is so because of the now-established medical fact, referred to above . . . that until the end of the first trimester mortality in abortion may be less than mortality in normal childbirth. It follows that, from and after this point, a State may regulate the abortion procedure to the extent that the regulation reasonably relates to the preservation and protection of maternal health. Examples of permissible state regulation in this area are requirements as to the qualifications of the person who is to perform the abortion; as to the licensure of that person; as to the facility in which the procedure is to be performed, that is, whether it must be a hospital or may be a clinic or some other place of less-than-hospital status; as to the licensing of the facility; and the like.

This means, on the other hand, that, for the period of pregnancy prior to this "compelling" point, the attending physician, in consultation with his patient, is free to determine, without regulation by the State, that, in his medical judgment, the patient's pregnancy should be terminated. If that decision is reached, the judgment may be effectuated by an abortion free of interference by the State.

With respect to the State's important and legitimate interest in potential life, the "compelling" point is at viability. This is so because the fetus then presumably has the capability of meaningful life outside the mother's womb. State regulation protective of fetal life after viability thus has both logical and biological justifications. If the State is interested in protecting fetal life after viability, it may go so far as to proscribe abortion during that period, except when it is necessary to preserve the life or health of the mother.

Measured against these standards, Art. 1196 of the Texas Penal Code, in restricting legal abortions to those "procured or attempted by medical advice for the purpose of saving the life of the mother," sweeps too broadly. The statute makes no distinction between abortions performed early in pregnancy and those performed later, and it limits to a single reason, "saving" the mother's life, the legal justification for the procedure. The statute, therefore, cannot survive the constitutional attack made upon it here.

This conclusion makes it unnecessary for us to consider the additional challenge to the Texas statute asserted on grounds of vagueness. . . .

MR. JUSTICE WHITE, with whom MR. JUSTICE REHNQUIST joins, dissenting.

At the heart of the controversy in these cases are those recurring pregnancies that pose no danger whatsoever to the life or health of the mother but are, nevertheless, unwanted for any one or more of a variety of reasons—convenience, family planning, economics, dislike of children, the embarrassment of illegitimacy, etc. The common claim before us is that for any one of such reasons, or for no reason at all, and without asserting or claiming any threat to life or health, any woman is entitled to an abortion at her request if she is able to find a medical advisor willing to undertake the procedure.

The Court for the most part sustains this position: During the period prior to the time the fetus becomes viable, the Constitution of the United States values the convenience, whim, or caprice of the putative mother more than the life or potential life of the fetus; the Constitution, therefore, guarantees the right to an abortion as against any state law or policy seeking to protect the fetus from an abortion not prompted by more compelling reasons of the mother.

With all due respect, I dissent. I find nothing in the language or history of the Constitution to support the Court's judgment. The Court simply fashions and announces a new constitutional right for pregnant mothers and, with scarcely any reason or authority for its action, invests that right with sufficient substance to override most existing state abortion statutes. The upshot is that the people and the legislatures of the 50 states are constitutionally disentitled to weigh the relative importance of the continued existence and development of the fetus, on the one hand, against a spectrum of possible impacts on the mother, on the other hand. As an exercise of raw judicial power, the Court perhaps has authority to do what it does today; but in my view its judgment is an improvident and extravagent exercise of the power of judicial review that the Constitution extends to this Court.

The Court apparently values the convenience of the pregnant mother more than the continued existence and development of the life or potential life that she carries. Whether or not I might agree with that marshaling of values, I can in no event join the Court's judgment because I find no constitutional warrant for imposing such an order of priorities on the people and legislatures of the States. In a sensitive area such as this, involving as it does issues over which reasonable men may easily and heatedly differ, I cannot accept the Court's exercise of its clear power of choice by interposing a constitutional barrier to state efforts to protect human life and by investing mothers and doctors with the constitutionally protected right to exterminate it. This issue, for the most part, should be left with the people and to the political processes the people have devised to govern their affairs.

It is my view, therefore, that the Texas statute is not constitutionally infirm because it denies abortions to those who seek to serve only their convenience rather than to protect their life or health. Nor is this plaintiff, who claims no threat to her mental or physical health, entitled to assert the possible rights of those women whose pregnancy assertedly implicated their health. This, together with *United States* v. *Vuitch,* 402 U.S. 62 (1971), dictates reversal of the judgment of the District Court. . . .

ABORTION AND THE LAW

BARUCH A. BRODY

One of the most frustrating aspects of discussions about abortion is the way in which they rapidly turn into a discussion of the status of the foetus and of whether destroying the foetus constitutes the taking of a human life. Since these latter questions seem difficult, if not impossible, to resolve upon rational grounds, frustration results. It therefore seems desirable to find aspects of the abortion problem that can be resolved independently of the status of the foetus problem. One such possibility is the question of whether there should be a law against abortions performed by licensed physicians upon the request of the mother (or perhaps the parents). There are, after all, many people who, while opposed to abortion on the grounds that it involves the taking of a human life, maintain that it would still be wrong (or at least inappropriate) for a state to legislate against such abortions. If their claim can be shown to be right, then we could at least resolve the legal problem about abortion. This paper is an attempt to assess their claim.

Such a claim cannot, of course, be considered without considering, at the same time, the more general question of when it is right (or appropriate) for a state to legislate against some action. It is hoped that our discussion of this problem in the context of a specific issue may shed light upon this fundamental general problem about the law.

I

The claim that we will be considering consists of the joint assertion that

1. *It is wrong for* x *to perform an abortion upon* y, *even when* x *is a licensed physician and* y *a consenting mother (who may also have the consent of the father), and it is wrong because this act would be the taking of an innocent human life.*

and

2. *It is wrong (or at least inappropriate) for the state, in the circumstances we find ourselves in now, to have a law prohibiting such abortions.*

Reprinted by permission of the author and publisher from B. Brody, "Abortion and the Law," *The Journal of Philosophy, 68,* No. 12 (1971), pp. 357–369.

These two claims are not, by themselves, clearly incompatible. But it might be claimed that when a correct principle about what type of wrong actions ought to be prohibited by law is conjoined to these two claims, we get an inconsistent triad. If this is correct, then the position we are considering is indefensible. But is it correct? What is the principle?

One principle that would certainly do the job is

3. *It is right (or appropriate) for the state to have a law prohibiting any action that is wrong.*

The trouble with this principle is that it seems too strong for two reasons. To begin with, it implies that such laws ought to be passed even if the wrong action produces no bad consequences, or at least none for anyone but the person who is doing the action. This implication about crimes without victims certainly casts familiar doubts upon (3). But even more significantly, (3) leaves out the whole question whether the passing of a law is the right (or appropriate) means for discouraging or preventing the performance of that wrong action. Would it be right to pass a law prohibiting such actions if, because of special circumstances in the case at hand, the existence of the law would increase the performance of these wrong actions? Or would it be right to pass a law prohibiting such actions if, because of special circumstances in the case at hand, the existence of the law has side effects worse than the continued performance of the wrong actions?

Much more plausible than (3) is another principle that would do the job, the principle that

4. *It is right (or appropriate) for the state to have a law prohibiting any wrong action that results in bad consequences for someone other than the performer of the action (or someone other than him and anyone else who has voluntarily consented to his doing the action).*

(4) is much more plausible than (3) because it avoids the problem about crimes without victims. (4) does the job of destroying the claim we are considering because it, (1) and (2), and certain additional very plausible assumptions (the foetus does not consent to the abortion, and the taking of a human life is an action that has bad consequences for him whose life is being taken) do form an inconsistent set of propositions. The trouble with (4), however, is that it does not avoid the second objection against (3). Even if the wrong action does have these bad consequences, it might still be wrong (or inappropriate) to have a law prohibiting these actions if, for example, the existence of the law would, because of special circumstances in the case at hand, increase the performance of these wrong actions or produce very undesirable side effects.

A similar problem arises even for the extremely plausible principle that

5. *It is right (or appropriate) for the state to have a law prohibiting any wrong action that involves the taking of a human life.*

This principle, (1), and (2) do form an inconsistent triad. But one cannot consent even to this general principle. After all, the existence of the law still might, because of special circumstances in the case at hand, increase the performance of these wrong actions or produce overriding side effects like the loss of even more human lives.

What rapidly emerges from a consideration of principles like (3)–(5) is that questions about the rightness (or appropriateness) of a law prohibiting certain actions cannot be settled by decisions about the rightness or wrongness of the action, or even by such decisions coupled with decisions about why the action is wrong. This is true even in the extreme case where it is decided that the action is wrong because it involves the taking of an innocent human life.

We are now in a position to see that our original objection to the joint assertability of (1) and (2) will not do. It, after all, requires some true principle of the form, "It is right (or appropriate) for the state to have a law prohibiting any action of type *A*" (where abortion, if it is the taking of an innocent human life, would necessarily be of the type in question). We have now seen that all such principles are wrong, and so our initial challenge is not sustainable. In examining the question of whether there should be a law against abortions, therefore, we must go beyond the fact, if it is a fact, that abortion is the taking of a human life, to consider the results of having such a law in a given society.

II

Before turning to such considerations, however, we must first consider an important claim about law and society which, if true, might serve as the basis for the justification of the joint assertion of (1) and (2). This is the claim that citizens of a pluralistic society must forego the use of the law as a method of enforcing their private moral point of view. On the basis of this claim, it might well be argued that, in our pluralistic society, in which there are serious disagreements about the status of the foetus and about the resulting rightness or wrongness of abortion, it would be wrong (or inappropriate) to have laws against abortion.

Such claims are difficult to evaluate because of their imprecision. So let us first try to formulate some version of them more carefully. Consider the following general claim:

6. *When the citizens of a society disagree about the rightness or wrongness of a given type of action, and a considerable number think it is right, then it is*

wrong (or inappropriate) for that society to have a law prohibiting that action, even if the majority of the citizens think that the action is wrong.

There are a variety of reasons that might be offered in support of such a claim. One appeals to the right of the minority to follow its own conscience and not to be forced into following the conscience of the majority. Another appeals to the inappropriateness of a majority's doing something to a minority that it would oppose were it the minority and were that action being done to it. Still another appeals to the detrimental consequences to a society of a strong feeling on the part of a significant minority that the state has passed unjust laws and that the law is being used by the majority to coerce the minority. All these weighty considerations make it seem certain that some principle like (6) is true.

If, however, (6) is true, it seems easy to offer a defense of the joint assertability of (1) and (2). All we need are the additional obvious truths that the citizens of our society strongly disagree about the rightness and wrongness of abortion and that a significant minority think that, in at least many cases, the right thing to do is to have an abortion. From these truths and (6), (2) follows even if (1) is true.

The trouble with this argument is that it depends upon (6). I quite agree that, because of the considerations mentioned above, something like (6) must be true. But (6) is much too broad to be defensible. Consider, after all, a society in which a significant minority thinks it is OK, and perhaps even obligatory, to kill blacks, jews, etc., because they are not really human. It would seem to follow from (6) that the majority of that society, even though they thought these actions were very evil, should not pass a law prohibiting such actions, should not use the law as a means of preventing the minority from killing blacks, jews, etc. Surely this is wrong. Even if citizens of a pluralistic society should forego passing certain laws out of deference to the views of those minorities who think that the action in question is not wrong, there are some cases where, because of the extent of the evil of the action in question according to the majority conception, the law must be passed anyway. It is easy to see why this is so. If, according to the majority conception, the actions in question produce bad enough results and infringe upon the rights of enough innocent people, then the possibility of preventing this by passing and enforcing a law against such actions may well override the rights of that minority to follow its conscience.

The truth in (6) seems to be captured by a weaker principle like:

7. *When the citizens of a society strongly disagree about the rightness or wrongness of a given type of action, and a considerable number think that it is right, then it is wrong (or inappropriate) for that society to have a law prohibiting that action, even if the majority of the citizens think that the action is wrong, unless the action in question is so evil that the desirability*

of its legal prohibition outweighs the desirability of granting to the minority the right to follow their own consciences on this issue.

(7) is, of course, rather vague. In particular, its last clause needs further clarification. But (7) is clear enough so that we can see that it cannot be used as the basis for a simple justification of the joint assertability of (1) and (2). (7), conjoined with the obvious truths that the citizens of our society strongly disagree about the rightness and wrongness of abortion and that a significant minority think that, in at least many cases, the right thing to do is to have an abortion, does not yield (2) if (1) is true. After all, if (1) is true, then the action in question is the taking of an innocent human life. Therefore, it might well fall under the last clause of (7). After all, according to (1), the destruction of a foetus is not unlike the killing of a black or a jew. They are all cases of the taking of an innocent human life.

It would seem, therefore, that an adequate account of the relation between law and morality in a pluralistic society does not easily support the joint assertability of (1) and (2). Even in a pluralistic society, the minority cannot always have the right to follow its conscience. Whether it should, in a given case, have this right depends, as (7) indicates, both on the nature of the action in question and on the results of having such a law. After all, both of these help determine "the desirability of its legal prohibition." So we are left once more with the need to consider the effects of abortion laws, and not merely the rightness and wrongness of the abortion.

III

There are a large number of arguments that are commonly offered in defense of the legalization of abortions performed by doctors upon consenting mothers. These include:

(i) The only legitimate interest the state ever had in preventing abortion, the threat to the health of the mother, is no longer present because of the safety of modern abortion techniques.

(ii) Laws prohibiting abortions interfere with a woman's right to do what she wants to do with her body.

(iii) Laws prohibiting abortions are harmful to the practice of medicine, partially because they lead to state intrusion into what should be a private relation between the doctor and his patient and partially because they prevent the doctor from practicing medicine as he sees fit.

(iv) Abortion laws that allow abortions only in a few cases (e.g., when necessary to save the life of the mother) are necessarily vague and therefore result either in a doctor's avoiding performing abortions when they should be performed, in order to play safe, or in a doctor's having to gamble that the law will uphold his decision.

(v) Abortion is a method of population control, and, with our growing population problem, we have no business outlawing any method of population control.

(vi) Laws prohibiting abortions lead to the birth of unwanted children, children who often suffer psychologically because of this, and who therefore sometimes grow up to be social problems.

(vii) Legalizing abortions is the only way of avoiding tragedies for the pregnant woman and her family, tragedies due to the cause of the pregnancy (rape, incest) or to events occurring during the pregnancy (certain types of maternal illness or certain drugs taken by the mother), or to the family's circumstances (its inability to support another child, the mother's mental or physical health).

(viii) Laws prohibiting abortions are commonly disregarded, and, like prohibition laws, all that they produce is a disrespect for the law.

(ix) Laws prohibiting abortions force pregnant women who need abortions to go to quacks, and this results in many deaths and much harm to these women's physical and mental health.

(x) Laws prohibiting abortions discriminate against the poor, who, unlike the rich, cannot get legal or safe abortions.

We are not now concerned with an evaluation of these arguments. What we want to see is whether someone who believes (1) can still use any of them as an argument for (2), i.e., whether these arguments can be used as the basis for an argument for the joint assertability of (1) and (2). We will have to ask ourselves two questions. Are the premises of these arguments compatible with (1), the claim that abortion is wrong because it is the taking of an innocent human life? And if they are compatible, does the addition to them of (1) still leave us with a good argument for legalizing abortions, an argument of the form, "Although (1), there should be no law against abortion because . . ."? If we can answer yes to both of these questions in connection with one or more of the above-mentioned arguments, or in connection with some other, as yet unmentioned, argument, then we will have a good argument in defense of the joint assertion of (1) and (2).

A good example of an argument that is not open for our use is (i). For it is clear that one of its premises, the claim that the only interest the state has in abortions is protecting the mother from a dangerous operation, is not compatible with the view that abortion is the taking of an innocent human life. After all, if abortion is the taking of an innocent human life, it would seem as though the state, as part of its general role as a protector of human life, has an interest in prohibiting abortions to protect the life of a child.

We could not use (i) because its premises are incompatible with (1). We cannot use (v) because when we conjoin (1) to its premises, the resulting statement, while consistent, is not the basis for a sound argument for legalizing abortion. Even if we do need methods of population control, and even if it's too late to talk about birth control after the woman is pregnant, the fact, if it is a fact, that abortion is the taking of an innocent human life seems to rule it out as a method of population control. At least it does, so long as we do not also allow infanticide, or even just plain killing of randomly selected adults, as methods of population control.

Argument (ii) won't do for slightly more complicated reasons. As it stands, it seems that one of its premises is that the destruction of a foetus is nothing more than the destruction of some part of a woman's body. As such, argument (ii) has a premise incompatible with (1). But I sometimes suspect that the people offering this argument (the so-called "woman's liberation argument") really only want to be arguing that women, like all other people, should be free to do what they want to do. If this is what the argument is, it still is wrong. People do not, and should not, have the right to do *anything* they want to do, and if (1) is true, then it looks as though abortion is the sort of thing that no one should have the right to do.

Argument (iii) is like argument (v). Although we may want to grant the truth of its premises that a doctor should be free to practice medicine as he sees fit and that the doctor–patient relation should be private, and although these premises are compatible with (1), the argument does not go through when we conjoin (1) with its original premises. We would not want to say, I believe, that although abortion is the taking of an innocent human life, we should have no law against it, in order to preserve the rights of the doctor and the privacy of the doctor–patient relationship. There are, after all, some limitations that will have to be placed upon these rights, like all other rights, and certainly the most obvious case for this is when these limitations are necessary to save an innocent human life.

Argument (iv) has a special type of shortcoming. It presupposes that any abortion law would allow for at least some cases of legalized abortion. It then goes on to argue that such an allowance will make any such law unfairly (and in the context of American law, unconstitutionally) vague, and concludes from this that all abortions should therefore be legal. But given (1), why should any abortions be allowed? Can we take one innocent life, the life of the child, to save another innocent life, the life of the mother? And even if we grant that we should allow this special case, it seems odd to claim that we should therefore allow all cases just to avoid a vague law. Given the problem of justifying any abortions if (1) is true, wouldn't it be more reasonable to live with this unfortunately vague law, or not allow any legal abortions at all, rather than to allow all of them?

It therefore seems to me that arguments (i)–(v) cannot be used by a believer in (1) to support (2). But these were clearly the weaker arguments, even if they are often used. The far stronger arguments are (vi) and (vii), arguments that are often used to support even the morality of abortions, and (viii)–(x). All these arguments deal with the bad consequences of not having legalized abortions; they are, therefore, the type of argument that, as was argued earlier, is most likely to support the joint assertability of (1) and (2). We turn therefore to a consideration of them.

IV

Arguments (vi) and (vii) seem to have the following structure: There are certain social problems (the problem of the unwanted child, the family that can't afford another child, the young girl pregnant because she was raped, etc.) for which abortion offers a solution; therefore, we must legalize abortion so as to make this solution available when needed. The premises of these arguments, which assert that these social problems exist and that abortion would solve them, are, of course, compatible with (1). But will we still have a good argument if we conjoin (1) to these premises; i.e., would we want to argue that, although abortion is the taking of an innocent human life, it should, by being legalized, be made available as a solution to these problems?

It might be thought that the answer to this question is unequivocally no, that we shouldn't solve our social problems by allowing some people to take the lives of other innocent people. It would seem that the sanctity of life must take precedence over the solution to these problems, and that abortions should not be allowed even if no alternative solutions are available. But, in any case, they often are available. The family who cannot afford another child can be helped financially; the unwanted child can be put up for adoption. And when an alternative solution is not available, as in the case of the young girl who has been raped, we can at least alleviate that problem by providing her in her pregnancy with an environment in which she can continue her studies, lead her life without being ridiculed or made to feel shame, and in which she can receive psychological and social counseling. It would seem, therefore, that if (1) is true, arguments (vi) and (vii) won't do.

But perhaps we are going too quickly. After all, aren't there many things that we allow, as the best solution to a problem or simply because it is convenient, which are responsible for the taking of innocent human lives? We could, for example, save many innocent human lives by banning private automobiles, and yet we don't. And surely the problems we solve by allowing them are far less significant than those we can solve by legalizing abortion. This is not, however, a very persuasive analogy. In the automobile case, what we allow is, in itself, perfectly OK. It is the abuse of the right to drive a car that results in the loss of innocent human lives. But in the case of abortion, if (1) is right, then the very thing that we are allowing is the taking of innocent human lives, and this is a very different matter.

We turn now to (viii). It must be noted, to begin with, that (viii) grossly exaggerates when it claims that the only result of our abortion laws is a disrespect for the law. It surely is the case that a good percentage of contemplated abortions are not carried out just because they are illegal (and, therefore, unsafe and expensive), even though a good percentage are carried out even though they are illegal. So what (viii) must really be arguing is that the disrespect for the law engendered by abortion laws is so great and poses such a problem that we must legalize abortions, even at the expense of an increase in the number of abortions due to the actual performance of at least some of those abortions now contemplated but not performed because of their illegality. The question we must now consider is whether one can still argue this way if (1) is true. This is indeed a difficult matter, but I am inclined to say that one cannot. After all, one must suppose that abortion is now seriously considered by a great many women who then reject it because of its illegality but who would have the abortion if it were legal. So legalized abortion would result, if (1) is right, in a great many additional cases of the taking of innocent human lives. It seems hard to believe, however, that the problem of disrespect for the law produced by abortion laws is so great that we must pay so drastic a price for its solution. The situation would be very different, of course, if (viii) were literally true, and few, if any, abortions were prevented by their being illegal—but this seems highly unlikely.

Argument (viii) derives much of its plausibility, I suppose, from the comparison with prohibition. But these two situations are very different. We today feel that prohibition attempted to solve a social problem at much too great a price in terms of disrespect for the law because we feel that the only problem about drinking is excessive drinking. As serious a problem as that may be, it is surely far less serious than the problem posed by abortion, the problem of the taking of innocent human lives, if (1) is right. And, secondly, it seems likely that the drinking laws may have even increased the number of drinkers, and not decreased them; but this can hardly be said for the abortion laws.

(ix) is, in a way, the strongest of the arguments. It claims that we must abolish laws prohibiting abortions so as to save the lives of those unfortunate women killed by abortion quacks. It would also seem that the sanctity of human life requires the abolishment of abortion laws. But I am afraid that even this argument will not do if we grant the truth of (1). If (1) is true, this argument would be calling upon us to legalize abortions so that, by increasing the number of innocent human lives taken by abortionists, we can save the lives of some of those people, the mothers, guilty of the offense of taking these innocent human lives. This surely seems wrong. Even if, as is very unlikely, the number of human lives saved is greater if we legalize abortions, surely the innocent lives (the foetuses) should take precedence over the lives of those (the mothers) who would take them. And if, as is far more likely, the number of human lives saved would be greater if we kept the abortion laws, then this argument makes no sense at all.

Before leaving arguments (vi)–(ix), I would like to add one additional point. All I have been trying to show is that, if (1) is true, then the problems raised in (vi)–(ix) won't justify the legalization of abortions. This is not to say, however, that these are not serious problems. They are indeed extremely serious, and we must do our best to solve them, but they are not serious enough to justify the taking of innocent lives in order to solve them.

We are left with (x). (x), as a claim about the hypocrisy of society, is very powerful. There are far too many people who oppose (or do nothing for) the legalization of abortion laws because they know that the problems created by these laws will never disturb them–they are wealthy or influential enough to be able to get for themselves or for their wives, girl friends, etc. a legal or safe, if illegal, abortion. But as an argument for legalizing abortion even though (1) is true, (x) will not do. It is then, after all, arguing that, since there is an inequality about who gets away with murder, we should allow everyone to get away with it. The obvious alternative to be pursued, even if it is difficult to achieve, is the abolishment of this inequality by no one's getting away with it. Even if this can never be achieved, however, this argument will not do: no man can claim an equal right to do something if that thing is a thing, like the taking of an innocent human life, which no man has a right to do. . . .

VI

Let me end by saying a few words about how the problem now stands. It was not my intention, in this paper, to consider the truth of (1), that so difficult question which permeates the abortion issue. What was considered in this paper was the possibility of arguing for the legalization of abortion even if (1) is true. But, unfortunately, no such argument seems forthcoming; so it looks as if the legal problem about abortion cannot be resolved independently of the status of the foetus problem.

I should like to suggest, however, that there may be other aspects of the abortion problem that are resolvable independently of the status of the foetus problem. One such possible aspect is the moral justification of abortions in cases where they are necessary to save the life of the mother. But this is an issue that we will have to consider on another occasion.

A DEFENSE OF ABORTION[1]

JUDITH JARVIS THOMSON

Most opposition to abortion relies on the premise that the fetus is a human being, a person, from the moment of conception. The premise is argued for, but, as I think, not well. Take, for example, the most common argument. We are asked to notice that the development of a human being from conception through birth into childhood is continuous; then it is said that to draw a line, to choose a point in this development and say "before this point the thing is not a person, after this point it is a person" is to make an arbitrary choice, a choice for which in the nature of things no good reason can be given. It is concluded that the fetus is, or anyway that we had better say it is, a person from the moment of conception. But this conclusion does not follow. Similar things might be said about the development of an acorn into an oak tree, and it does not follow that acorns are oak trees, or that we had better say they are. Arguments of this form are sometimes called "slippery slope arguments"—the phrase is perhaps self-explanatory—and it is dismaying that opponents of abortion rely on them so heavily and uncritically.

I am inclined to agree, however, that the prospects for "drawing a line" in the development of the fetus look dim. I am inclined to think also that we shall probably have to agree that the fetus has already become a human person well before birth. Indeed, it comes as a surprise when one first learns how early in its life it begins to acquire human characteristics. By the tenth week, for example, it already has a face, arms and legs, fingers and toes; it has internal organs, and brain activity is detectable.[2] On the other hand, I think that the premise is false, that the fetus is not a person from the moment of conception. A newly fertilized

---

"A Defense of Abortion," by Judith Jarvis Thomson, *Philosophy and Public Affairs*, vol. 1, no. 1 (copyright © 1971 by Princeton University Press), pp. 47–66. Reprinted by permission of Princeton University Press.

[1] I am very much indebted to James Thomson for discussion, criticism, and many helpful suggestions.

[2] Daniel Callahan, *Abortion: Law, Choice and Morality* (New York, 1970), p. 373. This book gives a fascinating survey of the available information on abortion. The Jewish tradition is surveyed in David M. Feldman, *Birth Control in Jewish Law* (New York, 1968), Part 5, the Catholic tradition in John T. Noonan, Jr., "An Almost Absolute Value in History," in *The Morality of Abortion*, ed. John T. Noonan, Jr. (Cambridge, Mass., 1970).

ovum, a newly implanted clump of cells, is no more a person than an acorn is an oak tree. But I shall not discuss any of this. For it seems to me to be of great interest to ask what happens if, for the sake of argument, we allow the premise. How, precisely, are we supposed to get from there to the conclusion that abortion is morally impermissible? Opponents of abortion commonly spend most of their time establishing that the fetus is a person, and hardly any time explaining the step from there to the impermissibility of abortion. Perhaps they think the step too simple and obvious to require much comment. Or perhaps instead they are simply being economical in argument. Many of those who defend abortion rely on the premise that the fetus is not a person, but only a bit of tissue that will become a person at birth; and why pay out more arguments than you have to? Whatever the explanation, I suggest that the step they take is neither easy nor obvious, that it calls for closer examination than it is commonly given, and that when we do give it this closer examination we shall feel inclined to reject it.

I propose, then, that we grant that the fetus is a person from the moment of conception. How does the argument go from here? Something like this, I take it. Every person has a right to life. So the fetus has a right to life. No doubt the mother has a right to decide what shall happen in and to her body; everyone would grant that. But surely a person's right to life is stronger and more stringent than the mother's right to decide what happens in and to her body, and so outweighs it. So the fetus may not be killed; an abortion may not be performed.

It sounds plausible. But now let me ask you to imagine this. You wake up in the morning and find yourself back to back in bed with an unconscious violinist. A famous unconscious violinist. He has been found to have a fatal kidney ailment, and the Society of Music Lovers has canvassed all the available medical records and found that you alone have the right blood type to help. They have therefore kidnapped you, and last night the violinist's circulatory system was plugged into yours, so that your kidneys can be used to extract poisons from his blood as well as your own. The director of the hospital now tells you, "Look, we're sorry the Society of Music Lovers did this to you—we would never have permitted it if we had known. But still, they did it, and the violinist now is plugged into you. To unplug you would be to kill him. But never mind, it's only for nine months. By then he will have recovered from his ailment, and can safely be unplugged from you." Is it morally incumbent on you to accede to this situation? No doubt it would be very nice of you if you did, a great kindness. But do you *have* to accede to it? What if it were not nine months, but nine years? Or longer still? What if the director of the hospital says, "Tough luck, I agree, but you've now got to stay in bed, with the violinist plugged into you, for the rest of your life. Because remember this. All persons have a right to life, and violinists are persons. Granted you have a right to decide what happens in and to your body, but a person's right to life outweighs your right to decide what happens in and to your body. So you cannot ever be unplugged from him." I imagine you would regard this as outrageous, which suggests that something really is wrong with that plausible-sounding argument I mentioned a moment ago.

In this case, of course, you were kidnapped; you didn't volunteer for the operation that plugged the violinist into your kidneys. Can those who oppose abortion on the ground I mentioned make an exception for a pregnancy due to rape? Certainly. They can say that persons have a right to life only if they didn't come into existence because of rape; or they can say that all persons have a right to life, but that some have less of a right to life than others, in particular, that those who came into existence because of rape have less. But these statements have a rather unpleasant sound. Surely the question of whether you have a right to life at all, or how much of it you have, shouldn't turn on the question of whether or not you are the product of a rape. And in fact the people who oppose abortion on the ground I mentioned do not make this distinction, and hence do not make an exception in case of rape.

Nor do they make an exception for a case in which the mother has to spend the nine months of her pregnancy in bed. They would agree that would be a great pity, and hard on the mother; but all the same, all persons have a right to life, the fetus is a person, and so on. I suspect, in fact, that they would not make an exception for a case in which, miraculously enough, the pregnancy went on for nine years, or even the rest of the mother's life.

Some won't even make an exception for a case in which continuation of the pregnancy is likely to shorten the mother's life; they regard abortion as impermissible even to save the mother's life. Such cases are nowadays very rare, and many opponents of abortion do not accept this extreme view. All the same, it is a good place to begin: a number of points of interest come out in respect to it.

1. Let us call the view that abortion is impermissible even to save the mother's life "the extreme view." I want to suggest first that it does not issue from the argument I mentioned earlier without the addition of some fairly powerful premises. Suppose a woman has become pregnant, and now learns that she has a cardiac condition such that she will die if she carries the baby to term. What may be done for her? The fetus, being a person, has a right to life, but as the mother is a person too, so has she a right to life. Presumably they have an equal right to life. How is it supposed to come out that an abortion may not be performed? If mother and child have an equal right to life, shouldn't we perhaps flip a coin? Or should we add to the mother's right to life her right to decide what happens in and to her body, which everybody seems to be ready to grant—the sum of her rights now outweighing the fetus' right to life?

The most familiar argument here is the following. We are told that performing the abortion would be directly killing[3] the child, whereas doing nothing would not be killing the mother, but only letting her die. Moreover, in killing the child, one would be killing an innocent person, for the child has committed no crime, and is not aiming at his mother's death. And then there are a variety of ways in which this might be continued. (1) But as directly killing an innocent person is always and absolutely impermissible, an abortion may not be performed. Or, (2) as directly killing an innocent person is murder, and murder is always and absolutely impermissible, an abortion may not be performed.[4] Or, (3) as one's duty to refrain from directly killing an innocent person is more stringent than one's duty to keep a person from dying, an abortion may not be performed. Or, (4) if one's only options are directly killing an innocent person or letting a person die, one must prefer letting the person die, and thus an abortion may not be performed.[5]

Some people seem to have thought that these are not further premises which must be added if the conclusion is to be reached, but that they follow from the very fact that an innocent person has a right to life.[6] But this seems to me to be a mistake, and perhaps the simplest way to show this is to bring out that while we must certainly grant that innocent persons have a right to life, the theses in (1) through (4) are all false. Take (2), for example. If directly killing an innocent person is murder, and thus is impermissible, then the mother's directly killing the innocent person inside her is murder, and thus is impermissible. But it

[3] The term "direct" in the arguments I refer to is a technical one. Roughly, what is meant by "direct killing" is either killing as an end in itself, or killing as a means to some end, for example, the end of saving someone else's life. See note 6, below, for an example of its use.

[4] Cf. *Encyclical Letter of Pope Pius XI on Christian Marriage*, St. Paul Editions (Boston, n.d.), p. 32: "however much we may pity the mother whose health and even life is gravely imperiled in the performance of the duty allotted to her by nature, nevertheless what could ever be a sufficient reason for excusing in any way the direct murder of the innocent? This is precisely what we are dealing with here." Noonan (*The Morality of Abortion*, p. 43) reads this as follows: "What cause can ever avail to excuse in any way the direct killing of the innocent? For it is a question of that."

[5] The thesis in (4) is in an interesting way weaker than those in (1), (2), and (3): they rule out abortion even in cases in which both mother *and* child will die if the abortion is not performed. By contrast, one who held the view expressed in (4) could consistently say that one needn't prefer letting two persons die to killing one.

[6] Cf. the following passage from Pius XII, *Address to the Italian Catholic Society of Midwives:* "The baby in the maternal breast has the right to life immediately from God.–Hence there is no man, no human authority, no science, no medical, eugenic, social, economic or moral 'indication' which can establish or grant a valid juridical ground for a direct deliberate disposition of an innocent human life, that is a disposition which looks to its destruction either as an end or as a means to another end perhaps in itself not illicit.–The baby, still not born, is a man in the same degree and for the same reason as the mother" (quoted in Noonan, *The Morality of Abortion*, p. 45).

cannot seriously be thought to be murder if the mother performs an abortion on herself to save her life. It cannot seriously be said that she *must* refrain, that she *must* sit passively by and wait for her death. Let us look again at the case of you and the violinist. There you are, in bed with the violinist, and the director of the hospital says to you, "It's all most distressing, and I deeply sympathize, but you see this is putting an additional strain on your kidneys, and you'll be dead within the month. But you *have* to stay where you are all the same. Because unplugging you would be directly killing an innocent violinist, and that's murder, and that's impermissible." If anything in the world is true, it is that you do not commit murder, you do not do what is impermissible, if you reach around to your back and unplug yourself from that violinist to save your life.

The main focus of attention in writings on abortion has been on what a third party may or may not do in answer to a request from a woman for an abortion. This is in a way understandable. Things being as they are, there isn't much a woman can safely do to abort herself. So the question asked is what a third party may do, and what the mother may do, if it is mentioned at all, is deduced, almost as an afterthought, from what it is concluded that third parties may do. But it seems to me that to treat the matter in this way is to refuse to grant to the mother that very status of person which is so firmly insisted on for the fetus. For we cannot simply read off what a person may do from what a third party may do. Suppose you find yourself trapped in a tiny house with a growing child. I mean a very tiny house, and a rapidly growing child—you are already up against the wall of the house and in a few minutes you'll be crushed to death. The child on the other hand won't be crushed to death; if nothing is done to stop him from growing he'll be hurt, but in the end he'll simply burst open the house and walk out a free man. Now I could well understand it if a bystander were to say, "There's nothing we can do for you. We cannot choose between your life and his, we cannot be the ones to decide who is to live, we cannot intervene." But it cannot be concluded that you too can do nothing, that you cannot attack it to save your life. However innocent the child may be, you do not have to wait passively while it crushes you to death. Perhaps a pregnant woman is vaguely felt to have the status of house, to which we don't allow the right of self-defense. But if the woman houses the child, it should be remembered that she is a person who houses it.

I should perhaps stop to say explicitly that I am not claiming that people have a right to do anything whatever to save their lives. I think, rather, that there are drastic limits to the right of self-defense. If someone threatens you with death unless you torture someone else to death, I think you have not the right, even to save your life, to do so. But the case under consideration here is very different. In our case there are only two people involved, one whose life is threatened, and one who threatens it. Both are innocent: the one who is threatened is not threatened because of any fault, the one who threatens does not threaten because of any fault. For this reason we may feel that we bystanders cannot intervene. But the person threatened can.

In sum, a woman surely can defend her life against the threat to it posed by the unborn child, even if doing so involves its death. And this shows not merely that the theses in (1) through (4) are false; it shows also that the extreme view of abortion is false, and so we need not canvass any other possible ways of arriving at it from the argument I mentioned at the outset.

2. The extreme view could of course be weakened to say that while abortion is permissible to save the mother's life, it may not be performed by a third party, but only by the mother herself. But this cannot be right either. For what we have to keep in mind is that the mother and the unborn child are not like two tenants in a small house which has, by an unfortunate mistake, been rented to both: the mother *owns* the house. The fact that she does adds to the offensiveness of deducing that the mother can do nothing from the supposition that third parties can do nothing. But it does more than this: it casts a bright light on the supposition that third parties can do nothing. Certainly it lets us see that a third party who says "I cannot choose between you" is fooling himself if he thinks this is impartiality. If Jones has found and fastened on a certain coat, which he needs to keep him from freezing, but which Smith also needs to keep him from freezing, then it is not impartiality that says "I cannot choose between you" when Smith owns the coat. Women have said again and again "This body is *my* body!" and they have reason to feel angry, reason to feel that it has been like shouting into the wind. Smith, after all, is hardly likely to bless us if we say to him, "Of course it's your coat, anybody would grant that it is. But no one may choose between you and Jones who is to have it."

We should really ask what it is that says "no one may choose" in the face of the fact that the body that houses the child is the mother's body. It may be simply a failure to appreciate this fact. But it may be something more interesting, namely the sense that one has a right to refuse to lay hands on people, even where it would be just and fair to do so, even where justice seems to require that somebody do so. Thus justice might call for somebody to get Smith's coat back from Jones, and yet you have a right to refuse to be the one to lay hands on Jones, a right to refuse to do physical violence to him. This, I think, must be granted. But then what should be said is not "no one may choose," but only "*I* cannot choose," and indeed not even this, but "*I* will not *act,*" leaving it open that somebody else can or should, and in particular that anyone in a position of authority, with the job of securing people's rights, both can and should. So this is no difficulty. I have not been arguing that any given third party must accede to the mother's request that he perform an abortion to save her life, but only that he may.

I suppose that in some views of human life the mother's body is only on loan to her, the loan not being one which gives her any prior claim to it. One who held this view might well think it impartiality to say "I cannot choose." But I shall simply ignore this possibility. My own view is that if a human being has any just, prior claim to anything at all, he has a just, prior claim to his own body. And perhaps this needn't be argued for here anyway, since, as I mentioned, the arguments against abortion we are looking at do grant that the woman has a right to decide what happens in and to her body.

But although they do grant it, I have tried to show that they do not take seriously what is done in granting it. I suggest the same thing will reappear even more clearly when we turn away from cases in which the mother's life is at stake, and attend, as I propose we now do, to the vastly more common cases in which a woman wants an abortion for some less weighty reason than preserving her own life.

3. Where the mother's life is not at stake, the argument I mentioned at the outset seems to have a much stronger pull. "Everyone has a right to life, so the unborn person has a right to life." And isn't the child's right to life weightier than anything other than the mother's own right to life, which she might put forward as ground for an abortion?

This argument treats the right to life as if it were unproblematic. It is not, and this seems to me to be precisely the source of the mistake.

For we should now, at long last, ask what it comes to, to have a right to life. In some views having a right to life includes having a right to be given at least the bare mimimum one needs for continued life. But suppose that what in fact *is* the bare minimum a man needs for continued life is something he has no right at all to be given? If I am sick unto death, and the only thing that will save my life is the touch of Henry Fonda's cool hand on my fevered brow, then all the same, I have no right to be given the touch of Henry Fonda's cool hand on my fevered brow. It would be frightfully nice of him to fly in from the West Coast to provide it. It would be less nice, though no doubt well meant, if my friends flew out to the West Coast and carried Henry Fonda back with them. But I have no right at all against anybody that he should do this for me. Or again, to return to the story I told earlier, the fact that for continued life that violinist needs the continued use of your kidneys does not establish that he has a right to be given the continued use of your kidneys. He certainly has no right against you that *you* should give him continued use of your kidneys. For nobody has any right to use your kidneys unless you give him such a right; and nobody has the right against you that you shall give him this right—if you do allow him to go on using your kidneys, this is a kindness on your part, and not something he can claim from you as his due. Nor has he any right against anybody else that *they* should give him continued use of your kidneys. Certainly he had no right against the Society of Music Lovers that they should plug him into you in the first place. And if you now start to unplug yourself, having learned that you will otherwise have to spend nine years in bed with him, there is nobody in the world who must try to prevent you, in order to see to it that he is given something he has a right to be given.

Some people are rather stricter about the right to life. In their view, it does not include the right to be given anything, but amounts to, and only to, the right not to be killed by anybody. But here a related difficulty arises. If everybody is to refrain from killing that violinist, then everybody must refrain from doing a great many different sorts of things. Everybody must refrain from slitting his throat, everybody must refrain from shooting him—and everybody must refrain from unplugging you from him. But does he have a right against everybody that they shall refrain from unplugging you from him? To refrain from doing this is to allow him to continue to use your kidneys. It could be argued that he has a right against us that *we* should allow him to continue to use your kidneys. That is, while he had no right against us that we should give him the use of your kidneys, it might be argued that he anyway has a right against us that we shall not now intervene and deprive him of the use of your kidneys. I shall come back to third-party interventions later. But certainly the violinist has no right against you that *you* shall allow him to continue to use your kidneys. As I said, if you do allow him to use them, it is a kindness on your part, and not something you owe him.

The difficulty I point to here is not peculiar to the right to life. It reappears in connection with all the other natural rights; and it is something which an adequate account of rights must deal with. For present purposes it is enough just to draw attention to it. But I would stress that I am not arguing that people do not have a right to life—quite to the contrary, it seems to me that the primary control we must place on the acceptability of an account of rights is that it should turn out in that account to be a truth that all persons have a right to life. I am arguing only that having a right to life does not guarantee having either a right to be given the use of or a right to be allowed continued use of another person's body—even if one needs it for life itself. So the right to life will not serve the opponents of abortion in the very simple and clear way in which they seem to have thought it would.

4. There is another way to bring out the difficulty. In the most ordinary sort of case, to deprive someone of what he has a right to is to treat him unjustly. Suppose a boy and his small brother are jointly given a box of chocolates for Christmas. If the older boy takes the box and refuses to give his brother any of the chocolates, he is unjust to him, for the brother has been given a right to half of them. But suppose that, having learned that otherwise it means nine years in bed with that violinist, you unplug yourself from him. You surely are not being unjust to him, for you gave him no right to use your kidneys, and no one else can have given him any such right. But we have to notice that in unplugging yourself, you are killing him; and violinists, like everybody else, have a right to life, and thus in the view we were considering just now, the right not

to be killed. So here you do what he supposedly has a right you shall not do, but you do not act unjustly to him in doing it.

The emendation which may be made at this point is this: the right to life consists not in the right not to be killed, but rather in the right not to be killed unjustly. This runs a risk of circularity, but never mind: it would enable us to square the fact that the violinist has a right to life with the fact that you do not act unjustly toward him in unplugging yourself, thereby killing him. For if you do not kill him unjustly, you do not violate his right to life, and so it is no wonder you do him no injustice.

But if this emendation is accepted, the gap in the argument against abortion stares us plainly in the face: it is by no means enough to show that the fetus is a person, and to remind us that all persons have a right to life—we need to be shown also that killing the fetus violates its right to life, i.e., that abortion is unjust killing. And is it?

I suppose we may take it as a datum that in a case of pregnancy due to rape the mother has not given the unborn person a right to the use of her body for food and shelter. Indeed, in what pregnancy could it be supposed that the mother has given the unborn person such a right? It is not as if there were unborn persons drifting about the world, to whom a woman who wants a child says "I invite you in."

But it might be argued that there are other ways one can have acquired a right to the use of another person's body than by having been invited to use it by that person. Suppose a woman voluntarily indulges in intercourse, knowing of the chance it will issue in pregnancy, and then she does become pregnant; is she not in part responsible for the presence, in fact the very existence, of the unborn person inside her? No doubt she did not invite it in. But doesn't her partial responsibility for its being there itself give it a right to the use of her body?[7] If so, then her aborting it would be more like the boy's taking away the chocolates, and less like your unplugging yourself from the violinist—doing so would be depriving it of what it does have a right to, and thus would be doing it an injustice.

And then, too, it might be asked whether or not she can kill it even to save her own life: If she voluntarily called it into existence, how can she now kill it, even in self-defense?

The first think to be said about this is that it is something new. Opponents of abortion have been so concerned to make out the independence of the fetus, in order to establish that it has a right to life, just as its mother does, that they have tended to overlook the possible support they might gain from making out that the fetus is *dependent* on the mother, in order to establish that she has a

[7] The need for a discussion of this argument was brought home to me by members of the Society for Ethical and Legal Philosophy, to whom this paper was originally presented.

special kind of responsibility for it, a responsibility that gives it rights against her which are not possessed by any independent person—such as an ailing violinist who is a stranger to her.

On the other hand, this argument would give the unborn person a right to its mother's body only if her pregnancy resulted from a voluntary act, undertaken in full knowledge of the chance a pregnancy might result from it. It would leave out entirely the unborn person whose existence is due to rape. Pending the availability of some further argument, then, we would be left with the conclusion that unborn persons whose existence is due to rape have no right to the use of their mothers' bodies, and thus that aborting them is not depriving them of anything they have a right to and hence is not unjust killing.

And we should also notice that it is not at all plain that this argument really does go even as far as it purports to. For there are cases and cases, and the details make a difference. If the room is stuffy, and I therefore open a window to air it, and a burglar climbs in, it would be absurd to say, "Ah, now he can stay, she's given him a right to the use of her house—for she is partially responsible for his presence there, having voluntarily done what enabled him to get in, in full knowledge that there are such things as burglars, and that burglars burgle." It would be still more absurd to say this if I had had bars installed outside my windows, precisely to prevent burglars from getting in, and a burglar got in only because of a defect in the bars. It remains equally absurd if we imagine it is not a burglar who climbs in, but an innocent person who blunders or falls in. Again, suppose it were like this: people-seeds drift about in the air like pollen, and if you open your windows, one may drift in and take root in your carpets or upholstery. You don't want children, so you fix up your windows with fine mesh screens, the very best you can buy. As can happen, however, and on very, very rare occasions does happen, one of the screens is defective; and a seed drifts in and takes root. Does the person-plant who now develops have a right to the use of your house? Surely not—despite the fact that you voluntarily opened your windows, you knowingly kept carpets and unholstered furniture, and you knew that screens were sometimes defective. Someone may argue that you are responsible for its rooting, that it does have a right to your house, because after all you *could* have lived out your life with bare floors and furniture, or with sealed windows and doors. But this won't do—for by the same token anyone can avoid a pregnancy due to rape by having a hysterectomy, or anyway by never leaving home without a (reliable!) army.

It seems to me that the argument we are looking at can establish at most that there are *some* cases in which the unborn person has a right to the use of its mother's body, and therefore *some* cases in which abortion is unjust killing. There is room for much discussion and argument as to precisely which, if any. But I think we should sidestep this issue and leave it open, for at any rate the argument certainly does not establish that all abortion is unjust killing.

5. There is room for yet another argument here, however. We surely must all grant that there may be cases in which it would be morally indecent to detach a person from your body at the cost of his life. Suppose you learn that what the violinist needs is not nine years of your life, but only one hour: all you need do to save his life is to spend one hour in that bed with him. Suppose also that letting him use your kidneys for that one hour would not affect your health in the slightest. Admittedly you were kidnapped. Admittedly you did not give anyone permission to plug him into you. Nevertheless it seems to me plain you *ought* to allow him to use your kidneys for that hour–it would be indecent to refuse.

Again, suppose pregnancy lasted only an hour, and constituted no threat to life or health. And suppose that a woman becomes pregnant as a result of rape. Admittedly she did not voluntarily do anything to bring about the existence of a child. Admittedly she did nothing at all which would give the unborn person a right to the use of her body. All the same it might well be said, as in the newly emended violinist story, that she *ought* to allow it to remain for that hour–that it would be indecent in her to refuse.

Now some people are inclined to use the term "right" in such a way that it follows from the fact that you ought to allow a person to use your body for the hour he needs, that he has a right to use your body for the hour he needs, even though he has not been given that right by any person or act. They may say that it follows also that if you refuse, you act unjustly toward him. This use of the term is perhaps so common that it cannot be called wrong; nevertheless it seems to me to be an unfortunate loosening of what we would do better to keep a tight rein on. Suppose that box of chocolates I mentioned earlier had not been given to both boys jointly, but was given only to the older boy. There he sits, stolidly eating his way through the box, his small brother watching enviously. Here we are likely to say "You ought not to be so mean. You ought to give your brother some of those chocolates." My own view is that it just does not follow from the truth of this that the brother has any right to any of the chocolates. If the boy refuses to give his brother any, he is greedy, stingy, callous–but not unjust. I suppose that the people I have in mind will say it does follow that the brother has a right to some of the chocolates, and thus that the boy does act unjustly if he refuses to give his brother any. But the effect of saying this is to obscure what we should keep distinct, namely the difference between the boy's refusal in this case and the boy's refusal in the earlier case, in which the box was given to both boys jointly, and in which the small brother thus had what was from any point of view clear title to half.

A further objection to so using the term "right" that from the fact that A ought to do a thing for B, it follows that B has a right against A that A do it for him, is that it is going to make the question of whether or not a man has a right to a thing turn on how easy it is to provide him with it; and this seems not merely unfortunate, but morally unacceptable. Take the case of Henry Fonda again. I said earlier that I had no right to the touch of his cool hand on my fevered brow, even though I needed it to save my life. I said it would be frightfully nice of him to fly in from the West Coast to provide me with it, but that I had no right against him that he should do so. But suppose he isn't on the West Coast. Suppose he has only to walk across the room, place a hand briefly on my brow—and lo, my life is saved. Then surely he ought to do it, it would be indecent to refuse. Is it to be said "Ah, well, it follows that in this case she has a right to the touch of his hand on her brow, and so it would be an injustice in him to refuse"? So that I have a right to it when it is easy for him to provide it, though no right when it's hard? It's rather a shocking idea that anyone's rights should fade away and disappear as it gets harder and harder to accord them to him.

So my own view is that even though you ought to let the violinist use your kidneys for the one hour he needs, we should not conclude that he has a right to do so—we should say that if you refuse, you are, like the boy who owns all the chocolates and will give none away, self-centered and callous, indecent in fact, but not unjust. And similarly, that even supposing a case in which a woman pregnant due to rape ought to allow the unborn person to use her body for the hour he needs, we should not conclude that he has a right to do so; we should conclude that she is self-centered, callous, indecent, but not unjust, if she refuses. The complaints are no less grave; they are just different. However, there is no need to insist on this point. If anyone does wish to deduce "he has a right" from "you ought," then all the same he must surely grant that there are cases in which it is not morally required of you that you allow that violinist to use your kidneys, and in which he does not have a right to use them, and in which you do not do him an injustice if you refuse. And so also for mother and unborn child. Except in such cases as the unborn person has a right to demand it—and we were leaving open the possibility that there may be such cases—nobody is morally *required* to make large sacrifices, of health, of all other interests and concerns, of all other duties and commitments, for nine years, or even for nine months, in order to keep another person alive.

6. We have in fact to distinguish between two kinds of Samaritan: the Good Samaritan and what we might call the Minimally Decent Samaritan. The story of the Good Samaritan, you will remember, goes like this:

> A certain man went down from Jerusalem to Jericho, and fell among thieves, which stripped him of his raiment, and wounded him, and departed, leaving him half dead.
>
> And by chance there came down a certain priest that way; and when he saw him, he passed by on the other side.
>
> And likewise a Levite, when he was at the place, came and looked on him, and passed by on the other side.

> But a certain Samaritan, as he journeyed, came where he was; and when he saw him he had compassion on him.
>
> And went to him, and bound up his wounds, pouring in oil and wine, and set him on his own beast, and brought him to an inn, and took care of him.
>
> And on the morrow, when he departed, he took out two pence, and gave them to the host, and said unto him, "Take care of him; and whatsoever thou spendest more, when I come again, I will repay thee." (*Luke 10:30–35*)

The Good Samaritan went out of his way, at some cost to himself, to help one in need of it. We are not told what the options were, that is, whether or not the priest and the Levite could have helped by doing less than the Good Samaritan did, but assuming they could have, then the fact they did nothing at all shows they were not even Minimally Decent Samaritans, not because they were not Samaritans, but because they were not even minimally decent.

These things are a matter of degree, of course, but there is a difference, and it comes out perhaps most clearly in the story of Kitty Genovese, who, as you will remember, was murdered while thirty-eight people watched or listened, and did nothing at all to help her. A Good Samaritan would have rushed out to give direct assistance against the murderer. Or perhaps we had better allow that it would have been a Splendid Samaritan who did this, on the ground that it would have involved a risk of death for himself. But the thirty-eight not only did not do this, they did not even trouble to pick up a phone to call the police. Minimally Decent Samaritanism would call for doing at least that, and their not having done it was monstrous.

After telling the story of the Good Samaritan, Jesus said "Go, and do thou likewise." Perhaps he meant that we are morally required to act as the Good Samaritan did. Perhaps he was urging people to do more than is morally required of them. At all events it seems plain that it was not morally required of any of the thirty-eight that he rush out to give direct assistance at the risk of his own life, and that it is not morally required of anyone that he give long stretches of his life—nine years or nine months—to sustaining the life of a person who has no special right (we were leaving open the possibility of this) to demand it.

Indeed, with one rather striking class of exceptions, no one in any country of the world is *legally* required to do anywhere near as much as this for anyone else. The class of exceptions is obvious. My main concern here is not the state of the law in respect to abortion, but it is worth drawing attention to the fact that in no state in this country is any man compelled by law to be even a Minimally Decent Samaritan to any person; there is no law under which charges could be brought against the thirty-eight who stood by while Kitty Genovese died. By contrast, in most states in this country women are compelled by law to be not merely Minimally Decent Samaritans, but Good Samaritans to unborn persons inside them. This doesn't by itself settle anything one way or the other, because it may well be argued that there should be laws in this country—as there are in many European countries—compelling at least Minimally Decent Samaritanism.[8] But it does show that there is a gross injustice in the existing state of the law. And it shows also that the groups currently working against liberalization of abortion laws, in fact working toward having it declared unconstitutional for a state to permit abortion, had better start working for the adoption of Good Samaritan laws generally, or earn the charge that they are acting in bad faith.

I should think, myself, that Minimally Decent Samaritan laws would be one thing, Good Samaritan laws quite another, and in fact highly improper. But we are not here concerned with the law. What we should ask is not whether anybody should be compelled by law to be a Good Samaritan, but whether we must accede to a situation in which somebody is being compelled—by nature, perhaps—to be a Good Samaritan. We have, in other words, to look now at third-party interventions. I have been arguing that no person is morally required to make large sacrifices to sustain the life of another who has no right to demand them, and this even where the sacrifices do not include life itself; we are not morally required to be Good Samaritans or anyway Very Good Samaritans to one another. But what if a man cannot extricate himself from such a situation? What if he appeals to us to extricate him? It seems to me plain that there are cases in which we can, cases in which a Good Samaritan would extricate him. There you are, you were kidnapped, and nine years in bed with that violinist lie ahead of you. You have your own life to lead. You are sorry, but you simply cannot see giving up so much of your life to the sustaining of his. You cannot extricate yourself, and ask us to do so. I should have thought that—in light of his having no right to the use of your body—it was obvious that we do not have to accede to your being forced to give up so much. We can do what you ask. There is no injustice to the violinist in our doing so.

7. Following the lead of the opponents of abortion, I have throughout been speaking of the fetus merely as a person, and what I have been asking is whether or not the argument we began with, which proceeds only from the fetus's being a person, really does establish its conclusion. I have argued that it does not.

But of course there are arguments and arguments, and it may be said that I have simply fastened on the wrong one. It may be said that what is important is not merely the fact that the fetus is a person, but that it is a person for whom the woman has a special kind of responsibility issuing from the fact that she is its mother. And it might be argued that all my analogies are therefore irrelevant—for you do not have that special kind of responsibility for that violinist, Henry Fonda does not have that special kind of responsibility for me. And our

[8] For a discussion of the difficulties involved, and a survey of the European experience with such laws, see *The Good Samaritan and the Law*, ed. James M. Ratcliffe (New York, 1966).

attention might be drawn to the fact that men and women both *are* compelled by law to provide support for their children.

I have in effect dealt (briefly) with this argument in section 4 above; but a (still briefer) recapitulation now may be in order. Surely we do not have any such "special responsibility" for a person unless we have assumed it, explicitly or implicitly. If a set of parents do not try to prevent pregnancy, do not obtain an abortion, and then at the time of birth of the child do not put in out for adoption, but rather take it home with them, then they have assumed responsibility for it, they have given it rights, and they cannot *now* withdraw support from it at the cost of its life because they now find it difficult to go on providing for it. But if they have taken all reasonable precautions against having a child, they do not simply by virtue of their biological relationship to the child who comes into existence have a special responsibility for it. They may wish to assume responsibility for it, or they may not wish to. And I am suggesting that if assuming responsibility for it would require large sacrifices, then they may refuse. A Good Samaritan would not refuse—or anyway, a Splendid Samaritan, if the sacrifices that had to be made were enormous. But then so would a Good Samaritan assume responsibility for that violinist; so would Henry Fonda, if he is a Good Samaritan, fly in from the West Coast and assume responsibility for me.

8. My argument will be found unsatisfactory on two counts by many of those who want to regard abortion as morally permissible. First, while I do argue that abortion is not impermissible, I do not argue that it is always permissible. There may well be cases in which carrying the child to term requires only Minimally Decent Samaritanism of the mother, and this is a standard we must not fall below. I am inclined to think it a merit of my account precisely that it does *not* give a general yes or a general no. It allows for and supports our sense that, for example, a sick and desperately frightened fourteen-year-old schoolgirl, pregnant due to rape, may *of course* choose abortion, and that any law which rules this out is an insane law. And it also allows for and supports our sense that in other cases resort to abortion is even positively indecent. It would be indecent in the woman to request an abortion, and indecent in a doctor to perform it, if she is in her seventh month, and wants the abortion just to avoid the nuisance of postponing a trip abroad. The very fact that the arguments I have been drawing attention to treat all cases of abortion, or even all cases of abortion in which the mother's life is not at stake, as morally on a par ought to have made them suspect at the outset.

Secondly, while I am arguing for the permissibility of abortion in some cases, I am not arguing for the right to secure the death of the unborn child. It is easy to confuse these two things in that up to a certain point in the life of the fetus it is not able to survive outside the mother's body; hence removing it from her body guarantees its death. But they are importantly different. I have argued that you are not morally required to spend nine months in bed, sustaining the life of that violinist; but to say this is by no means to say that if, when you unplug yourself, there is a miracle and he survives, you then have a right to turn around and slit his throat. You may detach yourself even if this costs him his life; you have no right to be guaranteed his death, by some other means, if unplugging yourself does not kill him. There are some people who will feel dissatisfied by this feature of my argument. A woman may be utterly devastated by the thought of a child, a bit of herself, put out for adoption and never seen or heard of again. She may therefore want not merely that the child be detached from her, but more, that it die. Some opponents of abortion are inclined to regard this as beneath contempt—thereby showing insensitivity to what is surely a powerful source of despair. All the same, I agree that the desire for the child's death is not one which anybody may gratify, should it turn out to be possible to detach the child alive.

At this place, however, it should be remembered that we have only been pretending throughout that the fetus is a human being from the moment of conception. A very early abortion is surely not the killing of a person, and so is not dealt with by anything I have said here.

THE MORALITY OF ABORTION[1]

RICHARD B. BRANDT

The term "abortion" is used in this discussion to refer to deliberate removal (or deliberate action to cause the expulsion) of a fetus[2] from the womb of a human female, at the request or through the agency of the mother, so as in fact to result in the death of the fetus but with insignificantly small risk to the life or health of the mother. The question I raise is roughly whether abortion in that sense is morally wrong. I am not raising the question whether abortion should be prohibited by law. That is a very different question, and I am confining myself to the moral issue. There is another question I am not raising: whether a fetus should be removed, *irrespective* of the preferences of the mother, when there is good reason to think the child will be seriously defective, mentally or physically. Since it is a grave responsibility to bring a human being into the world, I think this latter is an important question; but I do not propose to discuss it here.

The question must be made more precise. First, let us distinguish the question whether abortion is morally wrong from the question whether a person who does it is morally blameworthy, that is, whether even if and when abortion is morally wrong there can be considerations which morally excuse the agent. (Some persons might use alternative language, saying that, even if abortion is morally wrong, it might be that a person who performed an abortion has sometimes not, in the total circumstances, committed a "sin"; but I prefer to avoid this term, with its theological overtones.) Let me explain how this might be. Suppose a physician performs an abortion in the sincere belief that in the circumstances it is his moral obligation to do so, and at some risk to his professional standing; I would think it highly likely that he is morally blameless

Taken from, *Abortion: Pro and Con,* R. L. Perkins, ed., (Cambridge, Mass: Schenkman Publishing Company, 1974), pp. 151–169. This article is a revision of one of the same title which first appeared in *The Monist, 56* (1972), pp. 504–526. Reprinted by permission of the author, Schenkman Publishing Co., R. L. Perkins, and the editors of *The Monist.*

[1] The following paper is a revision of one of the same title, published in *The Monist 56* (1972), pp. 504–26. The first part is substantially identical with the first part of the *Monist* paper; the second part has been completely altered and rewritten, although the most important idea is the same. I am grateful to the editors of the *Monist* for their permission to republish the parts of that article that reappear here.

[2] I shall use this term in a general sense to denote an unborn, potential human being, at any stage from the moment of conception until birth.

(unless his having the belief is itself culpable), even though the abortion was wrong. Again, the temptations for an unwed mother to commit an abortion on herself might be so strong that we might say her act was morally blameless, even though in fact wrong–since an action can hardly be morally blameworthy unless it shows a defect of character, and her act might not show a defect of character. Now I do not wish to discuss the question when a person may be morally blameworthy for performing an action. What I do wish to discuss is whether acts of abortion are morally wrong in the sense that their agents *would* be morally blameworthy *unless* their actions were *excused* (not justified, but excused) in some way. This question seems to me the most controverted one today.

The question I wish to discuss, however, is even narrower than this. There are some persons who are absolutists about abortion; that is, they hold that abortion in the sense explained above is *universally* and *unconditionally* morally wrong. This position is widely held, especially by writers in the Catholic Church. I do mean to discuss this issue by implication, but the thesis I wish to discuss is a weaker thesis than this one. I am going to question the weaker thesis, and *a fortiori* I shall be questioning the stronger thesis. The thesis I wish to discuss is the thesis that abortion as above explained is *prima facie* morally wrong. Let me explain. Persons who hold this weaker thesis are prepared to concede that, although abortion is usually and normally morally wrong, there are circumstances in which it is not. For instance, it is sometimes held that if an abortion is necessary to save the life of the mother, or if it is highly likely that the infant will be seriously defective, or if the mother already has crippling responsibilities so that her other children will suffer if another child is born (etc.), then an abortion is morally justified. But they hold that an abortion is *prima facie* wrong; or, in other words, they hold that there is a *prima facie* obligation not to bring about an abortion. (What I mean here by saying there is a "*prima facie* obligation" to do *A*, is that there is an obligation to do *A*, everything considered, except in the presence of a *prima facie* obligation to do something which is incompatible with doing *A*.) Thus the position is that there is always a *prima facie* moral obligation not to cause the death of a fetus, and that, *unless* there is a stronger *prima facie* moral obligation to do something which is incompatible with meeting the obligation not to kill the fetus, it is morally wrong to perform the abortion. So, generally, it is morally wrong to perform an abortion except where there is some moral obligation which can be met only by taking such action. Evidently, the stronger view that it is unconditionally and morally wrong to perform an abortion is untenable unless one believes it is at least morally wrong to do so and there are no conflicting and stronger moral obligations which can be met only by performing an abortion.

I wish to discuss the question whether it is *prima facie* morally wrong to perform an abortion (or *prima facie* obligatory not to do so), viz., whether abortion is wrong except when it is called for by *moral obligations*. Since the preferences of the mother may not constitute a moral obligation, what I am asking is, in part, whether there is anything morally objectionable about performing an abortion for no other reason than that the mother wants it, or, perhaps better, does not want to continue a pregnancy or have the responsibility of rearing a child.

I shall confine my attention to abortions not later than the third month, partly because I think such abortions are the primary object of contemporary controversy, and partly because I think it may make a difference if the abortion occurs, say, in the six or seventh month.

In what follows I shall discuss the issue on two levels. In the first section I shall make no assumption about what moral principles are or about how to decide whether a particular moral principle is acceptable; I shall argue in a very commonsense way. I shall maintain that there is not an unrestricted *prima facie* obligation not to kill, but only a *prima facie* obligation not to kill in certain types of cases; and I shall tentatively suggest a general formulation of a restricted principle which would have the effect of not implying that there is a *prima facie* obligation not to cause an abortion. In the second section I shall attempt to answer my question about abortion on a deeper level, at which explicit assumptions are being made about what a moral principle is and how the acceptability of one may be decided, by inquiring whether a rule prohibiting abortions, except where there are conflicting other moral rules, would be a part of a rational moral system.

It is convenient to begin with comments on the views of Pope Pius XI, as expressed in his encyclical *Casti Connubii* (1930), although he is there asserting not just the *prima facie* wrongness of abortion but its absolute wrongness. He affirms:

> But can any reason ever avail to excuse the direct killing of the innocent? For this is what is at stake. The infliction of death whether upon mother or upon child is against the commandment of God and the voice of nature: "Thou shalt not kill." The lives of both are equally sacred, and no one, not even public authority, can ever have the right to destroy them. It is absurd to invoke against innocent human beings the right of the State to inflict capital punishment, for this is valid only against the guilty. Nor is there any question here of the right of self-defense, even to the shedding of blood, against an unjust assailant, for none could describe as an unjust assailant an innocent child. Nor, finally, does there exist any so called right of extreme necessity which could extend to the direct killing of an innocent human being.

Pope Pius presents his moral principle as being both a commandment of God and the expression of the "voice of nature." His reason for saying that the prohibition is a commandment of God seems to be simply that the Bible asserts that God has said, "Thou shalt not kill." It does not, however, follow from agreement that there is a commandment not to kill that there is a divine commandment not to cause abortion. The question remains: Was this kind of killing intended? The Commandment as reported in the Bible is a very summary statement, requiring interpretation. As it stands, it might be taken to forbid killing of animals, capital punishment, action necessary for self-defense, euthanasia to relieve extreme pain, and suicide. The Church has not, I believe, proposed understanding of the Commandment as forbidding *all* killing. Exceptions must be admitted. But which ones? The question whether God prohibits abortion is not settled by appeal to the Biblical Commandment until it has been shown whether abortion falls among the admissible exceptions.

Abortion, incidentally, is not explicitly mentioned in the New Testament, although it is forbidden as early as the *Didache* (c. 80 A.D.) and by the early Church Fathers. Prohibition of abortion as being morally wrong is not part of the Jewish tradition, and it is far from universal among human societies.

Pope Pius XI also said that abortion is contrary to the "voice of nature"; this contention is too obscure and difficult to justify discussion in this brief essay.

If, like the Pope, we accept the injunction, "Thou shalt not kill," because it is believed to be God's commandment, the question of how it should be interpreted must presumably be answered by arguments to determine what God really does command, or how God intends his commandment to be interpreted. Actually, the Pope interprets the principle (1) as not applying to the killing of animals, (2) as applying to the killing of a human fetus, (3) as not applying to the killing of an "uninnocent" human being, where "uninnocent" is apparently construed as meaning "guilty of a capital crime," and (4) as not applying to killing in self-defense.

Let us now set aside the contention that the general injunction "Do not kill" should be accepted because it is God's commandment; we can then ignore matters of exegesis of Biblical texts, theological arguments, etc. Let us rather consider how the injunction must be construed, if it is to be viewed as a principle of *prima facie* moral obligation and is to be restricted so as to be compatible with what appear to be firm intuitive moral convictions.

One might ask, "Why should we discuss this more general principle at all, when all we want to know is whether it is *prima facie* wrong to perform an abortion?" To this the answer is that the above injunction, in some interpretation, is probably accepted by everyone, whereas no principle directly about abortions enjoys such status. One way to show that abortion is *prima facie* wrong would be to show that the only reasonable construction of the above principle is such that it would be inconsistent to accept it and at the same time to condone abortion as not even being *prima facie* wrong.

There is, of course, a difficulty in using the above injunction to settle the abortion question. For anyone who denies that abortion is *prima facie* wrong will presumably insist on an interpretation of the general injunction about killing, such a prohibition of abortion will not follow from it. So, if an argument from a general injunction against killing to a prohibition of abortion is to succeed, the defender of abortion must somehow be brought to accept a relatively unrestricted injunction against killing.

It can be and may properly be argued that a proponent of permitting abortion is going to find it much harder to show reasonable grounds for restricting a general principle about killing human beings intentionally, when construed as only a principle of *prima facie* moral obligation, then when construed as an absolute principle. For instance, probably most people today would accept the view that a fetus may be aborted if absolutely necessary to save the life of the mother; this view, if accepted, is a counterexample to the view that it is unconditionally and always wrong to perform an abortion. But this view is by no means necessarily a counterexample to the thesis that abortion is only *prima facie* wrong. In this case there presumably is a *prima facie* moral obligation to save the life of the mother; hence it could be argued that there is a *prima facie* obligation to avoid the abortion, but because of the stronger *prima facie* obligation to save the life of the mother, an abortion would not be wrong. How might one show that there is not an unrestricted *prima facie* obligation to avoid abortion?

Let me begin by pointing out that it *may* be that any plausible *prima facie* principle about killing will require restriction. To see this, it is convenient to consider the matter of keeping promises. It has been maintained that there is a *prima facie* moral obligation to keep *all* promises as such, although there are of course promises which, everything considered, morally ought to be broken, e.g., promises of a rather trivial sort which could be kept only at the cost of serious injury to someone. But, on reflection, it seems that some promises have not even a *prima facie* force (corresponding, perhaps, to contracts which courts regard as essentially null and void). Consider a promise made under duress (when the person had no right to try to extract the promise), or a promise made as a result of a deliberate misrepresentation of the facts to the promise for the purpose of extracting the promise, or an old promise made to a person who has since violated all his obligations to the promisor. It is at least plausible to say that such promises have no moral force at all; there is not even a *prima facie* obligation to keep them—surely one need not discover that one has a stronger moral obligation to do something incompatible with keeping the promise, in order to be free of an obligation to keep such promises. So, if we are going to formulate a sound general principle about a *prima facie* obligation to keep promises, the principle must be restricted to promises that do not have certain voiding properties.

The very same thing could be true about killing. It could be that there are types of situations, and kinds of human beings, such that it is not wrong to destroy them, even in the absence of a *prima facie* obligation to do something which requires that. Are there, or are there not? I propose one example that seems clear.

Suppose a human being has suffered massive brain damage in an accident. He is unconscious, and it is quite clear he will never regain consciousness–his brain is beyond repair. His body, however, can be kept alive by means known to science, more or less indefinitely. Is there a *prima facie* obligation to keep this being alive, or to refrain from terminating its existence? I believe there is no such thing.

Critics may say that this instance is no support for the claim that there is no unrestricted *prima facie* obligation not to kill–indeed not even support for the claim that there is no unrestricted absolute obligation not to kill. For, they may say, all that is here in question is the withdrawal of artificial means of sustaining life; when these are withdrawn, nature is allowed to take its course, and the man dies. He is not *killed.*

This criticism is, however, without force. (1) If it seems convincing to some because an abstention from providing help seems very different from actively killing, we must point out that acts and abstentions are morally not different; morally it is as wrong to allow a man to die when one could have saved his life by lifting a finger, as it is to kill him. In both cases one is responsible for the death. (2) If it seems convincing because "allowing nature to take its course" is very different from "interfering with nature"–what happens in an abortion–we must point out that interference with a natural process has never been shown to have a moral standing different from abstention from using some artificial means. We certainly think we must use artificial means to save an ill person pain; in fact, all surgery and medication might be viewed as interfering with nature.

(3) In any case the criticism can be met by pointing out that we believe a moral proposition to which it clearly does not apply: that in case the injured man's vital organs are needed for another person, it is permissible to operate and remove them, even when this terminates his life there and then. In other words, we think it permissible directly to bring about the death of a person in some circumstances, and not merely to allow nature to take its course. It may be said that all this retort shows is that the obligation not to kill is not absolute, but only *prima facie.* But I suggest we feel that there is *no moral issue* about a lethal operation in this total situation. (I admit, however, that some persons apparently feel qualms, and hence the device is used of defining "medically dead" so that the injured man is declared to be already dead.)

I suggest, then, that it is at least quite plausible to say there is no *prima facie* obligation not to bring about the death of a man injured in this way. Hence any plausible general statement about the *prima facie* obligation not to kill must be restricted so as to take this fact into account.[3]

[3]We would feel differently if we knew that, after nine months' treatment, the man would regain consciousness and live a normal life. One might say this difference of view is relevant to abortion. All I want to say here, however, is that the case as described is an exception to any unrestricted rule about a *prima facie* obligation not to kill.

Can we formulate a more general restriction, of which the above example may be viewed as an instance, on the *prima facie* obligation not to kill? One logically possible view is one apparently espoused by William James: that there is no moral claim where there is no *sentience.* James wrote:

> Take any demand, however slight, which any creature, however weak, may make. Ought it not, for its own sake, to be satisfied? If not, prove why not. The only possible kind of proof you could adduce would be the exhibition of another creature who should make a demand that ran the other way. The only possible reason there can be why any phenomenon ought to exist is that such a phenomenon actually is desired.

The implication of James's principle seems to be: it is *prima facie* wrong to kill a being that wants to live, or that has desires which can be satisfied only if it is alive. But the fetus is not sentient; it has no wish or desire, is insufficiently developed to be capable of any. Since the fetus has no desires at all, it is not *prima facie* wrong to destroy it. (Moreover, if the mother wishes that she not bear a child, or have the lifelong responsibilities that go with bringing up a child, that *is* a reason, according to James, why what she wants should be done.) So, if we follow James, we might assert: It is *prima facie* wrong to kill human beings, except those which are not sentient and have no desires, and except in reasonable defense of self or others against unjust assault. Doubtless this suggestion is only a rough first approximation, which needs amendment and complication—as it surely does, since we do not wish to hold that it is not wrong to kill a person who is asleep or in a coma (and in that sense is not "sentient"). It seems clear, however, that a restricted principle can be formulated along the suggested line, which is not obviously a less acceptable principle than the one which affirms an unrestricted *prima facie* obligation not to cause the death of any human being.

It is sometimes argued, in criticism of the restriction of the right to life to sentient, desiring beings, that if one accepts such a restriction one then has no reason in principle not to withdraw the right to life from babies, the elderly, the feeble-minded, and indeed anyone it is socially inconvenient to have around. But, it is said, such a withdrawal of a *prima facie* right to life is morally intolerable. What is necessary in order to avoid this dread slippery slope is to espouse the right of human beings to life in the most unrestricted way.

But in what sense is there "no reason in principle" not to withdraw the right to life from babies, etc.? A child has sentience, desires, and memory, even if in rather undeveloped form at first. An old person has all the feelings and awareness any of us has (and persons of any age would feel anxiety if they knew they were up for execution beyond a certain age). The feeble-minded, again, have some sort of life with value, at least on the same level as the higher subhuman species. So what could be meant by saying there is "no reason in principle"?

It would, of course, be a complete mistake for anyone to argue that the restriction proposed could not be accepted because then *no line* could be drawn between this restriction and the unacceptable ones. A line can be drawn; we have already drawn it by distinguishing the types of cases. Indeed, one can draw lines rather more sharply, if one wishes. Since it is a matter of observation and inference when "sentience" begins, some finer lines in the form of behavioral criteria need to be drawn, which I do not know how to draw. Some writers draw a line at implantation, a proposal that would permit morning-after pills but not abortion thereafter. The Church has sometimes drawn a line at the moment of "quickening." My point is that there is no problem about distinguishing the proposed restriction of the right to life from other quite different restrictions and also that the restriction I have suggested can be, and needs to be, formulated somewhat more sharply.

Another argument is sometimes used: as a practical matter the permission of abortions will weaken respect for human life generally, so in effect we encourage murder in general by permitting abortion even in the early stages. This argument is a claim about empirical fact, and one would like to see evidence which supports it. I should suppose that actually we make a very sharp distinction between a virtually formless fetus in its early stages and a sentient human being; surely a woman who requests an abortion might not dream of killing an adult. The difference is so great than an explanation is needed why we should think that attitudes toward the killing of a fetus might generalize to the killing of the already born.

Some kind of Jamesian principle about when it is *prima facie* wrong to kill a human being, then, seems not open to some objections that have been raised. Of course, I have only put it forward as a suggestion; all I have done by way of support for it is to have offered an example which is a counterexample to an unrestricted right to life of "innocent" persons not engaged in an unjust assault. We shall now approach the question of the *prima facie* wrongness of abortion on a more fundamental level where, if we are fortunate, we may hope to find more decisive reasoning. This effort will at least disclose some problems which are intriguing.

I shall do this by assuming, as a useful meaning to be assigned to the expression "*A* is *prima facie* wrong," a definition the value of at least some variant of which has grown on moral philosophers in recent years. The proposal is that "*A* is *prima facie* morally wrong" be assigned the meaning:

> *A* would be prohibited by a rule of the moral code the currency of which for their society would be preferred over all others by all persons who
> (A) expected to live a lifetime in that society,
> (B) were rational at least in the sense that their preferences were fully guided by all relevant available knowledge, and
> (C) had some general benevolence, the degree and extent being compatible with the restriction that the preference of each be rational.

There are some alternative proposals which some philosophers might prefer to substitute for (C). For instance, we might have

> (C′) were impartial in the sense that (and this is a restriction on (B)) their preferences were uninfluenced by information which would enable them to choose a system which would especially advantage them.

I shall conduct my discussion in terms of (C), but it would be interesting for the reader to follow the implications of a choice of (C′),[4] which is a somewhat stronger assumption than (C), since it appears to be equivalent to requiring a degree of benevolence toward all others equal to the degree of the person's "self-love," or interest in his own future welfare.

It will naturally be asked: Why is the foregoing proposed as a "useful" meaning to be assigned to "*A* is *prima facie* wrong"? A full reply would have to be lengthy, but it may be helpful to mention three points. (1) If "wrong" is construed in this way, then intelligent persons must regard moral principles not just as traditional prejudices or commitments but as restrictions or requirements with real point in terms of the welfare of sentient beings. Such persons will tend to respect and support an actual morality to the extent to which it conforms with true moral principles understood in the explained way. (2) If moral principles are construed in the way suggested, intelligent people will be generally disposed to adjudicate conflicts of interest by appeal to them and to recognize that they cannot in decency argue publicly that a claim of theirs deserves public support if it does not comport with such principles. (3) If we identify right and wrong actions in the way the proposed definition suggests, then it is in principle possible to find out, by methods or ordinary reasoning and evidence (but doubtless not in a very simple way), which actions are right or wrong. These considerations seem important in deciding on a useful meaning for moral words; indeed, I think we would be prepared to alter our formulation if we found it could not have the support of these considerations.

There is an ambiguity in the above formulation which needs to be resolved. If what is right or wrong is determined, in the end, by the preferences of "all persons" who meet certain conditions, one will wonder who is to count as a "person." Does this include the mentally defective? children? small babies? fetuses, whom the recent Supreme Court decision has ruled are not persons within the meaning of the law at least during the early months of development? chimpanzees? intelligent dogs? any sentient creatures whatever? The reason for the interest in these questions is that, for instance, it might be that dogs would

[4] The original article in the *Monist*, of which the present paper is a reformulation, adopted this reading, and considered also a further possible reading. See the *Monist* *56* (1972), pp. 513 ff.

not prefer the currency of a moral system which committed stray dogs to a gas chamber, whereas such a system might be preferred if dogs do not "have a vote." I point this out, because a person hardly wants to accept the claim that a proposed definition of a moral term is "useful" until he understands it sharply enough to see something of the moral principles it may commit him to.

I propose to construe "persons" to mean "adult human beings with reasonable intelligence." This is not very precise and, if one wanted, one could spell it out as "human beings at least eighteen years of age, and with an I.Q. at the time of at least 110, and not temporarily in a psychotic or neurotic or even a highly emotional state of mind." An advantage of such a construction of "persons" is that our definition then clearly excludes any such bizarre conception as babies (or, worse, animals or fetuses) *expecting* to live a lifetime in a society, and so on. The restricted conception also, I think, does not really have implications different from those of a less restricted conception which in effect "gives a vote" to animals, etc. The reason for this is that if these adults have the benevolence specified in (C), they will presumably take an interest in the welfare of babies, animals, etc. and will prefer a moral system which will give them the same protection these would give themselves if they "had a vote."

What is meant by saying that a person has "benevolence"? The Shorter Oxford Dictionary defines "benevolent" as "desirous of the good of others." Notice that it is defined as desire for the good *of others,* not some abstract goodness or value at large. So far the Dictionary reminds us of the New Testament, which enjoins us to love "thy neighbor as thyself" but does not enjoin loving any abstract good.

But we need to be more specific about "the good" of others, and it is easiest to be so by considering first the notion of "self-love." We all of us have what have been called "particular" desires, like the desire to eat a certain French pastry or to be praised for a certain achievement. There is also a desire of a different kind, just to be in *some state or other* which one likes (notice how one can motivate a child by promising him "something nice" without telling him what it is to be), and a desire not to be in any state one dislikes such as fatigue, feeling hungry, feeling lonely, or in general being uncomfortable. Let us call this desire "self-love." Obviously one way to gratify one's self-love is by gratifying one's other particular desires; for if we do this we do not suffer the discomfort of hunger or longing, and we get to enjoy the consummatory experience associated with these desires.

We can think of benevolence as a desire like self-love, but directed at the inner states of *other* persons. Being benevolent is not sharing another person's particular desires; if another person wants a French pastry, benevolence does not imply that I shall be motivated by the prospect of his eating a French pastry in the way in which he is. Like self-love, benevolence is a desire for another person to be in a state he likes, one of joy or satisfaction or comfort and not in a state he dislikes, like fatigue, hunger, or other distress. Benevolence, then, is a positive interest in another's being in a state which he likes and a negative interest in his being in a state which he dislikes.

States of joy and distress are not limited to human beings; correspondingly people who are benevolent take an interest in the joy and distress, are motivated at least to relieve the misery of animals, and to some extent to satisfy them.

What is meant by the term "general" in clause (C)? This is a requirement that the benevolence be directed not merely to specific preferred individuals like one's own wife or children or even compatriots. How far must benevolence reach in order to be "general"? We probably do better not to try to make any further definitional specification of this for the purpose of application of clause (C). It is perhaps enough that (C) already specifies that the degree and extent are to be compatible with the preferences of the individual being rational (guided by full available information).

After all these preliminaries about a proposed meaning for "*A* is *prima facie* morally wrong" (or "is *prima facie* morally obligatory"), it is time to get back to our question about the morality of abortion. Our question whether abortion is *prima facie* morally wrong may now be construed as the question whether abortion as defined (destruction of a fetus within the first three months when this is wanted or undertaken by the mother) would be forbidden by the rules of the moral system which rational persons, with the benevolence specified, would prefer to any other moral system. This question may not be easy to answer.

At first it may seem that of course benevolent persons would not want a fetus destroyed, and hence that they would prefer a moral system which forbade this. But matters are not quite that simple.

There is an important distinction to be made. It seems very plausible that rational persons with some benevolence will take a favorable interest in the well-being of future generations, in that of persons who live in the future. (Obviously intelligent benevolent persons do, as we can notice from all the present concern about the environment and conservation of resources.) But it is one thing to take an interest in persons, living now or in the future, being deprived of happiness or brought joy; it is quite another thing to take an interest in the being *brought into being* of a sentient creature which, *if* produced, will be happy and enjoy well-being. It is not impossible for one to be strongly interested in the welfare of all actual persons, present or future, but be totally indifferent to whether a person *is produced* who, *if* produced, will be happy. We must remember that *no person is frustrated or made unhappy or miserable by not coming to exist.* No one is *deprived* by non-birth as a sentient being. So it is possible that a person take a strong interest in the welfare of actual (present or future) persons, but that he be left cold by the idea of producing more happy people. Producing somebody who would be unhappy (say, because of physical defects) would presumably be unacceptable to a benevolent person; but *failing* to produce a happy person would seem not necessarily unacceptable to a benevolent person.

The implication of this distinction for the morality of abortion is obvious. Suppose we construe abortion as the prevention of the *coming into being* of a person (perhaps one who will be happy, if he is brought into being). Then it would be a mistake to infer from the fact that rational benevolent people will presumably want a moral system which protects the life and happiness and welfare of sentient creatures that they will want a moral system which forbids the *prevention of the coming into being* of some sentient creatures.

It may be objected that the preceding paragraph misunderstands the nature of abortion. It may be said that a fetus is an *already existing person,* so that its destruction is not merely preventing the coming into being of a person, but is incompatible with the protection of the happiness and welfare of sentient beings, just as much as the murder of an adult. I shall come back to this—and argue that the preceding paragraph does not misunderstand the nature of abortion.

First I want to consider how we might argue that rational benevolent persons would want a moral system which did *not* forbid abortion (or, if you like, would want a system which forbids killing human beings but with an exceptive clause about fetuses in the first three months). How will rational benevolent people make up their minds on a point of this sort? Presumably they will count up all the probable harms to people and the benefits to people, if one rule or its opposite is adopted. They will ask: Will it do more good to sentient creatures, in terms of happiness in the long run, to have current a rule forbidding abortions or not to have such a rule? Now the first thing they will notice is that "abortion" has been defined as the destruction of a fetus *instigated by the mother.* If it is so instigated, there will normally, or always, be a reason. Mothers do not want abortions for no reason at all. Either they cannot afford another child or another child is a risk to the health of the mother or a child will be an unwanted burden and so on. It will be agreed that it can be assumed that if a mother wants an abortion, then it is her judgment that from her point of view there will be more harm than good in the arrival of the offspring. That fact will be a harm, something which counts against having a rule forbidding abortion. Is there usually a *public* benefit from the arrival of another child? Doubtless in a country with a manpower shortage, the more children the better. But at the present time, for most of the world, there are too many people already. Another mouth to feed, or someone to pollute the air by driving another car, is not a benefit to others. It looks as if other people would be better off if those who are not wanted anyway did not make it to existence. Average welfare is obviously related to population size; there is an optimum size at which the average welfare is maximal. The addition of people beyond this optimal point has the effect of reducing the welfare of somebody else. In most countries the population is already such that the optimal population has been exceeded. So, will benevolent people want to insist that a fetus be permitted to develop and be born, when the nonexistence of the extra person would bring the population nearer the optimal? It looks as if, from the point of view of the general public, there is no reason to prefer a moral rule forbidding abortion.

Someone might, unwisely, reply to this as follows: "In consistency you should permit a few murders. After all, a murder does somebody good–the murderer, or else he would not be doing it. And it does bring the population down." The answer to this objection is, however, obvious. If murders are permitted, then nobody can feel secure. Even if there would be some public benefit from certain murders, they cannot be permitted without rendering human life so precarious as to be of no account. The situation is very different in the case of abortion. The fetus has not the foresight to feel insecure; it suffers no pain, has no worries. It is *not yet* a sentient, intelligent being. So there is not the reason against a policy of terminating a fetus that there is against permitting a murder.

Nevertheless, it may be claimed that the above line of argument, counting the harm or good to persons arising from a moral rule about abortions, has overlooked an important point: the future happy life of the fetus itself. Suppose it *is* true that another person will slightly reduce the welfare of persons who will exist whether this person, whose existence is being deliberated, comes into being or not. Must not *its* happiness be counted too? Possibly that is a good to be considered so great that its development must be allowed, even if an additional person will somewhat diminish the well-being of other persons.

Should the supposititious happiness of the prospective extra person be counted or not? I have questioned above whether benevolent persons *would* count it, on the ground that no *person* is harmed, or deprived of happiness, if the fetus is not permitted to *become* a human person. So we are brought back to the controversial point I raised at the conclusion of making what I said was an "important distinction": whether, in some sense appropriate to our issue, the fetus is already a person. What we want to know is whether somebody is being deprived or harmed by the act of abortion, which admittedly prevents the fetus from developing into a human being with a possibly happy life. I said I would return to this question. I shall now try to meet it head on.

I am going to argue that it is mistaken to suppose that the fetus is, in an appropriate sense, *the same person* as the child who will occur after the development of the fetus, and birth. And, if it is not the same person as the person which begins at some later time, then it is not, in an appropriate sense, now a person at all–and hence we cannot say that somebody is harmed or deprived by the abortion.

This claim may seem puzzling, and it may seem initially that any argument to support it is bound to be contrived. I think, however, that we can see the force of the claim by considering a possible situation which is closely parallel. Suppose I were seriously ill, and were told that, for a sizable fee, an operation to save "my life" could be performed, of the following sort: my brain would be removed to another body which could provide a normal life, but the unfortunate result of the operation would be that my memory and learned abilities would be wholly erased, and that the forming of memory brain traces must begin again from scratch, as in a newborn baby. Now, how large a fee would I be willing to pay for this operation, when the alternative is my peaceful demise? My own answer would be: None at all. I would take no interest in the continued existence of "myself" in that sense, and I would rather add the sizable fee to the inheritance of my children. It is true that in some sense "I" would continue to exist; obviously my *brain* would. But it looks as if something is missing which is necessary for personal identity in an important sense. The important thing that is missing seems to be *memory;* if the latter person cannot remember my present and earlier experiences at all, then he is as good as a new person from my point of view. And I cannot see the point of forfeiting my children's inheritance in order to start off a person who is brand new except that he happens to enjoy the benefit of having my present brain, without the memory traces. It appears that some continuity of memory is a necessary condition for personal identity in an important sense.[5]

How does the above example bear on the situation of the fetus? Let us compare the physical components. In the case of the operation, one part of the first body (the brain) is transferred relatively intact (except for memory traces) into a different body; in the case of the fetus, one whole body at the earlier stage grows, through ingestion of new materials, into a structure quite different at the later stage (although a structure already determined by the genetic components there at the earlier stage). In both cases the total body at the first stage is qualitatively very different from the total body at the second stage. But in the operation case one part is identical in the two bodies at different times, whereas in the case of the fetus there is development, involving ingestion of new materials, from one stage into a very different later stage (although we say, in view of the continuity of development, that it is "the same body").

In a most important respect, however, the two cases are exactly similar; in both there is a complete break between the two body-stages as far as *memory* is

[5] The following question has been put to me: Suppose you knew that the operation was going to be performed, and you have a choice whether a very severe pain was going to occur to this "descendant" of you, or to someone entirely different. Would you prefer the pain to occur to the totally different person?–the suggestion being that, if the answer is affirmative, we do after all think there is some important sense of identity between this later self and our present self. I am inclined to think that the right answer to the question is "no preference," and that any tendency to give a different answer arises from failing to get the hypothesized situation clearly in mind.

concerned. The person who exists after birth will not recall any events of the first three months of fetal life, just as, after the operation, the new memory will contain no recollections of experiences up to the time of the operation. In saying this, of course, I am affirming as a fact what conceivably might not be true. I believe that (a) there are *no conscious experiences* at all in the fetus in the first three months, or else that (b) even if there are, they cannot be recalled by the baby which gets born. (I would also defend a stronger position about the intermediate stages, but there seems no point in doing so here.) Thus I think that if a fetus could consider the importance to it of the existence of the baby after birth, that existence would seem no more important than the existence of the life of the person with "my brain" after the hypothesized operation seems important to me. If I am right in my suggestion above that some continuity of memory is a necessary condition for personal identity in an important sense, then that important condition for personal identity is lacking in the case of the fetus just as in the case of the operation.

As a result, I would say I was not being harmed, or deprived, if it was decided just to let me die in peace rather than perform the operation transferring my brain to another body. For similar reasons it appears to make no sense to say that the fetus is harmed, or deprived, if steps are taken so that it does not develop into a normal baby. In that sense, we can say that the fetus is not now a person at all, in the important sense. (We might say that since it is not the same person as the later baby, in the important sense, and since it is obviously not *some other* person, it follows that it is not a person at all in any important sense.)

How does this conclusion bear on our question? What we were trying to decide is whether rational, somewhat benevolent persons would prefer a moral system which would prohibit the destruction of a fetus when the abortion is wanted by the mother. Now, we agreed that benevolent persons take an interest in the present and future happiness and well-being of other persons. But we earlier raised the question whether it followed from this that they would take an interest in the preservation of the fetus—since this seems to be an interest in the *addition of another person.* It seemed to be one thing to want the future happiness and welfare of a person who does or will exist, and another to want to produce another being which, if produced, would be happy and enjoy well-being. Our most recent discussion about the conditions of personal identity has been such as to sustain this distinction. It appears from this discussion that no person is being harmed or deprived by the termination of a fetus, nor being benefited by its continuance so that a baby is born. So it is one thing to want all persons to be happy and not miserable; it is quite another thing to want that there develop, from non-persons such as fetuses or ova or whatever, persons who will be happy or unhappy. Whereas it is clear that a benevolent person wants all persons to be happy and not miserable, now or in the future, it is not obvious that a benevolent person wants sentient beings *brought into being* at all, much less that he will want them to be brought into being at a cost to others who already do or will exist, such as the mother or other persons whose standard of living may be reduced by the appearance of another human being.

It is not impossible that there should be some other reason for thinking that, after all, benevolent beings would want fetuses protected from destruction because of the happiness that would come into the world if they are so protected, or even that benevolent beings would want as many sentient creatures produced as possible, provided that if they were born their lives would be happy, or lives of positive welfare, that is, above the indifference point.[6]

[6] I do not believe that anything I have said shows that it is impossible that there be some other reason supporting conclusions of this sort.

ABORTION

THE AVOIDABLE MORAL DILEMMA

JAMES M. HUMBER

The recent Supreme Court decision in *Roe* v. *Wade* could hardly be said to have solved the problem of abortion.[1] Grave religious and moral issues remain, and until these are settled there will be continued assaults upon the law as promulgated by the Supreme Court. Now I agree with those who hold that the Supreme Court ruling in *Roe* v. *Wade* gives legal sanction to immoral conduct. At the same time, I would urge those presently engaged in trying to change that law to give up this line of attack. There are two reasons for this. First, at this particular point in time, all avenues for change seem blocked. Given its present constitution, the Supreme Court is not going to modify its ruling; and because the majority of citizens of this country favor abortion in at least some cases,[2] movements for constitutional amendment seem unlikely to succeed. Second, even if the law could somehow be changed, it is quite certain that it would be (as it was in the past) virtually ineffectual. As long as people remain convinced that abortion is both moral and beneficial, they will continue to make use of that procedure regardless of legal sanctions. To attack the law is to attack a symptom, not the disease; and if anti-abortionists are to have any hope of eradicating abortion, they must find some way of convincing abortion advocates that this procedure is morally wrong. Now despite the many failures of the past, I think that this can be done. There are several reasons for my optimism here, but primary among them is this: unless I am mistaken, opponents of abortion have been unable to develop convincing arguments, only because they have failed to discern the true nature of their adversaries' position. Indeed, it is my belief that no party to the dispute has ever seen the abortion controversy for what it is, and that once this is done, a much more persuasive case can be made for the

---

This paper is a revised version of an essay of the same title, published in *The Journal of Value Inquiry*, Vol. IX (Winter, 1975), pp. 282–302. The only substantial difference between the two papers lies in the discussion of Professor Brandt's article, "The Morality of Abortion." I am indebted to the editors of *The Journal of Value Inquiry* for their permission to republish my essay.

[1] *Roe* v *Wade*, 410 U.S. 113 (1973).

[2] Herman Schwartz, "The Parent or the Fetus," *Humanist*, Vol. 27 (1967), p. 126.

immorality of abortion. In order to substantiate this belief, then, I propose to argue in the following way:

First, in sections I and II of this essay, the major defenses of abortion will be examined in detail and shown to be unsound. Further, it will be argued that abortion must always be the taking of a human life, for if we are to take seriously our ordinary ways of speaking, human life must be acknowledged to begin at conception. Next, in section III, a case will be made for the view that the arguments of the pro-abortionists are all so poor that they sould not be accepted at face value, but rather should be seen as after-the-fact rationalizations for beliefs held to be true on other grounds. In addition, an attempt will be made to uncover these "other grounds." Finally, in section IV it will be argued that the true basis for the abortion advocates' moral position is such that this view must be rejected out of hand. At the same time, the emotive force of the pro-abortion stance will be recognized, and the present impossibility of assuring moral behavior acknowledged. Taking these facts into consideration, then, new programs for anti-abortion action will be outlined.

I

Most of those arguing for and against abortion see the controversy as being one which can be resolved only by determining the proper use of *human*. Those opposed to the procedure, for example, usually offer into evidence biological data which, they say, clearly indicate the fetus' humanity. If prenatal beings are known to be human, they then continue, such organisms must be seen as having the right to life. And since abortion is always the violation of that right, the procedure must be considered immoral.[3] In opposition to this "conservative" position, those favoring abortion use various ploys. Some try to show that *human* is ordinarily being used in such a way that it excludes at least some prenatal beings (e.g., zygotes and/or embryos) from its extension. Unfortunately, all such attempts fail, for with each definition offered, some commonly recognized group of human beings is denied human status.[4] Recognizing this fact, other abortion advocates seek to defend their view by appealing to the arbitrary character of definition. Basically, the argument takes two forms. The

[3] Two of the chief proponents of this view are: Paul Ramsey, "The Morality of Abortion," in *Life or Death: Ethics and Options* (Seattle: 1968), and John T. Noonan Jr., "Abortion and the Catholic Church: A Summary History," *Natural Law Forum*, Vol. 12 (1967). For a complete discussion of the failures of these analyses, see D. Callahan, Abortion: *Law Choice, and Morality* (London: 1970), pp. 378–384.

[4] For example, Herman Schwartz (*op. cit.*, p. 126), contends that abortion is moral because the fetus is not "a rational creature, with unique emotions and feelings, intellect and personality, a being with whom we can identify." But this definition excludes (at least) newborn babies, and thus allows for infanticide.

most radical position is represented by Mr. Garret Hardin: "Whether the fetus is or is not a human being is a matter of definition, not fact; and we can define any way we wish. In terms of the human problem involved, it would be unwise to define the fetus as human."[5]

The difficulty with this view is all too apparent. That is, if definition is *purely* arbitrary, we may classify any group of persons as non-human, just so long as we believe the procedure is warranted by the presence of some "human problem." Alcoholics, the senile, those on welfare, *any* group legitimately may be classed as nonhuman and dealt with as we please. Hardin recognizes the problem, but refused to acknowledge its force: "This is, of course, the well-known argument of 'the camel's nose' –which says that if we let the camel put his nose in the tent, we will be unable to keep him from forcing his whole body inside. The argument is false. It is *always* possible to draw arbitrary lines *and enforce them.*"[6]

Seemingly, Hardin wants prenatal beings *alone* to be arbitrarily classed as non-human. But upon what basis can this preference be supported? It is obvious that he could not seriously assert that embryonic organisms constitute the only group of beings posing a "human problem." What we are left with, then, is simply Hardin's "bare feeling" that arbitrary definition is proper if and only if the beings to be classified are *in utero.* But what makes this feeling more proper than the feeling, say, that alcoholics should be classed as non-human and exterminated? Hardin gives us no answer.

A less radical version of the arbitrary definition defense of abortion has been developed by Glanville Williams. Sensing Hardin's problem, Williams attempts to give rational support for Hardin's "feeling" that it is with prenatal beings alone that arbitrary non-human classification is legitimate: "Do you wish to regard the microscopic fertilized ovum as a human being? You can if you want to. . . . But there are most important social arguments for *not* adopting this language. Moreover, *if you look at actual beliefs and behavior, you will find almost unanimous rejection of it.*"[7]

In effect, the arbitrary classification of fetuses is now held to be proper, not merely because there are social arguments which make such action desirable, but even more importantly, because society's behavior indicates that the majority does not believe such entities are human anyway (e.g., women do not mourn the loss of spontaneously aborted zygotes as they do the death of children, etc.)

The first thing which must be noted about Williams' attenuated version of the arbitrary definition defense is that it is not entirely clear. Is Williams

[5] Garrett Hardin, "Abortion–or Compulsory Pregnancy?" *Journal of Marriage and the Family,* Vol. 30 (May, 1968), pp. 250–251.

[6] Garrett Hardin, "Semantic Aspects of Abortion," *ETC,* (Sept., 1967), p. 264.

[7] Glanville Williams, "The Legalization of Medical Abortion," *The Eugenics Review,* Vol. 56 (April, 1964), p. 21.

claiming that, society's attitudes and behavior being what they are, fertilized ova have *already* been classified as non-human, and that it would be improper to disagree with majority opinion? Or is his view merely that the classificatory status of such organisms is in doubt, and that the available "social arguments," together with majority consent, provide us with good reasons for grouping prenatal beings as non-human? Let us examine each possibility in turn.

If the first interpretation of Williams' meaning is accepted his position must be rejected out of hand, for what it amounts to is simply the assertion that the majority is always right. But whether or not the proposition "*x* is human" is true or not, is not something to be resolved by an appeal to majority opinion. In the late Middle Ages, for example, the majority's "actual beliefs and behavior" were such that children were not considered fully human.[8] Would Williams want to admit that one who killed a troublesome child in those days was acting in a morally acceptable manner? And what of the Salem witch hunts? Was witch killing "right for them," but not for us? Surely not; indeed, if the fact that the majority of the residents of Salem thought witch killing proper indicates anything, it is only that the majority of the people of that city were ignorant. The conclusion, then, seems clear: if Williams wants to show that fertilized ova are non-human (i.e., if he wants to prove that it is improper today to disagree with the majority concerning the status of prenatal beings), he must provide us with some reasoned argument which demonstrates that fact. But no such argument is offered. Consequently, this interpretation of his position must be rejected.

Although the second construction of Williams' argument is stronger than the first, there seem to be two good reasons for concluding that it is unsound. First, if Williams takes majority doubt concerning classification plus the availability of "social arguments" as together providing a moral justification for dealing with certain groups of beings as non-human, various undesirable consequences follow. In both word and deed, for instance, we everyday illustrate that the majority view in this country is that there are at least some criminals who are subhuman. Rapists and murderers are often referred to as "animals" and "mad dogs"; and when these men are caught, they are housed like animals in a zoo. Then too, the social arguments for getting rid of such misanthropes are numerous. Would Williams want to use these insights to develop a new argument for capital punishment?[9]

Although it is not likely that he would do so, it could be that Williams would be willing to accept the logical implications of his argument, and admit that capital punishment, euthanasia, the killing of mental defectives, etc. are

[8] See Phillippe Aries, *Centuries of Childhood,* trans., Robert Baldick (New York: 1965), pp. 38–39.

[9] My choice of criminals here was purely arbitrary. The argument would apply with equal force to the senile, children born with mental defects, etc.

indubitably moral. But even if this were done, Williams' defense of abortion must be rejected as it stands, for unless it is modified in some significant way, it can only be seen as fostering morally irresponsible action. To illustrate: let us say that we are deer hunting with some friends. One of our colleagues sees a movement in the bushes and fires. When we ask him why he acted as he did, he replies: (1) although he was not sure what had moved in the bushes, it was much more probable that it was a deer than a man, and (2) if it were a deer he did not want it to get away, for his family needed the food, and the deer in this region had overpopulated and were destroying the crops of local farmers. Now even if we grant the truth of (1) and (2), I doubt that anyone would accept these facts as a justification for our trigger-happy friend's behavior. True, we would not call our hunter immoral; but we *would* think him careless, and insist that he not be allowed to carry a loaded gun. And in the same vein, a woman who was not sure that the fertilized ovum within her was human, and thus decided to have an abortion simply because she wanted no more children, or would be subject to economic and/or pyschological hardship given the child's birth, should not be held responsible enough to make that life and death decision.

In order to buttress Williams' argument, defenders of abortion could object to our hunter analogy on two grounds: (a) It could be said that it does not cover two very important cases, viz., that in which the life of the mother, and that in which the lives of the mother and her fetus are endangered; or (b) it could be objected that it fails to note that there is a special relationship between a woman and the organism developing within her. Even if modified in the ways suggested, however, Williams' argument must be rejected.

First, *if* it were true that we could never be sure that prenatal beings were human, abortion would have to be considered morally acceptable whenever the conditions of (a) were in evidence. But the reason abortion would be permitted in these two special instances is not what Williams would have us believe. That is to say, in neither case would we arbitrarily define individual fetuses as non-human in order to legitimatize particular abortions. On the contrary, if only the mother's life were in jeopardy we would reason that we *have* to act in order to save her, and that since the fetus is not known to be human, we would be justified in "playing the odds," hoping all the while that our actions were not destructive of a human being. If, on the other hand, the lives of both the mother and her fetus were endangered, the argument for abortion would be stronger—indeed, abortion in this case might be warranted even if the fetus were known to be human, for our only choice here would be between saving one life, or letting two persons die. If one were to accept Williams' version of the argument, however, he would be led to one of two ridiculous results: either fetuses would be non-human (in the two cases mentioned) and not non-human (in all other cases); or, because there would sometimes be good reasons for classifying fetuses as non-human, it could be argued that they should always be so categorized.

Now the first alternative is just blatantly inconsistent. As for the second, it leads to moral absurdities. For example, an advocate of this view would have to claim that since a *starving* hunter is justified in arbitrarily classifying anything moving in the bushes as non-human and shooting, *all* movements in bushes may be thought of as being due to non-human causes, and all hunters may "fire at will."

Before a proper evaluation of (b) can be made, we must first get as clear to the exact nature of the "special relationship" to which it alludes. Surely it is not simply that of mother to child, for if our hunter analogy is amended so that the careless hunter is a woman who, at the moment of shooting, knew that the motion in the bushes must have been caused either by a deer or her child, we would think her *more* careless rather than less so in firing. If there is anything "special" about the relationship, then, it can only be that the embryonic organism exists as a parasite, and that when an abortion is effected, conditions which are both necessary for the continued life of the fetus, and in some sense possessed (or "owned") by the mother, are removed.

Having clarified the nature of the relationship referred to in (b), a second question arises: how could abortion advocates use knowledge of this relationship's existence in order to demonstrate that persons seeking abortions are not morally irresponsible? Two alternatives present themselves. First, it could be argued that the parasitic status of the conceptus shows that it is not a "separate human being endowed with human rights."[10] Now although it may well be true that the fetus' total dependence upon the mother is one factor causing some to doubt its humanity, dependency can hardly be taken to prove that the fetus is non-human and without the right to life. To hold otherwise, one should have to accept the contention that it would be morally proper for a mother to kill her newborn baby anytime prior to the severing of its umbilical cord. But this is clearly ridiculous. If knowledge of the parasitic status of the fetus is of any importance at all, then, it can only be in helping us to understand why some people doubt prenatal organisms' humanity. But our hunter analogy has all along assumed that the status of fetal beings is dubious; and the reasons for this doubt are (at least at this point in the development of our thesis) of no consequence. As a result, the first interpretation of (b) contains nothing which could allow Williams to avoid the charge of fostering moral irresponsibility.

The second version of (b) is better founded than the first, and it has been defended in detail by Professor Judith Jarvis Thomson.[11] Women have the right to abort, she says, because they have the right to control their own bodies. And *even if the fetus is a human and possessed of human rights,* "having a right to life

[10] Although their arguments are not clear, when feminists argue for abortion on demand, they seem to be reasoning along these lines.

[11] Judith Jarvis Thomson, "A Defense of Abortion," *Philosophy and Public Affairs,* Vol. 1 (1971), pp. 47–66.

does not guarantee having either a right to be given the use of or a right to be allowed continued use of another person's body—even if one needs it for life itself."[12]

Because Thomson assumed from the first that prenatal beings are human, I am, by conjoining her argument with that of Williams, strengthening it considerably. In addition, this action gives us a new defense of abortion—one which, I believe, makes the strongest possible case for abortion's morality. If, as Williams holds, we could never be sure that fetuses were human, and if, as Thomson says, women have a right to control their bodies, then *perhaps* an appeal to the right to control one's body would justify abortion. Quite frankly, I am not sure. Luckily, we need not face this issue, for the combined thesis may be invalidated on other grounds. Specifically, Williams' assumption concerning the uncertain classificatory status of embryonic organisms can be shown to be false; and with this demonstration, the two theses will be separated, Thomson's argument then having to stand or fall on its own.

As previously noted, most of the discussion surrounding abortion has centered upon attempts to determine the proper use of "human." Not only has this procedure been particularly unproductive, it has also failed to explain why those who oppose abortion usually wish to hold that human life begins at conception, rather than at some other point in gestation.[13] Both these failures can be overcome, however, if only one is willing to shift his focus of attention. What, after all, is the meaning of "conception"? As employed in ordinary language, the term appears to have three uses: (1) Sometimes it is used to mean "beginning," "start," or "creation" (as when we say, "The design has been faulty since its conception."). (2) Sometimes it is used to mean "act of conceiving," where "conceiving" means "to imagine" or "to form a notion or idea" (e.g., "Conception of his meaning was possible for me only after he gave an example."). (3) And finally, "conception" is often used to mean "notion" or "idea" (e.g., "I now have a conception of what must have happened yesterday."). Now which of these three uses is being employed when we discuss human conception? Surely it is the first; for when I say (p) "My conception occurred approximately nine months before I was born," I do not mean to say (as with use 2) that I imagined or thought of something at that time. Similarly, I do not mean to hold (as use 3 would require) that someone else—presumably my parents—had an idea or notion of me. Well then, if uses (2) and (3) are excluded, what else could (p) mean than that I had my beginning, start, or creation about nine months before I was born? And note, it is *I* who got his start at the point

[12] *Ibid.*, p. 56.

[13] This preference is evident even among those who admit that there is no proof that human life begins at conception. See Paul Ramsey, "Abortion: A Review Article," *The Thomist*, Vol. XXXVII (1973).

thus denominated. This is extremely important, for if it is essential for me to be me that I be human (as surely it is), then what I am asserting when I assert (p) is that my creation *as a human* occurred nine months before my birth. And if this is so, the rationale for the anti-abortion position becomes clear: one can deny that human beings are created at conception, only by denying that they begin to exist when they begin to exist.

Having demonstrated that from conception on embryonic organisms are human, we can only conclude that Williams is in error when he holds that the classificatory status of these beings is uncertain. This being so, the combined Williams–Thomson argument must be rejected, and we are left with Thomson's argument alone. Now, can that reasoning stand by itself once the humanity of organisms *in utero* is admitted? In arguing that it can, Thomson constructs the following example: Assume, she says, that you are kidnapped by the Society of Music Lovers and connected *via* some medical equipment to a sick, unconscious virtuoso violinist. The violinist has a rare, potentially fatal disease, which can be cured only if he remains connected to you for nine months. Only you can perform the life-giving function, because you alone have the proper blood type. Given this as the situation, then, Thomson asks: "Is it morally incumbant on you to accede to this situation? No doubt it would be very nice of you if you did. . . . But do you *have* to accede to it? What if it were not nine months, but nine years? Or longer still? . . . I imagine you would regard this as outrageous. . . ."[14]

The conclusion seems clear: if the right to control one's body justifies "unplugging" the violinist, it must also legitimatize abortion—and this holds true regardless of whether the fetus is human or not.

Although Thomson's argument has immediate appeal, it is a relatively easy matter to show that her reasoning rests upon a confusion.[15] Consider the following counterexample: Let us say that I am involved in a shipwreck. After being thrown overboard, I manage to tie myself securely to a large piece of flotsam. As I am bobbing around in the water, a non-swimmer grabs my arm and asks that I help him get onto the piece of floating debris to which I have tied myself. To this I answer, "Having a right to life does not guarantee having either a right to be given the use of or a right to be allowed continued use of another

[14] Thomson, *op. cit.*, p. 49.

[15] Baruch Brody has challenged Professor Thomson's position by pointing out that she fails to distinguish between our duty to save someone's life, and our duty not to take it (B. Brody, "Thomson on Abortion," *Philosophy and Public Affairs,* Vol. 1 (1972), pp. 339–340.) There is some doubt, however, that this distinction is ever properly made in morals (See R.B. Brandt, "The Morality of Abortion," p. 157, in *Abortion: Pro and Con,* ed. by Robert L. Perkins (Cambridge: 1974), pp. 151–169). Brandt's article is a revision of one of the same title, published in *The Monist,* Vol. 56 (1972), pp. 504–526. All references in this paper will be to the revised essay.

person's body." With that, I shake him loose from my arm and watch him go under for the third time.

Surely no one will doubt that in my example, I acted in an immoral manner. But why does the immorality show up so clearly here, and not in Thomson's paradigm? The answer, I think, lies in the degree of hardship being imposed upon the persons whose bodies are being used. In my example, it would have required very little effort for the one shipwrecked person to have saved the life of the other. In Thomson's analogy, however, we are asked to consider ourselves bedridden for months, even years. And as Aristotle long ago realized, anyone can "break" under pressure and do something which he realizes is morally wrong. Now in certain cases (viz., when the pressure is so great that the average person could not reasonably be expected to withstand it), the man who "breaks" and acts immorally is said to have an excuse for his actions. But to excuse an act is not to say that it is morally right. Indeed, just the opposite is true, for unless an act is morally wrong, it hardly stands in need of an excuse. And if this is granted, two conclusions seem mandated: First, if Thomson's reasoning has some "convincing power," it is only because the reader has followed her in failing to distinguish between excused acts, and acts which are morally right. And second, if Thomson's argument shows anything at all about abortion, it is only that it is a morally wrong act which, like all other morally wrong acts, may sometimes be excused.

II

Professor Thomson's argument is not the only defense of abortion which proceeds upon the assumption that fetal beings are human. Some pro-abortionists admit this fact, but argue that there is a distinction to be drawn between "human" and "human person"[16] That is, even if human life begins at conception, these people say, no person is present until sometime late in gestation. And if this is so, ". . . then under some circumstances the welfare of actually existing pérsons might supersede the welfare of developing human tissue."[17]

Now there are two reasons why any defense constructed along these lines must fail. First, if there are formidable difficulties involved in trying to define "human," these problems must simply re-arise in attempts to define "human person;" and if this is so, it must be virtually impossible to distinguish a human from a human person. Even if we ignore this apparently irresolvable problem,

[16] Thomas L. Hayes, "A Biological View," *Commonweal,* Vol. 85 (March, 1967); Rudolph Ehrensing, "When is it Really Abortion?" *The National Catholic Reporter* (May, 1966).

[17] Ehrensing, *op. cit.,* p. 4.

however, what possibly could serve to justify the belief that an individual's rights may "supersede" those of a human? That is, since the right to life is a human right rather than a personal one, fetuses must be seen as possessing that right even if they are held to be non-persons. Why, then, do these proponents of abortion feel that a human's right to life may be negated whenever it conflicts with some right or rights possessed by a person? There seems to be only one answer, viz., these pro-abortionists must be assuming that it is more valuable, important, or worthy, to be a human person than to be human. But how could this presupposition be shown to be true? In some circumstances, perhaps, (as when the mother-to-be is, say, a doctor on the verge of discovering a cure for cancer), a good case could be made that this woman should be allowed to abort rather than having to risk her life in childbirth. But what if the expectant mother is an alcoholic, on welfare, and a general burden to society? Would the abortion advocate allow us to turn the argument around and insist that this woman should not be allowed to abort her "innocent" fetus, even if her own life were in jeopardy? This seems highly unlikely. But if this is so, the question simply re-arises: what is there about being a person which, in itself, makes one better or more worthy than a being who is merely human? To the best of my knowledge, no pro-abortionist has faced this issue head on. However, there is one defender of abortion whose arguments could be interpreted as replies to our question. It is to this defense, then, that we must now turn our attention.

Professor R. B. Brandt defends abortion with two separate arguments.[18] Neither argument is intended by Brandt to be an answer to the question we have posed, but both lend themselves to such an application. First, we have argued that even if it is legitimate to distinguish between "human" and "person," this distinction will be of no use to the pro-abortionist unless he can show that persons are more valuable or worthy than human non-persons. However, in arguing in this fashion we have assumed that all humans possess a right to life; and it is precisely this assumption which Brandt's arguments call into doubt. For Brandt, there is not an unrestricted *prima facie* obligation not to kill humans, but only a *prima facie* obligation not to kill certain *kinds* of humans. In his first argument, Brandt contends that there are two kinds of human beings which it is not *prima facie* wrong to kill: "It is *prima facie* wrong to kill human beings, except those which are not sentient and have no desires, and except in reasonable defense of self or others against unjust assault."[19]

Given this principle, there are two ways in which Brandt could attempt to justify abortion. ("Abortion" is defined by Brandt as "destruction of the fetus within the first three months, when this is wanted or undertaken by the mother.") First, he could claim that abortion may sometimes be justified as a

[18] Brandt, *op. cit.*
[19] *Ibid.*, p. 158.

"reasonable defense of self or others against unjust assault." But fetuses are non-sentient, and can intend no action. Thus, when a pregnant woman's life is threatened by the fetus growing within her, this "attack" cannot be characterized as an *unjust* assault. Even accepting Brandt's principle, then, it is not possible to justify abortion in terms of self-defense.[20]

The second way Brandt could use his principle to argue for abortion would be to claim that the procedure is justified by the fact that prenatal beings in the first three months of life are "not sentient and have no desires." In fact, this is the tack taken by Brandt; but the flaw in the argument is obvious: how can one argue in this way and consistently maintain that it is morally wrong to kill a person who is unconscious or in a coma? Brandt realizes that his thesis may be objected to in this way, and he replies that although his principle obviously needs amendment, "It seems clear . . . that a restricted principle can be formulated along the suggested line . . ." (p. 158). Now this statement of confidence is heartwarming, but it offers little in the way of practical help for one who wishes to amend the principle so as to allow for abortion without undesirable side effects. How could the necessary changes be made? Unless I am mistaken, it is a relatively easy matter to show that no amendment is possible.

First, one must get clear as to what Brandt is really trying to do. Rather than holding that all humans have a *prima facie* right to life and then arguing that to be human one must be sentient (as some have done), Brandt seeks to sidestep the discussion of "human" by using sentience (or the lack of it) as the defining characteristic for a special class of humans–viz., those to whom the right to life is denied. But this is clearly illegitimate. After all, the right to life is ordinarily spoken of as a *human* right, i.e., it is not a personal right, nor is it a right which one possesses by virtue of his being a member of a special group or class of humans. To hold otherwise not only makes a mockery of ordinary language, it also undercuts morals completely, potentially justifying all sorts of horrors. In short, one cannot, as Brandt wishes to do, make the question concerning whether or not one has the right to life rest upon a determination of the particular type of human being that that human is. If one is human he has, by definition, *all* human rights (including a *prima facie* right to life). And this remains true regardless of the *kind* of human being he is.

Brandt senses that an objection of this sort can be brought against his thesis, and in order to demonstrate that there are "kinds of humans, such that it is not wrong to destroy them," he constructs the following example: "Suppose a human being has suffered massive brain damage in an accident. He is unconscious, and it is quite clear he will never regain consciousness–his brain is beyond repair. His body, however, can be kept alive by means known to science,

[20] For a more detailed version of this argument, see B. Brody, "Abortion and the Sanctity of Human Life," *American Philsophical Quarterly*, Vol. X (1973).

more or less indefinitely. Is there a *prima facie* obligation to keep this being alive, or to refrain from terminating its existence? I believe there is no such thing."[21]

Now I have no doubts that most of us feel that we have no *prima facie* duty to keep alive the being described by Brandt. But why is this? Is it because, as Brandt suggests, we see the unconscious being as a human of a special type (viz., unsentient), one to whom the right to life is no longer applicable? Or is it because we think the unconscious being is, for one reason or another, not fully human? I think the weight of evidence is clearly in favor of the latter view. For instance, do we not refer to beings of the sort Brandt has described as "vegetables"? And do we not say that entities of this kind are "as good as dead"? This being so, there should be nothing mysterious about why we believe that it would be proper to let Brandt's brain damaged being die. Taking that being's non-human status for granted, we reason that there is nothing immoral about letting "it" die, for "it" is not human and does not possess the right to life. If added evidence is needed, one need only refer to Brandt's own example. If it is true, as Brandt believes, that his paradigm illustrates that there are kinds of humans to whom we deny the right to life, why does he end the quoted passage by asking: "Is there an obligation . . . to refrain from terminating *its* existence?" I submit that the reasons for this statement are clear, and that they confirm the view that we *never* consider a being to be without the right to life, unless we have sincere doubts concerning his or her humanity.

If our analysis of Brandt's first argument is correct, Brandt has not shown that some humans lack a *prima facie* right to life. Thus, his attempt to justify the killing of human fetuses by appealing to the fact that these beings are "non-sentient and have no desires," is at best otiose. But Brandt has a second argument to show that it is not *prima facie* wrong to kill some humans. As a first step toward proving his point, Brandt offers the following as a definition of "*A* is *prima facie* morally wrong":

> *A* would be prohibited by a rule of the moral code the currency of which for their society would be preferred over all others by all persons who: (A) expected to live a lifetime in that society, (B) were rational at least in the sense that their preferences were fully guided by all relevant available knowledge, and (C) had some general benevolence, the degree and extent being compatible with the restriction that the preferences of each be rational.[22]

Accepting this construction of meaning, it is a relatively easy matter to determine whether or not it is *prima facie* morally wrong to kill fetuses in the first three months of gestation. One need only answer the question, "Would abortion in the first three months of pregnancy be forbidden by the rules of a moral

[21] Brandt, *op. cit.*, pp. 156–157.
[22] *Ibid.*, p. 160.

system which Brandt's rational, benevolent, moral code choosers would adopt?" Brandt feels that moral code choosers of the type specified would regard fetuses under three months of age as non-persons, and for this reason would not view them as possessing a *prima facie* right to life.

Obviously, whether Brandt's argument stands or falls is wholly dependent upon whether he is right in insisting that rational, benevolent, moral code choosers would opt to permit abortion. Brandt feels that abortion would be favored by his moral code choosers, because he thinks they would not extend personal identity back to the early stages of prenatal growth:

> I am going to argue that it is mistaken to suppose that the fetus is, in an appropriate sense, *the same person* as the child who will occur after the development of the fetus, and birth. And, if it is not the same person as the person which begins at some later time, then it is not, in an appropriate sense, now a person at all—and hence we cannot say that somebody is harmed or deprived by the abortion . . . we can see the force of the claim by considering a possible situation which is closely parallel. Suppose I were seriously ill, and were told that, for a sizeable fee, an operation to save "my life" could be performed, of the following sort: my brain would be removed to another body which could provide a normal life, but the unfortunate result of the operation would be that my memory and learned abilities would be wholly erased, and that the forming of memory brain traces must begin again from scratch, as in a newborn baby. Now, how large a fee would I be willing to pay for this operation, when the alternative is my peaceful demise? My own answer would be: None at all. . . . The important thing that is missing seems to be *memory;* if the latter person cannot remember my present and earlier experiences at all, then he is as good as a new person. . . . It appears that some continuity of memory is a necessary condition for personal identity in an important sense.[23]

Just as Brandt would take no interest in his being reborn in the manner described, so too rational, benevolent, moral code choosers would take no interest in the welfare of fetuses under three months of age. Because no one recalls any events within the first three months of his fetal life, every rational person would consider his existence in the first trimester of gestation as the existence of a non-person, that is to say, as the existence of a human which is not valuable or important enough to merit protection of life.

Essential to Brandt's position is the view that "some continuity of memory is a necessary condition for personal identity in an important sense." But how far back does Brandt conceive this continuity as extending? Try as I will, I cannot recall events before I was one year old. And although I do not doubt that Professor Brandt has a phenomenal memory, I still feel certain that he cannot remember his birth or the events immediately following. Thus, if Brandt is correct in holding that the failure of memory to extend back to the first trimester of gestation will cause moral code choosers to favor abortion, he must also hold that it will lead them to condone infanticide. Unless one is willing to

[23] *Ibid.*, p. 166.

accept the view that the killing of newborn babies is not even *prima facie* wrong, then, he must reject Brandt's thesis. Of course, Brandt may object that *some* persons do claim to be able to remember their births, and that for this reason benevolent moral code choosers would opt to "play it safe," and prohibit infanticide. But then there are some individuals who claim to be able to remember their existences in former lives too. Would not these claims cause Brandt's "careful" moral code choosers to extend personal identity back through gestation to some former life? And if this were done, would not Brandt's argument demand that all abortions be prohibited? I fail to see why the testimony of those who claim to remember their births should be considered more trustworthy by rational, benevolent beings, than the testimony of those who honestly believe that they remember their former lives.

III

If our analysis is correct, none of the major defenses of abortion succeeds in its purpose. Still, in reviewing the literature as we have done, one is struck not so much by the arguments' failures, as by the reasons for those failures. That is to say, if we have represented the pro-abortion arguments fairly, it is not just that they are unsound, they are unsound for simple, almost foolish reasons. But this is not the worst of it. Sometimes proponents of abortion are aware of their mistakes; and whenever this happens, reasoned inquiry simply ends with the abortion advocate dogmatically insisting that although the criticism in question may be, in principle, applicable to his argument, in practice it will never constitute a problem (e.g., above, p. 73). Now how can these facts be explained? One thing is clear. Unless one is willing to claim that abortion advocates are both inconsistent and confused, he must hold that members of this group do not believe as they do for the reasons cited in their arguments. But if this is so, a further question arises: what is the true basis (or bases) for the pro-abortion position? It is this question to which I will now address myself. In doing so, the thesis that I shall argue for is briefly this: although abortion advocates find it impossible to identify or empathize with a fetus, they can and do sympathize with a suffering mother-to-be, and hence simply *feel* that abortion is moral. This view is not without support. Consider the following facts.

First, if our identification hypothesis is accepted, a full explanation can be given for each of the abortion advocate's errors. In general, these errors flow from one of two causes: either the proponent of abortion allows his identifications and sympathies to affect his judgment concerning what is and what is not human, or he tacitly assumes that a human's value or worth is properly determined by another's feelings toward him. In order to see how these causes have been operative in determining abortion advocates' errors, we must briefly reconsider the various pro-abortion arguments discussed above.

Those who defend abortion by trying to show that "human" is ordinarily used in such a way that it excludes prenatal beings from its extension, obviously believe that fetal organisms are non-human. Yet, all of their attempts to specify the proper use of "human" fail. If consciousness is taken as the touchstone for humanity, persons in a coma may be killed. If abortion is warranted because the fertilized ovum is not "viable," then one may kill all those persons being kept alive on heart-lung machines. And so it goes. The failures are legion, but proponents of lexical definition remain undaunted. They "know" that fetuses are not human, and so they continue seeking linguistic support for their views, despite constant frustration. Now how can this be explained? Given our analysis of "conception," we know that the elusive definition they seek will never be found. If their position cannot be given factual support, then, the only possible explanation for the abortion advocate's unwavering belief in the non-human status of fetal organisms is that this view is founded, not upon reason or fact at all, but rather upon emotively distorted "fact." Simply put, proponents of lexical definition cannot bring themselves to *feel* that prenatal beings are human. No empathy, no identification, no sympathy for such beings is present; hence, this "developing tissue" is considered by them to be non-human and without the right to life.

Aside from the above, there is another fact which operates in favor of our view. That is, if our identification hypothesis is accepted, not only are we able to explain why proponents of lexical definition persist in their efforts despite repeated failure, we can also explain why it is that different abortion advocates have favored different definitions of "human." Those who identify with thinking organisms opt for a definition of terms of sentience or rationality. Those who empathize with beings who look human hold out for a definition in terms of some specific form, etc. In each case, however, the abortion advocate is guilty of confusing the *reason* he has for *feeling* that the term "human" may be applied to a certain being, with the *meaning* of the term itself. But these pro-abortionists should not be judged too harshly; others have fallen victim to the same error. Is it at all surprising, for instance, that the ancient Greeks, enamored of reason, found man's essence in his rationality? And why is it that doctors speak of viability, while anthropologists stress the importance of tool making, family organization, and cultural forms? Why is it not the other way around?

As we have seen, the main problem faced by those defenders of abortion who are proponents of arbitrary definition is to show that their "justification" of abortion cannot be extended so as to make legitimate acts which are commonly recognized as immoral. In order to avoid this trap, then, these pro-abortionists have argued that arbitrary non-human classification should be permitted only in the case of prenatal beings. Try as they will, however, no proponent of arbitrary definition has ever been able to give a reasoned defense of this view. But if this is the case, one can only wonder why it is that fetal organisms alone have been selected for "special" treatment. Accepting our identification hypothesis, a ready answer presents itself: proponents of arbitrary definition believe that fetuses alone should be classified as non-human, because they cannot help but *feel* that these beings really fall outside the class of things we call human. In short, pro-abortionists of this sort are not truly advocates of arbitrary definition at all; rather, they are continually assuming that one's classificatory status is properly determined by another's feelings toward him. And if this is so, there is no significant difference between the so-called arbitrary definition defenses of abortion, and those which attempt to argue from some ordinary use of "human."

The reasoning of Professor Thomson, as well as the arguments of those who seek to distinguish between "human" and "human person," reflect a second way in which pro-abortionists have allowed their passions to affect their moral judgments. In both these cases, the abortion advocates admit (at least tacitly) that fetuses are humans possessed of the right to life.[24] Still, they hold, this right may be "negated" when it comes into conflict with some personal right—say, an individual's right to control her body. But we ordinarily do not think in this way (e.g., we do not believe that a human may be killed simply because his existence gives someone else great psychological pain.) What, then, can be the process whereby these proponents of abortion arrive at their extreme conclusion? Surely the answer is clear: because they identify or empathize with a suffering mother-to-be, and not with the unconscious fetus within her, these defenders of abortion feel that the former is more valuable or important than the latter. Thus, whenever the rights of the two beings conflict, the mother's rights are held to "supersede" those of her child. If added evidence is needed for

[24] Although Brandt's second argument seems to turn upon the *human–person* distinction, Brandt does not allow that all humans possess a *prima facie* right to life. Hence, this argument must be distinguished from those which separate persons from humans, but at the same time acknowledge that it is *prima facie* wrong to kill anything human. In essence, both of Brandt's arguments are identical with those positions which argue that fetuses are nonhuman, and thus not possessed of the right of life. To be sure, Brandt does not quibble about the meaning of "human." But he does argue that fetuses lack qualities (sentience and personhood) which are necessary for any being to possess a *prima facie* right to life. Consequently, Brandt is faced with the same problem as those who seek to define "human"; namely, he cannot deny that it is *prima facie* morally wrong to kill fetuses, without also denying that it is *prima facie* wrong to kill other kinds of beings whom we ordinarily think it morally wrong to destroy, e.g., humans in a coma, or newborn babies. So far as I can see, that Brandt failed to draw this conclusion himself can only be explained in terms of his having the ability to empathize with babies and unconscious humans, but not with fetuses.

the truth of this hypothesis, it may be found simply by contrasting Thomson's violinist example with my case of the shipwrecked sailor. If this is done, it readily becomes apparent that the only difference between the two examples is that, in the latter case, the locus of sympathy has been shifted from the person whose body is being used to the one who is doing the using.[25]

IV

If the considerations argued for in the first three sections of this work are accepted, certain conclusions seem mandated. First, having uncovered the real reasons for the abortion advocates' belief, we must now reject that position as being without proper support. Surely no one will deny that it is essential to morality that our emotions be allowed to have some influence upon our determinations of right and wrong. But admitting this, it is equally true that in seeking to discriminate between right and wrong, one's passions should never be allowed to enter into his determinations of matters of fact. This kind of "method" for resolving ethical disputes has been used in the past; and invariably, the conclusions reached were anything but moral. When African blacks were enslaved and American Indians exterminated, for example, the justifications given for these actions were twofold: either these beings were said to be not fully human (savages), or they were held to be of little worth (uncivilized beings whose rights were able to be "superseded" whenever they came into conflict with our "more enlightened and worthy" desires). If Americans and Europeans found this kind of reasoning appealing, however, it was only because they could not identify with the life-styles of the beings in question. And when this lack of identification was coupled with the desire to exploit, the morality of slavery and Indian killing was assured. Today, the same kind of thing appears to be happening in the minds of abortion advocates. Of course, there is a difference in that those who defend abortion desire to help rather than to exploit some group. Still, the fact that these people operate from praiseworthy motives does nothing to guarantee the morality of their conclusions. And insofar as proponents of

[25] Although there are numerous other facts supporting my identification hypothesis, most cannot be discussed in an essay of this length. There is one, however, which does bear mentioning. That is, accepting our thesis, one can explain why it is that certain "extremist" members of the feminist movement are virtually alone in defending abortion on demand. Most abortion advocates do not favor this view because they find themselves able to identify with developed fetal organisms—e.g., seven-month-old fetuses hear sounds, move, hiccup, suck their thumbs, etc., and are thus thought to be human (they are often referred to as "unborn children," for instance). On the other hand, zealous devotion to their "cause" so biases some feminists that they find it impossible to empathize even with the most developed fetal beings.

abortion allow their inability to sympathize with a prenatal human being support their desire to kill it, they are just as mistaken in their reasoning as were those who earlier argued for slavery or Indian killing.

Second, because our analysis focuses upon "conception" rather than "human," the objectivity of our determination of the facts is guaranteed. As we have seen, "human" is an "emotionally charged" term, and it may well be that its proper use will forever remain a subject of dispute. "Conception," however, does not suffer from this difficulty; and in discussing its various uses, we need not worry that our subjective biases have influenced our view of the facts. Indeed, because it is true that we cannot identify with fertilized ova, the objectivity of our analysis is put beyond doubt. That is, with the possible exception of those who blindly defend Catholic dogma (and as against these a modified version of my argument could easily be directed) no one can really *feel* that zygotes are human. Then too, we all empathize with suffering mothers-to-be. If we were not bound by reason to do otherwise, then, our desires would incline us to defend abortion—at least abortion performed in the first trimester of pregnancy. What better guarantee of objectivity could we have?

Third, given that we have a full and complete demonstration of the fact that human life begins at conception, there is no escaping the conclusion that abortion can be justified only when the circumstances are such that we would think it proper to take the life of a human being. Now we have allowed that there is one such case, namely, abortion must be permitted whenever the continuation of pregnancy would put in danger the lives of both the mother and her fetus. Could a case be made for the morality of abortion in any other instance? Certainly there is nothing in the abortion advocates' arguments which would incline us to believe so. Further, Professor Baruch Brody has argued convincingly that it would be morally wrong to perform an abortion, even if it were necessary to save the life of the mother.[26] And if this is true, how could any of the weaker defenses succeed? But apart from these specific claims, there are more general considerations which lead us to the same conclusion. That is, whether we like it or not, the fetus is a human being of a special kind. Because he is human, he has all human rights; yet, because he can intend no action, it makes no sense to say that he has any of the duties or obligations possessed by the rest of us. He is much like a newborn baby, unconscious since birth, and kept in isolation from everyone save his mother. How could one hope to argue for the morality of killing such a being? True, there are some instances in which a mother who killed such a baby could hope to be excused (e.g., the infant had a disease which, though it caused the baby no harm, would kill the mother if she contracted it). But even here, no argument could be made for the view that the mother who killed such a child would have acted in a morally right way.

[26] B. Brody, "Abortion and the Sanctity of Human Life," *op. cit.*

Finally, if it is true that there is no rational or factual basis upon which one can construct a sound defense of abortion, it is also true that the strength of our emotions cannot be denied. For example, what man can honestly say that he would attempt to stop his wife from having an abortion if it were clear that she would die or become psychotic given the continuation of her pregnancy? In cases such as these we are caught in a real moral dilemma: reason tells us one thing, the passions another. And even if one cannot agree that the passions should rule reason in deciding moral issues, he must acknowledge that (in the case of abortion at least), the emotions *do* rule. Apparently some opponents of abortion have realized this. Attempting to combat the emotions' force, then, they ask their readers to empathize with the fetus—to "feel" its desire for life.[27] But all such injunctions fall on deaf ears, for it is simply impossible for a normal person to identify with a zygote or embryo. If this is granted, however, what possible hope is there for fostering moral conduct? Unless I am mistaken, the anti-abortionist has only one choice: he must find some way of convincing women that abortion is both undesirable and unnecessary. Can this be done? In order to support the belief that it can, I now wish to offer for consideration a general set of guidelines which, though admittedly incomplete and sketchy, do offer some hope for escape from the dilemma of abortion.

First, before any substantive changes can be affected, abortion advocates must be made to see that their moral judgment rests upon an emotionally distorted view of the facts; for unless this is done, there will be no impetus for change. After all, why should proponents of abortion not rest content with the *status quo*? They feel that their position is morally right, and now the law allows that it is legally right. If anything at all is to get done, then, these people must be made to see that abortion presents us with a moral dilemma of sorts—one in which we cannot help but *feel* that the immoral course of action is right, and the moral course of action wrong.

Once people are educated to the point that their desire to be rational will (hopefully) cause them to want to avoid the abortion dilemma, action on several fronts is possible. If the statistics are to be believed, most women who seek abortions do so because they find themselves with unwanted pregnancies.[28] And women who have unwanted pregnancies turn to abortion for one or more of the following three reasons: (1) The woman is unmarried, the child is illegitimate, and the mother wants to avoid showing her "sin" to society. (2) The woman (especially the married woman) would agree to have the child and give it up for adoption, except that there is a social stigma attached to such action. And (3), the mother simply does not want to put up with pregnancy and childbirth.

[27] Germain Grisez, *Abortion: The Myths, The Realities, and the Arguments* (New York: 1970), pp. 277–287.

[28] Callahan, *op. cit.*, pp. 292–294.

Given these as the primary causes for abortion, then, anti-abortion action along the following lines would seem to be indicated:

1. If it is true that we today consider illegitimacy a "sin," it need not be so. In certain Scandinavian countries, for example, illegitimacy is not frowned upon as it is here. And if our attitudes could be so modified, one of the prime motivations women now have for seeking abortions would be undercut. Indeed, if people could only be made aware of the fact that abortion is the killing of a human being, more should be able to be accomplished than a simple change in attitude concerning illegitimacy. The unwed mother who elected to carry her child to term could become a person of respect—one to whom praise was due. After all, for the sake of another human being, this woman is putting her life on the line. Why not praise her? Surely changes of this sort are not impossible. And if the Catholic Church is sincere in its opposition to abortion, it should do its utmost to see that they *do* come about.

2. If our attitudes concerning illegitimacy can be changed, so too can our feelings toward a mother who "gives away" her child. What better expression of motherly love could there be than to give up one's baby for adoption when it was apparent that such a course of action was in the best interest of the child? Is abortion a more moral course? Is is better to keep the child out of some sense of duty, and thereby condemn it to a life without love or proper care? Surely we cannot believe so. In fact, a mother who follows either of the latter two courses must be operating out of self-interest and not love. And if people could only be made to realize this, another of the major reasons women have for seeking abortions would be nullified.

3. Some women, it is true, simply do not want to suffer through pregnancy and childbirth. And where this is the case, our hope lies, not in changing social attitudes, but in doing our best to advance scientific research. At present, the process is being developed whereby a fertilized egg can be transplanted from one woman's womb into another's. If this procedure could be perfected so that it was both successful and relatively inexpensive, women—all women—would have a viable alternative to abortion.[29] Instead of abortion clinics, we could have transplant clinics. Women who wanted children could register, just as they now register for adoption; and when the need arose, they could be called up for service. Now what could be the objections to such a procedure? Certainly the formation of transplant clinics would raise various new moral and legal problems. But could the new moral problems be as grave as those we now face in

[29] There is one possible exception to this statement. If a woman wanted an abortion because she knew that her child would be born deformed, it is almost certain that she would find no one willing to be a transplant recipient. The only way we can rid ourselves of abortions sought on these grounds, then, is to find the causes of birth defects and eradicate them.

abortion? And why could we not simply look upon transplantation as early adoption, thus minimizing the need for new legislation? Of course, there are also practical matters to be considered. Before uterine transplants could be used, for example, it would have to be shown that the procedure was safe, not only for the transplanted egg, but also for both the women involved. But if we are willing to spend the money for research, such safety could be secured. There remains, then, the matter of cost. How expensive would such an operation be? Frankly, I do not know. But it need not be as inexpensive as an abortion, for the cost could be split between the donor and donee. Also, why should it not be possible to obtain a small federal subsidy for such procedures? All in all, there appears to be nothing which would make it impossible in principle for uterine transplants to serve as an alternative to abortion.

Although the above three changes seem to hold out the most hope of success for the anti-abortionist, action along other lines would also be efficacious. First, a concerted effort should be made to overcome the Catholic Church's ridiculous opposition to birth control.[30] And second, those who are sincere in their opposition to abortion should do their utmost to make sex education mandatory in the schools. And perhaps a "refresher" course could be required for all those wishing to take out a marriage license. Further, pressure should be brought to bear for the development of more effective male contraceptives. Work is currently being done in this area—a pill is being developed—but the more procedures the better.

I harbor no illusions concerning the difficulties involved in getting something like the above program put into effect. Yet, it does provide hope. Indeed, if only the process of uterine transplants could be perfected, the possibility of eradicating abortion would be much increased. If women knew that there was a viable alternative to abortion, they would find it much easier to follow the dictates of reason. And if this state of affairs ever came into being, the time *would* be right for opponents of abortion to press for legal change. Of course, whether this point will ever be reached is doubtful. But even if there are good grounds for pessimism, we must try to bring about the needed changes. After all, there are human lives at stake.

[30] I cannot use this forum to argue for the view that the Catholic Church's position on birth control is "ridiculous." I see no escape from this conclusion, however, and hold that except for those devices which operate as abortive mechanisms (e.g., intra-uterine devices), all contraceptive procedures should be condoned.

FURTHER READINGS

Books

Clifford Bajema, *Abortion and the Meaning of Personhood* (Grand Rapids, Michigan: Baker Book House, 1974); Daniel Callahan, *Abortion: Law, Choice, and Morality* (New York: Macmillan and Company, 1970); Michel Chartier, *Avortement et respect de la vie humaine* (Paris: Editions du Sueil, 1972); Committee on Psychiatry and Law of the Group for the Advancement of Psychiatry, *The Right to Abortion: A Psychiatric View* (New York: Charles Scribners); Robert E. Cooke, *The Terrible Choice: The Abortion Dilemma* (New York: Bantam Books, 1968); Charles E. Curran, *Contraception: Authority and Dissent* (New York: Herder and Herder, 1969); Donald R. Cutler, *Updating Life and Death: Essays in Ethics and Medicine* (Boston: Beacon Press, 1969); Martin Ebon, *Every Woman's Guide to Abortion* (New York: Universe Books, 1971); David M. Feldman, *Birth Control in Jewish Law* (New York: University Press, 1968); Joseph Fletcher, *Morals and Medicine* (Boston: Beacon Press, 1960); Claude A. Frazier, *Should Doctors Play God?* (Nashville: Broadman Press); R. T. J. Galbally, *The Right to be Born* (Melbourne: A.C.T.S. Publications, 1970); David Granfield, *The Abortion Decision* (New York: Doubleday, 1969); Germain G. Grisez, *Abortion: The Myths, The Realities and Arguments* (New York: Corpus Books, 1970); Alan Frank Guttmacher, *The Case for Legalized Abortion Now* (Berekeley, California: Diablo Press, 1967); Robert E. Hall, *Abortion in a Changing World,* two volumes (New York: Columbia University Press, 1970); Anthony Horden, *Legal Abortion, The English Experience* (Oxford: Pergamon Press, 1971); Robert E. Joyce and Mary Rosera, *Let Us Be Born; The Inhumanity of Abortion* (Chicago: Franciscan Herald Press, 1970); Jerome M. Kummer, *Legal and Illegal: A Dialogue between Attorneys and Psychiatrists,* second edition (Santa Monica: J. M. Kummer, 1969); Daniel H. Labby, *Life or Death, Ethics and Options* (Seattle: University of Washington Press, 1968); Lawrence Lader, *Abortion* (Indianapolis: Bobbs-Merrill, 1966), also by this author: *Abortion II: Revolution in the Making* (Boston: Beacon Press, 1973); Nancy Howell Lee, *The Search for an Abortionist* (Chicago: University of Chicago Press, 1969); David Lowe, *Abortion and the Law* (Pocket Books, 1966); C. J. McFadden, *Medical Ethics* (London, 1962); Demenico Mondrone, *Mamma, Why Did You Kill Us?,* Tr. Dino Soria (Baltimore, 1970); John Clover Monsma, *Religion and Birth Control* (New York: Doubleday, 1963); John T. Noonan, Jr., *The Morality of Abortion; Legal and Historal Perspectives* (Harvard University Press, 1970); Robert L. Perkins, *Abortion: Pro and Con* (Cambridge: Schenkman Publishing Company, 1974); James Rachels, *Moral Problems* (New York: Harper and Row, 1971); Bernard Jerome Ransil, *Abortion* (Paramus, N. J.: Paulist Press, 1969); Harold Rosen,

Abortion in American, originally published as *Therapeutic Abortion,* (Boston: Beacon Press, 1967); Diane Schulder and Florynce Kenndy, *Abortion Rap–Shirley Chisholm, Member of Congress* (New York: McGraw-Hill, 1971); Edwin M. Schur, *Crimes without Victims: Deviant Behavior and Public Policy: Abortion, Homosexuality, Drug Addiction* (Englewood Cliffs, N. J.: Prentice-Hall, 1965); Russel B. Shaw, *Abortion on Trial* (London: Hale, 1969); David T. Smith, *Abortion and the Law* (Cleveland: Western Reserve University, 1967); Walter A. Spitzer and Caryle Saylor, *Symposium on the Control of Human Reproduction* (Wheaton, Ill.: Tyndale House, 1969); N. St. John-Stevas, *The Right to Life* (London: Hodder and Soughton, 1963); Herman Peter Tarnesby, *Abortion Explained* (London: Sphere Books, 1969); David F. Walbert and Douglas J. Butler eds., *Abortion, Society and the Law* (Cleveland: The Press of Case Western Reserve University, 1973); Glanville Williams, *The Sanctity of Life and Criminal Law* (New York: Knopf, 1968); Margaret Wynn and Arthur Wynn, *Some Consequences of Induced Abortion to Children Born Subsequently* (London: Foundation for Education and Research in Childbearing, 1972); Ludwig Wittgenstein, *Philosophical Investigations,* Tr. G. E. M. Anscombe (New York, 1953).

Articles

Louis Beinaert, "L' avortement est-il infanticide," *Etudes, 337* (1970), pp. 547–61; J. Bennett, "Whatever the Consequences," *Analysis, 26* (1966), pp. 83–102; John C. Bennett *et al., Christianity and Crisis, 32* (1973), pp. 287–98; Judith Blake, "Abortion and Public Opinion: the 1960–1970 Decade," *Science, 171* (1971), pp. 540–49; Sissela Bok, "Ethical Problems of Abortion," in *Raising Children in Modern Urban America,* ed. Nathan Talbot M.D. (New York: Little, Brown and Co., forthcoming), also by this author: "The Leading Edge of the Wedge," *The Hastings Center Report, 3* (1971), pp. 9–11; Richard B. Brandt, "The Morality of Abortion," *The Monist, 56* (1972), pp. 504–26; B.A. Brody, "Abortion and the Law," *Philosophy Journal, 68* (1971), pp. 357–69, also by this author: "Abortion and the Sanctity of Human Life," *American Philosophical Quarterly, 10* (1973), pp. 133–40, and "Thomson on Abortion," *Philosophy and Public Affairs, 1* (1972), pp. 335–40; M. J. Buss, "Beginnings of Human Life as an Ethical Problem," *Journal of Religion, 47* (1967), pp. 244–55; Daniel Callahan, "The Sanctity of Life," *Updating Life and Death: Essays in Ethics and Medicine,* ed. Donald Cutler (Boston: Beacon Press, 1969), also by this author: "Contraception and Abortion: American Catholic Response," *Annals of the American Academy of Political and Social Science, 387* (1970), pp. 109–17; Danner K. Clouser, "Abortion, Classification, and Competing Rights,"

Christian Century (1971), pp. 626–28, also by this author: "The Sanctity of Life: An Analysis of a Concept," *Annals of Internal Medicine* (1973), pp. 119–25; Patrick J. Coffey, "Toward a Sound Moral Policy on Abortion," *The New Scholasticism, 47* (1973), pp. 105–12; Charles E. Curran, "Abortion: Law and Morality in Recent Roman Catholic Thought," *The Jurist, 33* (1973); Joseph Donceel, "Abortion: Mediate and Immediate Animation," *Continuum, 5* (1969), pp. 167–71; Henry Dyck, "Perplexities of the Would-be Liberal in Abortion," *Journal of Reproductive Medicine, 8* (1972), pp. 351–54; Tristram H. Engelhardt, Jr., "Viability, Abortion, and the Difference Between a Fetus and an Infant," *American Journal of Obstetrics and Gyecology, 116* (1973), pp. 429–34; John M. Finnis, "Three Schemes of Regulations," in *The Morality of Abortion,* ed. John T. Noonan, Jr., (Cambridge: Harvard University Press, 1970); Philippa Foot, "The Problem of Abortion and the Doctrine of the Double Effect," *Oxford Review, 5* (1967); D. Gerber, "Abortion: The Uptake Argument," *Ethics, 83* (1972); Rudof J. Gerber, "Abortion: Parameters for Decision," *International Philosophical Quarterly, 11* (1971), pp. 561–84; Germain G. Grisez, "Toward a Consistent Natural Law Ethics of Killing," *American Journal of Jurisprudence, 15* (1970), pp. 64–96; James W. Gustafson, "A Protestant Ethical Approach," in *The Morality of Abortion,* ed. John T. Noonan, Jr., (Cambridge: Harvard University Press, 1970); A. F. Guttmacher and H. F. Pilpee, "Abortion and the Unwanted Child," *Family Planning Perspectives, 2* (1970), pp. 16–24; G. Hardin, "Abortion or Compulsory Pregnancy," *Marriage and the Family, 30* (1968), pp. 246–51; Bernard Haring, "A Theological Evaluation," *The Morality of Abortion,* ed. John T. Noonan, Jr. (Cambridge: Harvard University Press, 1970), pp. 123–45; D. M. Heer, "Abortion, Contraception, and Population Policy in the Soviet Union," *Soviet Studies, 17* (1965), pp. 76–83; James Humber, "Abortion: The Avoidable Moral Dilemma," *The Journal of Value Inquiry, 9* (1975), also by this author: "The Case Against Abortion," *The Thomist, 39* (1975), pp. 65–84; I. M. Ingram, "Abortion Games: An Inquiry into the Working of the Act," *The Lancet* (1971), pp. 969–70; Marvin Kohl, "Abortion and the Argument from Innocence," *Inquiry, 14,* pp. 147–51; Z. Leavy and J. Kummer, "Criminal Abortion: Human Hardship and Unyielding Laws," *Southern California Law Review, 35* (1962), pp. 123–28; Zigmond M. Leebensohn, "Abortion, Psychiatry and the Quality of Life," *American Journal of Psychiatry, 128,* pp. 946–54; C. Eric Lincoln, "Why I Reversed My Stand on Laissez-Faire Abortion," *Christian Century* (1973), pp. 477–79; David W. Louisell and John T. Noonan, Jr., "Constitutional Balance," in *The Morality of Abortion,* ed. John T. Noonan, Jr. (Cambridge: Harvard University Press, 1970), pp. 220–60; Joseph Margolis, "Abortion," *Ethics, 84* (1973), pp. 51–61; Richard A. McCormick, S.J., "Past Church Teaching on Abortion," *Proceedings of the Catholic Theological Society of America* (Yonkers, New York, 1968), *23,* pp. 131–51, also by this author: "Notes on Moral Theology," *Theological Studies, 33* (1972), pp. 68–78; D. S. Mileti and L. D. Barnett, "Nine Demographic Factors and Their Relationship toward Abortion Legalization," *Social Biology, 19* (1972), pp. 43–50; Giles Milhaven, "The Abortion Debate; An Epistemological Interpretation," *Theological Studies, 31* (1970); John T. Noonan, Jr., "An Almost Absolute Value in History," *The Morality of Abortion* (Cambridge: Harvard University Press, 1970), also by this author: "Abortion and the Catholic Church: A Summary History," *National Law Forum, 12* (1967), pp. 85–131; Michael Novak, "Abortion is Not Enough," *Christian Century, 84* (1967), pp. 430–31; John O'Conner, "On Humanity and Abortion," *National Law Forum, 13* (1968), pp. 127–33; Ralph Potter, "The Abortion Debate," *Updating Life and Death: Essays in Ethics and Medicine,* ed. Donald R. Cutler (Boston: Beacon Press, 1969); Eugene Quay, "Justifiable Abortion," *Georgetown Law Journal, 49* (1961); Paul Ramsey, "The Sanctity of Life," *Dublin Review, 241* (1967), pp. 3–21, also by this author: "Feticide/Infanticide upon Request," *Religion in Life, 39* (1970), pp. 170–86, and "Reference Points in Deciding about Abortion," *The Morality of Abortion,* ed. John T. Noonan, Jr. (Cambridge: Harvard University Press, 1970), and "The Morality of Abortion," *Moral Problems* (New York: Harper and Row, 1971), and "Abortion: A Review Article," *The Thomist, 37* (1973), pp. 174–226; Fred Rosner, "The Jewish Attitude Toward Abortion," *Tradition, 10* (1968), pp. 48–71; Schwartz, "The Parent or the Fetus," *Humanist, 27* (1967), p. 126; Bruno Schuller, "Typen ethischer Argumentution in der katholischen Moraltheologie," *Theologie und Philosophie, 45* (1970), pp. 526–50; Natalie Shainess, "Abortion: Inalienable Right," *New York State Journal of Medicine* (1972), pp. 1772–1775; Edward Shils, "The Sanctity of Life," *Life and Death: Ethics and Options* (Seattle: University of Washington Press, 1968), pp. 2–38; M. Simms, "Abortion Act after Three Years," *Political Quarterly, 42* (1971), pp. 269–86; Harmon L. Smith, "Abortion, Death, and the Sanctity of Life," *Social Science and Medicine, 5* (1971), pp. 211–18, also by this author: "Life as Relationship: Insight on Abortion," *Christian Advocate, 14* (1970), pp. 7–8, and "Religious and Moral Aspects of Population Control," *Religion in Life, 34* (1970), pp. 193–204, and "Abortion and the Right to Life," *Ethics and the New Medicine* (Nashville: Abingdon Press, 1970), pp. 17–54, and H. L. Smith and Louis W. Hodges, "The Human Shape of Life," *The Christian and His Decisions* (Nashville: Abingdon Press, 1969), pp. 233–52; Alan A. Stone, "Abortion and the Supreme Court: What Now?" *Modern Medicine* (1973), pp. 30–37; Judith Jarvis Thomson, "A Defense of Abortion," *Philosophy and Public Affairs* (Princeton University Press, 1971), pp. 47–66; Michael Tooley, "A Defense of Abortion and Infanticide," *Philosophy and Public Affairs, 2* (1972), pp. 37–65, also by this author: "A Defense of Abortion and Infanticide," *The Problem of Abortion,* ed. Joel

Feinberg, (Belmont, California, 1973), and "Michael Tooley Replies," Philosophy and Public Affairs, *2* (1973); Cornelius Ban der Poel, "The Principle of Double Effect," *Absolutes in Moral Theology?* ed. Charles E. Curran (Washington: Corpus Books, 1968), pp. 186–210; Robert Veatch, "When Does Life Begin?" *Face to Face, 1* (1969), pp. 18–19, also by this author: "Taxing of Childbearing: Can It Be Just?" Hastings-on-Hudson, New York: Institute of Society, Ethics and the Life Sciences Working Paper Series, No. 4 (1971), and "What About Abortion on Demand?" *Social Action* (1971), pp. 26–34; Rachel Conrad Walberg, "The Woman and the Fetus," *New Theology,* 10, eds. Martin E. Marty and Dean Peerman (New York: Macmillian, 1973); Mary Warren, "On the Moral and Legal Status of Abortion," *The Monist, 57* (1973), pp. 43–61; Roger Wertheimer, "Understanding the Abortion Argument," *Philosophy and Public Affairs, 1* (1971), pp. 67–95; G. H. Williams, "The Sacred Condominium," *The Morality of Abortion,* ed. John T. Noonan, Jr. (Cambridge: Harvard University Press, 1970), pp. 146–71, also by this author: "Religious Residues and Presuppositions in the American Debate on Abortion," *Theological Studies, 31* (1970), pp. 10–75; Ruth Jane Zuckerman, "Abortion and the Constitutional Rights of Minors," *American Civil Liberties Union Reports,* Juvenile Rights Project, (1973).

For additional lists of sources see: Danner K. Clouser, *Abortion and Euthanasia: An Annotated Bibliography* (with Arthur Zucker) Society for Health and Human Values (Philadelphia, 1974); Charles Dollen, *Abortion in Context: A Select Bibliography* (Metuchen: Scarecrow Press, 1970); William M. Pinson, *Resource Guide to Current Social Issues* (Waco: Word Books, 1968).

PART II: MENTAL ILLNESS

INTRODUCTION

The conceptual perplexity encountered in any serious attempt to understand the nature of mental illness, as well as the unique moral and legal problems resulting from the ways we deal with the mentally ill, invites close examination of both the nature of mental illness and the ways society customarily "treats" the mentally ill. The essays assembled in this chapter focus on the more prominent conceptual, moral, and legal problems met in any reasonably adequate discussion on mental illness and its proper treatment. Moreover, it seems clear that the legal problems, as well as the moral problems, attending our treatment of the mentally ill derive in no small measure from an apparant lack of agreement on the nature of mental illness itself. Accordingly, a good deal of this chapter is weighted toward the conceptual problems involved in defining the nature of mental illness.

Leon Eisenberg's "Psychiatric Intervention" depicts what is essentially the majority opinion on the nature and proper treatment of mental illness. But in "The Myth of Mental Illness" Thomas Szasz attacks the majority view and argues that mental illness (as we have come to know it) is a myth because mental illness is not a medical problem at all. For Szasz, terms generally used to designate various forms of mental illness only designate forms of behavior deviating from currently acceptable social, moral, or legal norms. In addition, it is essential to Szasz's view that although such deviant ways of coping with stress may require the counsel of "the psychotherapist," still there is no reason for thinking that the behavior is irresponsible. Although Szasz's thesis on the nature and treatment of mental illness is frequently accorded the status of a minority opinion in the field of psychiatry, Ruth Macklin's essay, "Mental Health and Mental Illness: Some Problems of Definition and Concept Formation," would appear to lend substantial support to the view that the opinion of the majority on the nature of mental illness is by no means free of deep-seated conceptual difficulties. She also offers her own evaluation of the merits of Szasz's general position.

In "Involuntary Mental Hospitalization: A Crime Against Humanity" Szasz further probes the moral implications of the practice of involuntary commitment and argues strenuously against the practice as part of the myth of mental illness. He also discusses other moral problems relating to current ways of treating the mentally ill.

Confusion over the nature of mental illness and, consequently, the extent to which the mentally ill may or may not be held responsible for their behavior, creates serious problems for legal practice with respect to the mentally ill. In his "Psychiatrists and the Adversary Process" Judge David Bazelon traces the history of his court rulings on the nature of the insanity plea as it relates to criminal responsibility. He also discusses the reasons for, as well as the problems relevant to, the court's recent attempts to subject expert psychiatric testimony to cross-examination. The difficulties involved in substantiating a plea of insanity, along with Szasz's urging that the plea be dropped altogether from the law, raises serious questions about the viability of the plea and its place in any system of jurisprudence.

PSYCHIATRIC INTERVENTION

LEON EISENBERG

The course and outcome of the major mental disorders of man have been profoundly altered by the advent of drug treatment and by changes in the methods of delivering health care. These gains have been made in spite of our continuing ignorance of the basic causes and mechanisms of mental disorders. The remarkable efficacy of chemotherapy has provided a major spur to basic research into the biochemical and genetic mechanisms underlying psychiatric illness, and recent discoveries hold great promise for new and better means of diminishing the misery associated with disorders of the mind. The potential power of this developing "psychotechnology" is, however, creating concern about unwarranted intrusions into personal privacy and individual rights.

Even though large gains have been made, there remain major areas that await the impetus of new ideas and better methods. Psychotherapy (the psychological treatment of mental disorders) has many sources, but in the U.S. it took root in the soil provided by psychoanalysis. Whatever its future evolution, dynamic psychotherapy has had a powerful humanizing influence on medicine. It requires the physician to listen and to try to understand the patient rather than merely to categorize his foibles while remaining indifferent to his suffering.

The majority of patients with neuroses (disorders characterized by anxiety or psychic defenses that seek to ward off anxiety) describe themselves as being improved after psychotherapy. The symptomatic changes, however, appear to be nonspecific. Similar rates of improvement are found following treatments based on theories in complete contradiction to one another, and sometimes following the mere anticipation of treatment.

Psychoanalysis has undergone many changes since Freud's original formulations. Its theories, however, rest on argument, in the philosophical sense, rather than on evidence that meets the canons of science. Behavior therapy (based on conditioning theory) has shaken psychiatric traditionalism. Its usefulness for particular symptoms such as phobias would now be acknowledged by most psychiatrists, but its more general applicability remains to be demonstrated.

Family-therapy methods have administered another jolt to conventional psychiatric thought. Its practitioners reformulate the problem from one "in" the patient to one "between" the patient and his family. The designated patient may merely be the scapegoat. The illness to be treated lies in distorted interpersonal relationships in the family. These concepts have broadened the psychiatrist's frame of reference. Family therapy, however, has spread more as a messianic movement and not because of convincing evidence from well-designed therapeutic trials.

Whether the psychological treatment of neurotic patients is an exclusive medical specialty is dubious. Such studies as we have provide no evidence that psychiatrists are no more effective therapists than psychologists, social workers and counselors. Therefore in this article I concentrate on the major responsibility of psychiatry: the care and treatment of serious mental illness. It must be emphasized, however, that psychological judgment and personal sensitivity are indispensable to the effective function of the psychiatrist as a physician.

The severe mental disorders we have learned to deal with more effectively are the psychoses, the most prominent of which are schizophrenia and manic-depressive psychosis. Psychoses are severe disorders characterized by profound and pervasive alterations of mood, disorganization of thought, and withdrawal from social interactions into fantasy. Schizophrenia is a psychosis with disturbances in the evaluation of reality and in conceptual thinking that are often accompanied by hallucinations and delusions. It usually becomes manifest in late adolescence or young adulthood. Manic-depressive psychosis is marked by severe disturbances in mood that are self-limited in time but are recurrent and frequently cyclic. Mania is manifested by psychic elation, increased motor activity, rapid speech, and the quick flight of ideas. The stigmata of depression are melancholia, the slowing of thought, unusual thought content (for example overwhelming guilt over imagined transgressions and delusions of rotting away), motor retardation, sleep disturbances, and preoccupation with bodily complaints.

The most obvious indicator of the extent of the recent change in psychiatric practices is evident in the number of resident patients in the state and county mental hospitals of the U.S. . . . The number of patients peaked at about 560,000 in 1955. Over the preceding decade the number of patients had increased at the rate of 3 percent per year, almost twice the rate of growth of the population. Then the trend reversed sharply. The resident state and county mental-hospital population fell to 276,000 by 1972, in spite of general population growth and increased rates for both first admissions and readmissions. From 1962 to 1969 first-admission rates rose from 130,000 to 164,000 and readmission rates from 150,000 to 216,000 as the number of resident patients fell from 516,000 to 370,000. These figures reflect the dramatic decline in the average length of stay.

Although these data convey an overall picture of national trends, they fail to portray the extent of change in some areas. At present California has only some 5,400 patients in its state hospitals, a reduction of 80 percent since 1961, and it plans to eliminate all state-hospital beds for mental patients by 1977. Lest this be mistaken for the elimination of mental illness in California, or even the elimination of inpatient care as a mode of treatment, it should be noted that the state plan projects the transfer of care of mental patients to the counties. California counties now operate, in more or less adequate fashion, programs for the mentally ill, including psychiatric inpatient units in general hospitals and beds in nursing homes.

The total number of patient-care episodes (inpatient plus outpatient) in the U.S. increased from 1,675,000 in 1955 to 4,038,000 in 1971. In that period the number of inpatient episodes rose from 1,296,000 to 1,721,000 and the number of outpatient episodes zoomed from 379,000 to 2,317,000. Placed in relation to population growth, inpatient episodes per 100,000 population rose marginally from 799 to 847, whereas outpatient episodes increased from 234 to 1,134. The locus of care has shifted from the isolated and neglected wards of the state hospital to newly created but not always adequate facilities in the community. There is growing evidence that some of the former hospital patients are not cared for by anyone; they live in single-room-occupancy units, kinless and friendless, subsisting marginally on welfare allotments. Given what most state mental hospitals once were and what many still are, most patients are better off out of them than in them. This, however, does not excuse our failure to provide for the patients lost in the shuffle from one pattern of care to another.

The net change in patient treatment has been enormous, with the number of patients in state and county mental hospitals reduced by half, with the great majority of patients spending less time in the hospital for a given episode of illness, and with far fewer of those admitted being condemned to an endless hospital stay. What factors account for these dramatic changes? Although a complete explanation is lacking, two important transformations in psychiatric care have played the major roles. They are the rediscovery of the principles of moral (that is, humane) treatment and the development of effective psychotropic drugs.

Responding to the humanistic ideas of the revolutionary era of the 18th century, Vincenzo Chiarugi of Italy, Philippe Pinel of France and William Tuke of England pioneered in the recognition that the way mental patients are treated affects the way they behave. Chiarugi wrote into the regulations of the Bonifacio hospital in Florence in 1788 the statement: "It is a supreme moral duty and medical obligation to respect the insane person as an individual." The famous engraving of Pinel striking the chains from the insane at the Bicêtre (an accomplishment he modestly credited to his nonmedical hospital governor, Jean-Baptiste Pussin) symbolizes, if it also mythologizes, his accomplishment. In

Pinel's words: "In lunatic hospitals, as in despotic governments, it is no doubt possible to maintain, by unlimited confinement and barbarous treatment, the appearance of order and loyalty. The stillness of the grave and the silence of death, however, are not to be expected in a residence consecrated for the reception of madmen. A degree of liberty, sufficient to maintain order, dictated not by weak but enlightened humanity and calculated to spread a few charms over the unhappy existence of maniacs, contributes in most instances to diminish the violence of symptoms and in some to remove the complaint altogether."

In England, Tuke and the Society of Friends founded The Retreat at York. They chose the name to suggest "a quiet haven in which the shattered bark might find the means of reparation or of safety." Samuel Tuke, grandson of the founder, wrote in a description of the moral treatment of the insane: "If it be true, that oppression makes a wise man mad, is it to be supposed that stripes, and insults, and injuries, for which the receiver knows no cause, are calculated to make a madman wise? Or would they not exasperate his disease, and excite his resentment?"

These quotations convey the extent to which convictions about the centrality of human liberty gave rise to a therapeutic philosophy that came to replace confinement and punishment. Succeeding waves of neglect, reform, and neglect of institutional patient treatment led a century and a half later to state mental hospitals that were unmanageably large, poorly staffed and grossly underfunded. These self-contained worlds became dedicated to self-perpetuation rather than to the patients to whom they were ostensibly dedicated.

In the late 1940s and early 1950s there were two significant developments. First, sociologists began to examine the nature of mental institutions and the interactions of patients and staff as these influenced the behavior of the patient. Second, innovative hospital superintendents, first in Britain and then in the U.S., opened locked doors, introduced a measure of patient self-government, and re-established bonds with the surrounding community. It soon became evident that the apparently deteriorated behavior of chronic mental patients was not a simple result of the mental disorder that had led to admission but of what has been termed the social-breakdown syndrome, that is, the alienation and dehumanization produced by the "total institution" the mental hospital had become. With a reconceptualization of the hospital as a therapeutic community (the new name for moral treatment), many of the chronic inpatients were able to be returned to the community, even after years of continuous hospitalization, and far fewer first admissions became chronic.

The reversal of the continuous growth of the inpatient population began with the reordering of the institutional environment, the development of community-care alternatives, changes in administrative policy favoring early discharge, and hard-won battles for community acceptance of the mentally ill. Although the first decrease of the inpatient population preceded the large-scale introduction of drug treatment, the continued and much higher rate of change would not have been possible unless effective chemotherapeutic means of managing acute psychotic disorders had become available at the same time.

The history of psychopharmacology can readily be made to fit Horace Walpole's parable of the "Three Princes of Serendip," but it may be more instructive to recall Louis Pasteur's aphorism that chance favors the prepared mind. Chlorpromazine, the prime example of the new psychotherapeutic drugs, resulted from efforts to synthesize a more effective antihistamine. In 1949 Henri-Marie Laborit of France used chlorpromazine to produce a "hibernation syndrome" in patients undergoing prolonged surgery, and he noticed that a side effect was a striking indifference to environmental stimuli. Then in 1952 Jean Delay and Pierre Deniker of France discovered the drug's remarkable effectiveness in aborting acute psychotic episodes, both in schizophrenia and in the manic phase of manic-depressive disorders. The confirmation of chlorpromazine's quite extraordinary value by controlled clinical trials led to its widespread use throughout the world and to an organized search for related families of compounds, a search that required close interaction of organic chemists, pharmacologists, and clinicians. New classes of drugs now termed major tranquilizers or neuroleptics were discovered.

In spite of the similarity of these drugs in pharmacologic properties, they vary in chemical structure, milligram-for-milligram potency, in frequency of side effects and in their utility for the treatment of clinical subtypes within the schizophrenic spectrum. (For example, trifluoperazine is thought to be more useful for the inhibited schizophrenic and chlorpromazine for the overactive schizophrenic.) Extensive clinical research has documented the effectiveness of the phenothazine class of drugs (which includes chlorpromazine and trifluoperazine) in terminating an episode of schizophrenia. The natural history of the disorder, however, indicates a substantial risk of recurrence and little residue of benefit from prior treatment. Studies of maintenance therapy, when the patient is willing and able to take an appropriate drug over a long period of time, indicate a definite attenuation in the rate of subsequent attacks. Unfortunately unpleasant side effects (sedation and symptoms resembling Parkinson's disease) are a problem for some patients and serious toxicity (persistent rhythmical, involuntary movements of tongue and face, abnormal pigmentation, low white-cell count, and jaundice) afflicts a substantial minority.

Ayurvedic medicine in India had for thousands of years made use of the snakeroot plant, *Rauwolfia serpentina*, for its sedative properties. When rauwolfia was introduced as an antihypertensive agent in the late 1940s and early 1950s, its active principle, reserpine, was identified. One serious side effect was the production of severe depression in some patients. The introduction of reserpine into psychiatry led to the discovery of the drug's effectiveness in aborting acute psychotic episodes in schizophrenia. Once widely used, reserpine has been replaced by the phenothiazines because of the greater frequency of reserpine's serious side effects. It continues, however, to play an important role in the experimental pharmacology of the psychoses.

In the same year that chlorpromazine was introduced into psychiatry alert clinicians treating tuberculosis with iproniazid noted that many of their patients displayed a marked euphoria well before there was any major improvement in their pulmonary disease. Subsequent trials in depressed patients established the power of this agent in relieving depression. It was also discovered that iproniazid acted to inhibit the enzyme monamine oxidase. (Monamines, such as serotonin, norepinephrine, and dopamine, are believed to act as chemical transmitters between nerve cells in the brain.) At about the same time organic chemists synthesized the drug imipramine by replacing the sulfur atom in the phenothiazine structure with a dimethyl bridge, thereby changing a six-member ring into a seven-member ring. Tests with animals had suggested that imipramine should have properties similar to the parent compound, but the drug proved relatively ineffective in schizophrenics.

In 1957, however, R. Kuhn of Switzerland, observing that depressed schizophrenics did seem to show some improvement on being given imipramine, demonstrated that patients suffering from psychotic depression displayed marked improvement. Imipramine was the prototype of the tricyclic antidepressants. The chemical structures of these two classes of antidepressants–the monamine oxidase inhibitors and the tricyclics–are quite different. . . . The drugs are effective in the management of from 70 to 80 percent of depressive episodes and have much reduced the reliance on electroconvulsive therapy, which was once the only known effective treatment for depressive illness. In Australia J. F. J. Cade was pursuing the hypothesis that mania was the result of a toxic state that he sought to detect by searching for metabolic products in the urine. Having employed lithium because of the solubility of its urate salt, he noted that the lithium given to laboratory animals acted as a sedative. He then used it in clinical trials that indicated the element could control mania. Cade's lead was ignored for more than a decade because of concern about lithium toxicity, which had become disturbingly evident when its salts were imprudently administered in high doses as a substitute for sodium in the treatment of hypertension. With proper attention to dosage and careful monitoring of levels in the blood serum, lithium has now been shown to be a safe agent for the treatment of mania, although it is not as prompt in its action as chlorpromazine. The most exciting development has been the demonstration by the Danish group headed by Mogens Schou that lithium is an effective prophylactic agent against the recurrence of psychotic episodes in patients with manic-depressive disease. For the first time we have an agent with acceptably low levels of toxicity (under close medical supervision), low cost, and ease of administration that prevents to a significant degree the otherwise inevitable recurrent attacks of a major psychosis.

Studies of the biochemistry of the brain, which were stimulated in large part by the desire to understand how the psychoactive drugs worked, indicate that selective errors in the metabolism of the monamines in specific areas of the brain may be the pathophysiological basis for psychotic disorders. Chemical and histological studies indicate that these substances are located in certain nerve cells and the ends of their long fibers. The biogenic amines (including serotonin, norepinephrine, and dopamine) most likely function in specific regions of the central nervous system as chemical transmitters between nerve cells or as modulators of transmission by other substances (such as acetylcholine).

There appears to be a close relation between the effects of the psychoactive drugs on monamine levels and on affective and behavioral states. Drugs that result in an inactivation and depletion of norepinephrine in the brain (such as reserpine and lithium) produce sedation and depression. Drugs such as imipramine, together with monamine oxidase inhibitors that increase or potentiate norepinephrine, are associated with behavioral stimulation and usually have an antidepressant effect in man. Therefore abnormally high or low levels of norepinephrine in functionally specialized areas of the brain may be responsible for at least some types of elation and depression.

There is growing evidence that the metabolic substance cyclic AMP mediates some of the effects of norepinephrine and dopamine in the brain. For example, in the brain structure called the caudate nucleus dopamine stimulates the activity of adenylate cyclase, the enzyme that catalyzes the synthesis of cyclic AMP from adenosine triphosphate. Cyclic AMP is present in large concentrations in the central nervous system and is known to be important in the induction of enzymes and the synthesis of other proteins. Recent experiments have indicated that protein synthesis may play an important role in long-term memory. When chemicals that inhibit protein synthesis are injected into the brain of an experimental animal, the formation of long-term memories is suppressed. Drugs that release or enhance norepinephrine in the brain counteract this suppression. These findings, made by Seymour S. Kety of the Harvard Medical School, support a tentative model of memory in which protein synthesis stimulated by the release of norepinephrine serves as the biochemical basis for the consolidation of learning. Although there are no completely adequate theories to account for the correlation between biochemical change, nerve-cell activity, and behavior, the exciting pace of interchange between clinical studies and laboratory investigations promises major gains in our understanding of the fundamental mechanisms.

Evidence has been accumulating that points toward a genetic basis for schizophrenia and for manic-depressive disorders. The risk of schizophrenia for a child with one schizophrenic parent is about 10 times greater than it is in the general population, and the risk for a child with two schizophrenic parents is about 40 times greater. This empirical finding can be explained on either a genetic basis or an environmental one. Research on twins, however, has shown that schizophrenia in both twins is much commoner (with a range of 35 to 70 percent) in monozygotic (identical) pairs than it is (with a range of 10 to 26 percent) in dizygotic (fraternal) pairs.

A continuing American-Danish study conducted by Kety, D. Rosenthal, P. H. Wender, and F. Schulsinger has provided independent evidence for a hereditary basis for schizophrenia. They have been studying people who as children were adopted and as adults became schizophrenic. A comparison of the prevalence of schizophrenia among the biological relatives of such people and among the adopting families makes it possible to separate the genetic component from the environmental one. The investigators have found more instances of schizophrenia and related disorders in biological relatives than in the adopting families, which offers strong support for genetic (although not Mendelian) transmission.

The emergence of these mental disorders probably requires a predisposition involving more than one gene (polygenic transmission) and still unspecified environmental precipitants. Studies of manic-depressive disorders in families indicate the need to differentiate bipolar disorders (with both manic and depressive episodes) from unipolar ones (with only depressive episodes). The prevalence of such disorders in the parents and extended family of bipolar patients is significantly greater than it is in the families of unipolar patients. The rates for families in both groups, however, are higher than they are for the general population.

Psychiatry, like the representation of the Roman god Janus, has two faces, one represented by treatment at the psychosocial level and the other by treatment at the pharmacologic level. The two forms of treatment often act to reinforce each other. Yet on the one hand the responsiveness of psychotic states to factors in the social environment has been taken to support contemporary views that mental illness is a "myth" arising from labels applied by psychiatrists in order to rationalize the segregation of people who exhibit disturbing (rather than disturbed) behavior. On the other, the power of psychotherapeutic drugs has been used to support a vulgarized medical model that views psychoses as nothing more than biochemical derangements to be corrected by a realignment of the biochemical machinery. Neither conclusion is warranted. The fundamental task of psychiatry is to understand behavior that results from interaction of stimulus (environmental) conditions and the response (psychobiological) capacities of the organism, capacities that reflect the organism's genetic endowment as well as its history (experience).

The notion that psychiatrists create insanity simply by labeling objectionable people as being insane has a certain charm. It implies that we can legislate mental illnesses out of existence by abolishing the myth that they are real.

However, either this myth in one form or another has mysteriously proved to be necessary in every society or mental illnesses are in fact endemic in every population known to us. Studies of the lifetime expectancy for schizophrenia in Switzerland, Iceland, and Japan reveal rates that vary only slightly: from .73 percent to 1.23 percent. The modal figure for all nations reporting is 1 ± .2 percent.

Some have argued that the frantic pace of industrialization increases psychotic behavior. H. Goldhammer and A. W. Marshall of the Rand Corporation compared the hospital rates for the psychoses of early and middle adult life in Massachusetts from 1840 to 1940. They found no substantial change.

The major psychoses are identifiable in preindustrial societies, although the local terms for them and the theories on which the terms are based differ sharply from our own. Both psychiatrists with Western training and native healers nonetheless identify a person with a psychosis as being abnormal. Mental disorders are identifiable even when the indigenous language has no verbal label for them and even when they are not defined as illness, as in senility. The signs and symptoms of mental deterioration are known to the community, but they are ascribed to the natural course of aging. The absence of the label does not abolish the behavior, but it does reflect a difference in social management; the patient's family, rather than some other community institution, assumes the burden of the affected person's care and protection.

Whatever the cause of a psychotic disorder, be it biological or psychogenic, the mental content of the psychosis must reflect the input to the mind and how that input is refracted by the mind's history and functional state. A patient suffering from a psychosis secondary to drug intoxication can only verbalize his hallucinations in the language he knows; he employs the images of his culture in expressing his terror. A schizophrenic with delusions of grandeur will think himself Jesus in one society and Mohammed in another. This does not say that the delusions were caused by the society but rather that the metaphors for grandeur are social in their origin. Moreover, if the psychotic abandons his insistence on being the deity in response to sympathy and support, that in itself does not tell us anything about the pathogenesis of the autistic thought.

Whether the patient is wildly delusional or rigidly catatonic, he remains a person in need of human satisfactions and he is always responsive to cues in his environment. Physicians quite regularly observe that the perceived pain of cancer, the visible tremor of neurological disease, the symptomatic malfunction of the lungs in emphysema increase or decrease in response to social transactions, which have awesome power to diminish or augment stress. The capacity of sedatives to diminish anxiety by damping physiological oscillations does not deny that the source of the anxiety lies in a psychological threat to personal esteem. The signs and symptoms of disordered function are a result of the state of the organism and the psychobiological forces acting on it.

The interactions among drug effects, psychobiological state, and social setting indicate the complexity of the determinants of human behavior. Amphetamines, which for most of us are stimulants and euphoriants, are almost totally ineffective in the treatment of psychotic depression. Tricyclics, in spite of their potency in relieving depressed patients, produce nothing more than sedation and unpleasant side effects (dry mouth, blurring of vision, and the like) in normal people. Lithium, which limits the swings of mood in manic-depressives, is without detectable effects in normal volunteers. An acutely psychotic patient may require for therapeutic benefit an amount of chlorpromazine capable of producing coma in a normal person. The perception of side effects, such as the dry mouth a patient has been warned to expect, may convince him of the potency and predictability of the drug and enhance favorable outcomes. For others a side effect such as sedation may arouse fears of losing self-control and result in refusal to continue treatment.

Social context affects a person's subjective response to drugs. Hallucinogens evoke different experiences when they are given as a hallowed act in a religious ceremony, when they are part of a carefully monitored laboratory experiment, or when they are taken in the search for "expanded consciousness." The effect depends on what the seeker expects and on whether he is alone or with others (the set and setting). Epinephrine (adrenaline) has been shown to produce irritation and anger when it is injected into a volunteer who does not know what he is receiving and is then exposed to a provocative and unpleasant situation. It generates feelings of well-being when it is injected into a person who is then exposed to a situation that is cheerful and humorous.

What transforms a troubled person into a patient (someone identified by himself, or his community, and a physician as being sick) is a complex social process influenced by cultural attitudes, medical knowledge and the availability of treatment facilities. It has long been observed that proximity to a state mental hospital, belief in the efficacy of psychiatric treatment, and administrative policy facilitating easy access all lead to higher admission rates. Conversely, shame about mental illness, guilt about its cause, and fear of long-term confinement are deterrents to acknowledging the need for help. Neither set of factors alters the actual prevalence of mental disorder, but they have a marked influence on the cases officially tabulated. Community-wide surveys always find disturbed individuals who are maintaining a marginal social existence; they indicate that statistics based only on hospital and clinic records underestimate the actual prevalence of mental disorders in the general population. A person is as likely to seek psychiatric help because of a family crisis, an economic misfortune, or a chance encounter with a public agency as because of the nature of his disorder. Thus single, divorced, widowed, or otherwise isolated persons are more likely to be hospitalized than those with family support. Life stress as a precipitant of perceived illness and prolonged hospitalization is a phenomenon of medical and surgical, as well as mental, illnesses.

With medical progress disorders once crudely lumped together as insanity have been separated into discrete entities. One type of insanity, a dementia accompanied by general paralysis, is now known to be the result of pathological changes in the central nervous system caused by the spirochete of syphilis; it can be prevented by the effective treatment of primary syphilis. The complex of mental disturbances associated with pellagra was shown to be the result of a dietary deficiency of B-complex vitamins (tryptophan, nicotinic acid, and pyridoxine). Similar disturbances in the absence of dietary inadequacy have now been shown to be the result of an inherited metabolic defect (Hartnup's disease) that leads to an inability to absorb tryptophan from the gut.

The fact that the remedies for schizophrenia and manic-depressive disorders are different has given added significance to the diagnostic differentiation of the two. Studies have shown that American and British psychiatrists have tended to make quite different diagnoses when they are given the same set of symptoms for disorders that could be schizophrenic or manic-depressive. A joint American-British diagnostic study has made it evident that the term *schizophrenia* has been used much more broadly by American psychiatrists. With more precise attention to the details of the psychiatric examination, because of the therapeutic importance of the diagnostic differentiation, higher degrees of replicability among psychiatrists can be attained. As long as we must rely on clinical judgments made by psychiatrists, in contrast to more objective laboratory indicators, differences of opinion will be inevitable.

It is these characteristics of a mental disorder—its changing definitions, the variability of its course, its responsiveness to the social environment, and imprecision in its diagnosis—that make psychiatry a center of controversy. If the recognition and appropriate treatment of psychosis can benefit the afflicted, it is also true that error can harm them. Hospitalization, particularly when it is involuntary, deprives the patient of his personal liberties. It can be misused by civil authorities and by disaffected families to remove unwanted persons from the community.

In the past five years commitment laws in many states have been rewritten to protect civil rights, but the laws will be only as effective as the vigilance of the community, the courts, and physicians guarantees. There is, however, a nagging question that should not be forgotten: Are we legislating a justification for indifference to human welfare? If we wish to avoid a paternalistic society, must we settle for an atomistic one? When someone is grossly disturbed but refuses to seek help, what is our responsibility toward him? The law recognizes the legitimacy of intervention when the patient is suicidal or homicidal. Under what other circumstances is intervention warranted? That is a major issue for parents with a disturbed adolescent, for children with a disturbed parent, and for friends helplessly watching a life being wasted. It may be that such casualties are the necessary price for the benefits of individual freedom, but the matter deserves more thought than it is being given.

Current psychiatric knowledge is compatible with a stress-diathesis model for the genesis of psychoses: Psychobiological stress acts on an individual with a genetic predisposition to psychosis and eventually leads to abnormal metabolic processes that cause disorders of mood and thought. Predisposition probably varies on a continuum. Although genetic factors are a prerequisite, it is unlikely that they are sufficient to evoke psychosis in themselves. The behavior of the mentally disturbed person may cause others to reject him, so that he experiences additional severe psychological stress.

At any given time the power of psychiatric remedies and their accessibility to the patient are major determinants of the outcome of an acute mental breakdown. Therapeutic drugs not only presumably restore metabolic equilibrium but also may directly affect the patient's perception of stress. The probability of further personality deterioration and the duration of the disorder is a function of the social environment provided for the patient by the health-care system. The degree of organization or disorganization of the community where a psychotic person finds himself will influence his mood and thought directly.

Our ability to identify the significant psychobiological stressors is limited, and it is still not possible to estimate degrees of genetic predisposition accurately enough to take preventive measures. There have nonetheless been major advances in our capacity for controlling or aborting mental disorders when they do occur. Moreover, we appear to be on the verge of answering some of the basic questions about the functioning of the brain. Psychiatric research is a fragile enterprise, yet it is currently argued in ruling government circles that it should receive not more public support but less. It would be the height of folly to stifle fundamental investigation just at the time when it holds so much promise for revealing the basic mechanisms of psychotic disorders and for developing rational therapies aimed at the causes of psychoses and not merely at their symptoms.

THE MYTH OF MENTAL ILLNESS

THOMAS SZASZ

I

At the core of virtually all contemporary psychiatric theories and practices lies the concept of mental illness. A critical examination of this concept is therefore indispensable for understanding the ideas, institutions, and interventions of psychiatrists.

My aim in this essay is to ask if there is such a thing as mental illness and to argue that there is not. Of course, mental illness is not a thing or physical object; hence it can exist only in the same sort of way as do other theoretical concepts. Yet, to those who believe in them, familiar theories are likely to appear, sooner or later, as "objective truths" or "facts." During certain historical periods, explanatory concepts such as deities, witches, and instincts appeared not only as theories but as *self-evident causes* of a vast number of events. Today mental illness is widely regarded in a similar fashion, that is, as the cause of innumerable diverse happenings.

As an antidote to the complacent use of the notion of mental illness—as a self-evident phenomenon, theory, or cause—let us ask: What is meant when it is asserted that someone is mentally ill? In this essay I shall describe the main uses of the concept of mental illness, and I shall argue that this notion has outlived whatever cognitive usefulness it might have had and that it now functions as a myth.

II

The notion of mental illness derives its main support from such phenomena as syphilis of the brain or delirious conditions—intoxications, for instance—in which persons may manifest certain disorders of thinking and behavior. Correctly speaking, however, these are diseases of the brain, not of the mind.

This paper originally appeared in vol. 15 of *The American Psychologist* and is reprinted here by permission of the author and *The American Psychological Association.*

According to one school of thought, *all* so-called mental illness is of this type. The assumption is made that some neurological defect, perhaps a very subtle one, will ultimately be found to explain all the disorders of thinking and behavior, Many contemporary physicians, psychiatrists, and other scientists hold this view, which implies that people's troubles cannot be caused by conflicting personal needs, opinions, social aspirations, values, and so forth. These difficulties—which I think we may simply call *problems in living*—are thus attributed to physicochemical processes that in due time will be discovered (and no doubt corrected) by medical research.

Mental illnesses are thus regarded as basically similar to other diseases. The only difference, in this view, between mental and bodily disease is that the former, affecting the brain, manifests itself by means of mental symptoms; whereas the latter, affecting other organ systems—for example, the skin, liver, and so on—manifests itself by means of symptoms referable to those parts of the body.

In my opinion, this view is based on two fundamental errors. In the first place, a disease of the brain, analogous to a disease of the skin or bone, is a neurological defect, not a problem in living. For example, a *defect* in a person's visual field may be explained by correlating it with certain lesions in the nervous system. On the other hand, a person's *belief*—whether it be in Christianity, in Communism, or in the idea that his internal organs are rotting and that his body is already dead—cannot be explained by a defect or disease of the nervous system. Explanations of this sort of occurrence—assuming that one is interested in the belief itself and does not regard it simply as a symptom or expression of something else that is more interesting—must be sought along different lines.

The second error is epistemological. It consists of interpreting communications about ourselves and the world around us as symptoms of neurological functioning. This is an error not in observation or reasoning, but rather in the organization and expression of knowledge. In the present case, the error lies in making a dualism between mental and physical symptoms, a dualism that is a habit of speech and not the result of known observations. Let us see if this is so.

In medical practice, when we speak of physical disturbances we mean either signs (for example, fever) or symptoms (for example, pain). We speak of mental symptoms, on the other hand, when we refer to a patient's communications about himself, others, and the world about him. The patient might assert that he is Napoleon or that he is being persecuted by the Communists. These would be considered mental symptoms only if the observer believed that the patient was *not* Napoleon or that he was *not* being persecuted by the Communists. This makes it apparent that the statement "*X* is a mental symptom" involves rendering a judgment that entails a covert comparison between the patient's ideas, concepts, or beliefs and those of the observer and the society in which they live. The notion of mental symptom is therefore inextricably tied to the social, and particularly the ethical, context in which it is made, just as the notion of bodily symptom is tied to an anatomical and genetic context.[1]

To sum up: For those who regard mental symptoms as signs of brain disease, the concept of mental illness is unnecessary and misleading. If they mean that people so labeled suffer from diseases of the brain, it would seem better, for the sake of clarity, to say that and not something else.

III

The term "mental illness" is also widely used to describe something quite different from a disease of the brain. Many people today take it for granted that living is an arduous affair. Its hardship for modern man derives, moreover, not so much from a struggle for biological survival as from the stresses and strains inherent in the social intercourse of complex human personalities. In this context, the notion of mental illness is used to identify or describe some feature of an invididual's so-called personality. Mental illness—as a deformity of the personality, so to speak—is then regarded as the cause of human disharmony. It is implicit in this view that social intercourse between people is regarded as something inherently harmonious, its disturbance being due solely to the presence of "mental illness" in many people. Clearly, this is faulty reasoning, for it makes the abstraction "mental illness" into a cause of, even though this abstraction was originally created to serve only as a shorthand expression for, certain types of human behavior. It now becomes necessary to ask: What kinds of behavior are regarded as indicative of mental illness, and by whom?

The concept of illness, whether bodily or mental, implies deviation from some clearly defined norm. In the case of physical illness, the norm is the structural and functional integrity of the human body. Thus, although the desirability of physical health, as such, is an ethical value, what health is can be stated in anatomical and physiological terms. What is the norm, deviation from which is regarded as mental illness? This question cannot be easily answered. But whatever this norm may be, we can be certain of only one thing: namely, that it must be stated in terms of psychosocial, ethical, and legal concepts. For example, notions such as "excessive repression" and "acting out an unconscious impulse" illustrate the use of psychological concepts for judging so-called mental health and illness. The idea that chronic hostility, vengefulness, or divorce are indicative of mental illness is an illustration of the use of ethical norms (that is, the desirability of love, kindness, and a stable marriage relationship). Finally, the

[1] See Szasz, T. S.: *Pain and Pleasure: A Study of Bodily Feelings* (New York: Basic Books, 1957), especially pp. 70–81; "The problem of psychiatric nosology." *Amer. J. Psychiatry, 114*:405–13 (Nov.), 1957.

widespread psychiatric opinion that only a mentally ill person would commit homicide illustrates the use of a legal concept as a norm of mental health. In short, when one speaks of mental illness, the norm from which deviation is measured is a *psychosocial and ethical* standard. Yet, the remedy is sought in terms of *medical* measures that–it is hoped and assumed–are free from wide differences of ethical value. The definition of the disorder and the terms in which its remedy are sought are therefore at serious odds with one another. The practical significance of this covert conflict between the alleged nature of the defect and the actual remedy can hardly be exaggerated.

Having identified the norms used for measuring deviations in cases of mental illness, we shall now turn to the question, Who defines the norms and hence the deviation? Two basic answers may be offered: First, it may be the person himself–that is, the patient–who decides that he deviates from a norm; for example, an artist may believe that he suffers from a work inhibition; and he may implement this conclusion by seeking help *for himself* from a psychotherapist. Second, it may be someone other than the "patient" who decides that the latter is deviant–for example, relatives, physicians, legal authorities, society generally; a psychiatrist may then be hired by persons other than the "patient" to do something *to him* in order to correct the deviation.

These considerations underscore the importance of asking the question, Whose agent is the psychiatrist? and of giving a candid answer to it. The psychiatrist (or non-medical mental health worker) may be the agent of the patient, the relatives, the school, the military services, a business organization, a court of law, and so forth. In speaking of the psychiatrist as the agent of these persons or organizations, it is not implied that his moral values, or his ideas and aims concerning the proper nature of remedial action, must coincide exactly with those of his employer. For example, a patient in individual psychotherapy may believe that his salvation lies in a new marriage; his psychotherapist need not share this hypothesis. As the patient's agent, however, he must not resort to social or legal force to prevent the patient from putting his beliefs into action. If his *contract* is with the patient, the psychiatrist (psychotherapist) may disagree with him or stop his treatment, but he cannot engage others to obstruct the patient's aspirations.[2] Similarly, if a psychiatrist is retained by a court to determine the sanity of an offender, he need not fully share the legal authorities' values and intentions in regard to the criminal, nor the means deemed appropriate for dealing with him; such a psychiatrist cannot testify, however, that the accused is not insane, but that the legislators are–for passing the law that decrees the offender's actions illegal.[3] This sort of opinion could be voiced, of

[2] See Szasz, T. S.: *The Ethics of Psychoanalysis: The Theory and Method of Autonomous Psychotherapy* (New York: Basic Books, 1965).

[3] See Szasz, T. S.: *Law, Liberty, and Psychiatry: An Inquiry into the Social Uses of Mental Health Practices* (New York: Macmillan, 1963).

course—but not in a courtroom, and not by a psychiatrist who is there to assist the court in performing its daily work.

To recapitulate: In contemporary social usage, the finding of mental illness is made by establishing a deviance in behavior from certain psychosocial, ethical, or legal norms. The judgment may be made, as in medicine, by the patient, the physician (psychiatrist), or others. Remedial action, finally, tends to be sought in a therapeutic—or covertly medical—framework. This creates a situation in which it is claimed that psychosocial, ethical, and legal deviations can be corrected by medical action. Since medical interventions are designed to remedy only medical problems, it is logically absurd to expect that they will help solve problems whose very existence have been defined and established on non-medical grounds.

IV

Anything that people *do*—in contrast to things that *happen* to them[4]—takes place in a context of value. Hence, no human activity is devoid of moral implications. When the values underlying certain activities are widely shared, those who participate in their pursuit often lose sight of them altogether. The discipline of medicine—both as a pure science (for example, research) and as an applied science or technology (for example, therapy)—contains many ethical considerations and judgments. Unfortunately, these are often denied, minimized, or obscured, for the ideal of the medical profession as well as of the people whom it serves is to have an ostensibly value-free system of medical care. This sentimental notion is expressed by such things as the doctor's willingness to treat patients regardless of their religious or political beliefs. But such claims only serve to obscure the fact that ethical considerations encompass a vast range of human affairs. Making medical practice neutral with respect to some specific issues of moral value (such as race or sex) need not mean, and indeed does not mean, that it can be kept free from others (such as control over pregnancy or regulation of sex relations). Thus, birth control, abortion, homosexuality, suicide, and euthanasia continue to pose major problems in medical ethics.

Psychiatry is much more intimately related to problems of ethics than is medicine in general. I use the word "psychiatry" here to refer to the contemporary discipline concerned with problems in living, and not with diseases of the brain, which belong to neurology. Difficulties in human relations can be analyzed, interpreted, and given meaning only within specific social and ethical contexts. Accordingly, the psychiatrist's socioethical orientations will influence his ideas on what is wrong with the patient, on what deserves comment or interpretation, in what directions change might be desirable, and so forth. Even

[4] Peters, R.S.: *The Concept of Motivation* (London: Routledge & Kegan Paul, 1958), especially pp. 12–15.

in medicine proper, these factors play a role, as illustrated by the divergent orientations that physicians, depending on their religious affiliations, have toward such things as birth control and therapeutic abortion. Can anyone really believe that a psychotherapist's ideas on religion, politics, and related issues play no role in his practical work? If, on the other hand, they do matter, what are we to infer from it? Does it not seem reasonable that perhaps we ought to have different psychiatric therapies—each recognized for the ethical positions that it embodies—for, say, Catholics and Jews, religious persons and atheists, democrats and Communists, white supremacists and Negroes, and so on? Indeed, if we look at the way psychiatry is actually practiced today, especially in the United States, we find that the psychiatric interventions people seek and receive depend more on their socioeconomic status and moral beliefs than on the "mental illnesses" from which they ostensibly suffer.[5] This fact should occasion no greater surprise than that practicing Catholics rarely frequent birth-control clinics, or that Christian Scientists rarely consult psychoanalysts.

V

The position outlined above, according to which contemporary psychotherapists deal with problems in living, not with mental illnesses and their cures, stands in sharp opposition to the currently prevalent position, according to which psychiatrists treat mental diseases, which are just as "real" and "objective" as bodily diseases. I submit that the holders of the latter view have no evidence whatever to justify their claim, which is actually a kind of psychiatric propaganda: their aim is to create in the popular mind a confident belief that mental illness is some sort of disease entity, like an infection or a malignancy. If this were true, one could *catch* or *get* a mental illness, one might *have* or *harbor* it, one might *transmit* it to others, and finally one could *get rid* of it. Not only is there not a shred of evidence to support this idea, but, on the contrary, all the evidence is the other way and supports the view that what people now call mental illnesses are, for the most part, *communications* expressing unacceptable ideas, often framed in an unusual idiom.

This is not the place to consider in detail the similarities and differences between bodily and mental illnesses. It should suffice to emphasize that whereas the term "bodily illness" refers to psysicochemical occurrences that are not affected by being made public, the term "mental illness" refers to sociopsychological events that are crucially affected by being made public. The psychiatrist thus cannot, and does not, stand apart from the person he observes, as the pathologist can and often does. The psychiatrist is committed to some picture of

[5] Hollingshead, A. B. and Redlich, F. C.: *Social Class and Mental Illness* (New York: Wiley, 1958).

what he considers reality, and to what he thinks society considers reality, and he observes and judges the patient's behavior in the light of these beliefs. The very notion of "mental symptom" or "mental illness" thus implies a covert comparison, and often conflict, between observer and observed, psychiatrist and patient. Though obvious, this fact needs to be re-emphasized, if one wishes, as I do here, to counter the prevailing tendency to deny the moral aspects of psychiatry and to substitute for them allegedly value-free medical concepts and interventions.

Psychotherapy is thus widely practiced as though it entailed nothing other than restoring the patient from a state of mental sickness to one of mental health. While it is generally accepted that mental illness has something to do with man's social or interpersonal relations, it is paradoxically maintained that problems of values–that is, of ethics–do not arise in this process. Freud himself went so far as to assert: "I consider ethics to be taken for granted. Actually I have never done a mean thing."[6] This is an astounding thing to say, especially for someone who had studied man as a social being as deeply as Freud had. I mention it here to show how the notion of "illness"–in the case of psychoanalysis, "psychopathology," or "mental illness"–was used by Freud, and by most of his followers, as a means of classifying certain types of human behavior as falling within the scope of medicine, and hence, by fiat, outside that of ethics. Nevertheless, the stubborn fact remains that, in a sense, much of psychotherapy revolves around nothing other than the elucidation and weighing of goals and values–many of which may be mutually contradictory–and the means whereby they might best be harmonized, realized, or relinquished.

Because the range of human values and of the methods by which they may be attained is so vast, and because many such ends and means are persistently unacknowledged, conflicts among values are the main source of conflicts in human relations. Indeed, to say that human relations at all levels–from mother to child, through husband and wife, to nation and nation–are fraught with stress, strain, and disharmony is, once again, to make the obvious explicit. Yet, what may be obvious may be also poorly understood. This, I think, is the case here. For it seems to me that in our scientific theories of behavior we have failed to accept the simple fact that human relations are inherently fraught with difficulties, and to make them even relatively harmonious requires much patience and hard work. I submit that the idea of mental illness is now being put to work to obscure certain difficulties that at present may be inherent–not that they need to be unmodifiable–in the social intercourse of persons. If this is true, the concept functions as a disguise: Instead of calling attention to conflicting human needs, aspirations, and values, the concept of mental illness provides an amoral and impersonal "thing"–an "illness"–as an explanation for problems in living. We may recall in this connection that not so long ago it was devils and

[6] Quoted in Jones, E.: *The Life and Work of Sigmund Freud* (New York: Basic Books, 1957), Vol. III, p. 247.

witches that were held responsible for man's problems in living. The belief in mental illness, as something other than man's trouble in getting along with his fellow man, is the proper heir to the belief in demonology and witchcraft. Mental illness thus exists or is "real" in exactly the same sense in which witches existed or were "real."

VI

While I maintain that mental illnesses do not exist, I obviously do not imply or mean that the social and psychological occurrences to which this label is attached also do not exist. Like the personal and social troubles that people had in the Middle Ages, contemporary human problems are real enough. It is the labels we give them that concern me, and, having labeled them, what we do about them. The demonologic concept of problems in living gave rise to therapy along theological lines. Today, a belief in mental illness implies—nay, requires—therapy along medical or psychotherapeutic lines.

I do not here propose to offer a new conception of "psychiatric illness" or a new form of "therapy." My aim is more modest and yet also more ambitious. It is to suggest that the phenomena now called mental illnesses be looked at afresh and more simply, that they be removed from the category of illnesses, and that they be regarded as the expressions of man's struggle with *the problem of how he should live.* This problem is obviously a vast one, its enormity reflecting not only man's inability to cope with his environment, but even more his increasing self-reflectiveness.

By problems in living, then, I refer to that explosive chain reaction that began with man's fall from divine grace by partaking of the fruit of the tree of knowledge. Man's awareness of himself and of the world about him seems to be a steadily expanding one, bringing in its wake an even larger *burden of understanding.*[7] This burden is to be expected and must not be misinterpreted. Our only rational means for easing it is more understanding, and appropriate action based on such understanding. The main alternative lies in acting as though the burden were not what in fact we perceive it to be, and taking refuge in an outmoded theological view of man. In such a view, man does not fashion his life and much of his world about him, but merely lives out his fate in a world created by superior beings. This may logically lead to pleading non-responsibility in the face of seemingly unfathomable problems and insurmountable difficulties. Yet, if man fails to take increasing responsibility for his actions, individually as well as collectively, it seems unlikely that some higher power or being would

[7] In this connection, see Langer, S. K.: *Philosophy in a New Key* (1942) (New York: Mentor Books, 1953), especially Chaps. 5 and 10.

assume this task and carry this burden for him. Moreover, this seems hardly a propitious time in human history for obscuring the issue of man's responsibility for his actions by hiding it behind the skirt of an all-explaining conception of mental illness.

VII

I have tried to show that the notion of mental illness has outlived whatever usefulness it may have had and that it now functions as a myth. As such, it is a true heir to religious myths in general, and to the belief in witchcraft in particular. It was the function of these belief-systems to act as social tranquilizers, fostering hope that mastery of certain problems may be achieved by means of substitutive, symbolic-magical, operations. The concept of mental illness thus serves mainly to obscure the everyday fact that life for most people is a continuous struggle, not for biological survival, but for a "place in the sun," "peace of mind," or some other meaning or value. Once the needs of preserving the body, and perhaps of the race, are satisfied, man faces the problem of personal significance: What should he do with himself? For what should he live? Sustained adherence to the myth of mental illness allows people to avoid facing this problem, believing that mental health, conceived as the absence of mental illness, automatically insures the making of right and safe choices in the conduct of life. But the facts are all the other way. It is the making of wise choices in life that people regard, retrospectively, as evidence of good mental health!

When I assert that mental illness is a myth, I am not saying that personal unhappiness and socially deviant behavior do not exist; what I am saying is that we categorize them as diseases at our own peril.

The expression "mental illness" is a metaphor that we have come to mistake for a fact. We call people physically ill when their body-functioning violates certain anatomical and physiological norms; similarly, we call people mentally ill when their personal conduct violates certain ethical, political, and social norms. This explains why many historical figures, from Jesus to Castro, and from Job to Hitler, have been diagnosed as suffering from this or that psychiatric malady.

Finally, the myth of mental illness encourages us to believe in its logical corollary: that social intercourse would be harmonious, satisfying, and the secure basis of a good life were it not for the disrupting influences of mental illness, or psychopathology. However, universal human happiness, in this form at least, is but another example of a wishful fantasy. I believe that human happiness, or well-being, is possible—not just for a select few, but on a scale hitherto unimaginable. But this can be achieved only if many men, not just a few, are willing and able to confront frankly, and tackle courageously, their ethical, personal, and social conflicts. This means having the courage and integrity to forego waging battles on false fronts, finding solutions for substitute problems—for instance, fighting the battle of stomach acid and chronic fatigue instead of facing up to a marital conflict.

Our adversaries are not demons, witches, fate, or mental illness. We have no enemy that we can fight, exorcise, or dispel by "cure." What we do have are problems in living—whether these be biologic, economic, political, or socio-psychological. In this essay I was concerned only with problems belonging in the last-mentioned category, and within this group mainly with those pertaining to moral values. The field to which modern psychiatry addresses itself is vast, and I made no effort to encompass it all. My argument was limited to the proposition that mental illness is a myth, whose function it is to disguise and thus render more palatable the bitter pill of moral conflicts in human relations.

MENTAL HEALTH AND MENTAL ILLNESS

SOME PROBLEMS OF DEFINITION AND CONCEPT FORMATION

RUTH MACKLIN

1. Introduction

In recent years there has been considerable discussion and controversy concerning the concepts of mental health and mental illness. The controversy has centered around the problem of providing criteria for an adequate conception of mental health and illness, as well as difficulties in specifying a clear and workable system for the classification, understanding, and treatment of psychological and emotional disorders. In this paper I shall examine a cluster of these complex and important issues, focusing on attempts to define "mental health" and "mental illness"; diverse factors influencing the ascription of the predicates "is mentally ill" and "is mentally healthy"; and some specific problems concerning these concepts as they appear in various theories of psychopathology. The approach here will be in the nature of a survey, directed at the specification of a number of problems—conceptual, methodological, and pragmatic—as they arise in various attempts to define and to provide criteria for applying the concepts of mental health and mental illness.

Closely related to the above-noted issues—indeed, an integral part of them—is a concern which has engendered much discussion and controversy among professionals in the past decade: the appropriateness of the "medical model," or health–disease conception of psychological disorders and emotional problems. The question arises as to whether or not the medical model of physical health and disease is the appropriate, correct, or most useful model for classifying, understanding, and treating the various emotional difficulties and behavior disorders which persons manifest. Space does not permit a detailed examination

This essay originally appeared in *Philosophy of Science*, vol. 39 (September 1972), and is reprinted with the kind permission of the author and the editor of *Philosophy of Science*.

here of this controversial and interesting issue, but some remarks will be addressed to the problem in the final section of the paper.

In the words of one recent writer, "There is hardly a term in current psychological thought as vague, elusive, and ambiguous as the term "mental health" ([5], p. 3).[1] There appears to be increasing recognition—even on the part of those who are not unhappy with the health-disease model—that some conceptual clarification is needed. Nevertheless, one psychologist claims that "The definition of mental illness is not especially difficult, though a number of problems need to be considered in certain peripheral areas of breakdown in normal human behavior. Ordinarily, we think of mental illness as an unusually persistent pattern of behavior over which the individual has little or no voluntary control; it differentiates him from his fellows; it incapacitates him; it interferes with his normal participation in life" ([1], p. 37).

It is immediately obvious that this conception of mental illness is too broad, since it fails to distinguish mental illness from physical illness—a task which most writers in this area consider crucial. It should be noted that attempts to provide an acceptable definition of these concepts and a systematic set of criteria for their application have implications beyond the concerns of psychological or psychiatric theory. There are consequences, as well, for the direction of social and institutional practices such as therapy, civil commitment proceedings, legal and judicial concerns with the "criminally insane," and other social issues concerned with antisocial and deviant behavior. In addition, given a situation where need exceeds availability of psychotherapeutic time and personnel (both inside and outside hospital and institutional settings), it seems important to have a precise conception according to which comparative judgments ("is sicker than," "is more healthy than") can be made. Such judgments—if made according to some systematic and uncontroversial professional conception of health and illness—would facilitate decision-making problems in the area of assigning priorities for private, outpatient, and inpatient treatment of persons afflicted with psychological disorders of various sorts. So there seem to be crucial practical concerns, as well as more narrowly theoretical issues, which could benefit from greater clarification and systematization in this domain.

A word should be said at this point about the enterprise of defining. The difficulties of attempting to provide a definition of a term in natural language or a set of necessary and sufficient conditions for its application are well known, and there is much philosophical literature on the subject. The generally acknowledged "open texture" of terms in natural languages (and even in scientific theories), in addition to multiple and possibly conflicting criteria of application, have led some philosophers to give up all attempts to provide such definitions.

[1] Numbers in brackets refer to references listed in the section at the end of this paper, p. 149.

But psychologists and psychiatrists do not seem to have given up, even if their efforts are bent towards abandoning old conceptions or models in favor of new ones ([11], [16]). In addition, the pragmatic necessity for the enterprise in the area of mental illness and psychotherapy is dictated by the need to identify persons for treatment and to decide when they are "well," and to have some more precise criteria for commitment to and release from mental hospitals. The existing state of affairs is that in general, "mental illness is regarded usually as a residual category for deviant behavior having no clearly specified label" ([10], p. 26). So a partial justification for the inquiry conducted here—the objections of some philosophers and psychiatrists notwithstanding—lies in the stated needs and attempts on the part of many theorists and clinicians at further clarification and provision of a set of criteria for an adequate conception of mental health and mental illness.

One chief difficulty lies in the vagueness and ambiguity of the concept of mental illness as construed even by professional groups presumably committed to similar theoretical orientations and diagnostic procedures. In one recent study, the authors write:

> One need only glance at the diagnostic manual of the American Psychiatric Association to learn what an elastic concept mental illness is. It ranges from the massive functional inhibition characteristic of one form of catatonic schizophrenia to those seemingly slight aberrancies associated with an unstable personality, but which are so close to conduct in which we all engage as to define the entire continuum involved.... And, because of the unavoidably ambiguous generalities in which the American Psychiatric Association describes its diagnostic categories, the diagnostician has the ability to shoehorn into the mentally diseased class almost any person he wishes, for whatever reason, to put there ([6], p. 80).

The authors point out that different norms of adjustment are employed by different users of the term "mental illness," and that "usually the use of the phrase 'mental illness' effectively masks the actual norms being applied" ([6], p. 80). The acuteness of the situation is evidenced by the recognition that the group in question here is the American Psychiatric Association, whose members we might expect, presystematically, to share a body of norms and a roughly similar theoretical background. That this is not the case is readily apparent, and the above-noted authors claim that "the usual reason for variance in diagnosis is a variance in the theoretical orientation of the diagnosticians" ([6], p. 80, n. 18). So if such variation exists even within a circumscribed group of professionals, it is not surprising that a great deal of confusion and inconsistency obtains in the entire field of mental health workers, including clinical psychologists, social psychologists, social workers, and psychiatric nurses, as well as medically trained psychiatrists and psychoanalysts. When we consider the further fact that at least some professionals in all these groups (psychiatrists included) explicitly reject the medical model and the attendant notion of "disease entities," the hope for unanimity or even some widespread agreement on the definition of "mental illness" becomes slim indeed.

In the light of these and other difficulties (a number of which will be discussed in the subsequent sections), Karl Menninger favors a "nonspecific, essentially unitary concept of mental illness" ([11], p. 87). In a brief historical review, Menninger notes that two systems of classification developed—the specific entity concept of mental illness and the unitary concept, the former of which prevailed in American Psychiatry. He writes:

> In the minds of many young doctors and in the minds of vast numbers of laymen, mental illness and particularly schizophrenia is a definite, specific, evil thing which invades the unsuspecting like a fungus or a tapeworm. The word schizophrenia becomes a damning designation. . . . A name implies a concept; and if this concept is unsound, the diagnosis can ruin the treatment; the very naming it can damage the patient whom we essay to help ([11], p. 88).

In formulating his unitary concept of mental illness, Menninger emphasizes the degree of disorganization of the ego and its course or trend of development; his system is based largely on the theoretical concept of the ego—its failures and attempts at survival and optimal adaptation under stress. He holds that there are no *natural* mental disease entities, but that "an ordering of clinical phenomena on the basis of the economics of adaptation does justice to the essential unity of sickness and health; at the same time it leaves room for recognizing the latent potentials of every individual" (p. 89).[7]

While Menninger departs from the official psychiatric nosology, still he retains the conception of psychological disorders as instances of illness—a fact which marks him as an adherent of the medical model. Other psychiatrists and many psychologists argue for a more radical departure from the traditional schema, holding that we need to drop the notions of health and illness altogether in forming a conception of emotional problems and behavior disorders. Limitations of space in this paper preclude a detailed study of these increasing efforts to abandon the medical model, but a thorough account must include an examination of the arguments for and against the model, as well as a study of alternative approaches currently in practice (e.g., behavior modification therapy, existential psychiatry, and others).

In the next four sections I shall examine some problems related to definition and concept formation, as falling under the following categories: attempts to define "mental health" in terms of the notion of mental illness; attempts to define "mental health" and "mental illness" in terms of normality and abnormality; obstacles to a clear conception of health or illness arising within or between specific theories of psychopathology; and, finally, problems with the conception of psychological disorders and malfunctioning as "disease" or "illness." There is an additional set of problems of a pragmatic sort: the identification of specific cases as instances of mental illness; conflicts between lay and professional definitions; the fact that different groups within society operate on the basis of different conceptions of mental health and illness; and the surprising fact that "the basic decision about illness is usually made by community members and *not professional personnel*. . . . Community persons are brought to the hospital on the basis of lay definitions, and once they arrive, their appearance alone is usually regarded as sufficient evidence of 'illness'" ([10], p. 27). Although these practical issues are interesting and relevant to the conceptual and theoretical concerns of this paper, unfortunately space does not permit an examination of them here.

Before proceeding to examine the problems falling under the categories cited above, one general point needs to be noted. There is an ambiguity in the terms "sick" and "healthy" as they are employed in contexts of physical as well as mental health and illness. Marie Jahoda characterizes this ambiguity as follows. The concept of mental health can be defined "as a *relatively constant and enduring function of personality,* leading to predictable differences in behavior and feelings depending on the stresses and strains of the situation in which a person finds himself; or as a *momentary function of personality and situation*" ([5], p. 8). According to the first conception, individuals will be classified as more or less healthy. On the second conception, actions will be classified as more or less healthy. There is an analogue in the context of physical health, as Jahoda notes: according to the first conception, a strong man with a bad cold is healthy; according to the second conception, he is sick. Other examples can be given to illustrate the distinction, and the point is an important one in avoiding sources of error that are often overlooked. Jahoda claims that "much of the confusion in the area of mental health stems from the failure to establish whether one is talking about mental health as an enduring attribute of a person or as a momentary attribute of functioning" ([5], p. 8). One area in which the relevance and importance of this distinction is crucial is that of the legal defense of persons charged with a criminal offense on grounds of insanity, temporary insanity, and the like. Still further implications exist for the selection of criteria according to which persons are committed to and released from mental hospitals. We now turn to an examination of some specific conceptual and theoretical problems in the area of mental health and illness.

2. *Attempts to Define "Health" in Terms of "Disease"*

The concept of disease is itself problematic, in somatic medicine as well as in the realm of psychological disorders. Two preliminary considerations will be noted briefly before examining specific problems in attempts to define the notion of health in terms of disease. Firstly, within the accepted framework of the medical model itself, there may be some difficulty in specifying relevant similarities in degree and kind between acknowledged cases of physical disease and putative instances of mental illness. This is the familiar problem of extending a concept from its standard or accepted usage to cover a new range of cases—a task first faced by Freud and others who were engaged in the process of noting affinities between the traditional cases of physical disease and the new cases (e.g., hysteria) which were being subsumed under a new, broadened concept of illness. But the familiarity of the enterprise of extending a concept does not render the issue any the less problematic (see the discussion in section 5 below).

Secondly, if one chooses to abandon or to circumvent the medical model, there is still the problem of specifying a different set of criteria from those which proceed by analogy with physical disease. While there are some guidelines according to which we can classify mental illness or psychological disorders—antisocial or socially deviant behavior, malfunctioning behavior, self-destructive behavior, and the like—these guidelines are extremely rough and may result in a category that is too *broad* to specify what we want presystematically to count as mental illness or personality disorder (cf. Albee's definition cited above). For example, an attempt along these lines may fail to distinguish between the mentally ill and the criminal. Or, it may result in classes of individuals—who by other psychological criteria would be considered "healthy," or "normal"—being classed as psychologically aberrant or emotionally disturbed, e.g., recluses, civil disobedients, radical revolutionaries, etc. On the other side of the coin, appeal to a specific theory of psychopathology might result in a definition of "mental illness" that is too *narrow* to cover the range of cases which ought to be included under the category.

It should be noted, in this connection, that even among theorists and practitioners operating within the medical model, the current emphasis is on behavior disorders rather than on internal states of illness or on mental disease entities. In an influential and widely used textbook for students and practitioners of psychiatry, the authors write:

> In older texts and in current lay parlance, psychiatry is often defined as the science dealing with mental diseases and illnesses of the mind or psyche. Since these are terms reminiscent of the metaphysical concepts of soul and spirit, we prefer to speak of behavior disorder. Behavior refers to objective data that are accessible to observation, plausible inference, hypothesis-making, and experimentation. The term disorder, although vague, is descriptive of malfunctioning of behavior without specifying etiology or underlying mechanisms. Only some of the behavior disorders are caused by diseases of the brain or are accompanied by somatic reactions. Whereas many cerebral diseases produce a behavior disorder, and while we believe that cerebral processes must be related fundamentally to behavior, *medically* recognizable diseases of the brain cannot, for the most part, be demonstrated in behavior disorders ([13], p. 2).

This position appears eminently sound, both in its emphasis on behavior and also in its presupposition that brain processes and other neurophysiological events underlie molar behavior. Behaviorists and other opponents of the medical model would do well to note the former point, while tough-minded philosophers who reject Cartesian dualism or other forms of "mentalism" (especially proponents of the neural identity thesis or modern materialism) should recognize the physicalist assumptions of at least some contemporary scientifically oriented psychiatrists. The fact that *medically* recognizable diseases of the brain cannot be demonstrated in most behavior disorders at the present time is no barrier to future progress in discovering such correlations and developing systematic psychophysical laws.

Turning now to the relationship between mental health and mental illness, we find that there is some disagreement on the matter. One writer points out the obvious fact that "consensus regarding positive mental health (or even mental abnormality) is far from unanimous" ([14], p. 3). He concludes, from an examination of several different conceptions, that the "criterion for mental health thus is simply the absence of mental illness" ([14], pp. 3–4). This view might be considered the standard conception, even among professionals in the field, as reflected in the words of Kenneth E. Appel, the 1954 president of the American Psychiatric Association: "mental illness is the opposite side of the coin of mental health" ([1], p. 38).

On the other hand, theorists such as Marie Jahoda, who are concerned to provide a workable *positive* conception of mental health, explicitly reject any attempt to define "mental health" as "the absence of mental illness or disease." She notes that at present "knowledge about deviations, illness, and malfunctioning far exceeds knowledge of healthy functioning" ([5], p. 6), and it is apparent that the emphasis in this area has been on the study of disease and malfunctioning with the result that the health–disease model has prevailed, influencing theoretical developments and providing a framework according to which treatment and therapy have proceeded. In Jahoda's view, the assumption that health and illness are different only in degree needs to be tested. She is, herself, a proponent of the view that "mental health" and "mental illness" are *not* correlative terms, each denoting a state of the organism to be understood in terms of the absence of the other. She claims that a definition of psychological health as the absence of mental disease "is based on the assumption that health is the opposite of disease, or that health and disease form the extreme poles of a continuum. What if this assumption should turn out to be unjustified and misleading? Some psychiatrists now speak of different health potentials in seemingly equally sick patients, as if they were dealing with two qualitatively different continua. . ."([5], p. 14).

Jahoda believes that this issue requires a good deal of further research, especially since there is difficulty in clearly circumscribing the notion of mental disease itself. So it would seem to be more fruitful to concentrate on "the concept of mental health in its more positive connotation, noting, however, that the absence of disease may constitute a necessary, but not a sufficient criterion for mental health" (p. 15).[15]

Among the theorists whose conceptions of mental health are not formulated in terms of the absence of disease are a number of self-realization or self-actualization theorists, as well as proponents of existential psychiatry. Rollo May, for one, claims that the carryover of concepts from physical to psychological science is often unsuitable, as in the term "health." "In the common popular meaning, physical health means the absence of infection and organic damage. In psychological therapy the term should mean something very different: "health" refers to dynamic processes; a person is healthy psychologically and emotionally to the extent that he can use all his capacities in day-to-day living" ([9], p. 167). May stresses the *active* role of the individual in relating to himself and to others in his environment–a view which supports a conception of mental health as being something different from or over and above the mere absence of illness or disease.

Jahoda cites a number of theorists who oppose the traditional view ([5], pp. 73–75), one of whom holds that the category of positive health "applies when there is evidence that the individual fully utilizes a capacity or is working in that direction" (p. 74). We shall return later in another connection to a brief examination of self-realization and self-actualization theories of psychopathology. The point to note in the present context is that the notion of *positive* mental health appears to be embedded in some theoretical conceptions while it is absent from others. To this extent, we might expect divergent views to be held on the relationship between mental health and illness by adherents of different theoretical systems in psychology and psychiatry. Jahoda points out that the relation between mental health and mental disease is still exceedingly complex, despite recent efforts at clarification, and that this relation remains one of the most urgent areas for future research.

3. Problems with Attempts to Define "Mental Health" and "Mental Illness" in Terms of Normality and Abnormality

In this category, a number of problematic issues can be delineated. As Redlich and Freedman point out, "The concepts of normality and abnormality are more complex in psychiatry than in general medicine, and some people have suggested abandoning the concepts of normal and abnormal behavior entirely because simple concepts of health and disease do not apply" ([13], p. 112). One of the sources of difficulty with these concepts is the familiar issue of cultural relativism. We turn now to an examination of that issue.

A. *Cultural Relativism*

Jahoda notes that the evidence presented by cultural anthropologists is sufficient to demonstrate the vast range of what can be regarded as normal. Cultural anthropologists "have convincingly demonstrated a great variety of social norms and institutions in different cultures in different parts of the world; and that in different cultures different forms of behavior are regarded as normal" ([5], p. 15). It should be added that even within our own culture, the different norms and standards of behavior vary greatly among different age groups, socio-economic classes, and sub-groups of the population including religious, racial, and ethnic minorities. Thus, what may be considered "normal" sexual behavior for a twenty-eight-year-old divorcée who is a professional woman living in New York City will be considered "abnormal" for her fourteen-year-old sister living in Hudson, Ohio. Similarly, what is "normal" behavior for a black youth living in a crime-infested ghetto will be considered abnormal for a white, upper-middle-class boy from Scarsdale.

Similar problems arise if we try to focus on the concept of abnormality as a criterion of mental *disease.* Again, Jahoda claims that anthropological studies throw doubt on the use of some symptoms for the diagnosis of mental disease: "According to Ruth Benedict (1934), the Kwakiutl Indians of British Columbia engage in behavior that is, by our standards, paranoid and megalomaniacal. Their view of the world is similar to a delusion of grandeur in our culture" ([5], p. 12). Other examples are given to support the view that "whereas identical observable symptoms are regarded in one culture as achievement, in another they are regarded as severe debility" ([5], p. 12). It is not clear, however, that the only conclusion to be drawn is that varying customs and accepted behavior in different cultures necessarily preclude a universally applicable conception of mental illness. Rather, it seems that in the absence of an overall psychological theory, or a well-developed personality theory that is acceptable to most, if not all, professionals and scientists in the field, no set of criteria can be agreed upon for making cross-cultural judgments of mental illness. The belief that a comprehensive psychological theory will be forthcoming someday may reflect an optimistic expression of faith in the progress and development of the science of psychology and related fields, but abundant evidence from the historical development of other sciences shows that such faith may not be wholly unwarranted. It is not unreasonable to hope that a more comprehensive and well-developed science of psychology (normal as well as abnormal) will provide us with a systematic approach to cultural and individual differences among people, enabling competent professionals to specify a workable set of parameters for formulating clear, univocal concepts of mental health and mental illness.

Redlich and Freedman address the issue of cultural relativism, claiming that abnormality depends on the cultural values of defining persons. They note that there is no agreement on what is normal drinking, that prostitution is accepted in some cultures but not in others, and that "there are remarkable differences in aggression, sexuality, and dependency needs in the different social classes of a single culture . . ." ([13], p. 114). They point out that despite the denial of relativism on the part of some social scientists, it is only extreme forms of behavior such as indiscriminate murder, cannibalism, or absolute disregard for property that are almost universally rejected. Severe behavior disorders are likely to be considered abnormal, no matter what the cultural setting. The authors write:

> In actual practice, psychiatrists use a composite approach; they diagnose behavior as clearly abnormal when it is seriously disabling, frustrating, deviates from established cultural norms, hence occurs relatively rarely; however, in borderline cases such an approach does not work well. . . .
>
> Only gross deviations are clearly recognized and agreed upon in all civilized societies; borderlines of normal and abnormal behavior are fuzzy and overlapping. Cultural relativism with respect to milder disorders is the rule. The judgments of psychiatrists cannot in reality be far removed from those of the common man of the societies and cultures in which psychiatrists and patients live. At present we cannot make precise statements about normal and abnormal. . . ([13], pp. 114–115).

Problems of cultural relativism constitute only a partial barrier to providing a clear and uncontroversial conception of normality and abnormality. Another set of difficulties lies in the way in which the terms *normality* and *abnormality* are to be construed, an inquiry to which we turn next.

B. Normality and Abnormality as Normative and as Statistical Concepts

It should be noted at the outset that if "mental health" and "mental illness" are not properly to be viewed as correlative terms (see section 2 above), then even *given* some acceptable account of normality and abnormality, we may be faced with the task of specifying an independent set of parameters for assessing types and degrees of mental health and mental illness. But apart from this, another set of problems emerges relating to the fact that normality is sometimes construed as a normative concept and sometimes as a statistical one. Jahoda rejects the attempt to provide a criterion for mental health based on normality on a number of grounds. One such ground is that of the problems presented by cultural relativity, as discussed above, and the second reason is as follows.

Noting that normality can be viewed either as a statistical frequency concept or as a normative idea of how people ought to function, Jahoda points out that a coincidence of statistical and normative correctness is, at best, fortuitous.

> To believe that the two connotations always coincide leads to the assertion that whatever exists in the majority of cases is right by virtue of its existence. The failure to keep the two connotations of normality separate leads straight back into an extreme cultural relativism according to which the storm trooper, for example, must be considered as the prototype of integrative adjustment in Nazi culture ([5], pp. 15–16).

The issue is now identified as the old problem in philosophical ethics: the is–ought gap. Although I am not concerned to argue here about whether there is or is not, or should or should not be such a gap, it seems that the significance of the distinction for the problem of defining "mental health" in terms of some conception of normality is clear. As Jahoda correctly points out, "insofar as normality is used in the normative sense, it is a synonym for mental health, and the problems of concept definition are, of course, identical" ([5], p. 16).

Another difficulty with "normality" as construed in the normative sense is that the concept tends to function as an "ideal type," so that the actual behavior of persons is, at best, an approximation to some optimal conditions. The problem is, then, that according to some psychological theoretical frameworks it may be extraordinarily difficult or even impossible to draw the line between normality and abnormality (and, consequently, following Jahoda's insight, impossible to draw the line between mental illness and mental health). This issue will be brought up again in connection with a problem to be discussed below: considerations *within* certain theories which preclude the possiblity of a precise definition of mental health and illness. Let it suffice to note at this point that an attempt to define "mental health" in terms of a normative conception of normality appears to lead either to circularity (as Jahoda claims), or else directly back to cultural relativism. We shall next examine the frequency concept of normality to see if it fares any better.

The most obvious difficulty with a statistical frequency concept of normality is that a majority of people may do many things we hesitate to call mentally healthy. Thus, "psychological health may, but need not be, the status of the majority of the people" ([5], p. 16). That this is so might be illustrated by considering the case of physical illness and health. No one would be likely to urge a definition of "physical health" based on statistical considerations, for it might turn out that a majority of the population is suffering from some form or other of physical ailment or disease (whether temporary or enduring). As Livermore, Malmquist, and Meehl point out in this connection:

> From a biological viewpoint, it is not inconsistent to assert that a sizable proportion–conceivably a majority–of persons in a given population are abnormal or aberrant. Thus if an epidemiologist found that 60% of the persons in a society were afflicted with plague or avitaminosis, he would (quite correctly) reject an argument that "Since most of them have it, they are okay, i.e. not pathological and not in need of treatment." It is admittedly easier to defend this nonstatistical, biological-fitness approach in the domain of physical disease, but its application in the domain of behavior is fraught with difficulties ([6], pp. 78–79, n. 11).

It is true, of course, that there are much more systematic and comprehensive biological and physiological theories on the basis of which the concept of physical or bodily health may be constructed in medicine than now exist in the realm of psychological or psychiatric theory. But the inadequacy of a statistical conception of normality for physical health provides an instructive comparison for present purposes. It may be objected here that the example just given presupposes the applicability of the medical model and a conception of mental health based on the analog of physical health. Although this is so, the reasons for questioning the adequacy of the frequency concept of normality for defining "mental health" do not depend on the analogy with physical health.

An additional difficulty with the statistical approach is noted by Redlich and Freedman. This difficulty constitutes a methodological problem rather than an objection in principle, but presents obstacles nevertheless. "Few exact data . . . are available on the frequency and distribution of behavior traits. Such an approach presupposes that behavior is quantifiable and measurable, but obviously many forms of behavior are not. . . . Few data . . . exist on the prevalence and the incidence of psychiatric symptoms, such as anxiety, hallucinations, phobias, and so forth" ([13], p. 113). The authors acknowledge that there are some good examples of the statistical approach in the data supplied by Kinsey et al. on sexual behavior, in the area of socio-economic data, and in the broad investigations of intelligence. However, the need for assessing the *relevance* of various sorts of behavioral data for the task of defining "mental health" and "mental illness" points to still another problem inherent in the statistical approach to normality and abnormality. This is the selection of a reference population–a procedure that involves nonstatistical considerations. Jahoda notes that "the choice of population inevitably contains, at least implicity, a nonstatistical concept of health" ([5], p. 17), a factor which indicates the inadequacy of an attempt to define "mental health" and "mental illness" on the basis of a purely statistical concept of normality.

Moreover, even when the relevant reference population has been delineated, equal weight would not be given to all measurable psychological functions in developing a set of norms against which to evaluate the mental health status of individuals. "We thus find again that some, at least tacit, nonstatistical considerations must precede the application of the statistical approach" ([5], pp. 17–18).

This is borne out once again by the example of physical health and disease where purely statistical considerations are insufficient for formulating a conception of health. In the domain of mental health, the most that a statistical approach can achieve is a specification of which behaviors and traits are "abnormal" in the general population. But we still require some nonstatistical parameters for deciding which "abnormalities" are to count as illness and which "normalities" should be construed as healthy. The selection of such parameters would appear to depend partly on considerations which are contingent upon the progress and development of a comprehensive psychological theory, and also on a range of value questions which, although relevant and important, cannot be gone into here (in this connection, see [3] and [8]). The above-noted difficulties with both the normative and statistical concepts of normality point to their inadequacy (at least if taken singly) as a basis for defining "mental health" and "mental illness." Redlich and Freedman identify one further approach to normality which will be discussed briefly in the next section: the clinical approach.

C. The Clinical Approach

It was noted in subsection *B* above (p. 132 ff.), that when construed in the normative sense, the concept of normality tends to function as an "ideal type" according to whatever theory is being employed. This is not the case, however, in the clinical approach, as Redlich and Freedman point out: "In general terms, clinical normality is not ideal performance but minimal performance, just above the level of pathological performance for a given individual" ([13], p. 113). But now the problem is to identify such "levels of performance," a task which is not only clinically difficult, but also depends on some theoretical assumptions on the part of the clinician. Indeed, according to Redlich and Freedman: "The clinical approach defines as abnormal anything that does not function according to its design. This approach is useful in somatic illness, including brain disease, but it is less helpful in behavior disorders, because all too often we do not know what design or function a certain behavior pattern serves" ([13], p. 113). Some criteria for normality which have been employed in the clinical approach are adaptation, maturity, "average expectable environment," and "predominance of conscious and preconscious motivations over unconscious motivation of behavioral acts" ([13], p. 113). But Redlich and Freedman find the concept of adaptation, "which is supposed to explain just about everything, only of very limited use in differentiating normal and abnormal behavior" ([13], p. 113); the conscious–unconscious criterion fails to apply to many forms of abnormal behavior determined by brain disease and ignores the fact that in many types of normal and socially desirable behavior, unconscious and preconscious motivations occur ([13], pp. 113–114). The criteria of "average expectable environment" and maturity are viewed more favorably, but the authors fail to note that "maturity" and "immaturity" are themselves value-laden terms, depending for their application not only on the theoretical orientation of the clinician but also on a set of cultural and sub-cultural norms espoused by him.

One final problem in connection with the clinical approach lies in the unwitting conflation of a number of different criteria on the part of professionals. These criteria may encompass those already cited here, in addition to the personal, subjective conceptions of the psychiatrist or clinical psychologist. The situation has been described as follows:

> It is especially tempting to the psychiatrist or clinical psychologist, given his usual clinical orientation, to slip unconsciously from the idea of "sickness," where treatment of a so-called "patient" is the model, to an application that justifies at most a statistical or ideological or psychological-adjustment usage of the word "norm." Probably the most pernicious error is committed by those who classify as "sick" behavior that is aberrant in *neither* a statistical sense *nor* in terms of any defensible biological or medical criterion, but solely on the basis of the clinician's personal ideology of mental health and interpersonal relationships. Examples might be the current psychiatric stereotype of what a good mother or a healthy family must be like, or the rejection as "perverse" of forms of sexual behavior that are not biologically harmful, are found in many infra-human mammals and in diverse human cultures, and have a high statistical frequency in our own society ([6], p. 79, n. 11).

This situation not only complicates the process of diagnosing mental illness in individual persons, but also, if widespread or typical, precludes the use of the clinical approach as an effective means of defining "normality" and "abnormality."

4. Obstacles to a Clear Conception Arising within or between Specific Theories of Psychopathology

This section will be devoted to an examination of some of the difficulties posed by different theories or theoretical conceptions—difficulties that stand in the way of a clear and precise concept of mental health or mental illness. The inquiry will be divided into two main areas: problems *within* theories (intratheoretical); and disparity or conflict *between* or *among* theories (intertheoretical).

A. Intratheoretical Problems

While it is likely that many different theories or theoretical systems contain tacit assumptions or explicit premises which create difficulties for attempts to provide criteria for the concepts of mental health and illness, we shall limit our inquiry here to two of these: Freudian and neo-Freudian psychoanalytic theories; and self-realization or self-actualization theories. It was noted above in connection with normality construed as a normative concept (p. 350), that according to some psychological theoretical frameworks it may be extraordinarily difficult or even impossible to draw the line between sickness and health. One such theory is the classical Freudian account, and one writer points out that if a psychiatrist is trained in the more orthodox psychoanalytic notions, "his belief system makes it impossible to determine the "sickness" or "wellness" of the patient, since the classical theories assume that all people have unconscious drives which interfere with optimal functioning, and no clear practical criteria are provided for judging the "sick" from the "well" ([10], p. 27). We shall not raise questions here about the nature of the theoretical entities to which Freudian theory is committed, nor about the testability of many propositions embedded in that theory. Although such questions are legitimate and interesting, they have been examined at great length by philosophers, psychologists, and psychiatrists alike, and are peripheral to the concerns of this paper. The question here is whether, *given* an initial acceptance of the Freudian notion of the Unconscious and all that it entails, a distinction can be made between healthy and unhealthy behavior. Jahoda seconds the view cited just above, noting that it has not been demonstrated that there are any human beings who are free from unconscious conflicts. "If it is reasonable to assume that such conflicts are universal, we are all sick in different degrees. Actually, the difference between anyone and a psychotic may lie in the way he handles his conflicts and in the appearance or lack of certain symptoms. If this is so, mental disease must inevitably be inferred from behavior. But, apart from extremes, there is no agreement on the types of behavior which it is reasonable to call "sick" ([5], p. 13). So according to this objection, the theory itself precludes any workable, clear distinction between instances of mental health and cases of mental illness. The most we can hope for, according to this theoretical framework, is the provision of *comparative* judgments of health and illness–a conclusion which would be welcomed by many Freudians and non-Freudians alike on this issue.

But this is not the end of the Freudian story. As Joseph Margolis contends, it appears that Freud himself employed a "mixed model that shows clear affinities with the models that obtain in physical medicine and at the same time with the models of happiness and well-being that obtain in the ethical domain" ([7], pp. 81–82). This further complicates the issue since the "mixed model" is really a combination of two different models, each having its own set of parameters along which health and disease, good and poor functioning, desirable and undesirable traits and behavior are identified. Margolis specifies further just how Freud's development of psychoanalytic medicine runs along two converging lines:

> In one, as in the studies of hysteria, Freud was extending case by case the medical concept of illness, by working out striking and undeniable affinities between physical illnesses and counterpart cases, for which the aetiology would have had to be radically different. And, in the other, Freud inevitably assimilated the concept of

> mental health to concepts of happiness–in particular, to his genital ideal. The result is that, *given* some version of this (or another such) ideal, deviation from the ideal tends to be viewed in terms of malady and disease, even though there are no strong analogical affinities between the pattern in question and clear-cut models of physical illness. Hence, patterns as significantly different as hysteria and homosexuality tend both to be assimilated to the concepts of health and disease ([7], pp. 75–76).

There is little doubt that Margolis's analysis here is correct. Indeed, on the Freudian account, the failure of an individual to pass successfully through the three developmental stages of sexuality in infancy and childhood (the oral phase, the anal phase, and the phallic phase) can lead to such diverse patterns of adult "illness" as those represented in the neuroses and those constituting the character disorders. In any case, successful passage (without "fixation," "regression," "arrest") through the early developmental stages is a necessary condition (although not sufficient) for the "healthy" adult, achieving or approximating the "genital ideal." The adult genital character is the *mature* individual; failure to attain the ideal results in varying degrees of "immaturity." Thus on the Freudian account, the concept of the "genital (ideal) character," which denotes the mature individual, is intimately bound up with the descriptive-explanatory theory of infant–childhood development. The interlacing of descriptive and normative components in this account, while not in itself pernicious, must be made explicit if there is to be any progress in providing criteria for the concepts of health and illness.

One need not, of course, accept the specific Freudian precepts concerning the developmental stages in infancy and childhood and the related notions of fixation at one level or another, regression to a previous stage, etc. Indeed, the refusal to accept this particular framework may result in different judgments as to the mental health or illness of a person who fails to live up to the "genital ideal." The fact that there are other ideals of health, happiness, and well-being which may be and are, in fact, postulated by different theorists leads not only to the presence of multiple criteria, but also to possibly conflicting ones. Even if we allow the legitimacy of a "mixed model" such as Freud's, it must be acknowledged that there exist multiple and possibly conflicting norms of health, happiness, and well-being. Consequently, deviation from one such norm might count as mental illness, while the same person might be termed mentally healthy according to some different set of norms whose ideal he fulfills or closely approximates.

We turn next to a brief look at some problems inherent in another set of influential theories: the self-realization or self-actualization theories. Mention was made earlier of these theories (above) in discussing the failure of attempts to define "mental health" in terms of the absence of disease or illness. The concepts of self-realization and self-actualization play a dual role in a number of psychopathological theories (notably, those of Karen Horney, Erich Fromm, Kurt Goldstein, and Abraham H. Maslow), functioning as a characterization of the healthy individual and also specifying the goal of psychotherapy (for a more detailed account, see [3]). Redlich and Freedman note that "in highly individualized cultures, self-actualization or self-realization is seen as the goal of certain psychotherapies. The goal of realizing one's human potential is encountered in Buddhism, particularly in the practices of Zen Buddhism, which fascinated and stimulated Karen Horney, Erich Fromm, and Alan Watts.... The behavioral changes ... should be defined in rationally and operationally verifiable terms. There is no place in psychiatry for mystical, irrational, or supra-rational approaches" ([13], p. 270).

Leaving aside the methodological issues connected with the goals and practice of psychotherapy, we may note that even in its role as characterizing the healthy person (a positive conception of mental health), the concept of self-realization is problematic. The difficulty is an old one in philosophy, harking back to Aristotelian notions of potentiality and essence. Indeed, Fromm cites Aristotle in his account of "activity" as "the exercise of the functions and capacities peculiar to man" and his emphasis on "the full development of our powers within the limitations set by the laws of our existence" ([3], p. 58).

Kurt Goldstein "speaks of a 'drive' that enables and impels the organism to actualize in further activities, according to its nature," emphasizing that "optimal self-actualization also means health" ([3], pp. 58–59). Karen Horney refers to "the real self as that central inner force common to all human beings and yet unique in each, which is the deep source of growth" ([3], p. 58). What all these theories–and others of this type–have in common is their emphasis on man's "inner nature" which he seeks to fulfill, his "potentialities" which need to be "actualized," and the "inner self" which develops and unfolds successfully through "self-realization." Both as goals of psychotherapy and as characterizations of positive mental health, these concepts are problematic. The assumptions about man's nature or essence cannot be accepted uncritically, and the difficulties of formulating a testable concept of man's "potentiality" (as a generic trait) are well-known. To the extent that a concept of mental health adopted by these theories rests on the foregoing assumptions, we cannot expect much in the way of criteria that meet the requirements of empirical testability and confirmability. Such theories may achieve a high degree of methodological adequacy in cases where a sensitive and insightful therapist can assess the specific "potentialities" or "natural inclinations" of a particular patient. But as attempts to provide a basis for constructing a clear and workable *concept* of mental health, the self-realization theories fail to satisfy the demands of conceptual clarity and empirical testability.

B. Intertheoretical Problems

A set of problems which naturally arises from the proliferation of theories in this area relates to the likelihood of multiple–possibly even conflicting–

criteria for mental health and mental illness. Especially among theories which are couched in terms of some normative ideal of health, happiness, or well-being, we may expect that deviations from the ideal might be construed as "sick" on one theory, while no decision might be forthcoming according to another. This raises the question of whether deviation from some *normative ideal* should properly be counted as sickness or illness, on the model of physical disease. In any case, an obvious consequence of the present situation is that the *general* concepts of mental health and mental illness can only be understood in terms of some *specific* theory of psychopathology.

Enough has been said in the preceding section to indicate the divergence between the concepts of health and illness as conceived by classical Freudian theorists, on the one hand, and self-realization theorists, on the other. Both conceptions are embedded in the theoretical systems themselves, and professional judgments concerning the health or illness of particular individuals may vary depending on which system the diagnostician espouses. The number of different theories is legion, and no attempt will be made here to survey them. Instead, a few general points will be noted.

Following Freud's tripartite division of the psyche into id, ego, and superego, much emphasis was placed by psychoanalytically oriented practitioners on the relationship among these and the relation of all of them to unconscious processes. Recent developments in the area known as ego psychology have tried to correct what some theorists held to be a bias in Freudian theory, attempting to replace the "one-sided emphasis on unconscious processes with a stronger acknowledgment of the importance of *conscious* experiences" ([3], p. xiii). It is apparent that with different norms of well-functioning and different emphases even in theories which accept many of the basic Freudian precepts, there is a broad scope of conceptions concerning what is to count as healthy or sick behavior.

Marie Jahoda favors a "multiple criterion" approach to problems of mental health. The value of this approach to the concepts of health and illness is that it has the requisite breadth and flexibility to comprehend a wide variety of human behavior without being so general as to become empty. Commenting on various criteria adopted by different theorists who propound different concepts of positive mental health and also on the various ways of using a multiple criterion, Jahoda writes: "There is no incompatibility between the idea of diverse types of health and the use of such a criterion. . . . At the present state of our knowledge it may well be best to combine the idea of various types of health with the use of a multiple criterion for each. The former will prevent over-generalizations; the latter will permit us to do justice to the complexity of human functioning" ([5], p. 73).

It is evident that the work of providing criteria for the concepts of mental health and illness, as well as the task of evaluating proposed definitions needs to be done by theorists and practioners in the fields of psychiatry and psychology. The task is partly conceptual and partly empirical, involving policy decisions as well as theoretical considerations. In this and the preceding sections, I have tried to specify a number of the problems which exist in this area and to note some of the issues that need clarification and careful scrutiny.

Finally, one large problem should be noted—a problem which might best be viewed as one that encompasses and gives rise to most of those already discussed. This is the absence of an overall scientific theory on which to base conceptions of mental health and illness, well-functioning and maladaptive behavior. It should be noted, however, that in this regard the concepts of health and disease face a number of similar problems in the domain of somatic medicine (i.e. the concepts are vague, there are multiple criteria for their application—criteria which may conflict occasionally, etc.). There is no general, well-integrated theory of the sort that exists in, say, physics, interconnecting the well-developed fields in medicine of physiology, anatomy, pathology, neurology, immunology, etc., with current developments in the biological sciences. The absence of bridging laws between these branches of medical and biological science, as well as the divergent theoretical and methodological approaches of experimental biologists, on the one hand, and medical scientists oriented towards pathology, on the other, all contribute to the present lack of systematization in the total field of biological science. So the absence of well-confirmed fundamental laws, from which other laws are derivable, and the absence of a systematic, general theory result in the situation that within medicine itself, there are no clear or precise formulations of the basic concepts of health and disease, and no set of necessary and sufficient conditions for their application.

These facts concerning the present stage of scientific development of somatic medicine may offer little consolation to the theorist in psychology or psychiatry who is looking to provide a clear, workable concept of mental health or mental illness. But it would seem that the clarity and precision of the basic concepts in this field—as in any other—go hand in hand with the related theoretical developments of formulating general laws, providing systematic interconnections among the various branches of the field, and relating all of these to existing, well-confirmed theories in the other sciences. The science of psychology is a long way from this goal, but there do not seem to be any good arguments that have been offered to show that the goal is unattainable, *in principle.* Some writers have attempted an operational reformulation of some of the basic psychoanalytic concepts [4], [15], and while the merits of such endeavors need to be assessed critically, these efforts point to the attempt to render more precise and testable concepts which have been found fruitful in psychotherapeutic practice. Moreover, it is not the case that all existing psychological and psychiatric theories should be viewed as competing with one another, or in some sense mutually incompatible. Rather, we might reasonably expect that with the further development of the science of psychology, bridging laws (perhaps of a very complex sort) will be formulated and theories at various levels (e.g. macro-behavior, physiological psychology, etc.) will be systematically integrated.

To show that the views expressed here do not reflect a philosopher's unwarranted optimism about the methods and goals of this field, it is appropriate to cite once again the views of Drs. Redlich and Freedman.

> As a technology based on the behavioral and biological sciences, psychiatry takes a deterministic point of view. This does not mean that all phenomena in our field can be explained, or that there is no uncertainty. It merely commits us to a scientific search for reliable and significant relationships. We assume causation–by which we mean that a *range* of similar antecedents in *both* the organism and environment produces a similar *set* of consequences. In general, we follow the procedures of basic sciences and attempt to determine the limits within which a range of antecedents has a high probability of producing similar results. . . .
>
> The principles and basic methods of studying normal and abnormal behavior in individuals and in populations, in clinical practice, as well as under experimental conditions, are the same as in other naturalistic sciences ([13], pp. 79–80).

To the extent that individual practitioners in the field–whether clinicians or theorists–depart from the above-noted principles and methods, psychiatry and psychology are not to be faulted, any more than physics is impugned by poor methodology or unsound theoretical conclusions on the part of some of its practitioners.

5. Problems with the Conception of Psychological Disorders as "Disease" or "Illness"

A number of arguments have been put forth by some influential psychologists and psychiatrists who hold that the conceptualization of psychological disorders in terms of illness represents an adherence to a mistaken model–the medical model of health and disease. These criticisms of the continued use of the medical model rest partly on conceptual and theoretical grounds and partly on pragmatic considerations relating to the consequences for the individual and society of adhering to this model. There is, however, a good deal of confusion surrounding these issues–confusion which stems largely from a tendency to conflate epistemological, conceptual, and pragmatic problems and attempts at solution to such problems. I shall concentrate on only a few of these issues here, specifically, those which relate most directly to the concerns of definition and concept formation.

The most outspoken opponent of the medical model is Dr. Thomas Szasz, a psychiatrist who holds the M.D. degree. Szasz claims that "although the notion of mental illness made good *historical* sense–stemming as it does from the historical identity of medicine and psychiatry–it made no *rational* sense. Although mental illness might have been a useful concept in the nineteenth century, today it is scientifically worthless and socially harmful" ([16], p. ix).

Thus he argues that it is inappropriate or logically mistaken to construe emotional problems and psychological disorders as a species of illness, on an analogy with bodily disease. On Szasz's view, there exists a *"major logical and procedural error in the evolution of modern psychiatry"* ([16], p. 26). One "error" lay in decreeing that some malingerers be called "hysterics," which led to obscuring the similarities and differences between organic neurological diseases and phenomena that only looked like them ([16], p. 26). But it does not follow from the fact that hysteria (and other psychological disorders) are called "illness" that the similarities and differences between organic and nonorganic illness cannot be duly noted and treated accordingly. Indeed, the very introduction of the notion of *mental* illness to cover phenomena such as hysteria marks a decision to treat a class of seeming bodily disorders as different in relevant respects from organic neurological disease. The labeling itself need not involve a failure to attend to the relevant similarities and differences for the purpose of diagnosis, explanation, or treatment.

In general, the precise nature of Szasz's objection to construing psychological disorders as illnesses is not always clear. Sometimes he writes as though the reclassification and introduction of a set of "new rules of the medical game" consist in a sort of logical or conceptual error: "During [the past sixty or seventy years] a vast number of occurrences were reclassified as 'illnesses.' We have thus come to regard phobias, delinquencies, divorce, homicide, addiction, and so on almost without limit as psychiatric illnesses. This is a colossal and costly mistake" ([16], p. 43).

In answer to the question, "from what point of view is it a mistake to classify non-illnesses as illnesses?" Szasz replies that "it is a mistake from the point of view of science and intellectual integrity." This would seem to imply that a proper scientific conception and a generally accepted classificatory schema preclude treating psychological disorders as illnesses. But there is no compelling evidence–either from Szasz's own account, or revealed in our inquiry in the preceding sections–to show that it is indeed the case that a clear "error" or "mistake" is involved in this type of classification. The consequence of the medical model approach and resulting reclassifications, according to Szasz, has been that although "some members of suffering humanity were promoted . . . to higher social rank, this was attained at the cost of obscuring the logical character of the observed phenomena" ([16], p. 295).

It is evident from the passages just cited that at least sometimes Szasz construes the reclassification of psychological disorders as illnesses to be a sort of error or mistake (logical or conceptual). At other times, however, he writes as though the change is "merely linguistic" and a matter of choice or preference of one classificatory schema rather than another. Thus he holds that it is "a matter of scientific and social choice whether we prefer to emphasize the similarities and, hence, place hysteria in the category of illness, or whether we prefer to emphasize the differences and place hysteria in a category of nonillness" ([16], p. 29). This view construes the issue as one of scientific and practical utility, rather than conceptual or logical error, and is borne out by Szasz's subsequent discussion. In a later passage, he reiterates this same view:

> From the standpoint of our present analysis, the entire change in renaming certain illnesslike forms of behavior from "malingering" to "hysteria" (and "mental illness") can be understood as nothing but a linguistic change employed for the purpose of achieving a new type of action-orientedness in the listener. The verbal change . . . served to command those charged with dealing with "hysterics" to abandon their moral-condemnatory attitude toward them and to adopt instead a solicitous and benevolent attitude, such as befitted the physician vis-a-vis his patient ([16], p. 132).

It appears from the above passages and others which could be cited, that Szasz's position is at the very least, unclear, and at worst, inconsistent, with regard to the question of what is wrong with classifying behavioral disorders and disabling psychological difficulties as forms of "illness." We now turn to a brief discussion of Szasz's own view of what properly constitutes illness and some criticisms of his charge against those who have been engaged in reclassifying certain nonbodily disorders as forms of illness.

It is at least an implicit assumption of Szasz's–one which he sometimes makes explicit–that the only proper candidates for the notion of disease are those which refer to genuine *bodily* (organic or functional) ailments or involve a physical lesion. We need to examine this assumption in order to evaluate Szasz's contention that it is a mistake or error to construe nonbodily disorders as illness. The issue then becomes, on what grounds does Szasz reject mental "illnesses" as instances of some sort of disease, and are those grounds justifiable? He writes:

> The adjectives "mental," "emotional," and "neurotic" are simply devices to codify–and at the same time obscure–the differences between two classes of disabilities or "problems" in meeting life. One category consists of bodily diseases–say, leprosy, tuberculosis, or cancer–which, by rendering imperfect the functioning of the human body as a machine, produce difficulties in social adaptation. In contrast to the first, the second category is characterized by difficulties in social adaptation not attributable to malfunctioning machinery but "caused" rather by the purposes the machine was made to serve . . . ([16], pp. 41–42).

This view sets up two mutually exclusive categories of disability, such that an instance of the one category can never be construed as falling also under the second category. An antireductionist bias is evident in Szasz's remarks here and elsewhere, and it is legitimate to ask whether a clear distinction can be made between "the functioning of the human body as a machine" and "the purposes the machine was made to serve."

Moreover, it is certainly true, as Szasz contends ([16], pp. 79 ff.), that Freud continued to seek organic or physico-chemical *causes* of the psychological disorders and malfunctioning which he observed in his patients. But the question remains, *even if* Freud was mistaken in his continued search for neurological or some other physical bases for these behavioral disorders, does it follow that such disorders cannot properly be construed as forms of disease or illness nonetheless? Szasz's position seems to be that the absence of identifiable or probable physiological causes disqualifies a disorder or disability from the category of disease. Consequently, construing nonorganically based behavioral and personality disorders as diseases turns out to be *both* a logical and a scientific error. It is a logical error because the two categories of problems in facing life are mutually exclusive; and it is a scientific mistake because it erroneously presupposes an organic or neurological cause for every psychological, social, or ethical problem resulting from the malfunctioning of persons. Szasz wishes, therefore, to eliminate the entire notion of mental illness, claiming that "mental illness is a myth. Psychiatrists are not concerned with mental illnesses and their treatments. In actual practice they deal with personal, social, and ethical problems in living" ([16], p. 296).

Whereas Szasz chooses to *close* the concept of disease or illness, requiring as a necessary condition that there be a known or probable physiological basis, another view of the matter holds that the labors of Freud and others resulted in a legitimate *extension* of the then existing concept of disease or illness. The strategy in replying to Szasz's position would thus consist in the following two-stage argument: (1) showing that Freud and his followers were not making a logical or conceptual *mistake* in treating psychological disorders as illnesses, but were rather engaged in the enterprise of extending or widening the concept of disease or illness; (2) showing that such extension in this case is *legitimate,* that is, can be justified by noting relevant and important similarities between cases of mental illness and cases of physical illness. I shall take the question, "Is it *ever* legitimate to extend or enlarge a concept?" as admitting of an uncontroversial affirmative answer. Accordingly, one reply to Szasz is given in the words of Joseph Margolis:

> Szasz is absolutely right in holding that Freud reclassified types of suffering. But what he fails to see is that this is a perfectly legitimate (and even necessary) maneuver. In fact, this enlargement of the concept of illness does not obscure the differences between physical and mental illness–and the differences themselves are quite gradual, as psychosomatic disorder and hysterical conversion attest. On the contrary, these differences are preserved and respected in the very idea of an *enlargement* of the concept of illness ([7], p. 73).

This passage serves not only to make the point about the legitimacy of enlarging the concept of illness to cover cases of mental or psychological illness, it also emphasizes that there is no clear and obvious line—as Szasz appears to think there is—between physical and mental illness, or between the "two categories" of problems in facing life. Indeed, it is apparent that Margolis has drawn this line in a different place from Szasz. While Szasz considers hysteria a nonbodily illness (hence, not an "illness" at all) on the grounds that it has no organic or neurological *causes,* Margolis construes hysterical conversion and other psychosomatic ailments at least as borderline cases, presumably on the grounds that such disorders are manifested in terms of observable and clear-cut bodily *symptoms* and *malfunctioning.* So it appears that there may be some genuine dispute as to the selection of criteria for an adequate or uncontroversial characterization of *physical* or *bodily* illness itself.

Once it is acknowledged that there are good reasons for construing Freud's maneuver as one of extending a concept rather than as a sort of logical or conceptual error, we may proceed to the second stage of the argument in reply to Szasz: the justification of the extension of the concept of illness to cover psychological problems and personality disorders. Margolis suggests the following, in answer to the question "Should mental 'disorders' be allowed, in a medical sense, to count as diseases or illnesses?"

> If I were to describe a condition in which a patient suffers great pain in walking and is quickly overcome by fatigue, a condition which lasts for several years, and we were to find that there is an organic cause for this pattern, we should be strongly inclined to regard what we have before us as a *physical illness.* Now, if we have the same sort of pattern but are unable to find any organic cause, and begin to suspect that, in some inexplained way, the condition is due to the emotional or psychical life of the patient, we may have a reason for insisting that the pattern is still a *pattern of illness* ([7], p. 74).

The reasonableness of this conclusion—Szasz's view notwithstanding—is shown by observing the affinities between the new cases and the standard ones. It should be noted, further, that the existence of unexplained phenomena and the absence at present of psycho-physical laws (covering normal as well as abnormal behavior) do not in themselves compel theoretical conclusions and conceptual decisions of the sort that Szasz is prone to make.

There is a line of argument different from that employed by Margolis which can serve to show the relevant similarities between cases of physical disease and cases of putative mental illness. Whereas Margolis's method is a case by case approach, proceeding by comparison of new (mental) cases to old (physical) instances of disease and noting the affinities between them, the alternative method distinguishes general *categories* of behavioral symptoms and demonstrates that these categories are common both to bodily diseases and to disorders commonly construed as mental illness. David Ausubel uses this approach in arguing that "the plausibility of subsuming abnormal behavioral reactions to stress under the general rubric of disease is further enhanced by the fact that these reactions include the same three principal categories of symptoms found in physical illness" ([2], p. 262). Ausubel characterizes these categories as manifestations of impaired functioning, adaptive compensation, and defensive overreaction, and cites examples of both physical and mental diseases falling under each category, noting the relevant similarities between them. He concludes that there is no inherent contradiction in regarding mental symptoms *both* as expressions of "problems in living" (Szasz's preferred locution) *and* as manifestations of illness. "The latter situation results when individuals are for various reasons unable to cope with such problems, and react with seriously distorted or maladaptive behavior" ([2], p. 265). So according to some opponents of Szasz, the position is taken that in order to qualify as a genuine manifestation of disease, a symptom need not reflect a physical lesion.

We may conclude from the inquiry in this section that there appears to be no compelling reason to adopt Szasz's view that mental illness is a "myth" and that personality disorders and psychological problems are inappropriately viewed as illness and properly to be construed as "problems in living." In sum, there appear to be no logical or conceptual reasons why such difficulties cannot or should not be subsumed under the category of "illness." Moreover, whatever is gained in terms of social utility by viewing these problems as "problems in living" is not precluded by viewing them *also* as manifestations of disease, as Ausubel suggests. So whatever merits Szasz's position may have in terms of pragmatic consequences, these same results can be achieved if we retain the concept of mental *illness* along with the present classificatory schema.

By way of summary and conclusion, it would be well to note where the rejection of Szasz's antimedical model position leads us. Most of the problems discussed in this paper can be seen to re-emerge upon consideration of a brief quotation from Ausubel's paper. Arguing specifically against Szasz's contention that to qualify as a genuine manifestation of disease a given symptom must be caused by a physical lesion, Ausubel writes: "Adoption of such a criterion would be arbitrary and inconsistent both with medical and lay connotations of the term "disease," which in current usage is generally regarded as including any marked deviation, physical, mental, or behavioral, from normally desirable standards of structural and functional integrity" ([2], p. 259).

While this statement may appear sufficiently general to escape controversy, upon closer analysis a range of familiar problems can be identified in connection with the phrase *normally desirable standards.* The immediate difficulty concerns whether the phrase is to be construed descriptively or normatively, but even if this question is decided, further problems remain.

Construed descriptively, *normally desirable standards* denotes those standards which people (the issue of just *which* people will be put aside for the moment) actually desire for themselves or others. The question of whether or not this is an adequate account of the general notion of desirability has been raised at least since John Stuart Mill wrote, in his essay *Utilitarianism:* "The only proof capable of being given that an object is visible, is that people actually see it. The only proof that a sound is audible, is that people hear it: and so of the other sources of our experience. In like manner, I apprehend, the sole evidence it is possible to produce that anything is desirable, is that people do actually desire it" ([12], p. 221).

Critical views contend that the notion of desirability is properly to be explicated in terms of what *ought* to be desired, or what it is *rational* to desire, and that Mill's purely descriptive account fails to capture the normative force of the term *desirable.*

Construed normatively (in terms of which standards ought to be adopted or are worthy of being maintained), the phrase generates a number of problems noted in detail above–chief among which is the cultural relativity of values. The questions "desirable *for* whom?" "desirable *according* to whom?" and "desirable for what ends or aims or purposes?" all need to be answered satisfactorily before Ausubel's general statement can serve to provide even a rough and ready criterion for the notions of mental health and illness. Moreover, it may emerge that there is no set of "normally desirable standards" that will be accepted without controversy by all individuals or groups because of the differing ideologies and value systems and differing background experiences in a society as large and diverse as ours (not to mention those of other societies and cultures).

It may be, however, that we need to emphasize the further phrase: "of structural and functional integrity" in analyzing Ausubel's statement. On this view, the appeal to "normally desirable standards of structural and functional integrity" presupposes some general theory which provides an account of an integrated, well-functioning system. Such an account is given, for the most part, in biological (anatomical and physiological) theories so that the notion of physical disease, although not without a number of problems, can be specified without engendering a great deal of controversy. With regard to mental health and illness, however, not only is there no generally accepted psychological or personality theory that can be presupposed, but the search for criteria of application for the basic concepts is itself an attempt to fill out such a theory and provide the very parameters which enable us to judge that a personality system possesses "structural and functional integrity."

In the words of Redlich and Freedman, "a completely acceptable super-theory on which psychiatry can generally rest its work does not exist" ([13], p. 79). But these authors would be quick to note that progress in the behavioral and biological sciences has been rapid and steadily advancing in recent years. So whatever pessimism may accrue to the observations made in this study about the concepts of mental health and mental illness as currently understood and employed by professionals and laymen alike, a measure of optimism exists in the belief that fruitful and systematic developments will continue to be forthcoming in the experimental, theoretical, and clinical areas of psychology and psychiatry.

References

[1] Albee, G. "Definition of Mental Illness and Mental Health Manpower Trends." *Psychopathology Today*. Edited by William S. Sahakian. Itasca, Ill.: F. E. Peacock Publishers, 1970.

[2] Ausubel, D. P. "Personality Disorder *Is* Disease." *Mental Illness and Social Processes.* Edited by Thomas J. Scheff. New York: Harper & Row, 1967.

[3] Buhler, C. *Values in Psychotherapy*. New York: The Free Press of Glencoe, 1962.

[4] Ellis, A. "An Operational Reformulation of Some of the Basic Principles of Psychoanalysis." *Minnesota Studies in the Philosophy of Science.* vol. 1. Edited by Herbert Feigl and Michael Scriven. Minneapolis: University of Minnesota Press, 1956.

[5] Jahoda, M. *Current Concepts of Positive Mental Health* New York: Basic Books, 1958.

[6] Livermore, J. M.; Malmquist, C. P.; and Meehl, P. E. "On the Justifications for Civil Commitment." *University of Pennsylvania Law Review 117* (1968): 75–96.

[7] Margolis, J. *Psychotherapy and Morality*. New York: Random House, 1966.

[8] Masserman, J. H., ed. *Psychoanalysis and Human Values.* New York: Grune and Stratton, 1960.

[9] May, R. "The Work and Training of the Psychological Therapist." *Psychology, Psychiatry, and the Public Interest.* Edited by Maurice H. Krout. Minneapolis: University of Minnesota Press, 1956.

[10] Mechanic, D. "Some Factors in Identifying and Defining Mental Illness." *Mental Illness and Social Processes.* Edited by Thomas J. Scheff. New York: Harper & Row, 1967.

[11] Menninger, K. "Unitary Concept of Mental Illness." *Psychopathology Today.* Edited by William S. Sahakian. Itasca, Ill.: F. E. Peacock Publishers, 1970.

[12] Mill, J. S. "Utilitarianism." *Essential Works of John Stuart Mill.* New York: Bantam Books, 1961.

[13] Redlich, F. C. and Freedman, D. X. *The Theory and Practice of Psychiatry.* New York: Basic Books, 1966.

[14] Sahakian, W. S., ed. *Psychopathology Today,* Itasca, Ill.: F. E. Peacock Publishers, 1970.

[15] Skinner, B. F. *Science and Human Behavior.* New York: Macmillan Company, 1953.

[16] Szasz, T. S. *The Myth of Mental Illness.* New York: Harper & Row, 1961.

INVOLUNTARY MENTAL HOSPITALIZATION

A CRIME AGAINST HUMANITY

THOMAS SZASZ

I

For some time now I have maintained that commitment—that is, the detention of persons in mental institutions against their will—is a form of imprisonment;[1] that such deprivation of liberty is contrary to the moral principles embodied in the Declaration of Independence and the Constitution of the United States;[2] and that it is a crass violation of contemporary concepts of fundamental human rights.[3] The practice of "sane" men incarcerating their "insane" fellow men in "mental hospitals" can be compared to that of white men enslaving black men. In short, I consider commitment a crime against humanity.

Existing social institutions and practices, especially if honored by prolonged usage, are generally experienced and accepted as good and valuable. For thousands of years slavery was considered a "natural" social arrangement for the securing of human labor; it was sanctioned by public opinion, religious dogma, church, and state;[4] it was abolished a mere one hundred years ago in the United States; and it is still a prevalent social practice in some parts of the world, notably in Africa.[5] Since its origin, approximately three centuries ago, commitment of the insane has enjoyed equally widespread support; physicians, lawyers,

Adapted from "Science and Public Policy: The Crime of Involuntary Mental Hospitalization," *Medical Opinion and Review, 4*:24–35 (May), 1968; reprinted by permission.

[1] Szasz, T. S.: "Commitment of the mentally ill: Treatment or social restraint?" *J. Nerv. & Ment. Dis. 125*:293–307 (Apr.–June), 1957.

[2] Szasz, T. S.: *Law, Liberty, and Psychiatry: An Inquiry into the Social Uses of Mental Health Practices* (New York: Macmillan, 1963), pp. 149–90.

[3] Ibid., pp. 223–55.

[4] Davis, D. B.: *The Problem of Slavery in Western Culture* (Ithaca, N. Y.: Cornell University Press, 1966).

[5] See Cohen, R.: "Slavery in Africa." *Trans-Action 4*:44–56 (Jan.–Feb.), 1967; Tobin, R. L.: "Slavery still plagues the earth." *Saturday Review*, May 16, 1967, pp. 24–25.

and the laity have asserted, as if with a single voice, the therapeutic desirability and social necessity of institutional psychiatry. My claim that commitment is a crime against humanity may thus be countered—as indeed it has been—by maintaining, first, that the practice is beneficial for the mentally ill, and second, that it is necessary for the protection of the mentally healthy members of society.

Illustrative of the first argument is Slovenko's assertion that "reliance solely on voluntary hospital admission procedures ignores the fact that some persons may desire care and custody but cannot communicate their desire directly."[6] Imprisonment in mental hospitals is here portrayed—by a professor of law!—as a service provided to persons by the state because they "desire" it but do not know how to ask for it. Felix defends involuntary mental hospitalization by asserting simply, "We *do* (his italics) deal with illnesses of the mind."[7]

Illustrative of the second argument is Guttmacher's characterization of my book *Law, Liberty, and Psychiatry* as "a pernicious book . . . certain to produce intolerable and unwarranted anxiety in the families of psychiatric patients."[8] This is an admission of the fact that the families of "psychiatric patients" frequently resort to the use of force in order to control their "loved ones," and that when attention is directed to this practice it creates embarrassment and guilt. On the other hand, Felix simply defines the psychiatrist's duty as the protection of society: "Tomorrow's psychiatrist will be, as is his counterpart today, one of the gatekeepers of his community."[9]

These conventional explanations of the nature and uses of commitment are, however, but culturally accepted justifications for certain quasi-medical forms of social control, exercised especially against individuals and groups whose behavior does not violate criminal laws but threatens established social values.

II

What is the evidence that commitment does not serve the purpose of helping or treating people whose behavior deviates from or threatens prevailing social norms or moral standards; and who, because they inconvenience their families, neighbors, or superiors, may be incriminated as "mentally ill"?

[6] Slovenko, R.: "The psychiatric patient, liberty, and the law." *Amer. J. Psychiatry, 121*:534–39 (Dec.), 1964, p. 536.
[7] Felix, R. H.: "The image of the psychiatrist: Past, present, and future." *Amer. J. Psychiatry, 121*:318–22 (Oct.), 1964, p. 320.
[8] Guttmacher, M. S.: "Critique of views of Thomas Szasz on legal psychiatry." *AMA Arch. Gen. Psychiatry, 10*:238–45 (March), 1964, p. 244.
[9] Felix, op. cit., p. 231.

1. The Medical Evidence

Mental illness is a metaphor. If by "disease" we mean a disorder of the physicochemical machinery of the human body, then we can assert that what we call functional mental diseases are not diseases at all.[10] Persons said to be suffering from such disorders are socially deviant or inept, or in conflict with individuals, groups, or institutions. Since they do not suffer from disease, it is impossible to "treat" them for any sickness.

Although the term *mentally ill* is usually applied to persons who do not suffer from bodily disease, it is sometimes applied also to persons who do (for example, to individuals intoxicated with alcohol or other drugs, or to elderly people suffering from degenerative disease of the brain). However, when patients with demonstrable diseases of the brain are involuntarily hospitalized, the primary purpose is to exercise social control over their behavior;[11] treatment of the disease is, at best, a secondary consideration. Frequently, therapy is nonexistent, and custodial care is dubbed "treatment."

In short, the commitment of persons suffering from "functional psychoses" serves moral and social, rather than medical and therapeutic, purposes. Hence, even if, as a result of future research, certain conditions now believed to be "functional" mental illnesses were to be shown to be "organic," my argument against involuntary mental hospitalization would remain unaffected.

2. The Moral Evidence

In free societies, the relationship between physician and patient is predicated on the legal presumption that the individual "owns" his body and his personality.[12] The physician can examine and treat a patient only with his consent; the latter is free to reject treatment (for example, an operation for cancer).[13] After death, "ownership" of the person's body is transferred to his heirs; the physician must obtain permission from the patient's relatives for a postmortem examination. John Stuart Mill explicitly affirmed that "each person

[10] See Szasz, T. S.: "The Myth of Mental Illness," *Medical Opinion and Review, 4*:12–24; *The Myth of Mental Illness: Foundations of a Theory of Personal Conduct* (New York: Hoeber-Harper, 1961); "Mental illness is a myth." *The New York Times Magazine*, June 12, 1966, pp. 30 and 90–92.

[11] See, for example, Noyes, A. P.: *Modern Clinical Psychiatry*, 4th ed. (Philadelphia: Saunders, 1956), p. 278.

[12] Szasz, T. S.: "The ethics of birth control; or, who owns your body?" *The Humanist, 20*:332–36 (Nov.–Dec.), 1960.

[13] Hirsch, B. D.: "Informed consent to treatment," in Averbach, A. and Belli, M. M., eds., *Tort and Medical Yearbook* (Indianapolis: Bobbs-Merrill, 1961), Vol. I, pp. 631–38.

is the proper guardian of his own health, whether bodily, or mental and spiritual."[14] Commitment is incompatible with this moral principle.

3. The Historical Evidence

Commitment practices flourished long before there were any mental or psychiatric "treatments" of "mental diseases." Indeed, madness or mental illness was not always a necessary condition for commitment. For example, in the seventeenth century, "children of artisans and other poor inhabitants of Paris up to the age of 25, . . . girls who were debauched or in evident danger of being debauched, . . ." and other "miserables" of the community, such as epileptics, people with venereal diseases, and poor people with chronic diseases of all sorts, were all considered fit subjects for confinement in the Hospital General.[15] And, in 1860, when Mrs. Packard was incarcerated for disagreeing with her minister-husband,[16] the commitment laws of the State of Illinois explicitly proclaimed that "married women . . . may be entered or detained in the hospital at the request of the husband of the woman or the guardian . . . , without the evidence of insanity required in other cases."[17] It is surely no coincidence that this piece of legislation was enacted and enforced at about the same time that Mill published his essay *The Subjection of Women.*[18]

4. The Literary Evidence

Involuntary mental hospitalization plays a significant part in numerous short stories and novels from many countries. In none that I have encountered is commitment portrayed as helpful to the hospitalized person; instead, it is always depicted as an arrangement serving interests antagonistic to those of the so-called patient.[19]

[14] Mill, J. S.: *On Liberty* (1859) (Chicago: Regnery, 1955), p. 18.

[15] Rosen, G.: "Social Attitudes to Irrationality and Madness in 17th and 18th Century Europe." *J. Hist. Med. & Allied Sciences, 18*:220–40 (1963), p. 223.

[16] Packard, E. W. P.: *Modern Persecution, or Insane Asylums Unveiled*, 2 Vols. (Hartford: Case, Lockwood, and Brainard, 1873).

[17] Illinois Statute Book, Sessions Laws 15, Section 10, 1851. Quoted in Packard, E. P. W.: *The Prisoner's Hidden Life* (Chicago: published by the author, 1868), p. 37.

[18] Mill, J. S.: *The Subjection of Women* (1869) (London: Dent, 1965).

[19] See, for example, Chekhov, A. P.: *Ward No. 6*, (1892), in *Seven Short Novels by Chekhov* (New York: Bantam Books, 1963), pp. 106–57; De Assis, M.: *The Psychiatrist* (1881–82), in De Assis, M., *The Psychiatrist and Other Stories* (Berkeley and Los Angeles: University of California Press, 1963), pp. 1–45; London, J.: The Iron Heel (1907) (New York: Sagamore Press, 1957); Porter, K. A.: *Noon Wine* (1937), in Porter, K. A., *Pale Horse, Pale Rider: Three Short Novels* (New York: Signet, 1965), pp. 62–112; Kesey, K.: *One Flew Over the Cuckoo's Nest* (New York: Viking, 1962); Tarsis, V.: *Ward 7: An Autobiographical Novel* (London and Glasgow: Gollins and Harvill, 1965).

III

The claim that commitment of the "mentally ill" is necessary for the protection of the "mentally healthy" is more difficult to refute, not because it is valid, but because the danger that "mental patients" supposedly pose is of such an extremely vague nature.

1. The Medical Evidence

The same reasoning applies as earlier: If "mental illness" is not a disease, there is no medical justification for protection from disease. Hence, the analogy between mental illness and contagious disease falls to the ground: The justification for isolating or otherwise constraining patients with tuberculosis or typhoid fever cannot be extended to patients with "mental illness."

Moreover, because the accepted contemporary psychiatric view of mental illness fails to distinguish between illness as a biological condition and as a social role,[20] it is not only false, but also dangerously misleading, especially if used to justify social action. In this view, regardless of its "causes"–anatomical, genetic, chemical, psychological, or social–mental illness has "objective existence." A person either has or has not a mental illness; he is either mentally sick or mentally healthy. Even if a person is cast in the role of mental patient against his will, his "mental illness" exists "objectively"; and even if, as in the case of the Very Important Person, he is never treated as a mental patient, his "mental illness" still exists "objectively"–apart from the activities of the psychiatrist.[21]

The upshot is that the term *mental illness* is perfectly suited for mystification: It disregards the crucial question of whether the individual assumes the role of mental patient voluntarily, and hence wishes to engage in some sort of interaction with a psychiatrist; or whether he is cast in that role against his will, and hence is opposed to such a relationship. This obscurity is then usually employed strategically, either by the subject himself to advance *his* interests, or by the subject's adversaries to advance *their* interests.

In contrast to this view, I maintain, first, that the involuntarily hospitalized mental patient is, by definition, the occupant of an ascribed role; and, second, that the "mental disease" of such a person–unless the use of this term is restricted to demonstrable lesions or malfunctions of the brain–is always the product of interaction between psychiatrist and patient.

[20] See Szasz, T. S.: Alcoholism: A Socio-ethical Perspective." *Western Medicine, 7*:15–21 (Dec.) 1966.

[21] See, for example, Rogow, A. A.: *James Forrestal: A Study of Personality, Politics, and Policy* (New York: Macmillan, 1964); for a detailed criticism of this view, see Szasz, T. S.: "Psychiatric classification as a strategy of personal constraint," *Medical Opinion and Review, 4*:190–217.

2. The Moral Evidence

The crucial ingredient in involuntary mental hospitalization is coercion. Since coercion is the exercise of power, it is always a moral and political act. Accordingly, regardless of its medical justification, commitment is primarily a moral and political phenomenon—just as, regardless of its anthropological and economic justifications, slavery was primarily a moral and political phenomenon.

Although psychiatric methods of coercion are indisputably useful for those who employ them, they are clearly not indispensable for dealing with the problems that so-called mental patients pose for those about them. If an individual threatens others by virtue of his beliefs or actions, he could be dealt with by methods other than "medical": if his conduct is ethically offensive, moral sanctions against him might be appropriate; if forbidden by law, legal sanctions might be appropriate. In my opinion, both informal, moral sanctions, such as social ostracism or divorce, and formal, judicial sanctions, such as fine and imprisonment, are more dignified and less injurious to the human spirit than the quasi-medical psychiatric sanction of involuntary mental hospitalization.[22]

3. The Historical Evidence

To be sure, confinement of so-called mentally ill persons does protect the community from certain problems. If it didn't, the arrangement would not have come into being and would not have persisted. However, the question we ought to ask is not *whether* commitment protects the community from "dangerous mental patients," but rather from precisely *what danger* it protects and by *what means?* In what way are prostitutes or vagrants dangerous in seventeenth-century Paris? Or married women in nineteenth-century Illinois?

It is significant, moreover, that there is hardly a prominent person who, during the past fifty years or so, has not been diagnosed by a psychiatrist as suffering from some type of "mental illness." Barry Goldwater was called a "paranoid schizophrenic";[23] Whittaker Chambers, a "psychopathic personality";[24] Woodrow Wilson, a "neurotic" frequently "very close to psychosis";[25] and Jesus, "a born degenerate" with a "fixed delusional system," and a "paranoid" with a "clinical picture (so typical) that it is hardly conceivable

[22] Szasz, T. S.: *Psychiatric Justice* (New York: Macmillan, 1965).

[23] "The Unconscious of a Conservative: A Special Issue on the Mind of Barry Goldwater." *Fact*, Sept.–Oct. 1964.

[24] Zeligs, M. A.: *Friendship and Fratricide: An Analysis of Whittaker Chambers and Alger Hiss* (New York: Viking, 1967).

[25] Freud, S., and Bullitt, W. C.: *Thomas Woodrow Wilson: A Psychological Study* (Boston: Houghton Mifflin, 1967).

that people can even question the accuracy of the diagnosis."[26] The list is endless.

Sometimes, psychiatrists declare the same person sane *and* insane, depending on the political dictates of their superiors and the social demand of the moment. Before his trial and execution, Adloph Eichmann was examined by several psychiatrists, all of whom declared him to be normal; after he was put to death, "medical evidence" of his insanity was released and widely circulated.

According to Hannah Arendt, "Half a dozen psychiatrists had certified him (Eichmann) as 'normal.'" One psychiatrist asserted, "his whole psychological outlook, his attitude toward his wife and children, mother and father, sisters and friends, was 'not only normal but most desirable.' . . . And the minister who regularly visited him in prison declared that Eichmann was "a man with very positive ideas."[27] After Eichmann was executed, Gideon Hausner, the Attorney General of Israel, who had prosecuted him, disclosed in an article in *The Saturday Evening Post* that psychiatrists diagnosed Eichmann as "'a man obsessed with a dangerous and insatiable urge to kill,' 'a perverted, sadistic personality.'"[28]

Whether or not men like those mentioned above are considered "dangerous" depends on the observer's religious beliefs, political convictions, and social situation. Furthermore, the "dangerousness" of such persons—whatever we may think of them—is not analogous to that of a person with tuberculosis or typhoid fever; nor would rendering such a person "non-dangerous" be comparable to rendering a patient with a contagious disease non-infectious.

In short, I hold—and I submit that the historical evidence bears me out—that people are committed to mental hospitals neither because they are "dangerous," nor because they are "mentally ill," but rather because they are society's scapegoats, whose persecution is justified by psychiatric propaganda and rhetoric.[29]

4. *The Literary Evidence*

No one contests that involuntary mental hospitalization of the so-called dangerously insane "protects" the community. Disagreement centers on the

[26] Quoted in Schweitzer, A.: *The Psychiatric Study of Jesus* (1913) transl. by Charles R. Joy (Boston: Beacon Press, 1956), pp. 37, 40–41.

[27] Arendt, H.: *Eichmann in Jerusalem: A Report on the Banality of Evil* (New York: Viking, 1963), p. 22.

[28] Ibid., pp. 22–23.

[29] For a full articulation and documentation of this thesis, see Szasz, T. S.: *The Manufacture of Madness: A Comparative Study of the Inquisition and the Mental Health Movement* (New York: Harper & Row, 1970).

nature of the threat facing society, and on the methods and legitimacy of the protection it employs. In this connection, we may recall that slavery, too, "protected" the community; it freed the slaveowners from manual labor. Commitment likewise shields the non-hospitalized members of society; first, from having to accommodate themselves to the annoying or idiosyncratic demands of certain members of the community who have not violated any criminal statutes; and, second, from having to prosecute, try, convict, and punish members of the community who have broken the law but who either might not be convicted in court, or, if they would be, might not be restrained as effectively or as long in prison as in a mental hospital. The literary evidence cited earlier fully supports this interpretation of the function of involuntary mental hospitalization.

IV

I have suggested that commitment constitutes a social arrangement whereby one part of society secures certain advantages for itself at the expense of another part. To do so, the oppressors must possess an ideology to justify their aims and actions; and they must be able to enlist the police power of the state to impose their will on the oppressed members. What makes such an arrangement a "crime against humanity"? It may be argued that the use of state power is legitimate when law-abiding citizens punish lawbreakers. What is the difference between this use of state power and its use in commitment?

In the first place, the difference between committing the "insane" and imprisoning the "criminal" is the same as that between the rule of man and the rule of law:[30] whereas the "insane" are subjected to the coercive controls of the state because persons more powerful than they have labeled them as "psychotic," "criminals" are subjected to such controls because they have violated legal rules applicable equally to all.

The second difference between these two proceedings lies in their professed aims. The principal purpose of imprisoning criminals is to protect the liberties of the law-abiding members of society.[31] Since the individual subject to commitment is not considered a threat to liberty in the same way as the accused criminal is (if he were, he would be prosecuted), his removal from society cannot be justified on the same grounds. Justification for commitment must thus rest on its therapeutic promise and potential: it will help restore the "patient" to "mental health." But if this can be accomplished only at the cost of robbing the

[30] Hayek, F. S.: *The Constitution of Liberty* (Chicago: University of Chicago Press, 1960), especially pp. 162–92.

[31] Mabbott, J. D.: "Punishment" (1939), in Olafson, F. A., ed., *Justice and Social Policy: A Collection of Essays* (Englewood Cliffs, N. J.: Prentice-Hall, 1961), pp. 39–54.

individual of liberty, "involuntary mental hospitalization" becomes only a verbal camouflage for what is, in effect, punishment. This "therapeutic" punishment differs, however, from traditional judicial punishment, in that the accused criminal enjoys a rich panoply of constitutional protections against false accusation and oppressive prosecution, whereas the accused mental patient is deprived of these protections.[32]

To support this view of involuntary mental hospitalization, and to cast it into historical perspective, I shall now briefly review the similarities between slavery and institutional psychiatry. (By the use of the term *institutional psychiatry* I refer generally to psychiatric interventions imposed on persons by others. Such interventions are characterized by the complete loss, on the part of the ostensible client or "patient," of control over his participation in his relationship with the expert. The paradigm "service" of institutional psychiatry is, of course, involuntary mental hospitalization.[33])

V

Suppose that a person wishes to study slavery. How would he go about doing so? First, he might study slaves. He would then find that such persons are generally brutish, poor, and uneducated, and he might accordingly conclude that slavery is their "natural" or appropriate social state. Such, indeed, have been the methods and conclusions of innumerable men throughout the ages.[34] Even the great Aristotle held that slaves were "naturally" inferior and were hence justly subdued. "From the hour of their birth," he asserted, "some are marked for subjection, others for rule."[35] This view is similar to the modern concept of "psychopathic criminality" and "schizophrenia" as genetically caused diseases.[36]

Another student, "biased" by contempt for the institution of slavery, might proceed differently. He would maintain that there can be no slave without a

[32] For documentation, see Szasz, T. S.: *Law, Liberty, and Psychiatry: An Inquiry into the Social Uses of Mental Health Practices* (New York: Macmillan, 1963); *Psychiatric Justice* (New York: Macmillan, 1965).

[33] For further discussion, and for a detailed exploration of the similarities between the Inquisition and institutional psychiatry, see Szasz, T. S., *The Manufacture of Madness, supra*, especially the Preface and Chap. 1–9.

[34] Davis, op. cit., passim.

[35] Ibid., p. 70.

[36] Stock, R. W.: "The XYY and the criminal," *The New York Times Magazine*, October 20, 1968, pp. 30–31, 90–104; Kallman, F. J.: "The Genetics of Mental Illness," in Arieti, S., ed., *American Handbook of Psychiatry* (New York: Basic Books, 1959), Vol. I, pp. 175–96.

master holding him in bondage; and he would accordingly consider slavery a type of human *relationship* and, more generally, a *social institution,* supported by custom, law, religion, and force. From this point of view, the study of masters is at least as relevant to the study of slavery as is the study of slaves.

The latter point of view is generally accepted today with regard to slavery, but not with regard to institutional psychiatry. "Mental illness" of the type found in psychiatric hospitals has been investigated for centuries, and continues to be investigated today, in much the same way as slaves had been studied in the ante-bellum South and before. Then, the "existence" of slaves was taken for granted; their biological and social characteristics were accordingly noted and classified. Today, the "existence" of "mental patients" is similarly taken for granted;[37] indeed, it is widely believed that their number is steadily increasing.[38] The psychiatrist's task is therefore to observe and classify the biological, psychological, and social characteristics of such patients.[39] This perspective is a manifestation, in part, of what I have called "the myth of mental illness,"[40] that is, of the notion that mental illnesses are similar to diseases of the body; and, in part, of the psychiatrist's intense need to deny the fundamental complementarity of his relationship to the involuntary mental patient. The same sort of complementarity obtains in all situations where one person or party assumes a superior or dominant role and ascribes an inferior or submissive role to another; for example, master and slave, accuser and accused, inquisitor and witch.

The fundamental parallel between master and slave on the one hand, and institutional psychiatrist and involuntarily hospitalized patient on the other, lies in this: in each instance, the former member of the pair *defines* the social role of the latter, and *casts* him in that role by force.

VI

Wherever there is slavery, there must be criteria for who may and who may not be enslaved. In ancient times, any people could be enslaved. Bondage was the usual consequence of military defeat. After the advent of Christianity, although the people of Europe continued to make war upon each other, they ceased enslaving prisoners who were Christians. According to Dwight Dumond,

[37] Caplan, G.: *Principles of Preventive Psychiatry* (New York: Basic Books, 1964).

[38] See, for example, Srole, L., Langer, T. S., Mitchell, S. T., Opler, M. K., and Rennie, T. A. C.: *Mental Health in the Metropolis: The Midtown Manhattan Study* (New York: McGraw-Hill, 1962).

[39] Noyes, A. P., and Kolb, L. C.: *Modern Clinical Psychiatry*, 5th ed. (Philadelphia: Saunders, 1958).

[40] Szasz, T. S., *The Myth of Mental Illness, supra.*

"the theory that a Christian could not be enslaved soon gained such wide endorsement as to be considered a point of international law."[41] By the time of the colonization of America, the peoples of the Western world considered only black men appropriate subjects for slave trade.

The criteria for distinguishing between those who may be incarcerated in mental hospitals and those who may not be are similar: poor and socially unimportant persons may be, the Very Important Persons my not be.[42] This rule is manifested in two ways: first, through our mental-hospital statistics, which show that the majority of institutionalized patients belong in the lowest socioeconomic classes;[43] second, through the rarity with which VIPs are committed.[44] Yet even sophisticated social scientists often misunderstand or misinterpret these correlations by attributing the low incidence of committed upper-class persons to a denial on their part, and on the part of those close to them, of the "medical fact" that "mental illness" can "strike" anyone.[45] To be sure, powerful people may feel anxious or depressed, or behave in an excited or paranoid manner; but that, of course, is not the point at all. This medical perspective, which defines all distressed and distressing behavior as mental illness—and which is now so widely accepted—only succeeds in confusing the observer's judgment of the quality of another person's behavior with the observer's power to cast that person in the role of involuntary patient. My argument here is limited to asserting that prominent and powerful persons are rarely cast into the role of involuntarily confined mental patient—and for obvious reasons: The degraded status of committed patient ill befits a powerful person. In fact, the two statuses are as mutually exclusive as those of master and slave.

VII

A basic assumption of American slavery was that the Negro was racially inferior to the Caucasian. "There is no malice toward the Negro in Ulrich Phillips' work," wrote Stanley Elkins about the author's book *American Negro Slavery*, a work sympathetic with the Southern position. "Phillips was deeply

[41] Dumond, D. L.: *Antislavery: The Crusade for Freedom* (Ann Arbor: University of Michigan Press, 1961), p. 4.

[42] Henderson, D., and Gillespie, R. D.: *A Textbook of Psychiatry*, 7th ed. (London: Oxford University Press, 1950), p. 684.

[43] Hollingshead, A. B., and Redlich, F. C.: *Social Class and Mental Illness* (New York: Wiley, 1958).

[44] See, for example, Rogow, op. cit.

[45] Ibid., pp. xxi, 44, 344–47.

fond of the Negroes as a people; it was just that he could not take them seriously as men and women; they were children."[46]

Similarly, the basic assumption of institutional psychiatry is that the mentally ill person is psychologically and socially inferior to the mentally healthy. He is like a child: he does not know what is in his best interests and therefore needs others to control and protect him.[47] Psychiatrists often care deeply for their involuntary patients, whom they consider—in contrast with the merely "neurotic" persons—"psychotic," which is to say, "very sick." Hence, such patients must be cared for as the "irresponsible children" they are considered to be.

The perspective of paternalism has played an exceedingly important part in justifying both slavery and involuntary mental hospitalization. Aristotle defined slavery as "an essentially domestic relationship"; in so doing, wrote Davis, he "endowed it with the sanction of paternal authority, and helped to establish a precedent that would govern discussions of political philosophers as late as the eighteenth century."[48] The relationship between psychiatrists and mental patients has been and continues to be viewed in the same way. "If a man brings his daughter to me from California," declares Braceland, "because she is in manifest danger of falling into vice or in some way disgracing herself, he doesn't expect me to let her loose in my hometown for that same thing to happen."[49] Indeed, almost any article or book dealing with the "care" of involuntary mental patients may be cited to illustrate the contention that physicians fall back on paternalism to justify their coercive control over the uncooperative patient. "Certain cases" (not individuals!)—writes Solomon in an article on suicide—"... must be considered irresponsible, not only with respect to violent impulses, but also in all medical matters." In this class, which he labels "The Irresponsible," he places "Children," "The Mentally Retarded," "The Psychotic," and "The Severely or Terminally Ill." Solomon's conclusion is that "repugnant though it may be, he (the physician) may have to act against the patient's wishes in order to protect the patient's life and that of others."[50] The fact that, as in the case of slavery, the physician needs the police power of the state to maintain his

[46] Elkins, S. M.: *Slavery: A Problem in American Institutional and Intellectual Life* (1959) (New York: Universal Library, 1963), p. 10.

[47] See, for example, Linn, L.: *A Handbook of Hospital Psychiatry* (New York: International Universities Press, 1955), pp. 420–22; Braceland, F. J.: Statement in *Constitutional Rights of the Mentally Ill* (Washington, D. C.: U. S. Government Printing Office, 1961), pp. 63–74; Rankin, R. S., and Dallmayr, W. B.: "Rights of Patients in Mental Hospitals," in *Constitutional Rights of the Mentally Ill, supra*, pp. 329–70.

[48] Davis, op. cit., p. 69.

[49] Braceland, op. cit., p. 71.

[50] Solomon, P.: "The burden of responsibility in suicide." *JAMA, 199*:321–24 (Jan. 30), 1967.

relationship with his involuntary patient does not alter this self-serving image of institutional psychiatry.

Paternalism is the crucial explanation for the stubborn contradiction and conflict about whether the practices employed by slaveholders and institutional psychiatrists are "therapeutic" or "noxious." Masters and psychiatrists profess their benevolence; their slaves and involuntary patients protest against their malevolence. As Seymour Halleck puts it: "the psychiatrist experiences himself as a helping person, but his patient may see him as a jailer. Both views are partially correct."[51] Not so. Both views are completely correct. Each is a proposition about a different subject: the former, about the psychiatrist's self-image; the latter, about the involuntary mental patient's image of his captor. In *Ward 7,* Valeriy Tarsis presents the following dialogue between his protagonist-patient and the mental-hospital physician: "This is the position. I don't regard you as a doctor. You call this a hospital. I call it a prison. . . . So now, let's get everything straight. I am your prisoner, you are my jailer, and there isn't going to be any nonsense about my health . . . or treatment."[52]

This is the characteristic dialogue of oppression and liberation. The ruler looks in the mirror and sees a liberator; the ruled looks at the ruler and sees a tyrant. If the physician has the power to incarcerate the patient and uses it, their relationship will inevitably fit into this mold. If one cannot ask the subject whether he likes being enslaved or committed, whipped or electroshocked—because he is not a fit judge of his own "best interests"—then one is left with the contending opinions of the practitioners and their critics. The practitioners insist that their coercive measures are beneficial; the critics, that they are harmful.

The defenders of slavery thus claimed that the Negro "is happier . . . as a slave, than he could be as a free man; this is the result of the peculiarities of his character";[53] that "it was actually an act of liberation to remove Negroes from their harsh world of sin and dark superstition";[54] and that "Negroes were better off in a Christian land, even as slaves, than living like beasts in Africa."[55]

Similarly, the defenders of involuntary mental hospitalization claim that the mental patient is healthier—the twentieth-century synonym for the nineteenth-century term *happier*—as a psychiatric prisoner than he would be as a free citizen; that "the basic purpose [of commitment] is to make sure that sick human

[51] Halleck, S. L.: *Psychiatry and the Dilemmas of Crime* (New York: Harper & Row, 1967), p. 230.

[52] Tarsis, V.: *Ward 7: An Autobiographical Novel* (London and Glasgow: Collins and Harvill, 1965), p. 62.

[53] Elkins, op. cit., p. 190.

[54] Davis, op. cit., p. 186.

[55] Ibid., p. 190.

beings get the care that is appropriate to their needs . . .";[56] and that "it is a feature of some illnesses that people do not have insight into the fact that they are sick. In short, sometimes it is necessary to protect them [the mentally ill] for a while from themselves. . . ."[57] It requires no great feat of imagination to see how comforting—indeed, how absolutely necessary—these views are to the advocates of slavery and involuntary mental hospitalization, even when they are contradicted by facts.

For example, although it was held that "a merrier being does not exist on the face of the globe than the Negro slave of the United States,"[58] there was an ever-lurking fear of Negro violence and revolt. As Elkins put it, "the failure of any free workers to present themselves for enslavement can serve as one test of how much the analysis of the 'happy slave' may have added to Americans' understanding of themselves."[59]

The same views and the same inconsistencies apply to involuntary psychiatric hospitalization. Defenders of this system maintain that committed patients are better off in hospitals, where they are contented and harmless; "most patients," declares Guttmacher, "when they get in a (mental) hospital are quite content to be there. . . ."[60] At the same time, such patients are feared for their potential violence, their escapes from captivity occasion intense manhunts, and their crimes are prominently featured in the newspapers. Moreover, as with slavery, the failure of citizens to present themselves for involuntary psychiatric hospitalization can serve as a test of how much the currently popular analysis of mental health problems has added to Americans' understanding of themselves.

The social necessity, and hence the basic value, of involuntary mental hospitalization, at least for some people, is not seriously questioned today. There is massive consensus in the United States that, properly used, such hospitalization is a good thing. It is thus possible to debate *who* should be hospitalized, or *how*, or for *how long*—but not whether *anyone should* be. I submit, however, that just as it is improper to enslave anyone—whether he is black or white, Moslem or Christian—so it is improper to hospitalize anyone without his consent—whether he is depressed or paranoid, hysterical or schizophrenic.

Our unwillingness to look at this problem searchingly may be compared to the unwillingness of the South to look at slavery. "(A) democratic people," wrote Elkins, "no longer 'reasons' with itself when it is all of the same mind.

[56] Ewalt, J.: Statement, in *Constitutional Rights of the Mentally Ill, supra*, pp. 74–89, p. 75.
[57] Braceland, op. cit., p. 64.
[58] Elkins, op. cit., p. 216.
[59] Ibid.
[60] Guttmacher, M.: Statement, in *Constitutional Rights of the Mentally Ill, supra*, pp. 143–60, p. 156.

Men will then only warn and exhort each other, that their solidarity may be yet more perfect. The South's intellectuals, after the 1830s, did really little more than this. And when the enemy's reality disappears, when his concreteness recedes, then intellect itself, with nothing more to resist it and give it resonance, merges with the mass and stultifies, and shadows become monsters."[61]

Our growing preoccupation with the menace of mental illness may be a manifestation of just such a process–in which "concreteness recedes . . . and shadows become monsters." A democratic nation, as we have been warned by Tocqueville, is especially vulnerable to the hazards of a surfeit of agreement; "The authority of a king is physical, and controls the actions of men without subduing their will. But the majority possesses a power that is physical and moral at the same time, which acts upon the will as much as upon the actions, and represses not only all contests, but all controversy."[62]

VIII

There are essential similarities in relationships between masters and subjects–no matter whether plantation owners and Negro slaves or institutional psychiatrists and committed mental patients.

To maintain a relationship of personal or class superiority, it is necessary, as a rule, that the oppressor keep the oppressed uninformed, especially about matters pertinent to their relationship. In America the history of the systematic efforts by the whites to keep the Negro ignorant is well known. A dramatic example is the law passed in 1824 by the Virginia Assembly that provided a $50 fine and two months' imprisonment for teaching *free* Negroes to read and write.[63] Nor was the situation very different in the North. In January 1833 Prudence Crandall admitted to her private school in Canterbury, Connecticut, a young lady of seventeen, the daughter of a highly respected Negro family. Miss Crandall was thereupon ostracized and persecuted by her neighbors: "They dumped a load of manure in her well. They refused to sell her supplies and threatened her father and brother with mob violence, fines, and imprisonment if they continued to bring her food from their nearby farm. They piled refuse from a slaughter house on her front porch."[64] She was also accused of, and tried for, breaking a law that forbade the harboring, boarding, or instruction in any

[61] Elkins, op. cit., p. 222.
[62] Tocqueville, A. de: *Democracy in America* (1835–40) (New York: Vintage Books, 1945), Vol. I, p. 273.
[63] Dumond, op. cit., p. 11.
[64] Ibid., p. 211.

manner of any person of color, and was convicted. Finally, her school was set on fire.

A similar effort to educationally degrade and psychologically impoverish their charges characterizes the acts of the managers of madhouses. In most prisons in the United States, it is possible for a convict to obtain a high-school diploma, to learn a trade, to become an amateur lawyer, or to write a book. None of these things is possible in a mental hospital. The principal requirement for an inmate of such an institution is to accept the psychiatric ideology of his "illness" and the things he must do to "recover" from it. The committed patient must thus accept the view that he is "sick" and that his captors are "well"; That his own view of himself is false and that of his captors true; and that to effect any change in his social situation he must relinquish his "sick" views and adopt the "healthy" views of those who have power over him.[65] By accepting himself as "sick," and his institutional environment and the various manipulations imposed on him by the staff as "treatment," the mental patient is compelled to authenticate the psychiatrist's role as that of a benevolent physician curing mental illness. The mental patient who maintains the forbidden image of reality that the institutional psychiatrist is a jailer is considered paranoid. Moreover, since most patients—as do oppressed people generally—sooner or later accept the ideas imposed on them by their superiors, hospital psychiatrists are constantly immersed in an environment where their identity as "doctors" is affirmed. The moral superiority of white men over black was similarly authenticated and affirmed through the association between slaveowners and slaves.

In both situations, the oppressor first subjugates his adversary and then cites his oppressed status as proof of his inferiority. Once this process is set in motion, it develops its own momentum and psychological logic.

Looking at the relationship, the oppressor will see his superiority and hence his well-deserved dominance, and the oppressed will see his inferiority and hence his well-deserved submission. In race relations in the United States, we continue to reap the bitter results of this philosophy, while in psychiatry we are even now sowing the seeds of this poisonous fruit whose eventual harvest may be equally bitter and long.

Convicts are entitled to fight for their "legal rights," but not involuntary mental patients. Like slaves, such patients have no rights except those granted them by their medical masters. According to Benjamin Apfelberg, Clinical Professor of Psychiatry and Medical Director of the Law-Psychiatry Project at New York University, "Our students come to realize that by fighting for a patient's *legal* rights they may actually be doing him a great disservice. They

65 Goffman, E.: *Asylums: Essays on the Social Situation of Mental Patients and Other Inmates* (Garden City, N. Y.: Doubleday Anchor, 1961).

learn that there is such a thing as a person's *medical* rights, the right to get treatment, to become well."[66]

The "medical right" to which Apfelberg refers is a euphemism for the *obligation* to remain confined in a mental institution, not the opportunity to *choose* between hospitalization and no hospitalization. But calling involuntary mental hospitalization a "medical right" is like calling involuntary servitude in ante-bellum Georgia a "right to work."

Oppression and degradation are unpleasant to behold and are, therefore, frequently disguised or concealed. One method for doing so is to segregate—in special areas, as in camps or "hospitals"—the degraded human beings. Another is to conceal the social realities behind the fictional facade of what we call, after Wittgenstein,[67] "language games." While psychiatric language games may seem fanciful, the psychiatric idiom is actually only a dialect of the common language of oppressors.[68] Thus slaveholders called the slaves "livestock," mothers "breeders," their children "increase," and gave the term "drivers" to the men set over them at work.[69] The defenders of psychiatric imprisonment call their institutions "hospitals," the inmates "patients," and the keepers "doctors"; they refer to the sentence as "treatment," and to the deprivation of liberty as "protection of the patient's best interests."

In both cases, the semantic deceptions are supplemented by appeals to tradition, to morality, and to social necessity. The proslavery forces in America argued that the abolitionists were wrong because "they were seeking to overthrow an ancient institution, one which was recognized by the Scriptures, recognized by the Constitution, and imbedded in the structure of southern society,"[70] Thus, an editorial in the Washington *Telegraph* in 1837 asserted, "As a man, a Christian, and a citizen, we believe that slavery is right; that the condition of the slave, as it now exists in slaveholding states, is the best existing organization of civil society";[71] while another proslavery author, writing in 1862, defended the institution on mainly religious grounds: "Slavery, authorized by God, permitted by Jesus Christ, sanctioned by the apostles, maintained by good men of all ages, is still existing in a portion of our beloved country."[72]

[66] Quoted in "Attorneys-at-Psychiatry," *Smith, Kline & French Psychiatric Reporter*, July–August 1965, p. 23.

[67] See Wittgenstein, L.: *Philosophical Investigations* (Oxford: Blackwell, 1953); and Hartnack, J.: *Wittgenstein and Modern Philosophy* (Garden City, N. Y.: Doubleday Anchor, 1965).

[68] On the language game of Nazi anti-Semitism, see Arendt, op. cit., especially pp. 80, 96, 141.

[69] Dumond, op. cit., p. 251.

[70] Ibid., p. 233.

[71] Elkins, op. cit., p. 36.

[72] Ibid.

One has only to scan present-day psychiatric journals, popular magazines, or daily newspapers to find involuntary mental hospitalization similarly extolled and defended.

The contemporary reader may find it difficult to believe how unquestioningly slavery was accepted as a natural and beneficial social arrangement. Even as great a liberal thinker as John Locke did not advocate its abolition.[73] Moreover, protests against the slave trade would have provoked the hostility of powerful religious and economic interests. Opposition to it, as Davis observed, would therefore have required "considerable independence of mind, since the Portuguese slave posts were closely connected with missionary establishments and criticism of the African slave trade might challenge the very ideal of spreading the faith."[74]

Indeed, the would-be critic or opponent of slavery would have found himself at odds with all the tradition and wisdom of Western civilization. "(O)ne could not lightly challenge," wrote Davis, "an institution approved not only by the Fathers and canons of the Church, but by the most illustrious writers of antiquity. . . . (T)he revival of classical learning, which may have helped to liberate the mind of Europe from bondage to ignorance and superstition, only reinforced the traditional justification for human slavery. . . . (H)ow could an institution supported by so many authorities and sanctioned by the general custom of nations be intrinsically unjust or repugnant to natural reason?"[75]

In Western nations and the Soviet Bloc alike, there are thus two contradictory views on commitment. According to the one, involuntary mental hospitalization is an indispensable method of medical healing and a humane type of social control; according to the other, it is a contemptible abuse of the medical relationship and a type of imprisonment without trial. We adopt the former view and consider commitment "proper" if we use it on victims of our choosing whom we despise; we adopt the latter view and consider commitment "improper" if our enemies use it on victims of their choosing whom we esteem.

IX

The change in perspective—from seeing slavery occasioned by the "inferiority" of the Negro and commitment by the "insanity" of the patient, to seeing each occasioned by the interplay of, and especially the power relation between, the participants—has far-reaching practical implications. In the case of slavery, it meant not only that the slaves had an obligation to revolt and emancipate

[73] Davis, op. cit., p. 121.
[74] Ibid., p. 187.
[75] Ibid., pp. 107, 115.

themselves, but also that the masters had an even greater obligation to renounce their roles as slaveholders. Naturally, a slaveholder with such ideas felt compelled to set his slaves free, at whatever cost to himself. This is precisely what some slaveowners did. Their action had profound consequences in a social system based on slavery.

For the individual slaveholder who set his slaves free, the act led invariably to his expulsion from the community—through economic pressure or personal harassment or both. Such persons usually emigrated to the North. For the nation as a whole, these acts and the abolitionist sentiments behind them symbolized a fundamental moral rift between those who regarded Negroes as objects or slaves, and those who regarded them as persons or citizens. The former could persist in regarding the slave as existing in nature; whereas the latter could not deny his own moral responsibility for creating man in the image, not of God, but of the slave-animal.

The implications of this perspective for institutional psychiatry are equally clear. A psychiatrist who accepts as his "patient" a person who does not wish to be his patient, defines him as a "mentally ill" person, then incarcerates him in an institution, bars his escape from the institution and from the role of mental patient, and proceeds to "treat" him against his will—such a psychiatrist, I maintain, creates "mental illness" and "mental patients." He does so in exactly the same way as the white man who sailed for Africa, captured the Negro, brought him to America in shackles, and then sold him as if he were an animal, created slavery and slaves.

The parallel between slavery and institutional psychiatry may be carried one step further: Denunciation of slavery and the renouncing of slaveholding by some slaveowners led to certain social problems, such as Negro unemployment, the importation of cheap European labor, and a gradual splitting of the country into pro- and anti-slavery factions. Similarly, criticisms of involuntary mental hospitalization and the renouncing by some psychiatrists of relationships with involuntary mental patients have led to professional problems in the past, and are likely to do so again in the future. Psychiatrists restricting their work to psychoanalysis and psychotherapy have been accused of not being "real doctors"—as if depriving a person of his liberty required medical skills; of "shirking their responsibilities" to their colleagues and to society by accepting only the "easier cases" and refusing to treat the "seriously mentally ill" patient—as if avoiding treating persons who do not want to be treated were itself a kind of malpractice; and of undermining the profession of psychiatry—as if practicing self-control and eschewing violence were newly discovered forms of immorality.[76]

[76] See, for example, Davidson, H. A.: "The image of the psychiatrist." *Amer. J. Psychiatry, 121*:329–33 (Oct.), 1964; Glaser, F. G.: "The dichotomy game: A further consideration of the writings of Dr. Thomas Szasz." *Amer. J. Psychiatry, 121*:1069–74 (May), 1965.

X

The psychiatric profession has, of course, a huge stake, both existential and economic, in being socially authorized to rule over mental patients, just as the slaveowning classes did in ruling over slaves. In contemporary psychiatry, indeed, the expert gains superiority not only over members of a specific class of victims, but over nearly the whole of the population, whom he may "psychiatrically evaluate."[77]

The economic similarities between chattel slavery and institutional psychiatry are equally evident: The economic strength of the slaveowner lay in the Negro slaves he owned. The economic strength of the institutional psychiatrist lies, similarly, in his involuntary mental patients, who are not free to move about, marry, divorce, or make contracts, but are, instead, under the control of the hospital director. As the plantation owner's income and power rose with the amount of land and number of slaves he owned, so the income and power of the psychiatric bureaucrat rise with the size of the institutional system he controls and the number of patients he commands. Moreover, just as the slaveholder could use the police power of the state to help him recruit and maintain his slave labor force, so can the institutional psychiatrist rely on the state to help him recruit and maintain a population of hospital inmates.

Finally, since the state and federal governments have a vast economic stake in the operation of psychiatric hospitals and clinics, the interests of the state and of institutional psychiatry tend to be the same. Formerly, the state and federal governments had a vast economic stake in the operation of plantations worked by slaves, and hence the interests of the state and of the slaveowning classes tended to be the same. The wholly predictable consequence of this kind of arrangment is that just as the coalition of chattel slavery and the state created a powerful vested interest, so does the coalition of institutional psychiatry and the state.[78] Moreover, as long as the oppressive institution has the unqualified support of the state, it is invincible. On the other hand, since there can be no oppression without power, once such an institution loses the support of the state, it rapidly disintegrates.

If this argument is valid, pressing the view that psychiatrists now create involuntary mental patients just as slaveholders used to create slaves is likely to lead to a cleavage in the psychiatric profession, and perhaps in society generally, between those who condone and support the relationship between psychiatrist and involuntary mental patient, and those who condemn and oppose it.

It is not clear whether, or on what terms, these two psychiatric factions could coexist. The practices of coercive psychiatry and of paternalistic psychia-

[77] See Menninger, W.: *A Psychiatrist for a Troubled World* (New York: Viking, 1967).
[78] See Davis, op. cit., p. 193.

trists do not, in themselves, threaten the practices of non-coercive psychiatry and of contracting psychiatrists. Economic relations based on slavery coexisted over long periods with relations based on contract. But the moral conflict poses a more difficult problem. For just as the abolitionists tended to undermine the social justifications of slavery and the psychological bonds of the slave, so the abolitionists of psychiatric slavery tend to undermine the justifications of commitment and the psychological bonds of the committed patient.

Ultimately, the forces of society will probably be enlisted on one side or the other. If so, we may, on the one hand, be ushering in the abolition of involuntary mental hospitalization and treatment; on the other, we may be witnessing the fruitless struggles of an individualism bereft of moral support against a collectivism proffered as medical treatment.[79]

XI

We know that man's domination over his fellow man is as old as history; and we may safely assume that it is traceable to prehistoric times and to pre-human ancestors. Perennially, men have oppressed women; white men, colored men; Christians, Jews. However, in recent decades, traditional reasons and justifications for discrimination among men—on the grounds of national, racial, or religious criteria—have lost much of their plausibility and appeal. What justification is there now for man's age-old desire to dominate and control his fellow man? Modern liberalism—in reality, a type of statism—allied with scientism, has met the need for a fresh defense of oppression and has supplied a new battle cry: Health!

In this therapeutic-meliorist view of society, the ill form a special class of "victims" who must, both for their own good and for the interests of the community, be "helped"—coercively and against their will, if necessary—by the healthy, and especially by physicians who are "scientifically" qualified to be their masters. This perspective developed first and has advanced farthest in psychiatry, where the oppression of "insane patients" by "sane physicians" is by now a social custom hallowed by medical and legal tradition. At present, the medical profession as a whole seems to be emulating this model. In the Therapeutic State toward which we appear to be moving, the principal requirement for the position of Big Brother may be an M.D. degree.

[79] Szasz, T. S.: "Whither psychiatry?" *Medical Opinion and Review, 4*:218–45.

PSYCHIATRISTS AND THE ADVERSARY PROCESS

DAVID L. BAZELON

In our society part of the task of scrutinizing the decisional process involving experts in a great many disciplines has fallen to the judiciary. With public issues increasingly conditioned by scientific discovery and technological change, courts often confront technical questions with legal and moral implications for society. As a judge on the United States Court of Appeals for the District of Columbia Circuit for the past 25 years, I have been exposed to almost every sensitive scientific and medical question that has legal, moral, and ethical implications for our society. Arguments before our court, where the Federal Government is often a party, have ranged over the spectrum from abortion and blood transfusions to the underground-nuclear-explosion experiment in the Aleutian Islands and the safety of nuclear reactors. It is my duty to approach these questions not as a surgeon, a physicist, an ecologist, or any other technical expert does but as one charged with monitoring the decisional process. In scrutinizing the decision-making process of experts I have seen our familiar judicial procedures attempt to bring the most arcane sciences and technologies under public surveillance.

Psychiatry, I suppose, is the ultimate wizardry. My experience has shown that in no case is it more difficult to elicit productive and reliable expert testimony than in cases that call on the knowledge and practice of psychiatry. Such cases turn up in every jurisdiction. They raise fundamental questions: Who can be morally convicted of a crime? Who can be ordered into a hospital for compulsory treatment, and for how long? What kinds of treatment can be imposed without the consent of the patient? These questions engage the overriding question of the balance of power between the state and the individual. The effort by the courts to strike that balance requires the knowledge and expertise of the experts in the behavioral sciences, particularly psychiatry.

The discipline of psychistry has direct relevance to cases involving human behavior. One might hope that psychistrists would open up their reservoirs of knowledge in the courtroom. Unfortunately in my experience they try to limit then testimony to conclusory statements couched in psychiatric terminology.

Thereafter they take shelter in a defensive resistance to questions about the facts that are or ought to be in their possession. They thus refuse to submit their opinions to the scrutiny that the adversary process demands.

Psychiatrists are not alone, of course, in their failure to comprehend the nature and the importance of the adversary process. A word of explanation about what I mean when I use the term is in order here. The adversary process is the central feature of the system of legal institutions and procedures set up by our society to resolve controversies that arise between contending interests, values, and ideologies. The adversity—to use the word in its old dictionary meaning—is supplied not by the process but by the parties to the conflict; the adversary process is merely the decisional mechanism for resolving their conflict. Decisions must be reached even in the absence of any source of perfect information or wisdom. We therefore arrange an orderly contest of the parties in the courtroom, in which adversary roles are assigned to reflect the reality of the underlying dispute. Those of us who are engaged in conducting these proceedings have an awed awareness of the risk of arriving at an imperfect decision often with enormous consequences for the individual and for our society.

Cases always present conflicts over both facts and values since they arise on petitions for redress of grievances. To find facts we rely on an exhaustive inquiry. Parties and counsel must make the best case for themselves; they must check and correct the material offered by their adversary. Specific rules make the system at once skeptical and objective. These rules of evidence presuppose that men are biased and that their testimony is invariably shaped by their background, personality, interests, and values. Cross-examination challenges witnesses' veracity, accuracy, and bias. The inquiry is conducted in the presence of fact finders—the jury—chosen by measures that it is hoped will ensure their impartiality. Counsel are expected to be sensitive to their own conflicts of interest and to take care they do not serve more than one master.

We might be able to develop better rules to bring out facts in the courtroom and to dispel the excesses in the proceedings that excite hostile charges from those who are subjected to rigorous cross-examination. With full awareness of its weaknesses as well as its strengths, however, we still rely on the adversary process to uncover as many of the relevant facts as possible.

The law must also reconcile competing values that seem irreconcilable. In this task it does not seek final solutions; it recognizes the ongoing nature of deep-rooted conflicts. A judge reviews and develops criteria for resolving each case as it comes before him. The criteria are made known to the public in written opinions in which the competing values are ventilated. The court's decisions are never fixed and frozen; they are altered in response to new information, new understanding and new public demands. The law of itself does not provide wisdom. It offers a method for seeking wisdom.

In this system the expert witness has a special role. He is the only witness who is allowed to testify to a conclusion ("The defendant was intoxicated") as well as to the facts ("The defendant had two drinks last night"). Precisely because the expert testifies to a conclusion, and to one that almost inevitably favors one side, he must be open to cross-examination by the other side on the facts and premises on which he rests his conclusion. To the court and to the great aim of the proceeding he owes all the insight and information that his knowledge and training can offer.

My first exposure to psychiatry inside the courtroom concerned the expert opinions that psychiatrists were asked to render in criminal trials. In the early 1950s psychiatry and the law were deadlocked on the same issue of criminal responsibility, the so-called insanity defense. The traditional legal test for insanity, in use in virtually every American jurisdiction, was the M'Naghten test. According to this holding of a British court in 1843, the accused could not be found morally responsible if it was shown that he was suffering such "a defect of reason, from disease of the mind, as not to know the nature and the quality of the act he was doing, or, if he did know it, that he did not know he was doing what was wrong."

The M'Naghten rule thus betrayed its origins in a period when the dominant perception of human behavior was that people, as rational beings, made free choices informed by conscious consideration. The psychiatric profession was outspokenly critical of this model of the human psyche. It ignored, they said, modern dynamic understanding of man as an integrated personality, manifesting nonrational and irrational as well as rational, compulsive as well as volitional, behavior. The M'Naghten rule recognized only one aspect of that personality—cognitive reasoning—as the determinant of conduct.

Psychiatrists declared that, if the law would let them, they could give a more adequate account of realities. They complained that the M'Naghten test forced the physician to testify on whether or not the accused knew right from wrong and hence to decide the ultimate issue of moral responsibility, which should be left to the jury. Some urged their colleagues to refuse to speak to the question of "right or wrong, by virtue of reason alone." If psychiatrists were to be qualified as experts in criminal cases, they should be allowed to address the issue of responsibility in terms appropriate to their medical discipline.

Now, the law recognizes that the question of guilt or innocence is essentially a moral one. I believe the morality of a person's actions cannot be determined solely by abstract philosophical principles without regard to the facts that condition human behavior in the real world. To obtain these facts I wrote the opinion in the *Durham* case, in 1954, in which our court formulated a new test of criminal responsibility. *Durham* held that an accused man is not criminally responsible if it is shown that his unlawful act was the product of a mental disease or defect. *Durham* was not based on any notion that psychiatrists know everything there is to know about behavior. Its purpose was to bring into the courtroom the knowledge they do have and to restore to the jury its traditional function of applying "our inherited ideas of moral responsibility" to those accused of crime.

The response of the psychiatric profession to the new rule was enthusiastic. Here was the chance they had asked for to bring their expertise into the courtroom. Karl Menninger described the decision as "more revolutionary" than the *Brown* decision, also given in 1954, that outlawed racial segregation in the public schools. If the *Durham* "revolution" comes with the same "deliberate speed" as the desegregation ordered by the United States Supreme Court in *Brown,* however, only our great-grandchildren will be able to validate the Menninger appraisal.

Gregory Zilboorg had this to say about *Durham:* "[The *Durham* rule] might require of the psychiatrist that he study offenders and examine them clinically with much greater care than has often been the custom. . . . But at least . . . psychiatry is permitted to take the witness stand with all the dignity, medical and professional, which is due it. . . . The old unpleasantness of the 'adversary proceeding' in which psychiatric expert testimony has so often been engulfed . . . is bound to disappear."

Zilboorg's appraisal can be evaluated now. He was wrong. Psychiatrists continued to limit their testimony to conclusions–applying conclusory labels to the defendant–without explaining the origin, development, or manifestation of a disease in terms comprehensible to the jury. They began to wage a war of words in the courtroom, arguing about whether a defendant had a "personality defect," a "personality problem," a "personality disorder," a "disease," an "illness," or simply a "type of personality." Even more disturbing, they began to speak conclusively to the question of whether the criminal act was the "product" of mental disorder. The word "product" in their testimony conveyed no more clinical meaning than "right" or "wrong."

From St. Elizabeths, the Federal Government's well-known mental hospital in the District of Columbia, from which many cases come before our court, there came this insight into the response of the profession to the *Durham* test.

"Our psychiatric staff became alarmed that the courts would equate personality disorders with the psychoses and because of this anxiety, we erroneously–although in good faith–decided to add the words 'without mental disorder' in parenthesis immediately following the diagnosis . . . of personality disorder."

What psychiatrists have not understood is that conclusory labels are no substitute in judicial proceedings for facts derived from disciplined investigation. Labeling a person "schizophrenic" does not make him so! Although the law must settle for an "educated guess," that guess is only as good as the investigation, the facts, and the reasoning that underpin it.

The sterility of the profession's response to *Durham,* I now conclude, was due to the fact that its observance was bound to make the psychiatrist's task in the courtroom much more demanding than before. The late Winfred Overholser, superintendent of St. Elizabeths, once told me that the breadth of information I envisioned being placed before the jury would require from 50 to 100 man-hours of interviewing and investigation; he declared that a public hospital simply could not afford it. If that were the case, I replied, psychiatrists should frankly explain on the witness stand that their opinions are thus qualified by lack of time and resources. It was no service to the administration of justice for them to create the false impression that they had learned substantially all that could be known about someone on the basis of study they knew was inadequate.

The jury is equally entitled to know the differences of opinion and outright conflicts involved in psychiatric diagnosis. Consider this case reported by a leading forensic psychiatrist:

"A 17-year-old boy, confined as a patient in a state hospital, strangled another patient. He was sent to another hospital for medical-legal observation. The diagnostic choices lay between schizophrenia and schizoid personality disorder. Which diagnosis was made would naturally have great bearing on his criminal responsibility. When the case was presented to the staff conference, there was a great division of opinion. . . . After repeated medical conferences over a period of a year, it was agreed that further observation was futile, and that a diagnosis would have to be agreed on. The vote was five to four in favor of psychopathy [that is, personality disorder]. A report was then sent to the court stating that this diagnosis had been made, following a year's observation, and that he was now sane and could stand trial. The wording of the report clearly implied that the diagnosis was definitely established and that there was full agreement."

In the words of the psychiatrist reporting this incident: "What we have here, of course, is nothing else than the familiar star-chamber proceeding. The hospital staff usurped the function of the jury . . . and the court was then deprived of the full evidence, the conflict of medical opinion."

Attempts by my court to obtain records or tapes of just such clinical conferences have consistently been opposed and thwarted by the psychiatric staff at St. Elizabeths. In asking the profession to open up its opinions and decisions, we were asking no more than that psychiatric expertise should submit to the process by which the shortcomings of all opinion evidence are tested. The status of court-appointed "impartial" expert is not open to any profession. The potential for bias, distortion, and deviation from the truth is unavoidable. Like any other man, a physician acquires an emotional identification with an opinion that comes down on one side of a conflict; he has an inescapable, prideful conviction in the accuracy of his own findings. These realities belie the myth that a medical expert can speak from above or outside the legal adversary system.

In the end, after 18 years, I favored the abandonment of the *Durham* rule because in practice it had failed to take the issue of criminal responsibility away from the experts. Psychiatrists continued to testify to the naked conclusion instead of providing information about the accused so that the jury could render the ultimate moral judgment about blameworthiness. *Durham* had secured little improvement over M'Naghten.

In 1972, in the *Brawner* case, our court unanimously set aside the *Durham* rule. Essentially we replaced it with the insanity test that had been proposed by the American Law Institute:

"A person is not responsible for criminal conduct if at the time of such conduct as a result of mental disease or defect he lacks substantial capacity either to appreciate the wrongfulness of his conduct or to conform his conduct to the requirements of the law."

The *Brawner* formulation was designed in large part to end the expert's dominance over the question of moral responsibility. Psychiatrists will nonetheless still be able to take away the jury's function by presenting conclusory testimony. Thus they will testify that a defendant lacked capacity "as a result," just as under *Durham* they testified to whether the act was "the product" of a mental disorder.

Although no phrase will magically solve the problem of expert dominance, my own separate opinion in *Brawner* suggested the jury be instructed that a defendant is not responsible "if at the time of his unlawful conduct his mental or emotional processes or behavior controls were impaired to such an extent that he cannot justly be held responsible." This approach envisions that the jury will be provided with a broad range of information about the accused from a variety of sources including but not necessarily limited to psychiatrists. Other disciplines with special skills and knowledge in the field of human behavior would not be precluded from the opportunity to show the relevance of their data in the courtroom. Moreover, experts will be less likely to address the ultimate issue: whether the accused can be "justly held responsible." Even the promise of this approach, however, will be broken unless means are provided for the defendant, who is virtually always indigent, to obtain and present the required broad spectrum of information.

The form of words by which we call the insantity defense is not crucial, but the provision to the jury of all arguably relevant information about the accused's freedom of choice is. As I said in my opinion in *Brawner*, "while [we] generals are designing inspiring new insignia for the standard, the battle is being lost in the trenches."

In the *Jenkins* case, in 1962, spokesmen for the psychiatric profession betrayed a further misunderstanding of the role of the psychiatric witness that must have been widely shared among their colleagues. The trial court in this case had excluded the testimony of highly qualified and certified clinical psychologists, on the ground that "a psychologist is not competent to give a medical opinion." In the appeal to our court the American Psychiatric Association supported the lower court's decision. It asserted that in medical problems medical opinion can be the only guide. It chose to overlook the fact that the problem of criminal responsibility is not the exclusive terrain of psychiatry. I wrote the opinion for our court rejecting such guild mentality.

The "right to treatment" cases, in the 1960s, brought the issues under consideration here into sharper focus. In the *Rouse* case, in 1966, a patient confined without his consent sued for release or for adequate treatment of his disability. For our court I wrote that the plaintiff had a right to treatment, founded on a statute passed by Congress for the District of Columbia. This placed a more difficult question before us: How to establish whether this theoretical right was violated in actual practice. We held that adequacy of treatment is reviewable in court. In view of the inherent lack of certainty in psychiatric (as in other scientific) decisions, we held that the question of adequacy must be weighed on the basis of the "state of the art." We imposed no artificial criteria of success or failure, nor did we suggest we would overturn informed medical judgments. Independent experts should be enlisted, we said, to establish the parameters of acceptable treatment, and we urged that the American Psychiatric Association, which had published standards of medical care, be consulted.

The American Psychiatric Association responded to *Rouse* with an adamant statement of what I must call professional mystique: "The definition of treatment and the appraisal of its adequacy are matters for medical determination." This declaration ignored the explicit message in *Rouse* that the court does not presume to assess the quality of anyone's performance, unless that performance is patently arbitrary and capricious. As in all administrative law, the task of the court is to ensure that the administrative process itself controls abuse of discretion, that a factual record is established, that alternatives are considered and that reasons for decisions are set forth.

The instant opposition to *Rouse* was a clue to a deeper discordance between the professional and the judicial outlook. Plainly the profession was blind to or was concealing the conflict between the imposition of treatment and the human and civil rights of patients. In the view of most members medical decisions are by definition made in the best interests of the patient. If a physician says a man is sick, he must be sick; if he says the man must be treated or confined, that must be what is best for him. Such bootstrap reasoning comes under scrutiny only when, in the case of psychiatric prescriptions, it calls for involuntary treatment. The patient's interest in release, in less restrictive confinement or in adequate treatment cannot be matters solely "for medical determination."

Bringing these matters into court does not impose an artificial adversary relationship between the patient and his keepers; it reflects an adversity that already exists. This proposition comes as a surprise, of course, to the psychiatrist engaged exclusively in office practice and to his voluntary patient. In the public sector the adversity of interests that confronts the psychiatrist and his involuntary patient—although it does not encompass the entire relationship—must be recognized as an inescapable reality.

What is more, it should be recognized that the physician–patient relationship in the public sector is compromised by the interests of third parties and the pressures from hidden agendas. At the state hospital in Napa, Calif., the superintendent told me a few years ago in a public meeting that the staff had "Sacramento looking over its shoulder." Psychiatric decisions—to confine or release—are influenced by public outcry for "law and order." In some hospitals the shortage of beds and manpower has been known to override medical determinations; in Veterans Administration hospitals the need to fill empty beds also presses medical determinations, but in the opposite direction. Psychiatrists have justified fudging their testimony on "dangerousness"—a ground for involuntary confinement—when they were convinced that an individual was too sick to seek help voluntarily.

What is disturbing about these situations is not that they impute venality or frailty to merely human practitioners, nor that conflicting societal interests can dictate different and not necessarily the best medical results. It is rather that the psychiatric profession should resist facing these conflicts in the open. Serious legal challenges have been needed to surface its hidden agendas.

The hazard implicit in professional resistance to public scrutiny is well illustrated by the use in the U.S.S.R. of psychiatric facilities for the suppression of political dissenters. I had occasion to read the Russian case studies as a member of an *ad hoc* committee set up by the American Psychiatric Association. The studies showed how the medical model of "sickness" could be perverted to encompass judgment of what is socially and politically unacceptable behavior. On the record the physicians did not seem to be acting in their patients' best interests, or even in their own direct self-interest, but were using psychiatric terminology and techniques in the service of state policy. Yet when I was in the U.S.S.R., the Russian psychiatrists steadfastly insisted that they follow the medical model, much as their American colleagues are known to do.

As I read these case studies, however, I was impressed to realize how, in many analogous situations in this country, I had had occasion to find psychiatrists making decisions for motives and under pressures from outside their professional role. Whenever psychiatrists enter the public sector to apply their knowledge in the service of public institutions—the military, state hospitals, schools and penal institutions, to name only a few—they face conflicts between the therapeutic interests of their patients and the institutional interests of their employers. Needless to say I found the leadership of the profession in America not nearly as eager to investigate such conflicts in their own ranks as they were to look into evidence of malpractice in the U.S.S.R. To make a not very pleasant story short, when our *ad hoc* committee turned its attention to the American scene, at first with official approval, its charter was soon revoked. The need for examination remains. The American Bar Association and the Institute of Medicine of the National Academy of Sciences plan studies of these conflicts. The American Psychiatric Association apparently recognizes the importance of acting itself; it now has commissioned an inquiry by the Institute of Society, Ethics and the Life Sciences in Hastings, N.Y.

Such guild self-protection is not, of course, peculiar to the psychiatric profession. Business enterprises, labor unions and government agencies all exhibit the same penchant for privacy. The usual counter to the call for public scrutiny is the promise of "self-regulation." Whereas peer review is a much praised and not much observed principle of the medical profession, it has been largely foreign to the practice of psychiatry. Its practitioners work alone with their patients, behind ritually locked doors. For medical practitioners of all kinds peer review is now required, however, for the validation of professional performance paid for by Medicaid and Medicare funds. Curiously, this requirement was imposed not out of professional agreement that self-evaluation is in order but because of Congressional concern at rising medical costs.

For monitoring the performance of a profession there is no substitute, in the end, for the adversary process. This discussion has focused on psychiatry because decisions grounded on this discipline and on counsel from its practitioners are employed by the state to confine people against their will and to treat people in ways they do not ask for. As such the discipline is of concern to those of us whose judicial duty is to scrutinize governmental intrusions on liberty. Much of what I have said, however, applies equally well to the public surveillance of other highly specialized professions on which the operation of our complex civilization depends. Today every profession is being challenged by those who believe that trust should rest not on mystique but rather on what the public knows about its exercise of its expertise. Challenging the expert and digging into the facts behind his opinion is the lifeblood of our legal system, whether it is a psychiatrist characterizing a mental disturbance, a physicist testifying on the environmental impact of a nuclear power plant or a Detroit engineer insisting on the impossibility of meeting legislated automobile exhaust-emission standards by 1975. It is the only way a judge or a jury—or the public—can decide whom to trust.

FURTHER READINGS

For extensive lists of sources see: Ralph Slovenko, *Psychiatry and Law,* first edition (Boston: Little, Brown, 1973), p. 611; and E. Fuller Torrey, *The Death of Psychiatry* (Radnor, Pa.: Chilton Book Co., 1974), p. 213.

Books

Comprehensive Mental Health, eds. N. S. Greenfield, M. H. Miller, and C. M. Roberts (Madison: University of Wisconsin Press, 1968); D. C. Cooper, *Psychiatry and Anti-Psychiatry* (London: Tavistock, 1967); A. M. Dershowitz, Joseph Goldstein, and Jay Katz, *Psychoanalysis, Psychiatry, and Law* (New York: Free Press, 1967); *Deviance and Respectability: The Social Construction of Moral Meanings,* ed. Jack D. Douglas (New York: Basic Books, 1970); M. Eliade, *Shamanism, Archaic Technique of Ectasy* (New York: Bollingen Foundation, 1964); Bruce Ennis, *Prisoners of Psychiatry* (New York: Harcourt, Brace, Jovanovich, 1972), and with Loren Siegel, *The Rights of Mental Patients* (New York: Avon Books, 1973); *Ethical Issues In Medicine: The Role of the Physician in Today's Society,* ed. E. F. Torrey (Boston: Little Brown, 1968); S. J. Fox, *Science and Justice: The Massachusetts Witchcraft Trials* (Baltimore: Johns Hopkins Press, 1968); Willard Gaylin, A. Carr, and H. Hendrin, *Psychoanalysis and Social Research* (Doubleday, 1965); Ervin Goffman, *Asylums* (New York: Doubleday Anchor, 1961); Group for the Advancement of Psychiatry, Criminal Responsibility and Psychiatric Expert Testimony, Rep. No. 26 (New York: Gap Publication Office, 1954); Seymour L. Halleck, *The Politics of Therapy* (New York: Science House, 1971); H. Hartmann, *Psychoanalysis and Moral Values* (New York: International Universities Press, 1960); Nicholas N. Kittrie, *The Right to Be Different: Deviance and Enforced Therapy* (Baltimore: Johns Hopkins Press, 1971); *Legal Rights of the Mentally Handicapped,* Three Vols., Criminal Law And Urban Problems Course Handbook Series No. 57 eds. Bruce Ennis, P. R. Friedman with Bonnie Gitlin (Practising Law Institute, The Mental Health Law Project, 1973); Perry London, *Modes and Morals of Psychotherapy* (New York: Holt, Rinehart & Winston, 1964); C. M. Lowe, *Value Orientation in Counseling and Psychotherapy: The Meaning of Mental Health* (San Francisco, Chandler, 1969); Joseph Margolis, *Psychotherapy and Morality* (New York: Random House, 1966); *Medical, Moral, and Legal Issues in Mental Health Care* ed. Frank J. Ayd, Jr. (Baltimore: Williams and Wilkins, 1974); Roy and Zhores Medvedev, *The Question of Madness* (New York: Knopf, 1971); Jessica Mitford, *Kind and Usual Punishment* (New York: Knopf, 1973); Stanley Pearlstein,

Psychiatry, The Law, and Mental Health (Dobbs Ferry, New York: Oceana Publications, 1967); V. R. Potter, *Bioethics: Bridge to the Future* (Englewood Cliffs: Prentice-Hall, 1971); Philip Rieff, *The Triumph of the Therapeutic: Uses of Faith After Freud* (New York: Harper Torchbooks, 1968); Jonas Robitscher, *Pursuit of Agreement: Psychiatry and the Law* (Philadelphia: Lippincott, 1966); G. Rosen, *Madness in Society: Chapters in the Historical Sociology of Mental Illness* (New York: Harper & Row, 1968); David J. Rothman, *The Discovery of the Asylum* (Boston: Little, Brown, 1971); David Rudovsky, *The Rights of Prisoners* (New York: Avon Books, 1973); T. J. Scheff, *Being Mentally Ill: A Sociological Theory* (Chicago: Aldine Publishing, 1966); C. G. Schoenfeld, *Psychoanalysis and the Law* (Springfield, Ill.: Thomas, 1973); B. F. Skinner, *Beyond Freedom and Dignity* (New York: Knopf, 1971); Ralph Slovenko, *Psychiatry and Law,* first edition (Boston: Little, Brown, 1973); Thomas S. Szasz, *The Ethics of Psychoanalysis* (New York: Basic Books, 1965), and also by this author: *Law, Liberty, and Psychiatry* (New York: Macmillan, 1963), and *The Manufacture of Madness: A Comparative Study of the Inquisition and the Mental Health Movement* (New York: Dell, 1970), and *Psychiatric Injustice* (New York: Macmillan, 1965); E. Fuller Torrey, *The Death of Psychiatry* (Radnor, Pa.: Chilton Book Co., 1974), and also *The Mind Game: Witchdoctors and Psychiatrists* (New York: Bantam Books, 1972).

Articles and Addresses

David L. Bazelon, "The Adversary Process in Psychiatry," Address, Southern California Psychiatric Society (April 21, 1973), and also by this author: "Follow the Yellow Brick Road," *American Journal of Orthopsychiatry 40*:4 (1970), pp. 562–67, and "Institutional Psychiatry–The Self-Inflicted Wound," Address, Conference on Mental Health and the Law, The Catholic University of America (Jan. 19, 1974), and "Psychiatry's Fear of Analysis," (Washington Post June 24, 1973), and "The Right to Treatment: The Court's Role," *Hospital and Community Psychiatry 20* (1969), pp. 129–35, and A Statement to the President of the American Psychiatric Association, the Board of Trustees of the APA, and to the Members of the Ad Hoc Committee on the Use of Psychiatric Institutions for the Commitment of Political Dissenters (April 30, 1972); M. Birbaum, "A Rationale for the Right," *Georgetown Law Journal* 57 (1969), p. 752, and also "The Right to Treatment," *A.B.A.J. 46* (1960), p. 499; P. Breggin, "Psychotherapy as Applied Ethics," *Psychiatry 34* (1971), pp. 59–74; Alex Capron, Book Review of N. N. Kittrie, "The Right to Be Different," *Columbia Law Review 73* (1973), pp. 893–913; Frank Carrera and P. L. Adams, "An Ethical Perspective on Operant Conditioning," *Journal of the American Academy of Child Psychiatry 9* (1970), pp. 607–23; Lloyd H. Cotter, "Operant Conditioning in a Vietnamese Mental Hospital," *American Journal of Psychiatry 124* (1967), pp. 23–28; W. J. Curran, "Community Mental Health and the Commitment Laws: A Radical New Approach Is Needed," *Am. J. Public Health 57* (1967), pp. 1565–70; C. S. Deschin, "Knowledge is Neither Neutral Nor Apolitical," *American Journal of Orthopsychiatry 41* (1971), pp. 344–47; Leon Eisenberg, "The Future of Psychiatry," *Lancet 7842* (1973), p. 1371; L. H. Farber, "Psychoanalysis and Morality," *Commentary 40* (1965), pp. 69–74; Willard Gaylin, "Psychiatry and the Law: Partners in Crime," *Columbia Forum* (1965), pp. 23–27, also by this same author: "What's Normal–Mental Illness and Public Office," *New York Times Sunday Magazine* April 1, 1973); James S. Gordon, "The Uses of Madness," *Social Policy 4* (Sep./Oct. 1973), pp. 37–43; Jack Himmelstein and Robert Michels, "Case Studies in Bioethics: The Right to Refuse Psychoactive Drugs," *Hastings Center Report 3* (June, 1973), pp. 8–11; Paul C. Horton, "Normality–Toward a Meaningful Construct," *Comprehensive Psychiatry 12* (1971), pp. 54–66; Jay Katz, "The Right to Treatment–An Enchanting Legal Fiction?" *University of Chicago Law Review 36* (1969), pp. 755–83; Y. Kunasaka *et al.*, "Criteria for Involuntary Hospitalization," *Archives of General Psychiatry 26* (1972), pp. 399–404; Aston Lazare, "Hidden Conceptual Models in Clinical Psychiatry," *New England Journal of Medicine 288* (1973), pp. 345–51; C. M. Lowe, "Value Orientations: An Ethical Dilemma," *American Psychologist 14* (1959), pp. 687–93; R. Maisel, "Decision-Making in a Commitment Court," *Psychiatry 33* (1970), pp. 352; Joseph Margolis, "The Question of Insanity," *Wayne Law Review XIX* (1973), pp. 1007–22 and also, "The Myths of Psychoanalysis," *The Monist LVI* (1972), pp. 361–75; J. Masserman, "Is Mental Illness a Medicosocial 'Myth'?" *Arch. Gen. Psychiatry 9* (1963), p. 175; A. R. Mathews, "Mental Illness and the Criminal Law: Is Community Mental Health an Answer?" *American Journal of Public Health 57*:9 (1967), pp. 1571–79; A. L. McGarry and M. Greenblatt, "Conditional Volentary Mental Hospital Admission," *N. Engl. J. Med. 287* (1972), p. 279; Louis A. McGarry and Honora A. Kaplan, "Overview: Current Trends In Mental Health Law," *American Journal of Psychiatry 130* (1973), pp. 621–30; S. C. McMorris, "Can We Punish for the Acts of Addiction?" *Bulletin on Narcotics 22*:3 (1970), pp. 43–48; D. M. Callahan and M. Lorr, "Therapist 'Type' and Patient Response to Psychotherapy," *Journal of Consulting Psychology 26* (1962), pp. 425–29; J. N. McNeil *et al.*, "Community Psychiatry and Ethics," *American Journal of Orthopsychiatry 40*:1 (1970), pp. 22–29; W. M. Mendel, "On the Abolition of the Psychiatric Hospital," in Comprehensive Mental Health, eds. C. M. Roberts, N. S. Greenfield and M. H. Miller (Madison: University of Wisconsin, 1968), and also "Responsibility in Health, Illness, and Treatment," *Archives of General Psychiatry 18* (1968), pp. 697–705; M. H. Miller and S. L. Halleck, "The Critics of Psychiatry: A Review of Contemporary Critical Attitudes," *Am. J. Psychiatry 119* (1963), p. 705; D. J. Muller, "Involuntary Mental Hospitalization," *Compr.*

Psychiatry 9 (1968), p. 187; L. R. Oelbaum, "Clinical and Moral Views of Madness: Laing and Goffman Compared," *Bull. Menninger Clin. 36* (1972), p. 487; H. Osmond, "The Medical Model in Psychiatry: Love It or Leave It," *Medical Annals of District of Columbia 41* (1972), pp. 171–75; "Physical Manipulation of the Brain," *The Hastings Center Report* Special Supplement 9 (March, 1973); E. Barrett Prettyman, "The Indeterminate Sentence and the Right to Treatment," *American Criminal Law Review 11* (1972), pp. 7–37; R. Restak, "The Promise and Peril of Psychosurgery," *Saturday Review/World 54* (Sep. 25, 1973); Jonas Robitscher, "Controversial Crusaders: The Mentally Ill and Psychiatric Reform," *Medical Opinion and Review 4* (1968), pp. 54–67, and also by this author: "Courts, State Hospitals and the Right to Treatment," *American Journal of Psychiatry 129* (1972), pp. 298–304, and "The Right to Treatment: A Social-Legal Approach to the Plight of the State Hospital Patient," *Villanova Law Review 18* (1972), pp. 11–36, and "Social Legal Psychiatry," Address to the Annual Meeting, American College of Legal Medicine (1972), and with Sister Madeleine Brady, "Case Studies in Bioethics: Informed Consent: When Can It Be Withdrawn?" *Hastings Center Report 2* (June, 1972), pp. 10–11; Milton Rokeach, "Long Range Experimental Modification of Values, Attitudes and Behavior," *American Psychologist 26* (1971), pp. 453–59; D. L. Rosenhan, "On Being Sane in Insane Places," *Science 179* (1973), pp. 250–58; S. Rosenzweig, "Compulsory Hospitalization of the Mentally Ill," *Am. J. Pub. Health 61* (1971), pp. 121; M. Sabshin, "Psychiatric Perspectives on Normality," *Archives of General Psychiatry 17*:3 (1967), pp. 258–64; T. R. Sarbin, "On the Futility of the Proposition that Some People Be Labeled Mentally Ill," *J. Consulting Psychol. 31* (1967), p. 447; Morton Schatzman, "Madness and Morals," *Radical Therapist 1* (Oct./Nov., 1970), pp. 11–15; Aaron Schreiber, "Indeterminate Therapeutic Incarceration of Dangerous Criminals: Perspectives and Problems," *Virginia Law Review 56* (1970), pp. 602–34; David Shakow, "Ethics for a Scientific Age: Some Moral Aspects of Psychoanalysis," *Psychoanalytic Review 52* (Fall, 1965), pp. 5–18; M. Siegler and H. Osmond, "Models of Madness," *British Journal of Psychiatry 112* (Dec., 1966), pp. 1193–1203, and also "Laing's Models of Madness," *British Journal of Psychiatry 115* (Aug., 1969), pp. 947–58; Roy G. Spece, Jr., "Conditioning and Other Technologies Used to 'Treat'? 'Rehabilitate'? 'Demolish'? Prisoners and Mental Patients," *Southern California Law Review 45* (Spring, 1972), pp. 616–84; Alan A. Stone, "Psychiatry and the Law," *Psychiatric Annals* (Oct., 1971), pp. 18–44; I. F. Stone, "Betrayal By Psychiatry," (The New York Review Feb. 19, 1972), p. 8; Thomas S. Szasz, "Criminal Responsibility and Psychiatry," in *Legal and Criminal Psychology*, ed. H. Toch (New York: Holt, Rinehart & Winston, 1961), and also the following list of articles: "The Ethics of Addiction," *American Journal of Psychiatry 128* (1971), pp. 541–46, and "Hospital Refusal to Release Mental Patients," *Clev.-Mar. L. Rev. 9* (1960), p. 220, and "Justice and Psychiatry,"

Atlantic 222 (1968), p. 127, and "The Problem of Psychiatric Nosology," *American Journal of Psychiatry 114*:5 (1957), pp. 405–13, and "Problems Facing Psychiatry: The Psychiatrist as Party to Conflict," in *Ethical Issues in Medicine: The Role of the Physician in Today's Society,* ed. E. F. Torrey (Boston: Little, Brown, 1968), and "Psychiatry as a Social Institution," in *Psychiatry and Responsibility* eds. H. Schoeck and J. W. Wiggins (Princeton, N. J.: Van Nostrand, 1962), and "Psychiatry, Psychotherapy, and Psychology," *Arch. Gen. Psychiatry 5* (1959), p. 455, and "Psychoanalysis and Suggestion," *Comprehensive Psychiatry 4* (1963), pp. 271–80, and "Psychoanalysis and Taxation," *American Journal of Psychotherapy 18* (1964), pp. 635–43, and "Psychoanalytic Treatment as Education," *Arch. Gen. Psychiatry 9*:1 (1963), pp. 46–52, and "The Sane Slave: Social Control and Legal Psychiatry," *Am. Crim. L. Rev. 10* (1972), p. 337, and "Involuntary Mental Hospitalization: An Unacknowledged Practise of Medical Fraud," *New England Journal of Medicine 287*:6 (1972), pp. 277–78, and "Wither Psychiatry?" *Social Research 33*:3 (1966), 439–62, and with G. J. Alexander, "Law, Property and Psychiatry," *American Journal of Orthopsychiatry 42*:4 (1972), pp. 610–26, and also with G. J. Alexander, "Mental Illness as Excuse for Civil Wrongs," *Journal of Nervous and Mental Disease 147* (1968), pp. 113–23, and with M. H. Hollender, "A Contribution to the Philosophy of Medicine," *Arch. Int. Med. 97* (1956), p. 585, and with M. H. Hollander and W. F. Knoff, "The Doctor–Patient Relationship and Its Historical Context," *American Journal of Psychiatry 115* (1958), pp. 522–28; Laurence Tancredi and Diana Clark, "Psychiatry and the Legal Rights of Patients," *American Journal of Psychiatry 129* (1972), pp. 320–28; L. J. Tapp, "Psychology and the Law: The Dilemma," American Bar Foundation Pamphlet No. 2 (1969); W. Weintraub and H. Aronson, "Is Classical Psychoanalysis a Danger Procedure?" *Journal of Nervous and Mental Disease 149* (1969), pp. 224–28; L. J. West, "Psychiatry, 'Brainwashing,' and the American Character," *American Journal of Psychiatry 120* (1964), pp. 842–50; Daniel B. Wexler, "Token Taboo: Behavior Modification, Token Economies and the Law," *California Law Review 61* (1973), pp. 81–109; F. Wiseman, "Psychiatry and Law: Use and Abuse of Psychiatry in a Murder Case," *American Journal of Psychiatry 118* (1961), pp. 289–99.

PART III: HUMAN EXPERIMENTATION

INTRODUCTION

In his now-classic piece, "Ethics and Clinical Research," Dr. Henry K. Beecher introduces the problem of human experimentation by citing twenty-two examples of unethical or questionably ethical studies with human subjects. Although Dr. Beecher affirms his belief in the soundness of American medicine, he contends that unless steps are taken to correct activities of the sort he has noted, great harm will be done to medicine in the United States.

In order to ensure that the proprieties are followed by an investigator in any experiment, Dr. Beecher proposes that professional journals refuse to publish data which have been improperly obtained. Still, he admits that the most reliable safeguard will always be the presence of an intelligent, informed, conscientious, compassionate, and responsible investigator.

Dr. Michael Shimkin represents the medical research worker's viewpoint. He rejects the idea that professional journals refuse to publish data which have been obtained under questionable circumstances, contending that "we learn by our mistakes as well as our successes, and there are few situations in which ignorance or the hiding of facts are desirable solutions."

Dr. Shimkin outlines a list of items for a protocol to be followed in investigations on human beings. In order to ensure that subjects' rights are not subverted, he suggests: (1) that a technically qualified and objective counselor act in a continued advisory capacity to the subject in an experiment, and (2) that there be a review board, not made up exclusively of medical personnel, to judge protocols of proposed research, justifications for experimental studies, etc. Still, he insists, members of review boards should be sympathetic to research, and should not include individuals having "emotional blocks" to human experimentation. Further, he contends that there are cases in which the subjects in an experiment should perhaps be given no information at all concerning the experiment, as where the possession of such information would bias or negate the results.

Finally, Dr. Shimkin insists upon a fact often overlooked in the literature, viz., the rights of the investigator. He notes four such rights, and spells out the need for mechanisms by which the investigator can defend himself and exact penalties for unfounded charges.

Prof. Hans Jonas argues that neither the social contract theory of the state, nor the moral law, imposes an obligation upon one to offer himself up as the subject of experimental study. But if individuals must be asked to commit themselves to participation in experimental studies, to whom should the appeal be addressed? Contrary to established practice, Professor Jonas argues for what may be called an "elitist" standard. For Jonas, it is proper that the issuer of the call, viz., the researcher himself, be the first addressee. Still, Jonas realizes that it is impossible to keep the issue of human experimentation within the research community: "Neither in numbers nor in variety of material would its [the research community's] potential suffice for . . . the continual attack on disease. . . ." Thus, the aid of other persons must be sought. And in soliciting such aid, Jonas contends, the qualifications of the members of the scientific fraternity should be taken as the general criteria for selection. That is to say, researchers should look for additional subjects among the most highly motivated, most highly educated, and least "captive" members of our society.

Professor Jonas realizes that his recommendations, if taken seriously by the scientific community, would restrict the pool of potential subjects for experimental study and so perhaps work for a slower rate of progress in the conquest of disease. But Jonas contends that progress in overcoming disease is only an optional goal—one which, if not achieved with the greatest possible haste, would not threaten society. For Jonas, a far greater danger to society lies in the possible erosion of moral values which could be one of the side effects of too ruthless a pursuit of scientific progress.

Professors Dyck and Richardson concern themselves with the question of whether or not it is morally permissible to experiment with human subjects when the research activities are known to be potentially harmful or risk bearing. They conclude that if research of this kind is to be justified, two conditions must be met: (1) The anticipated benefits from such research must more than offset all the potential, associated risks, and (2) such research must be in accord with certain inviolable moral values that can only be secured by satisfying the requirement of *informed consent.*

Although conditions (1) and (2) must both be satisfied if experimentation involving possible risk to human subjects is to be morally justified, Dyck and Richardson stress the primacy of (2). On their view, failure to secure informed consent undermines a set of structural values; and regardless of how great the benefit of any particular experiment may be, violation of structural values cannot be tolerated for "these values maintain the social systems that are the presupposition of all benefits."

Dyck and Richardson deal with the requirement of informed consent at some length, and conclude that there seem to be no insurmountable barriers to obtaining informed consent when human subjects are participants in a controlled experiment. However, more serious problems arise when they consider the relationship of informed consent to persons affected by the use of drugs, food additives, and environmental toxins such as pesticides. In considering this issue, the authors contend that there is a need for licensing boards to represent the "informed consent" of the public, and for compensation to persons who are injured by toxins which affect them outside the controlled conditions of the laboratory.

Regardless of how simple the process of obtaining informed consent may appear to be in theory, the actual practice of obtaining such consent is always complex. Rev. John Fletcher examines the actualities of giving informed consent in medical experiments. Conducting a series of interviews with patient-volunteers, Fletcher seeks to discover whether or not experimental subjects manage to maintain a sense of personhood throughout their experience, and whether they feel coerced in their decisions to submit to experimental study.

Fletcher's study leads him to two conclusions. First, he is convinced that the subjects of his interviews were able to maintain their relationship to personhood, i.e., these individuals did not feel as though they were being treated as "things." And second, although none of the persons interviewed felt that their consent decision was coerced, several gave signs of having to defend themselves against their consent being "engineered." That is, Fletcher finds that patients tend to tie their participation in experimentation with continued treatment of their disease. Further, everything that happens seems to be filtered through illness; and arrangements at the institution of study tend to diminish the patient's willingness to complain or question. In order to combat these "engineering" forces, then, Fletcher suggests that research institutions act to designate a "physician-friend," who would bear the responsibility of advocating for the patient. Such a step would be particularly desirable, he says, whenever especially necessitous groups such as prisoners, children, or the mentally ill were involved in research.

As John Fletcher observes, certain classes of humans have limited capacity to consent to their involvement in experimental activities; and where individuals of this sort are used in research, special problems arise. Realizing this, the Department of Health, Education, and Welfare, through the National Institutes of Health, has made several attempts to formulate guidelines for the protection of such subjects. As the last selection in our section on human experimentation, then, we have chosen to include portions of two working documents drafted by a special study group of the National Institutes of Health. The sections of the documents which we have incorporated into our text concern themselves with fetal organisms, and represent two attempts to formulate guidelines for the protection of prenatal beings used in fetal research.

ETHICS AND CLINICAL RESEARCH*

HENRY K. BEECHER‡

Human experimentation since World War II has created some difficult problems with the increasing employment of patients as experimental subjects when it must be apparent that they would not have been available if they had been truly aware of the uses that would be made of them. Evidence is at hand that many of the patients in the examples to follow never had the risk satisfactorily explained to them, and it seems obvious that further hundreds have not known that they were the subjects of an experiment although grave consequences have been suffered as a direct result of experiments described here. There is a belief prevalent in some sophisticated circles that attention to these matters would "block progress." But, according to Pope Pius XII,[1] ". . . science is not the highest value to which all other orders of values . . . should be subordinated."

I am aware that these are troubling charges. They have grown out of troubling practices. They can be documented, as I propose to do, by examples from leading medical schools, university hospitals, private hospitals, governmental military departments (the Army, the Navy and the Air Force), governmental institutes (the National Institutes of Health), Veterans Administration hospitals and industry. The basis for the charges is broad.[2]

Printed with permission of the author and *The New England Journal of Medicine, 274:* 1354–1360 (June 16, 1966). © 1966 by the Massachusetts Medical Society. From the Anaesthesia Laboratory of the Harvard Medical School at the Massachusetts General Hospital.

*From the Anaesthesia Laboratory of the Harvard Medical School at the Massachusetts General Hospital.

‡Dorr Professor of Research in Anaesthesia, Harvard Medical School.

[1] Pope Pius XII. Address. Presented at First International Congress on Histopathology of Nervous System, Rome, Italy, September 14, 1952.

[2] At the Brook Lodge Conference on "Problems and Complexities of Clinical Research" I commented that "what seem to be breaches of ethical conduct in experimentation are by no means rare, but are almost, one fears, universal." I thought it was obvious that I was by "universal" referring to the fact that examples could easily be found in *all* categories where research in man takes place to any significant extent. Judging by press comments, that was not obvious; hence, this note.

I should like to affirm that American medicine is sound, and most progress in it soundly attained. There is, however, a reason for concern in certain areas, and I believe the type of activities to be mentioned will do great harm to medicine unless soon corrected. It will certainly be charged that any mention of these matters does a disservice to medicine, but not one so great, I believe, as a continuation of the practices to be cited.

Experimentation in man takes place in several areas: in self-experimentation; in patient volunteers and normal subjects; in therapy; and in the different areas of *experimentation on a patient not for his benefit but for that, at least in theory, of patients in general.* The present study is limited to this last category.

Reasons for Urgency of Study

Ethical errors are increasing not only in numbers but in variety—for example, in the recently added problems arising in transplantation of organs.

There are a number of reasons why serious attention to the general problem is urgent.

Of transcendent importance is the enormous and continuing increase in available funds, as shown below.

Money Available for Research Each Year

	Massachusetts General Hospital	*National Institutes of Health*[a]
1945	$ 500,000[b]	$ 701,800
1955	2,222,816	36,063,200
1965	8,384,342	436,600,000

[a]National Institutes of Health figures based upon decade averages, excluding funds for construction, kindly supplied by Dr. John Sherman, of National Institutes of Health.

[b]Approximation, supplied by Mr. David C. Crockett, of Massachusetts General Hospital.

Since World War II the annual expenditure for research (in large part in man) in the Massachusetts General Hospital has increased a remarkagle 17-fold. At the National Institutes of Health, the increase has been a gigantic 624-fold. This "national" rate of increase is over 36 times that of the Massachusetts General Hospital. These data, rough as they are, illustrate vast opportunities and concomitantly expanded responsibilities.

Taking into account the sound and increasing emphasis of recent years that experimentation in man must precede general application of new procedures in therapy, plus the great sums of money available, there is reason to fear that these requirements and these resources may be greater than the supply of responsible investigators. All this heightens the problems under discussion.

Medical schools and university hospitals are increasingly dominated by investigators. Every young man knows that he will never be promoted to a tenure post, to a professorship in a major medical school, unless he has proved himself as an investigator. If the ready availability of money for condcting research is added to this fact, one can see how great the pressures are on ambitious young physicians.

Implementation of the recommendations of the President's Commission on Heart Disease, Cancer and Stroke means that further astronomical sums of money will become available for research in man.

In addition to the foregoing three practical points there are others that Sir Robert Platt[3] has pointed out: a general awakening of social conscience; greater power for good or harm in new remedies, new operations, and new investigative procedures than was formerly the case; new methods of preventive treatment with their advantages and dangers that are now applied to communities as a whole as well as to individuals, with multiplication of the possibilities for injury; medical science has shown how valuable human experimentation can be in solving problems of disease and its treatment; one can therefore anticipate an increase in experimentation; and the newly developed concept of clinical research as a profession (for example, clinical pharmacology)—and this, of course, can lead to unfortunate separation between the interests of science and the interests of the patient.

Frequency of Unethical or Questionably Ethical Procedures

Nearly everyone agrees that ethical violations do occur. The practical question is, how often? A preliminary examination of the matter was based on 17 examples, which were easily increased to 50. These 50 studies contained references to 186 further likely examples, on the average 3.7 leads per study; they at times overlapped from paper to paper, but this figure indicates how conveniently one can proceed in a search for such material. The data are suggestive of widespread problems but there is need for another kind of information, which was obtained by examination of 100 consecutive human studies

[3] Platt (Sir Robert), 1st part. *Doctor and Patient: Ethics, morals, government.* 87 pp. London: Nuffield provincial hospitals trust, 1963. Pp. 62 and 63.

published in 1964, in an excellent journal; 12 of these seemed to be unethical. If only one quarter of them is truly unethical, this still indicates the existence of a serious situation. Pappworth,[4] in England, has collected, he says, more than 500 papers based upon unethical experimentation. It is evident from such observations that unethical or questionably ethical procedures are not uncommon.

The Problem of Consent

All so-called codes are based on the bland assumption that meaningful or informed consent is readily available for the asking. As pointed out elsewhere,[5] this is very often not the case. Consent in any fully informed sense may not be obtainable. Nevertheless, except, possibly, in the most trivial situations, it remains a goal toward which one must strive for sociologic, ethical, and clear-cut legal reasons. There is no choice in the matter.

If suitably approached, patients will accede, on the basis of trust, to about any request their physician may make. At the same time, every experienced clinician investigator knows that patients will often submit to inconvenience and some discomfort, if they do not last very long, but the usual patient will never agree to jeopardize seriously his health or his life for the sake of "science."

In only 2 of the 50[6] examples originally compiled for this study was consent mentioned. Actually, it should be emphasized in all cases for obvious moral and legal reasons, but it would be unrealistic to place much dependence on it. In any precise sense statements regarding consent are meaningless unless one knows how fully the patient was informed of all risks, and if these are not known, the fact should also be made clear. A far more dependable safeguard than consent is the presence of a truly *responsible* investigator.

Examples of Unethical or Questionably Ethical Studies

These examples are not cited for the condemnation of individuals; they are recorded to call attention to a variety of ethical problems found in experimental medicine, for it is hoped that calling attention to them will help to correct abuses present. During ten years of study of these matters it has become apparent that thoughtlessness and carelessness, not a willful disregard of the patient's rights, account for most of the cases encountered. Nonetheless, it is

[4] Pappworth, M. H. Personal communication.

[5] Beecher, H. K. Consent in clinical experimentation: Myth and reality. *J.A.M.A. 195:* 34, 1966.

[6] Reduced here to 22 for reasons of space.

evident that in many of the examples presented, the investigators have risked the health or the life of their subjects. No attempt has been made to present the "worst" possible examples; rather, the aim has been to show the variety of problems encountered.

References to the examples presented are not given, for there is no intention of pointing to individuals, but rather, a wish to call attention to widespread practices. All, however, are documented to the satisfaction of the editors of the *Journal.*

Known Effective Treatment Withheld

Example 1. It is known that rheumatic fever can usually be prevented by adequate treatment of streptococcal respiratory infections by the parenteral administration of penicillin. Nevertheless, definitive treatment was withheld, and placebos were given to a group of 109 men in service, while benzathine penicillin G was given to others.

The therapy that each patient received was determined automatically by his military serial number arranged so that more men received penicillin than received placebo. In the small group of patients studied 2 cases of acute rheumatic fever and 1 of acute nephritis developed in the control patients, whereas these complications did not occur among those who received the benzathine penicillin G.

Example 2. The sulfonamides were for many years the only antibacterial drugs effective in shortening the duration of acute streptococcal pharyngitis and in reducing its suppurative complications. The investigators in this study undertook to determine if the occurrence of the serious nonsuppurative complications, rheumatic fever, and acute glomerulonephritis, would be reduced by this treatment. This study was made despite the general experience that certain antibiotics, including penicillin, will prevent the development of rheumatic fever.

The subjects were a large group of hospital patients; a control group of approximately the same size, also with exudative Group A streptococcus, was included. The latter group received only nonspecific therapy (no sulfadiazine). The total group denied the effective penicillin comprised over 500 men.

Rheumatic fever was diagnosed in 5.4 percent of those treated with sulfadiazine. In the control group rheumatic fever developed in 4.2 percent.

In reference to this study a medical officer stated in writing that the subjects were not informed, did not consent and were not aware that they had been involved in an experiment, and yet admittedly 25 acquired rheumatic fever. According to this same medical officer *more than 70* who had had known definitive treatment withheld were on the wards with rheumatic fever when he was there.

Example 3. This involved a study of the relapse rate in typhoid fever treated in two ways. In an earlier study by the present investigators chloramphenicol had been recognized as an effective treatment for typhoid fever, being attended by half the mortality that was experienced when this agent was not used. Others had made the same observations, indicating that to withhold this effective remedy can be a life-or-death decision. The present study was carried out to determine the relapse rate under the two methods of treatment; of 408 charity patients 251 were treated with chloramphenicol, of whom 20, or 7.97 percent, died. Symptomatic treatment was given, but chloramphenicol was withheld in 157, of whom 36, or 22.9 percent, died. According to the data presented, 23 patients died in the course of this study who would not have been expected to succumb if they had received specific therapy.

Study of Therapy

Example 4. TriA (triacetyloleandomycin) was originally introduced for the treatment of infection with gram-positive organisms. Spotty evidence of hepatic dysfunction emerged, especially in children, and so the present study was undertaken on 50 patients, including mental defectives or juvenile delinquents who were inmates of a children's center. No disease other than acne was present; the drug was given for treatment of this. The ages of the subjects ranged from thirteen to thirty-nine years. "By the time half the patients had received the drug for four weeks, the high incidence of significant hepatic dysfunction . . . led to the discontinuation of administration to the remainder of the group at three weeks." (However, only two weeks after the start of the administration of the drug, 54 percent of the patients showed abnormal excretion of bromsulfalein.) Eight patients with marked hepatic dysfunction were transferred to the hospital "for more intensive study." Liver biopsy was carried out in these 8 patients and repeated in 4 of them. Liver damage was evident. Four of these hospitalized patients, after their liver-function tests returned to normal limits, received a "challenge" dose of the drug. Within two days hepatic dysfunction was evident in 3 of the 4 patients. In 1 patient a second challenge dose was given after the first challenge and again led to evidence of abnormal liver function. Flocculation tests remained abnormal in some patients as long as five weeks after discontinuance of the drug.

Physiologic Studies

Example 5. In this controlled, double-blind study of the hematologic toxicity of chloramphenicol, it was recognized that chloramphenicol is "well known as a cause of aplastic anemia" and that there is a "prolonged morbidity and high mortality of asplastic anemia" and that "chloramphenicol-induced aplastic anemia can be related to dose. . . ." The aim of the study was "further definition of the toxicology of the drug. . . ."

Forty-one randomly chosen patients were given either 2 or 6 gm. of chloramphenicol per day; 12 control patients were used. "Toxic bone-marrow depression, predominantly affecting erythropoiesis, developed in 2 of 20 patients given 2.0 gm. and in 18 of 21 given 6 gm. of chloramphenicol daily." The smaller dose is recommended for routine use.

Example 6. In a study of the effect of thymectomy on the survival of skin homografts 18 children, three and a half months to eighteen years of age, about to undergo surgery for congenital heart disease, were selected. Eleven were to have total thymectomy as part of the operation, and 7 were to serve as controls. As part of the experiment, full-thickness skin homografts from an unrelated adult donor were sutured to the chest wall in each case. (Total thymectomy is occasionally, although not usually part of the primary cardiovascular surgery involved, and whereas it may not greatly add to the hazards of the necessary operation, its eventual effects in children are not known.) This work was proposed as part of a long-range study of "the growth and development of these children over the years." No difference in the survival of the skin homograft was observed in the 2 groups.

Example 7. This study of cyclopropane anesthesia and cardiac arrhythmias consisted of 31 patients. The average duration of the study was three hours, ranging from two to four and a half hours. "Minor surgical procedures" were carried out in all but 1 subject. Moderate to deep anesthesia, with endotracheal intubation and controlled respiration, was used. Carbon dioxide was injected into the closed respiratory system until cardiac arrhythmias appeared. Toxic levels of carbon dioxide were achieved and maintained for considerable periods. During the cyclopropane anesthesia a variety of pathologic cardiac arrhythmias occurred. When the carbon dioxide tension was elevated above normal, ventricular extrasystoles were more numerous than when the carbon dioxide tension was normal, ventricular arrhythmias being continuous in 1 subject for ninety minutes. (This can lead to fatal fibrillation.)

Example 8. Since the minimum blood-flow requirements of the cerebral circulation are not accurately known, this study was carried out to determine "cerebral hemodynamic and metabolic changes . . . before and during acute reductions in arterial pressure induced by drug administration and/or postural adjustments." Forty-four patients whose ages varied from the second to the tenth decade were involved. They included normotensive subjects, those with essential hypertension and finally a group with malignant hypertension. Fifteen had abnormal electrocardiograms. Few details about the reasons for hospitalization are given.

Signs of cerebral circulatory insufficiency, which were easily recognized, included confusion and in some cases a nonresponsive state. By alteration in the tilt of the patient "the clinical state of the subject could be changed in a matter of seconds from one of alertness to confusion, and for the remainder of the flow, the subject was maintained in the latter state." The femoral arteries were cannulated in all subjects, and the internal jugular veins in 14.

The mean arterial pressure fell in 37 subjects from 109 to 48 mm. of mercury, with signs of cerebral ischemia. "With the onset of collapse, cardiac output and right ventricular pressures decreased sharply."

Since signs of cerebral insufficiency developed without evidence of coronary insufficiency the authors concluded that "the brain may be more sensitive to acute hypotension than is the heart."

Example 9. This is a study of the adverse circulatory responses elicited by intra-abdominal maneuvers:

> When the peritoneal cavity was entered, a deliberate series of maneuvers was carried out [in 68 patients] to ascertain the effective stimuli and the areas responsible for development of the expected circulatory changes. Accordingly, the surgeon rubbed localized areas of the parietal and visceral peritoneum with a small ball sponge as discretely as possible. Traction on the mesenteries, pressure in the area of the celiac plexus, traction on the gallbladder and stomach, and occlusion of the portal and caval veins were the other stimuli applied.

Thirty-four of the patients were sixty years of age or older; 11 were seventy or older. In 44 patients the hypotension produced by the deliberate stimulation was "moderate to marked." The maximum fall produced by manipulation was from 200 systolic, 105 diastolic, to 42 systolic, 20 diastolic; the average fall in mean pressure in 26 patients was 53 mm. of mercury.

Of the 50 patients studied, 17 showed either atrioventricular dissociation with nodal rhythm or nodal rhythm alone. A decrease in the amplitude of the T wave and elevation or depression of the ST segment were noted in 25 cases in association with manipulation and hypotension or, at other times, in the course of anesthesia and operation. In only 1 case was the change pronounced enough to suggest myocardial ischemia. No case of myocardial infarction was noted in the group studied although routine electrocardiograms were not taken after operation to detect silent infarcts. Two cases in which electrocardiograms were taken after operation showed T-wave and ST-segment changes that had not been present before.

These authors refer to a similar study in which more alarming electrocardiographic changes were observed. Four patients in the series sustained silent myocardial infarctions; most of their patients were undergoing gallbladder surgery because of associated heart disease. It can be added further that in the 34 patients referred to above as being sixty years of age or older, some doubtless had heart disease that could have made risky the maneuvers carried out. In any event, this possibility might have been a deterrent.

Example 10. Starling's law—"that the heart output per beat is directly proportional to the diastolic filling"—was studied in 30 adult patients with atrial fibrillation and mitral stenosis sufficiently severe to require valvulotomy. "Continuous alterations of the length of a segment of left ventricular muscle were recorded simultaneously in 13 of these patients by means of a mercury-filled resistance gauge sutured to the surface of the left ventricle." Pressures in the left ventricle were determined by direct puncture simultaneously with the segment length in 13 patients and without the segment length in an additional 13 patients. Four similar unanesthetized patients were studied through catheterization of the left side of the heart transeptally. In all 30 patients arterial pressure was measured through the catheterized brachial artery.

Example 11. To study the sequence of ventricular contraction in human bundle-branch block, simultaneous catheterization of both ventricles was performed in 22 subjects; catheterization of the right side of the heart was carried out in the usual manner; the left side was catheterized transbronchially. Extrasystoles were produced by tapping on the epicardium in subjects with normal myocardium while they were undergoing thoracotomy. Simultaneous pressures were measured in both ventricles through needle puncture of this group.

The purpose of this study was to gain increased insight into the psysiology involved.

Example 12. This investigation was carried out to examine the possible effect of vagal stimulation on cardiac arrest. The authors had in recent years transected the homolateral vagus nerve immediately below the origin of the recurrent laryngeal nerve as palliation against cough and pain in bronchogenic carcinoma. Having been impressed with the number of reports of cardiac arrest that seemed to follow vagal stimulation, they tested the effects of intrathoracic vagal stimulation during 30 of their surgical procedures, concluding, from these observations in patients under satisfactory anesthesia, that cardiac irregularities and cardiac arrest due to vagovagal reflex were less common than had previously been supposed.

Example 13. This study presented a technic for determining portal circulation time and hepatic blood flow. It involved the transcutaneous injection of the spleen and catheterization of the hepatic vein. This was carried out in 43 subjects, of whom 14 were normal; 16 had cirrhosis (varying degrees), 9 acute hepatitis, and 4 hemolytic anemia.

No mention is made of what information was divulged to the subjects, some of whom were seriously ill. This study consisted in the development of a technic, not of therapy, in the 14 normal subjects.

Studies to Improve the Understanding of Disease

Example 14. In this study of the syndrome of impending hepatic coma in patients with cirrhosis of the liver certain nitrogenous substances were administered to 9 patients with chronic alcoholism and advanced cirrhosis: ammonium chloride, di-ammonium citrate, urea or dietary protein. In all patients a reaction that included mental disturbances, a "flapping tremor," and electroencephalographic changes developed. Similar signs had occurred in only 1 of the patients before these substances were administered:

> The first sign noted was usually clouding of the consciousness. Three patients had a second or a third course of administration of a nitrogenous substance with the same results. It was concluded that marked resemblance between this reaction and impending hepatic coma, implied that the administration of these [nitrogenous] substances to patients with cirrhosis may be hazardous.

Example 15. The relation of the effects of ingested ammonia to liver disease were investigated in 11 normal subjects, 6 with acute virus hepatitis, 26 with cirrhosis, and 8 miscellaneous patients. Ten of these patients had neurologic changes associated with either hepatitis or cirrhosis.

The hepatic and renal veins were cannulated. Ammonium chloride was administered by mouth. After this, a tremor that lasted for three days developed in 1 patient. When ammonium chloride was ingested by 4 cirrhotic patients with tremor and mental confusion the symptoms were exaggerated during the test. The same thing was true of a fifth patient in another group.

Example 16. This study was directed toward determining the period of infectivity of infectious hepatitis. Artifical induction of hepatitis was carried out in an institution for mentally defective children in which a mild form of hepatitis was endemic. The parents gave consent for the intramuscular injection or oral administration of the virus, but nothing is said regarding what was told them concerning the appreciable hazards involved.

A resolution adopted by the World Medical Association states explicitly: "Under no circumstances is a doctor permitted to do anything which would weaken the physical or mental resistance of a human being except from strictly therapeutic or prophylactic indications imposed in the interest of the patient." There is no right to risk an injury to 1 person for the benefit of others.

Example 17. Live cancer cells were injected into 22 human subjects as part of a study of immunity to cancer. According to a recent review, the subjects (hospitalized patients) were "merely told they would be receiving 'some cells'"—"... the word cancer was entirely omitted...."

Example 18. Melanoma was transplanted from a daughter to her volunteering and informed mother, "in the hope of gaining a little better understanding of cancer immunity and in the hope that the production of tumor antibodies might be helpful in the treatment of the cancer patient." Since the daughter died on the day after the transplantation of the tumor into her mother, the hope expressed seems to have been more theoretical than practical, and the daughter's condition was described as "terminal" at the time the mother volunteered to be a recipient. The primary implant was widely excised on the twenty-fourth day after it had been placed in the mother. She died from metastatic melanoma on the four hundred and fifty-first day after transplantation. The evidence that this patient died of diffuse melanoma that metastasized from a small piece of transplanted tumor was considered conclusive.

Technical Study of Disease

Example 19. During bronchoscopy a special needle was inserted through a bronchus into the left atrium of the heart. This was done in an unspecified number of subjects, both with cardiac disease and with normal hearts.

The technic was a new approach whose hazards were at the beginning quite unknown. The subjects with normal hearts were used, not for their possible benefit but for that of patients in general.

Example 20. The percutaneous method of catheterization of the left side of the heart has, it is reported, led to 8 deaths (1.09 percent death rate) and other serious accidents in 732 cases. There was, therefore, need for another method, the transbronchial approach, which was carried out in the present study in more than 500 cases, with no deaths.

Granted that a delicate problem arises regarding how much should be discussed with the patients involved in the use of a new method, nevertheless where the method is employed in a given patient for *his* benefit, the ethical problems are far less than when this potentially extremely dangerous method is used "in 15 patients with normal hearts, undergoing bronchoscopy for other reasons." Nothing was said about what was told any of the subjects, and nothing was said about the granting of permission, which was certainly indicated in the 15 normal subjects used.

Example 21. This was a study of the effect of exercise on cardiac output and pulmonary-artery pressure in 8 "normal" persons (that is, patients whose diseases were not related to the cardiovascular system), in 8 with congestive heart failure severe enough to have recently required complete bed rest, in 6 with hypertension, and 2 with aortic insufficiency, in 7 with mitral stenosis, and in 5 with pulmonary emphysema.

Intracardiac catheterization was carried out, and the catheter then inserted into the right or left main branch of the pulmonary artery. The brachial artery was usually catheterized; sometimes, the radial or femoral arteries were catheterized. The subjects exercised in a supine position by pushing their feet against weighted pedals. "The ability of these patients to carry on sustained work was severely limited by weakness and dyspnea." Several were in severe failure. This was not a therapeutic attempt but rather a physiologic study.

Bizarre Study

Example 22. There is a question whether ureteral reflux can occur in the normal bladder. With this in mind, vesicourethrography was carried out on 26 normal babies less than forty-eight hours old. The infants were exposed to x-rays while the bladder was filling and during voiding. Multiple spot films were made to record the presence or absence of ureteral reflux. None was found in this group, and fortunately no infection followed the catheterization. What the results of the extensive x-ray exposure may be, no one can yet say.

Comment on Death Rates

In the foregoing examples a number of procedures, some with their own demonstrated death rates, were carried out. The following data were provided by 3 distinguished investigators in the field and represent widely held views.

Cardiac catheterization: right side of the heart, about 1 death per 1000 cases; left side, 5 deaths per 1000 cases. "Probably considerably higher in some places, depending on the portal of entry." (One investigator had 15 deaths in his first 150 cases.) It is possible that catheterization of a hepatic vein or the renal vein would have a lower death rate than that of catheterization of the right side of the heart, for if it is properly carried out, only the atrium is entered en route to the liver or the kidney, not the right ventricle, which can lead to serious cardiac irregularities. There is always the possibility, however, that the ventricle will be entered inadvertently. This occurs in at least half the cases, according to 1 expert—"but if properly done is too transient to be of importance."

Liver biopsy: the death rate here is estimated at 2 to 3 per 1000, depending in considerable part on the condition of the subject.

Anesthesia: the anesthesia death rate can be placed in general at about 1 death per 2000 cases. The hazard is doubtless higher when certain practices such as deliberate evocation of ventricular extrasystoles under cyclopropane are involved.

Publication

In the view of the British Medical Research Council[7] it is not enough to ensure that all investigation is carried out in an ethical manner: it must be made unmistakably clear in the publications that the proprieties have been observed. This implies editorial responsibility in addition to the investigator's. The question rises, then, about valuable data that have been improperly obtained.[8] It is

[7] Great Britain, Medical Research Council. *Memorandum,* 1953.

[8] As far as principle goes, a parallel can be seen in the recent Mapp decision by the United States Supreme Court. It was stated there that evidence unconstitutionally obtained cannot be used in any judicial decision, no matter how important the evidence is to the ends of justice.

my view that such material should not be published (see footnote 7). There is a practical aspect to the matter: failure to obtain publication would discourage unethical experimentation. How many would carry out such experimentation if they *knew* its results would never be published? Even though suppression of such data (by not publishing it) would constitute a loss to medicine, in a specific localized sense, this loss, it seems, would be less important than the far-reaching moral loss to medicine if the data thus obtained were to be published. Admittedly, there is room for debate. Others believe that such data, because of their intrinsic value, obtained at a cost of great risk or damage to the subjects, should not be wasted but should be published with stern editorial comment. This would have to be done with exceptional skill, to avoid an odor of hypocrisy.

Summary and Conclusions

The ethical approach to experimentation in man has several components; two are more important than the others, the first being informed consent. The difficulty of obtaining this is discussed in detail. But it is absolutely essential to *strive* for it for moral, sociologic, and legal reasons. The statement that consent has been obtained has little meaning unless the subject or his guardian is capable of understanding what is to be undertaken and unless all hazards are made clear. If these are not known this, too, should be stated. In such a situation the subject at least knows that he is to be a participant in an experiment. Secondly, there is the more reliable safeguard provided by the presence of an intelligent, informed, conscientious, compassionate, responsible investigator.

Ordinary patients will not knowingly risk their health or their life for the sake of "science." Every experienced clinician investigator knows this. When such risks are taken and a considerable number of patients are involved, it may be assumed that informed consent has not been obtained in all cases.

The gain anticipated from an experiment must be commensurate with the risk involved.

An experiment is ethical or not at its inception; it does not become ethical *post hoc*—ends do not justify means. There is no ethical distinction between ends and means.

In the publication of experimental results it must be made unmistakably clear that the proprieties have been observed. It is debatable whether data obtained unethically should be published even with stern editorial comment.

SCIENTIFIC INVESTIGATIONS ON MAN

A MEDICAL RESEARCH WORKER'S VIEWPOINT

MICHAEL B. SHIMKIN

Scientific investigations on human beings have a rich tradition, have yielded many important advances for the betterment of man, and will continue to have an increasingly larger role in biomedical research.

The expanding dimensions of human investigations also have expanded the vexing but inescapable moral and legal problems such investigations entail. The timeliness and the importance of the problems are suggested by the recent medical writings such as those of Beecher,[1] their public recognition,[2] and the official governmental actions.[3] For purposes of this discussion, we must assume that the extent and seriousness of the problems are commensurate with the concern they have evinced.

Scientific research is usually directed toward obtaining precise information on specific, defined questions. Scientists become uneasy when they are asked to deal with generalities and abstractions. They, as well as the public, are also susceptible to emotional reactions when confronted with semantic subtleties and reactive labels. For this reason the word *investigation* appears to me to be a desirable substitute for its emotion-laden synonym, *experimentation.* But call it what we will, when the hiatus between morals, which embody the aspirations of man, and mores, the actual conduct of man, becomes too wide, it must be

---

From M. Shimkin, "Scientific Investigations on Man: A Medical Research Worker's Viewpoint," in *Use of Human Subjects in Safety Evaluation of Food Chemicals–Proceedings of a Conference.* National Academy of Sciences and National Research Council, Pub. 1491 (Wash., D.C., 1967), pp. 217–227. Reproduced with permission of the author and the National Academy of Sciences.

[1] H. K. Beecher, "Ethics and Clinical Research," *New Engl. J. Med., 274,* 1354–1360 (1966).

[2] P. Lear, "Experiments on People–The Growing Debate," *Saturday Rev.,* pp. 41–50 (July 2, 1966).

[3] W. H. Stewart, "Memorandum on Revised Procedure on Clinical Research and Investigation Involving Human Subjects," Public Health Service, Dept. of Health, Education and Welfare. Washington, D.C. (July 1, 1966).

narrowed in one way or another.[4] Such a hiatus seems to have appeared in the expanded investigations on man in the United States.

The problems as they now exist must be defined and then considered despite the absence of absolute, immutable criteria for human conduct. Scientific research, morals, and laws are creations of man and are thus defined as aspects of human behavior in our period of a space–time continuum.[5]

Resolutions to the problems, it is hoped, will emerge from discussions of specific questions, such as: How and when should human investigations be undertaken? What principles and rules should guide such investigations? What measures will reduce to a minimum the consequences of human frailty and misconduct, yet impose minimum restrictions upon useful investigations? The considerations of these and similar questions must be modified by experience and new information and relevant to specific conditions of each situation and its temporal occurrence. The aim is to seek workable methods of successive approximations, rather than nonexistent final solutions.

The Conduct of Investigations on Man

Existing moral and legal concepts governing investigations on man are derived to a great extent from past experiences in clinical research. The compilation of this material by Ladimer and Newman[6] is an essential basis for further discussion. It is evident that clinical considerations are also of direct relevance to investigations on healthy individuals and on human populations.

Table I outlines a functional sequence that applies to most projects for investigations involving human beings. The outline provides no surprises, for it includes essentially the same points and steps as are entailed in most biomedical research, although some are of particular importance when man is involved. Many of the points are interrelated, the value judgments are inescapable.

I. As in all research, perhaps the most important items are the primary question or hypothesis and the quality of the investigator who proposes to conduct the research.

The question must be valid and approachable by existent methods. The background requires documentation on the importance of the problem, scientifically and practically, and an analytical review of preceding basic and applied

[4] H. F. Pilpel, "Morals, Mores and Mandates," *Vassar Alumnae Magazine, 51* (5), 18–21 (June 1966).

[5] E. Goldberger, *How Physicians Think,* Charles C Thomas, Springfield, Ill. (1965).

[6] I. Ladimer, and R. W. Newman, editors, *Clinical Investigation in Medicine: Legal, Ethical, and Moral Aspects,* Law-Medicine Research Institute, Boston University, Boston, Mass. (1963).

Table I. Items for a Protocol Involving Investigation on Human Beings

I. The Question (Problem, Hypothesis)
   A. Is it valid and substantive?
   B. Is it approachable scientifically?
   C. Review of relevant knowledge, basic and applied
   D. Who proposes the question?
      1. Qualifications of investigator(s)
      2. Facilities and consultations available

II. Type of Research Proposed
   A. Observational
      1. Retrospective
      2. Prospective
   B. Manipulative (experimental)
      1. Episodic
      2. Repetitive or continual

III. Materials Needed
   A. Type of individuals
      1. With defined characteristics (e.g., patients)
      2. Normal individuals
      3. Contratests (controls)
   B. Type of facilities
      1. Field
      2. Institutional
      3. Special
   C. Type of instruments, laboratories

IV. Methods Proposed
   A. Type of data to be gathered
      1. Surveys of existing data
      2. Questionnaires, interviews
      3. Psychological tests
      4. Physical measurements
      5. Physiological measurements
         a. Basal
         b. Under stress (surgical, etc.)
      6. Pharmacological tests
   B. Type of specimens required
      1. Excretory (urine, etc.)
      2. Blood
      3. Tissues
      4. Organs

V. Investigative Procedure
   A. Pilot phase
      1. Safety of procedures

continued

Table I. continued.

2. Validation of data-procurement procedures
3. Validation of test procedures

B. Biometric considerations
1. Number of individuals required
2. Number and type of observations required
3. Randomization, double-blinding, sequential trials

C. Practical considerations
1. Cost
2. Time

D. Formal investigation

VI. Control and Analysis
A. During investigation
B. Final analysis and conclusions
C. Publication or other dissemination

research relating to the problem with special justification of its extension to man.

Particularly when human populations are involved, the topic must be substantive and of interest to the subject-participants as well as to the investigators, and to the community as well as to those involved in the study, in order that the investigation be initiated and pursued at all. It must also be able to attract and to hold capable scientific personnel, thus helping to assure competent planning and performance.

As for the investigator, there must be confidence in his judgment, integrity, and technical competence. This confidence must extend to his associates and his institution. It is necessarily based on past performance.

II. Research on man, like other scientific research, can be divided into two general types. One type is *observational,* in which findings are compared and analyzed, without active participation of, or procedures on, the subjects. Even in this type of inquiry, there will be differences of opinion regarding the proper use of records, especially those involving confidential information. Clinical records are the usual problem, but so are data on occupation, income, habits, or legal entanglements. Observational research may extend to data on the future status of the subjects, requiring examinations and follow-up procedures, or the gathering of specimens of excreta, blood, and various tissues. The latter type of research extends beyond simple passive observations and analyses for interrelations.

Our chief concern is with the second type of research, *manipulative* or experimental research, in which some deliberate procedure is imposed upon the subjects. This may involve the addition of an agent to or its removal from the environment, or some physiological or pharmacological procedure and the study of the effects of this manipulation on the subjects.

III. It is also necessary to categorize research on man by the type of subject to be used in the research. One type is exemplified by patients with disease upon whom new procedures are instituted for diagnostic or therapeutic aims, and are therefore of potential benefit to the patient. Most of us accept the thought that in patients with rapidly progressing fatal afflictions, more risk is justified than in, say, the treatment of an annoying allergy.

The other type of research involves either patients or normal individuals on whom procedures are contemplated that do not aim to be of direct benefit to that individual. The two categories may overlap, and the latter may be rationalized into the former. However, limiting human research to clinical aims is neither realistic nor desirable. Man's aspirations and interests include adventure, discovery, contribution to society, and the recognition and honors that may result. These drives are not limited to healthy individuals, and they exist in the potential subjects as well as in the potential investigators. Perhaps we should recognize the subjects in research more often as our co-workers. The initial participation of the investigator as one of the subjects in research reinforces the equality of their roles.

IV. The methods specify procedures to be performed on the human subjects. These extend from simple observations to complex physiological and pharmacological situations in which stress and risk are unavoidable. Risk must be commensurate with the importance of the potential discovery, and the estimate of the risk must be revised with experience. An extreme example is afforded by the current explorations of space, in which disaster is a possibility at each step, while the importance of potential discoveries remains unknown.

Possible effects of external environmental stresses are now equalled by the potential dangers of changes in man's internal milieu. Chemical agents with therapeutic effects may also transform genetic material, produce fetal abnormalities, block cellular metabolic pathways, and affect personality. In pharmacological investigations, the exposure of man should come only after a series of laboratory studies, culminating with animals, have been performed and evaluated. But at some point data on nonhuman material have to be transposed to man. Complex investigational systems have been devised to achieve proper therapeutic trials, yet, with each new drug, different problems arise. The consequences of chemical agents in trace amounts over long periods and the cumulative effects and interactions between such agents require much more laboratory and animal data, as well as observations on man.

V. When the question has been specified and the appropriate materials and methods selected, the next step is the experimental design. In modern times, this involves biometric considerations of the choice of the samples to be compared, the criteria by which individuals are selected, and the numbers required to establish differences or to test nondifferences at predetermined levels of statistical significance. It involves strict definitions and interpretative criteria for the observations to be made. Randomized assignment of individuals to the groups to be compared, double-blinding wherever possible, and sequential analysis of data are among techniques useful for economy and as safety measures. The words *randomization, double-blind,* and *control* sometimes meet emotional resistance, and their scientific meaning requires careful, specific explanation. In the case of control, the concept in human investigations is perhaps better expressed by a substitute word, *contratest.*[7]

Formal studies on man in almost all instances require a pilot phase, directed toward further establishing safety and feasibility for the investigation. The pilot phase also provides additional data for biometric design and final considerations of cost.

VI. The last item relates research to communications. Research should start and end in the library, and research is not completed if it is not reported and disseminated. The history of man is in his written records, and those who do not read history are condemned to repeating it. We learn by mistakes as well as by successes, and there are few situations in which ignorance or the hiding of facts are desirable solutions.

The Rights of the Human Subject

Recognized human rights are the unique feature that separates investigations on man from all other research. Table II suggests four such features. Again, all are relative and involve subjective reactions and judgments.

There appears to be a consensus that informed consent of the subjects is essential for manipulative investigations. But what is informed consent? How can one convey "all" relevant information, especially when the investigator himself cannot predict the course or the outcome of a research situation?

The usual practice in clinical research is to accept patients who are referred by other physicians. This, in effect, provides the subject with a professional counselor who assists him toward arriving at a decision. It has been suggested[8] that this may be developed into a more formal arrangement in which a "physician-friend" would act in a continued advisory capacity to the patient. Perhaps

[7] M. B. Shimkin, "The Numerical Method in Therapeutic Medicine," *Public Health Repts.*, *79*, 1–12 (1964).

[8] O. Guttentag, "The Problem of Experimentation on Human Beings: II. The Physician's Point of View," *Science, 117,* 207–210 (1953).

Table II. The Rights of the Human Subject

I.	Informed consent
II.	Volunteer participation
III.	Review of proposal and conduct by board of peers
IV.	Recourse to religious and other counsel

this can be usefully extended to nonclinical situations, by providing technically qualified but presumably objective counselors to the potential subject.

In many specific therapeutic situations detailed discussion with a patient is not called for and may even be harmful to the patient. For nontherapeutically oriented investigations, however, the procedure, the possible risks, and the aims should be available to the prospective subject and his counselors. How much to tell and how to convey the information depend on specific circumstances. This may vary from full participation in the actual research to no information in situations where the possession of such information would bias or negate the results.

There also appears to be a consensus regarding the principle that the subjects in manipulative investigations, whether they be patients or normal individuals, must be volunteers in a strict sense. This implies an untrammeled option to reject as well as to accept the proposition. All human beings have equal, inalienable rights, and their role as subjects or as investigators does not modify such rights.[9]

In order to assure these rights, prior review mechanisms as well as remedies in courts of law must be available for the protection of both the investigators and the subjects. In therapeutic research, this can be relatively simple and informal, perhaps by the investigator's peers in the institution or of the professional group. For human volunteers, sick or healthy, more rigorous and formal review may be required, depending on the potential dangers of the experimental design. Protocols of the proposed research, its justification, and its need appear unavoidable.

And here again we are faced with the Socratic question, "Who selects the judges?" For patients and for medical institutions, physicians must bear the responsibility. But for many other types of human investigations, an exclusively medical jury is parochial. Legal and religious viewpoints, as well as technical proficiency in the contemplated techniques, are desirable for the protection of the subjects and the investigators. Of course, members of review boards should

[9] M. B. Shimkin, "The Problem of Experimentation on Human Beings: The Research Worker's Point of View," *Science, 117,* 205–207 (1953).

be sympathetic to research, and should not include individuals with emotional blocks to human experimentation.

A sticky point confronts us regarding individuals who are legally incapable of giving consent, especially children or mental defectives. There seems to be no alternative to placing the burden of consent on the legal guardians. But this only begs the question, since the aspect of volunteer participation would still be lacking. This assigns additional responsibility to the judgment of the review mechanism.

The last item of Table II emphasizes the equality of human rights in all situations and man's need for recourse to forces beyond his limitations. The subjects of investigation, as well as the investigators, should have ready access to theological and other counsel.

All these precautions are susceptible to the formulation of memoranda, codes, and laws. How far these should go at any stage, and how detailed they should be, are crucial questions that will determine how human investigations will and should progress.

The Rights of the Investigator

This brings us to another aspect of the dynamic equilibrium between the subject, the investigator, and society that is represented in human investigations.

In order to perform investigations, there must be investigators. The investigators must be motivated to plan and to pursue the work. Most good research is self-motivated, meaning that the idea stems primarily from the investigator, is considered to be important by the investigator, and, despite frustrations and difficulties, remains rewarding to the investigator.

Truly novel ideas, the most important elements for scientific advance, are born in individual minds. New ideas are stimulated by an environment of sufficient freedom to allow unorthodoxy and possible error. They are not encouraged by reviews and restrictions.

Scientists do not differ from other individuals in their responsibilities and their commitments to human values. They also have rights, as real as those of the subjects who may be involved in the investigation. And research, with its yields to society, will not flourish unless these rights are also recognized and reasonably met.

Table III lists four such rights, which must be placed in correct relationship with the rights of the subjects, and of society, including its ethics and its laws.

Privacy of ideas and of preliminary investigations is the first item. It is unrealistic to expect investigators to divulge original ideas. Moreover, without actual data, somehow obtained, such ideas have an excellent chance of being

I. Privacy of own ideas and investigations
II. Freedom to reject ideas and assignments of others
III. Protection against excessive detail of plans, reports, records
IV. Appeal and defense, with confrontation, against accusations

rejected just because they are unsupported. To get out of this circular trap requires opportunity for the investigator to generate some data.

Plans presented in writing to committees, and examined by clerical and scientific assistants, seldom remain confidential. The better qualified the reviewers, the more opportunity there is for the transfer of ideas, consciously or subconsciously, with possible loss of priority and claim for discovery.

The second point is the right to reject ideas and assignments of others, be they other scientists, committees, industrial concerns, or governmental agencies. Such rejection, just as for the potential subject, should not carry implied penalties for the investigator. He should be free to voice his objections, but must allow identical freedom to others who might wish to proceed despite his criticisms.

Research on man should be substantive from scientific as well as from practical standpoints. It may be of considerable commercial importance to provide data on the safety and utility of some product. Whether an actual study is to be performed, however, requires consideration of its scientific interest and priority.

The most precious commodity of an investigator is his time. A substantial proportion of his time is already devoted to composing plans, forwarding reports, and filling out forms on fiscal and administrative matters. Operations research is highly desirable to develop methods that would meet these responsibilities with a minimum of paper work. Funding agencies, governmental or private, are not owners but stewards of funds that are in their custody to meet defined objectives. Mechanisms that preclude or make difficult the meeting of the objectives should be revised. Otherwise investigators best qualified to do the work will find other outlets for their drives, to the detriment of all.

The fourth point is seldom openly discussed. Review committees, funding agencies, and university faculties reach many of their decisions *in camera.* Rejection of a proposal may carry only theoretical right to reconsideration; seldom is there actual confrontation between the body that makes the decision and the investigator. The subjects of human investigation require protection, we all admit. Investigators who are accused of unethical practices, openly or by innuendo, also need mechanisms through which they can defend themselves and exact penalties for unfounded charges.

One of the problems facing us, in all aspects of life, is the proliferation of laws and their interpretations. Most of these mandates are promulgated for worthy reasons, but their effect *en masse* is not without drawbacks. Rules and regulations produce records and restrictions in a never-ending cycle. In an area as complex and as varied as research on human beings, all rules are imperfect and susceptible to many interpretations, and these are not made perfect by additional rules. But neither are they made useless by the presence of imperfections.

Discussions such as these should be helpful in the pragmatic resolution of the problems inherent in scientific investigations on human beings. Yet it is difficult to improve upon the conclusion written by Claude Bernard[10] almost a century ago: "So, among the experiments that may be tried on man, those that can only harm are forbidden, those that are innocent are permissible, and those that may do good are obligatory." Surely this is in keeping with the pivotal moral and ethical guide of our major religions, stated many times in the Bible: "Thou shalt love thy neighbor as thyself."[11]

Acknowledgment

I am indebted to Dr. Nathaniel I. Berlin, of Bethesda, Maryland, to Dr. Otto E. Guttentag and Mr. Stillman Drake of San Francisco, and to Mrs. Mary N. Shimkin for their thoughtful suggestions and comments.

[10] C. Bernard, *An Introduction to the Study of Experimental Medicine,* 1878, transl. by H. C. Greene, Henry Schuman, New York (1949), p. 102.

[11] *The Bible,* Leviticus, 19:18, Matthew, 22:39; Mark, 12:31.

PHILOSOPHICAL REFLECTIONS ON EXPERIMENTING WITH HUMAN SUBJECTS

HANS JONAS

Experimenting with human subjects is going on in many fields of scientific and technological progress. It is designed to replace the overall instruction by natural, occasional, and cumulative experience with the selective information from artificial, systematic experiment which physicial science has found so effective in dealing with inanimate nature. Of the new experimentation with man, medical is surely the most legitimate; psychological, the most dubious; biological (still to come), the most dangerous. I have chosen here to deal with the first only, where the case *for* it is strongest and the task of adjudicating conflicting claims hardest. When I was first asked[1] to comment "philosophically" on it, I had all the hesitation natural to a layman in the face of matters on which experts of the highest competence have had their say and still carry on their dialogue. As I familiarized myself with the material,[2] any initial feeling of moral rectitude that might have facilitated my task quickly dissipated before the awesome complexity of the problem, and a state of great humility took its place. The awareness of the problem in all its shadings and ramifications

---

Originally published in *Daedalus* 98 (Spring 1969), and in its present revised version, with a comment by Arthur J. Dyck, in *Experimentation with Human Subjects*, ed. Paul A. Freund (New York: Braziller, 1970). The article as it appears here is from *Philosophical Essays: From Ancient Creed To Technological Man*, ed. Hans Jonas (Englewood Cliffs, N.J.: Prentice-Hall, 1974). Reprinted by permission of *DAEDALUS*, Journal of The American Academy of Arts and Sciences, Boston, Mass. Spring, 1969, *Ethical Aspects of Experimentation with Human Subjects*.

1 The American Academy of Arts and Sciences invited me to participate in a conference on the Ethical Aspects of Experimentation on Human Subjects, sponsored by the Academy journal *Daedalus* and the National Institutes of Health. The conference was held September 26–28, 1968, in Boston, Massachusetts, and all papers were subsequently published in *Daedalus*. A previous conference of the same title and under the same auspices is documented in *Proceedings of the Conference on the Ethical Aspects of Experimentation on Human Subjects*, Nov. 3–4, 1967 (Boston; hereafter called *Proceedings*).

2 Since the time of writing this essay (1968), the literature on the subject has grown so much that listing what was then available to me would be of no more than historical interest.

speaks out with such authority, perception, and sophistication in the published discussions of the researchers themselves that it would be foolish of me to hope that I, an onlooker on the sidelines, could tell those battling in the arena anything they have not pondered themselves. Still, since the matter is obscure by its nature and involves very fundamental, transtechnical issues, anyone's attempt at clarification can be of use, even without novelty. And even if the philosophical reflection should in the end achieve no more than the realization that in the dialectics of this area we must sin and fall into guilt, this insight may not be without its own gains.

I. The Peculiarity of Human Experimentation

Experimentation was originally sanctioned by natural science. There it is performed on inanimate objects, and this raises no moral problems. But as soon as animate, feeling beings become the subject of experiment, as they do in the life sciences and especially in medical research, this innocence of the search for knowledge is lost, and questions of conscience arise. The depth to which moral and religious sensibilities can become aroused over these questions is shown by the vivisection issue. Human experimentation must sharpen the issue as it involves ultimate questions of personal dignity and sacrosanctity. One profound difference between the human experiment and the physical (beside that between animate and inanimate, feeling and unfeeling nature) is this: The physical experiment employs small-scale, artificially devised substitutes for that about which knowledge is to be obtained, and the experimenter extrapolates from these models and simulated conditions to nature at large. Something deputizes for the "real thing"—balls rolling down an inclined plane for sun and planets, electric discharges from a condenser for real lightning, and so on. For the most part, no such substitution is possible in the biological sphere. We must operate on the original itself, the real thing in the fullest sense, and perhaps affect it irreversibly. No simulacrum can take its place. Especially in the human sphere, experimentation loses entirely the advantage of the clear division between vicarious model and true object. Up to a point, animals may fulfill the proxy role of the classical physical experiment. But in the end man himself must furnish knowledge about himself, and the comfortable separation of noncommittal experiment and definitive action vanishes. An experiment in education affects the lives of its subjects, perhaps a whole generation of schoolchildren. Human experimentation for whatever purpose is always *also* a responsible, nonexperimental, definitive dealing with the subject himself. And not even the noblest purpose abrogates the obligations this involves.

This is the root of the problem with which we are faced: Can both that purpose and this obligation be satisfied? It not, what would be a just compromise? Which side should give way to the other? The question is inherently philosophical as it concerns not merely pragmatic difficulties and their arbitration, but a genuine conflict of values involving principles of a high order. May I put the conflict in these terms. On principle, it is felt, human beings *ought not* to be dealt with in that way (the "guinea pig" protest); on the other hand, such dealings are increasingly urged on us by considerations, in turn appealing to principle, that claim to override those objections. Such a claim must be carefully assessed, especially when it is swept along by a mighty tide. Putting the matter thus, we have already made one important assumption rooted in our "Western" cultural tradition: The prohibitive rule is, to that way of thinking, the primary and axiomatic one; the permissive counter-rule, as qualifying the first, is secondary and stands in need of justification. We must justify the infringement of a primary inviolability, which needs no justification itself; and the justification of its infringement must be by values and needs of a dignity commensurate with those to be sacrificed.

Before going any further, we should give some more articulate voice to the resistance we feel against a merely utilitarian view of the matter. It has to do with a peculiarity of human experimentation quite independent of the question of possible injury to the subject. What is wrong with making a person an experimental subject is not so much that we make him thereby a means (which happens in social contexts of all kinds), as that we make him a thing—a passive thing merely to be acted on, and passive not even for real action, but for token action whose token object he is. His being is reduced to that of a mere token or "sample." This is different from even the most exploitative situations of social life: there the business is real, not fictitious. The subject, however much abused, remains an agent and thus a "subject" in the other sense of the word. The soldier's case is instructive: Subject to most unilateral discipline, forced to risk mutilation and death, conscripted without, perhaps against, his will—he is still conscripted with his capacities to act, to hold his own or fail in situations, to meet real challenges for real stakes. Though a mere "number" to the High Command, he is not a token and not a thing. (Imagine what he would say if it turned out that the war was a game staged to sample observations on his endurance, courage, or cowardice.)

These compensations of personhood are denied to the subject of experimentation, who is acted upon for an extraneous end without being engaged in a real relation where he would be the counterpoint to the other or to circumstance. Mere "consent" (mostly amounting to no more than permission) does not right this reification. Only genuine authenticity of volunteering can possibly redeem the condition of "thinghood" to which the subject submits. Of this we shall speak later. Let us now look at the nature of the conflict, and especially at the nature of the claims countering in this matter those on behalf of personal sacrosanctity.

II. "Individual Versus Society" as the Conceptual Framework

The setting for the conflict most consistently invoked in the literature is the polarity of individual versus society—the possible tension between the individual good and the common good, between private and public welfare. Thus, W. Wolfensberger speaks of "the tension between the long-range interests of society, science, and progress, on one hand, and the rights of the individual on the other."[3] Walsh McDermott says: "In essence, this is a problem of the rights of the individual versus the rights of society."[4] Somewhere I found the "social contract" invoked in support of claims that science may make on individuals in the matter of experimentation. I have grave doubts about the adequacy of this frame of reference, but I will go along with it part of the way. It does apply to some extent, and it has the advantage of being familiar. We concede, as a matter of course, to the common good some pragmatically determined measure of precedence over the individual good. In terms of rights, we let some of the basic rights of the individual be overruled by the acknowledged rights of society—as a matter of right and moral justness and not of mere force or dire necessity (much as such necessity may be adduced in defense of that right). But in making that concession, we require careful clarification of what the needs, interests, and rights of society are, for society—as distinct from any plurality of individuals—is an abstract and, as such, is subject to our definition, and his basic good is more or less known. Thus the unknown in our problem is the so-called common or public good and its potentially superior claims, to which the individual good must or might sometimes be sacrificed, in circumstances that in turn must also be counted among the unknowns of our question. Note that in putting the matter in this way—that is, in asking about the right of society to individual sacrifice—the consent of the sacrificial subject is no necessary part of the *basic* question.

"Consent," however, is the other most consistently emphasized and examined concept in discussions of this issue. This attention betrays a feeling that the "social" angle is not fully satisfactory. If society has a right, its exercise is not contingent on volunteering. On the other hand, if volunteering is fully genuine, no public right to the volunteered act need be construed. There is a difference between the moral or emotional appeal of a cause that elicits volunteering and a right that demands compliance—for example, with particular reference to the social sphere, between the *moral claim* of a common good and society's *right* to that good and to the means of its realization. A moral claim cannot be met without consent; a right can do without it. Where consent is present anyway, the

[3] Wolfensberger, "Ethical Issues in Research with Human Subjects," *World Science 155* (Jan. 6, 1967), p. 48.
[4] *Proceedings,* p. 29.

distinction may become immaterial. But the awareness of the many ambiguities besetting the "consent" actually available and used in medical research[5] prompts recourse to the idea of a public right conceived independently of (and valid prior to) consent; and, vice versa, the awareness of the problematic nature of such a right makes even its advocates still insist on the idea of consent with all its ambiguities: an uneasy situation either way.

Nor does it help much to replace the language of "rights" by that of "interests" and then argue the sheer cumulative weight of the interest of the many over against those of the few or the single individual. "Interests" range all the way from the most marginal and optional to the most vital and imperative, and only those sanctioned by particular importance and merit will be admitted to count in such a calculus–which simply brings us back to the question of right or moral claim. Moreover, the appeal to numbers is dangerous. Is the number of those afflicted with a particular disease great enough to warrant violating the interests of the nonafflicted? Since the number of the latter is usually so much greater, the argument can actually turn around to the contention that the cumulative weight of interest is on *their* side. Finally, it may well be the case that the individual's interest in his own inviolability is itself a public interest, such that its publicly condoned violation, irrespective of numbers, violates the interest of all. In that case, its protection in *each* instance would be a paramount interest, and the comparison of numbers will not avail.

These are some of the difficulties hidden in the conceptual framework indicated by the terms *society-individual, interest,* and *rights.* But we also spoke of a moral call, and this points to another dimension–not indeed divorced from the social sphere, but transcending it. And there is something even beyond that: true sacrifice from highest devotion, for which there are no laws or rules except that it must be absolutely free. "No one has the right to choose martyrs for science" was a statement repeatedly quoted in the November 1967 *Daedalus* conference. But no scientist can be prevented from making himself a martyr for his science. At all times, dedicated explorers, thinkers, and artists have immolated themselves on the altar of their vocation, and creative genius most often pays the price of happiness, health, and life for its own consummation. But no one, not even society, has the shred of a right to expect and ask these things in the normal course of events. They come to the rest of us as a *gratia gratis data.*

III. The Sacrificial Theme

Yet we must face the somber truth that the *ultima ratio* of communal life is and has always been the compulsory, vicarious sacrifice of individual lives. The

[5]See M. H. Pappworth, "Ethical Issues in Experimental Medicine" in D. R. Cutler, ed., *Updating Life and Death* (Boston: Beacon Press, 1969), pp. 64–69.

primordial sacrificial situation is that of outright human sacrifices in early communities. These were not acts of blood-lust or gleeful savagery; they were the solemn execution of a supreme, sacral necessity. One of the fellowship of men had to die so that all could live, the earth be fertile, the cycle of nature renewed. The victim often was not a captured enemy, but a select member of the group: "The king must die." If there was cruelty here, it was not that of men, but that of the gods, or rather of the stern order of things, which was believed to exact that price for the bounty of life. To assure it for the community, and to assure it ever again, the awesome *quid pro quo* had to be paid over and over.

Far should it be from us to belittle, from the height of our enlightened knowledge, the majesty of the underlying conception. The particular *causal* views that prompted our ancestors have long since been relegated to the realm of superstition. But in moments of national danger we still send the flower of our young manhood to offer their lives for the continued life of the community, and if it is a just war, we see them go forth as consecrated and strangely ennobled by a sacrificial role. Nor do we make their going forth depend on their own will and consent, much as we may desire and foster these. We conscript them according to law. We conscript the best and feel morally disturbed if the draft, either by design or in effect, works so that mainly the disadvantaged, socially less useful, more expendable, make up those whose lives are to buy ours. No rational persuasion of the pragmatic necessity here at work can do away with the feeling, a mixture of gratitude and guilt, that the sphere of the sacred is touched with the vicarious offering of life for life. Quite apart from these dramatic occasions, there is, it appears, a persistent and constitutive aspect of human immolation to the very being and prospering of human society—an immolation in terms of life and happiness, imposed or voluntary, of the few for the many. What Goethe has said of the rise of Christianity may well apply to the nature of civilization in general: *"Opfer fallen hier, / Weder Lamm noch Stier, / Aber Menschenopfer unerhört."*[6] We can never rest comfortably in the belief that the soil from which our satisfactions sprout is not watered with the blood of martyrs. But a troubled conscience compels us, the undeserving beneficiaries, to ask: Who is to be martyred? In the service of what cause and by whose choice?

Not for a moment do I wish to suggest that medical experimentation on human subjects, sick or healthy, is to be likened to primeval human sacrifices. Yet something sacrificial is involved in the selective abrogation of personal inviolability and the ritualized exposure to gratuitous risk of health and life,

[6] "Victims do fall here, /Neither lamb nor steer, / Nay, but human offerings untold." –*Die Braut von Korinth.*

justified by a presumed greater, social good. My examples from the sphere of stark sacrifice were intended to sharpen the issues implied in that context and to set them off clearly from the kinds of obligation and constraint imposed on the citizen in the normal course of things or generally demanded of the individual in exchange for the advantages of civil society.

IV. The "Social Contract" Theme

The first thing to say in such a setting-off is that the sacrificial area is not covered by what is called the "social contract." This fiction of political theory, premised on the primacy of the individual, was designed to supply a rationale for the *limitation* of individual freedom and power required for the existence of the body politic, whose existence in turn is for the benefit of the individuals. The principle of these limitations is that their *general* observance profits all, and that therefore the individual observant, assuring this general observance for his part, profits by it himself. I observe property rights because their general observance assures my own; I observe traffic rules because their general observance assures my own safety; and so on. The obligations here are mutual and general; no one is singled out for special sacrifice. Moreover, for the most part, *qua* limitations of my liberty, the laws thus deducible from the hypothetical "social contract" enjoin me from certain actions rather than obligate me to positive actions (as did the laws of feudal society). Even where the latter is the case, as in the duty to pay taxes, the rationale is that I am myself a beneficiary of the services financed through these payments. Even the contributions levied by the welfare state, though not originally contemplated in the liberal version of the social contract theory, can be interpreted as a personal insurance policy of one sort or another–be it against the contingency of my own indigence, be it against the dangers of disaffection from the laws in consequence of widespread unrelieved destitution, be it even against the disadvantages of a diminished consumer market. Thus, by some stretch, such contributions can still be subsumed under the principle of enlightened self-interest. But no complete abrogation of self-interest at any time is in the terms of the social contract, and so pure sacrifice falls outside it. Under the putative terms of the contract alone, I cannot be required to die for the public good. (Thomas Hobbes made this forcibly clear.) Even short of this extreme, we like to think that nobody is entirely and one-sidedly the victim in any of the renunciations exacted under normal circumstances by society "in the general interest"–that is, for the benefit of others. "Under normal circumstances," as we shall see, is a necessary qualification. Moreover, the "contract" can legitimize claims only on our overt, public actions and not on our invisible, private being. Our powers, not our persons, are beholden to the common weal.

In one important respect, it is true, public interest and control do extend to the private sphere by general consent: in the compulsory education of our children. Even there, the assumption is that the learning and what is learned, apart from all future social usefulness, are also for the benefit of the individual in his own being. We would not tolerate education to degenerate into the conditioning of useful robots for the social machine.

Both restrictions of public claim in behalf of the "common good"—that concerning one-sided sacrifice and that concerning the private sphere—are valid only, let us remember, on the premise of the primacy of the individual, upon which the whole idea of the "social contract" rests. This primacy is itself a metaphysical axiom or option peculiar to our Western tradition, and the whittling away of its force would threaten the tradition's whole foundation. In passing, I may remark that systems adopting the alternative primacy of the community as their axiom are naturally less bound by the restrictions we postulate. Whereas we reject the idea of "expendables" and regard those not useful or even recalcitrant to the social purpose as a burden that society must carry (since their individual claim to existence is as absolute as that of the most useful), a truly totalitarian regime, Communist or other, may deem it right for the collective to rid itself of such encumbrances or to make them forcibly serve some social end by conscripting their persons (and there are effective combinations of both). We do not normally—that is, in nonemergency conditions—give the state the right to conscript labor, while we do give it the right to "conscript" money, for money is detachable from the person as labor is not. Even less than forced labor do we countenance forced risk, injury, and indignity.

But in time of war our society itself supersedes the nice balance of the social contract with an almost absolute precedence of public necessities over individual rights. In this and similar emergencies, the sacrosanctity of the individual is abrogated, and what for all practical purposes amounts to a near-totalitarian, quasi-communist state of affairs is *temporarily* permitted to pervail. In such situations, the community is conceded the right to make calls on its members, or certain of its members, entirely different in magnitude and kind from the calls normally allowed. It is deemed right that a part of the population bears a disproportionate burden of risk of a disproportionate gravity; and it is deemed right that the rest of the community accepts this sacrifice, whether voluntary or enforced, and reaps its benefits—difficult as we find it to justify this acceptance and this benefit by any normal ethical standards. We justify it transethically, as it were, by the supreme collective emergency, formalized, for example, by the declaration of a state of war.

Medical experimentation on human subjects falls somewhere between this overpowering case and the normal transactions of the social contract. On the one hand, no comparable extreme issue of social survival is (by and large) at stake.

And no comparable extreme sacrifice or foreseeable risk is (by and large) asked. On the other hand, what is asked goes decidedly beyond, even runs counter to, what it is otherwise deemed fair to let the individual sign over of his person to the benefit of the "common good." Indeed, our sensitivity to the kind of intrusion and use involved is such that only an end of transcendent value or overriding urgency can make it arguable and possibly acceptable in our eyes.

V. Health as a Public Good

The cause invoked is health and, in its more critical aspect, life itself–clearly superlative goods that the physician serves directly by curing and the researcher indirectly by the knowledge gained through his experiments. There is no question about the good served nor about the evil fought–disease and premature death. But a good to whom and an evil to whom? Here the issue tends to become somewhat clouded. In the attempt to give experimentation the proper dignity (on the problematic view that a value becomes greater by being "social" instead of merely individual), the health in question or the disease in question is somehow predicated of the social whole, as if it were society that, in the persons of its members, enjoyed the one and suffered the other. For the purposes of our problem, public interest can then be pitted against private interest, the common good against the individual good. Indeed, I have found health called a national resource, which of course it is, but surely not in the first place.

In trying to resolve some of the complexities and ambiguities lurking in these conceptualizations, I have pondered a particular statement, made in the form of a question, which I found in the *Proceedings* of the earlier *Daedalus* conference: "Can society afford to discard the tissues and organs of the hopelessly unconscious patient when they could be used to restore the otherwise hopelessly ill, but still salvageable, individual?" And somewhat later: "A strong case can be made that society can ill afford to discard the tissues and organs of the hopelessly unconscious patient; they are greatly needed for study and experimental trial to help those who can be salvaged."[7] I hasten to add that any suspicion of callousness that the "commodity" language of these statements may suggest is immediately dispelled by the name of the speaker, Dr. Henry K. Beecher, for whose humanity and moral sensibility there can be nothing but admiration. But the use, in all innocence, of this language gives food for thought. Let me, for a moment, take the question literally. "Discarding" implies proprietary rights–nobody can discard what does not belong to him in the first place. Does society then own my body? "Salvaging" implies the same and,

[7]*Proceedings*, pp. 50–51.

moreover, a use-value to the owner. Is the life-extension of certain individuals then a public interest? "Affording" implies a critically vital level of such an interest—that is, of the loss or gain involved. And "society" itself—what is it? When does a need, an aim, an obligation become social? Let us reflect on some of these terms.

VI. What Society Can Afford

"Can society afford . . .?" Afford what? To let people die intact, thereby withholding something from other people who desperately need it, who in consequence will have to die too? These other, unfortunate people indeed cannot afford not to have a kidney, heart, or other organ of the dying patient, on which they depend for an extension of their lease on life; but does that give them a right to it? And does it oblige society to procure it for them? What is it that *society* can or cannot afford—leaving aside for the moment the question of what it has a *right* to? It surely can afford to lose members through death; more than that, it is built on the balance of death and birth decreed by the order of life. This is too general, of course, for our question, but perhaps it is well to remember. The specific question seems to be whether society can afford to let some people die whose death might be deferred by particular means if these were authorized by society. Again, if it is merely a question of what society can or cannot afford, rather than of what it ought or ought not to do, the answer must be: Of course, it can. If cancer, heart disease, and other organic, noncontagious ills, especially those tending to strike the old more than the young, continue to exact their toll at the normal rate of incidence (including the toll of private anguish and misery), society can go on flourishing in every way.

Here, by contrast, are some examples of what, in sober truth, society cannot afford. It cannot afford to let an epidemic rage unchecked; a persistent excess of deaths over births, but neither—we must add—too great an excess of births over deaths; too low an average life expectancy even if demographically balanced by fertility, but neither too great a longevity with the necessitated correlative dearth of youth in the social body; a debilitating state of general health; and things of this kind. These are plain cases where the whole condition of society is critically affected, and the public interest can make its imperative claims. The Black Death of the Middle Ages was a *public* calamity of the acute kind; the life-sapping ravages of endemic malaria or sleeping sickness in certain areas are a public calamity of the chronic kind. Such situations a society as a whole can truly not "afford," and they may call for extraordinary remedies, including, perhaps, the invasion of private sacrosanctities.

This is not entirely a matter of numbers and numerical ratios. Society, in a subtler sense, cannot "afford" a single miscarriage of justice, a single inequity in the dispensation of its laws, the violation of the rights of even the tiniest minority, because these undermine the moral basis on which society's existence rests. Nor can it, for a similar reason, afford the absence or atrophy in its midst of compassion and of the effort to alleviate suffering–be it widespread or rare–one form of which is the effort to conquer disease of any kind, whether "socially" significant (by reason of number) or not. And in short, society cannot afford the absence among its members of *virtue* with its readiness for sacrifice beyond defined duty. Since its presence–that is to say, that of personal idealism–is a matter of grace and not of decree, we have the paradox that society depends for its existence on intangibles of nothing less than a religious order, for which it can hope, but which it cannot enforce. All the more must it protect this most precious capital from abuse.

For what objectives connected with the medico-biological sphere should this reserve be drawn upon–for example, in the form of accepting, soliciting, perhaps even imposing the submission of human subjects to experimentation? We postulate that this must be not just a worthy cause, as any promotion of the health of anybody doubtlessly is, but a cause qualifying for transcendent social sanction. Here one thinks first of those cases critically affecting the whole condition, present and future, of the community we have illustrated. Something equivalent to what in the political sphere is called "clear and present danger" may be invoked and a state of emergency proclaimed, thereby suspending certain otherwise inviolable prohibitions and taboos. We may observe that averting a disaster always carries greater weight than promoting a good. Extraordinary danger excuses extraordinary means. This covers human experimentation, which we would like to count, as far as possible, among the extraordinary rather than the ordinary means of serving the common good under public auspices. Naturally, since foresight and responsibility for the future are of the essence of institutional society, averting disaster extends into long-term prevention, although the lesser urgency will warrant less sweeping licenses.

VII. Society and the Cause of Progress

Much weaker is the case where it is a matter not of saving but of improving society. Much of medical research falls into this category. As stated before, a permanent death rate from heart failure or cancer does not threaten society. So long as certain statistical ratios are maintained, the incidence of disease and of disease-induced mortality is not (in the strict sense) a "social" misfortune. I hasten to add that it is not therefore less of a human misfortune, and the call for relief issuing with silent eloquence from each victim and all potential victims is of no lesser dignity. But it is misleading to equate the fundamentally human response to it with what is owed to society: it is owed by man to man—and it is thereby owed by society to the individuals as soon as the adequate ministering to these concerns outgrows (as it progressively does) the scope of private spontaneity and is made a public mandate. It is thus that society assumes responsibility for medical care, research, old age, and innumerable other things not originally of the public realm (in the original "social contract"), and they become duties toward "society" (rather than directly toward one's fellow man) by the fact that they are socially operated.

Indeed, we expect from organized society no longer mere protection against harm and the securing of the conditions of our preservation, but active and constant improvement in all the domains of life: the waging of the battle against nature, the enhancement of the human estate—in short, the promotion of progress. This is an expansive goal, one far surpassing the disaster norm of our previous reflections. It lacks the urgency of the latter, but has the nobility of the free, forward thrust. It surely is worth sacrifices. It is not at all a question of what society can afford, but of what it is committed to, beyond all necessity, by our mandate. Its trusteeship has become an established, ongoing, institutionalized business of the body politic. As eager beneficiaries of its gains, we now owe to "society," as its chief agent, our individual contributions toward its *continued* pursuit. I emphasize "continued pursuit." Maintaining the existing level requires no more than the orthodox means of taxation and enforcement of professional standards that raise no problems. The more optional goal of pushing forward is also more exacting. We have this syndrome: Progress is by our choosing an acknowledged interest of society, in which we have a stake in various degrees; science is a necessary instrument of progress; research is a necessary instrument of science; and in medical science experimentation on human subjects is a necessary instrument of research. Therefore, human experimentation has come to be of societal interest.

The destination of research is essentially melioristic. It does not serve the preservation of the existing good from which I profit myself and to which I am obligated. Unless the present state is intolerable, the melioristic goal is in a sense gratuitous, and this not only from the vantage point of the present. Our descendants have a right to be left an unplundered planet; they do not have a right to new miracle cures. We have sinned against them, if by our doing we have destroyed their inheritance—which we are doing at full blast; we have not sinned against them, if by the time they come around arthritis has not yet been conquered (unless by sheer neglect). And generally, in the matter of progress, as humanity had no claim on a Newton, a Michelangelo, or a St. Francis to appear, and no right to the blessings of their unscheduled deeds, so progress, with all our methodical labor for it, cannot be budgeted in advance and its fruits received as a due. Its coming-about at all and its turning out for good (of which we can never be sure) must rather be regarded as something akin to grace.

VIII. The Melioristic Goal, Medical Research, and Individual Duty

Nowhere is the melioristic goal more inherent than in medicine. To the physician, it is not gratuitous. He is committed to curing and thus to improving the power to cure. Gratuitous we called it (outside disaster conditions) as a *social* goal, but noble at the same time. Both the nobility and the gratuitousness must influence the manner in which self-sacrifice for it is elicited, and even its free offer accepted. Freedom is certainly the first condition to be observed here. The surrender of one's body to medical experimentation is entirely outside the enforceable "social contract."

Or can it be construed to fall within its terms—namely, as repayment for benefits from past experimentation that I have enjoyed myself? But I am indebted for these benefits not to society, but to the past "martyrs," to whom society is indebted itself, and society has no right to call in my personal debt by way of adding new to its own. Moreover, gratitude is not an enforceable social obligation; it anyway does not mean that I must emulate the deed. Most of all, if it was wrong to exact such sacrifice in the first place, it does not become right to exact it again with the plea of the profit it has brought me. If, however, it was not exacted, but entirely free, as it ought to have been, then it should remain so, and its precedence must not be used as a social pressure on others for doing the same under the sign of duty.

Indeed, we must look outside the sphere of the social contract, outside the whole realm of public rights and duties, for the motivations and norms by which we can expect ever again the upwelling of a will to give what nobody—neither society, nor fellow man, nor posterity—is entitled to. There are such dimensions in man with trans-social wellsprings of conduct, and I have already pointed to the paradox, or mystery, that society cannot prosper without them, that it must draw on them, but cannot command them.

What about the moral law as such a transcendent motivation of conduct? It goes considerably beyond the public law of the social contract. The latter, we saw, is founded on the rule of enlightened self-interest: *Do ut des*—I give so that I be given to. The law of individual conscience asks more. Under the Golden Rule, for example, I am required to give as I wish to be given to under like circumstances, but not in order that I be given to and not in expectation of return. **Reciprocity, essential to the social law, is not a condition of the moral law.** One subtle "expectation" and "self-interest," but of the moral order itself, may even then be in my mind: I prefer the environment of a moral society and can expect to contribute to the general morality by my own example. But even if I should always be the dupe, the Golden Rule holds. (If the social law breaks faith with me, I am released from its claim.)

IX. Moral Law and Transmoral Dedication

Can I, then, be called upon to offer myself for medical experimentation in the name of the moral law? *Prima facie,* the Golden Rule seems to apply. I should wish, were I dying of a disease, that enough volunteers in the past had provided enough knowledge through the gift of their bodies that I could now be saved. I should wish, were I desperately in need of a transplant, that the dying patient next door had agreed to a definition of death by which his organs would become available to me in the freshest possible condition. I surely should also wish, were I drowning, that somebody would risk his life, even sacrifice his life, for mine.

But the last example reminds us that only the negative form of the Golden Rule ("Do not do unto others what you do not want done unto yourself") is fully prescriptive. The positive form ("Do unto others as you would wish them to do unto you"), in whose compass our issue falls, points into an infinite, open horizon where prescriptive force soon ceases. We may well say of somebody that he ought to have come to the succor of B, to have shared with him in his need, and the like. But we may not say that he ought to have given his life for him. To have done so would be praiseworthy; not to have done so is not blameworthy. It cannot be asked of him; if he fails to do so, he reneges on no duty. But *he* may say of himself, and only he, that he ought to have given his life. *This* "ought" is strictly between him and himself, or between him and God; no outside party—fellow man or society—can appropriate its voice. It can humbly receive the superogatory gifts from the free enactment of it.

We must, in other words, distinguish between moral obligation and the much larger sphere of moral value. (This, incidentally, shows up the error in the widely held view of value theory that the higher a value, the stronger its claim and the greater the duty to realize it. The highest are in a region beyond duty and claim.) The ethical dimension far exceeds that of the moral law and reaches into the sublime solitude of dedication and ultimate commitment, away from all reckoning and rule—in short, into the sphere of the *holy.* From there alone can the offer of self-sacrifice genuinely spring, and this source of it must be honored religiously. How? The first duty here falling on the research community, when it enlists and uses this source, is the safeguarding of true authenticity and spontaneity.

X. The "Conscription" of Consent

But here we must realize that the mere issuing of the appeal, the calling for volunteers, with the moral and social pressures it inevitably generates, amounts even under the most meticulous rules of consent to a sort of *conscripting.* And some soliciting is necessarily involved. This was in part meant by the earlier remark that in this area sin and guilt can perhaps not be wholly avoided. And this is why "consent," surely a non-negotiable minimum requirement, is not the full answer to the problem. Granting then that soliciting and therefore some degree of conscripting are part of the situation, who may conscript and who may be conscripted? Or less harshly expressed: Who should issue appeals and to whom?

The naturally qualified issuer of the appeal is the research scientist himself, collectively the main carrier of the impulse and the only one with the technical competence to judge. But his being very much an interested party (with vested interests, indeed, not purely in the public good, but in the scientific enterprise as such, in "his" project, and even in his career) makes him also suspect. The ineradicable dialectic of this situation—a delicate incompatibility problem—calls for particular controls by the research community and by public authority that we need not discuss. They can mitigate, but not eliminate the problem. We have to live with the ambiguity, the treacherous impurity of everything human.

XI. Self-Recruitment of the Scientific Community

To whom should the appeal be addressed? The natural issuer of the call is also the first natural addressee: the physician-researcher himself and the scientific confraternity at large. With such a coincidence—indeed, the noble tradition with which the whole business of human experimentation started—almost all of the associated legal, ethical, and metaphysical problems vanish. If it is full, autonomous identification of the subject with the purpose that is required for the dignifying of his serving as a subject—here it is; if strongest motivation—here it is; if fullest understanding—here it is; if freest decision—here it is; if greatest integration with the person's total, chosen pursuit—here it is. With the fact of self-solicitation the issue of consent in all its insoluble equivocality is bypassed *per se.* Not even the condition that the particular purpose be truly important and the project reasonably promising, which must hold in any solicitation of others, need be satisfied here. By himself, the scientist is free to obey his obsession, to play his hunch, to wager on chance, to follow the lure of ambition. It is all part of the "divine madness" that somehow animates the ceaseless pressing against frontiers. For the rest of society, which has a deep-seated disposition to look with reverence and awe upon the guardians of the mysteries of life, the profession assumes with this proof of its devotion the role of a self-chosen, consecrated fraternity, not unlike the monastic orders of the past, and this would come nearest to the actual, religious origins of the art of healing.

It would be the ideal, but is not a real solution, to keep the issue of human experimentation within the research community itself. Neither in numbers nor in variety of material would its potential suffice for the many-pronged, systematic, continual attack on disease into which the lonely exploits of the early investigators have grown. Statistical requirements alone make their voracious demands; and were it not for what I have called the essentially "gratuitous" nature of the whole enterprise of progress, as against the mandatory respect for invasion-proof self-hood, the simplest answer would be to keep the whole population enrolled, and let the lot, or an equivalent of draft boards, decide which of each category will at any one time be called up for "service." It is not difficult to picture societies with whose philosophy this would be consonant. We are agreed that ours is not one such and should not become one. The specter of it is indeed among the threatening utopias on our own horizon from which we should recoil, and of whose advent by imperceptible steps we must beware. How then can our mandatory faith be honored when the recruitment for experimentation goes outside the scientific community, as it must in honoring another commitment of no mean dignity? We simply repeat the former question: To whom should the call be addressed?

XII. "Identification" as the Principle of Recruitment in General

If the properties we adduced as the particular qualifications of the members of the scientific fraternity itself are taken as general criteria of selection, then one should look for additional subjects where a maximum of identification, understanding, and spontaneity can be expected—that is, among the most highly motivated, and most highly educated, and the least "captive" members of the community. From this naturally scarce resource, a descending order of permissibility leads to greater abundance and ease of supply, whose use should become proportionately more hesitant as the exculpating criteria are relaxed. An inversion of normal "market" behavior is demanded here—namely, to accept the lowest quotation last (and excused only by the greatest pressure of need); to pay the highest price first.

The ruling principle in our considerations is that the "wrong" of reification can only be made "right" by such authentic identification with the cause that it is the subject's as well as the researcher's cause—whereby his role in its service is not just permitted by him, but *willed*. That sovereign will of his which embraces the end as his own restores his personhood to the otherwise depersonalizing context. To be valid it must be autonomous and informed. The latter condition can, outside the research community, only be fulfilled by degrees; but the higher the degree of the understanding regarding the purpose and the technique, the more valid becomes the endorsement of the will. A margin of mere trust inevitably remains. Ultimately, the appeal for volunteers should seek this free and generous endorsement, the appropriation of the research purpose into the person's own scheme of ends. Thus, the appeal is in truth addressed to the one, mysterious, and sacred source of any such generosity of the will—"devotion," whose forms and objects of commitment are various and may invest different motivations in different individuals. The following, for instance, may be responsive to the "call" we are discussing: compassion with human suffering, zeal for humanity, reverence for the Golden Rule, enthusiasm for progress, homage to the cause of knowledge, even longing for sacrificial justification (do not call that "masochism," please). On all these, I say, it is defensible and right to draw when the research objective is worthy enough; and it is a prime duty of the research community (especially in view of what we called the "margin of trust") to see that this sacred source is never abused for frivolous ends. For a less than adequate cause, not even the freest, unsolicited offer should be accepted.

XIII. The Rule of the "Descending Order" and Its Counter-Utility Sense

We have laid down what must seem to be a forbidding rule to the number-hungry research industry. Having faith in the transcendent potential of man, I do not fear that the "source" will ever fail a society that does not destroy it—and only such a one is worthy of the blessings of progress. But "elitistic" the rule is (as is the enterprise of progress itself), and elites are by nature small. The combined attribute of motivation and information, plus the absence of external pressures, tends to be socially so circumscribed that strict adherence to the rule might numerically starve the research process. This is why I spoke of a descending order of permissibility, which is itself permissive, but where the realization that it is a *descending* order is not without pragmatic import. Departing from the august norm, the appeal must needs shift from idealism to docility, from high-mindedness to compliance, from judgment to trust. Consent spreads over the whole spectrum. I will not go into the casuistics of this penumbral area. I merely indicate the principle of the order of preference: The poorer in knowledge, motivation, and freedom of decision (and that, alas, means the more readily available in terms of numbers and possible manipulation), the more sparingly and indeed reluctantly should the reservoir be used, and the more compelling must therefore become the countervailing justification.

Let us note that this is the opposite of a social utility standard, the reverse of the order by "availability and expendability": The most valuable and scarcest, the least expendable elements of the social organism, are to be the first candidates for risk and sacrifice. It is the standard of *noblesse oblige;* and with all its counter-utility and seeming "wastefulness," we feel a rightness about it and perhaps even a higher "utility," for the soul of the community lives by this spirit.[8] It is also the opposite of what the day-to-day interests of research clamor for, and for the scientific community to honor it will mean that it will have to fight a strong temptation to go by routine to the readiest sources of supply—the suggestible, the ignorant, the dependent, the "captive" in various senses.[9] I do not believe that heightened resistance here must cripple research, which cannot be permitted; but it may indeed slow it down by the smaller numbers fed into experimentation in consequence. This price—a possibly slower rate of progress—may have to be paid for the preservation of the most precious capital of higher communal life.

XIV. Experimentation on Patients

So far we have been speaking on the tacit assumption that the subjects of experimentation are recruited from among the healthy. To the question "Who is conscriptable?" the spontaneous answer is: least and last of all the sick—the most available of all as they are under treatment and observation anyway. That the afflicted should not be called upon to bear additional burden and risk, that they are society's special trust and the physician's trust in particular—these are elementary responses of our moral sense. Yet the very destination of medical research, the conquest of disease, requires at the crucial stage trial and verification on precisely the sufferers from the disease, and their total exemption would defeat the purpose itself. In acknowledging this inescapable necessity, we enter the most sensitive area of the whole complex, the one most keenly felt and most searchingly discussed by the practitioners themselves. No wonder, for it touches the heart of the doctor–patient relation, putting its most solemn obligations to the test. There is nothing new in what I have to say about the ethics of the

[8] Socially, everyone is expendable relatively—that is, in different degrees; religiously, no one is expendable absolutely: the "image of God" is in all. If it can be enhanced, then not by anyone being expended, but by someone expending himself.

[9] This refers to captives of circumstance, not of justice. Prison inmates are, with respect to our problem, in a special class. If we hold to some idea of guilt, and to the supposition that our judicial system is not entirely at fault, they may be held to stand in a special debt to society, and their offer to serve—from whatever motive—may be accepted with a minimum of qualms as a means of reparation.

doctor–patient relation, but for the purpose of confronting it with the issue of experimentation some of the oldest verities must be recalled.

A. *The Fundamental Privilege of the Sick*

In the course of treatment, the physician is obligated to the patient and to no one else. He is not the agent of society, nor of the interests of medical science, nor of the patient's family, nor of his cosufferers, or future sufferers from the same disease. The patient alone counts when he is under the physician's care. By the simple law of bilateral contract (analogous, for example, to the relation of lawyer to client and its "conflict of interest" rule), the physician is bound not to let any other interest interfere with that of the patient in being cured. But, manifestly, more sublime norms than contractual ones are involved. We may speak of a sacred trust; strictly by its terms, the doctor is, as it were, alone with his patient and God.

There is one normal exception to this—that is, to the doctor's not being the agent of society vis-à-vis the patient, but the trustee of his interests alone: the quarantining of the contagious sick. This is plainly not for the patient's interest, but for that of others threatened by him. (In vaccination, we have a combination of both: protection of the individual and others.) But preventing the patient from causing harm to others is not the same as exploiting him for the advantage of others. And there is, of course, the abnormal exception of collective catastrophe, the analog to a state of war. The physician who desperately battles a raging epidemic is under a unique dispensation that suspends in a nonspecifiable way some of the strictures of normal practice, including possibly those against experimental liberties with his patients. No rules can be devised for the waiving of rules in extremities. And as with the famous shipwreck examples of ethical theory, the less said about it the better. But what is allowable there and may later be passed over in forgiving silence cannot serve as a precedent. We are concerned with nonextreme, nonemergency conditions where the voice of principle can be heard and claims can be adjudicated free from duress. We have conceded that there are such claims, and that if there is to be medical advance at all, not even the superlative privilege of the suffering and the sick can be kept wholly intact from the intrusion of its needs. About this least palatable, most disquieting part of our subject, I have to offer only groping, inconclusive remarks.

B. *The Principle of "Identification" Applied to Patients*

On the whole, the same principles would seem to hold here as are found to hold with "normal subjects": motivation, identification, understanding on the part of the subject. But it is clear that these conditions are peculiarly difficult to satisfy with regard to a patient. His physical state, psychic preoccupation, dependent relation to the doctor, the submissive attitude induced by treatment—everything connected with his condition and situation makes the sick person inherently less of a sovereign person than the healthy one. Spontaneity of self-offering has almost to be ruled out; consent is marred by lower resistance or captive circumstance, and so on. In fact, all the factors that make the patient, as a category, particularly accessible and welcome for experimentation at the same time compromise the quality of the responding affirmation that must morally redeem the making use of them. This, in addition to the primacy of the physician's duty, puts a heightened onus on the physician-researcher to limit his undue power to the most important and defensible research objectives and, of course, to keep persuasion at a minimum.

Still, with all the disabilities noted, there is scope among patients for observing the rule of the "descending order of permissibility" that we have laid down for normal subjects, in vexing inversion of the utility order of quantitative abundance and qualitative "expendability." By the principle of this order, those patients who most identify with and are cognizant of the cause of research—members of the medical profession (who after all are sometimes patients themselves)—come first; the highly motivated and educated, also least dependent, among the lay patients come next; and so on down the line. An added consideration here is seriousness of condition, which again operates in inverse proportion. Here the profession must fight the tempting sophistry that the hopeless case is expendable (because in prospect already expended) and therefore especially usable; and generally the attitude that the poorer the chances of the patient the more justifiable his recruitment for experimentation (other than for his own benefit). The opposite is true.

C. Nondisclosure as a Borderline Case

Then there is the case where ignorance of the subject, sometimes even of the experimenter, is of the essence of the experiment (the "double blind"-control group-placebo syndrome). It is said to be a necessary element of the scientific process. Whatever may be said about its ethics in regard to normal subjects, especially volunteers, it is an outright betrayal of trust in regard to the patient who believes that he is receiving treatment. Only supreme importance of the objective can exonerate it, without making it less of a transgression. The patient is definitely wronged even when not harmed. And ethics apart, the practice of such deception holds the danger of undermining the faith in the bona fides of treatment, the beneficial intent of the physician—the very basis of the doctor–patient relationship. In every respect, it follows that concealed experiment on patients—that is, experiment under the guise of treatment—should be the rarest exception, at best, if it cannot be wholly avoided.

This has still the merit of a borderline problem. The same is not true of the other case of necessary ignorance of the subject—that of the unconscious patient. Drafting him for nontherapeutic experiments is simply and unqualifiedly impermissible; progress or not, he must never be used, on the inflexible principle that utter helplessness demands utter protection.

When preparing this paper, I filled pages with a casuistics of this harrowing field, but then scrapped most of it, realizing my dilettante status. The shadings are endless, and only the physician-researcher can discern them properly as the cases arise. Into his lap the decision is thrown. The philosophical rule, once it has admitted into itself the idea of a sliding scale, cannot really specify its own application. It can only impress on the practitioner a general maxim or attitude for the exercise of his judgment and conscience in the concrete occasions of his work. In our case, I am afraid, it means making life more difficult for him.

It will also be noted that, somewhat at variance with the emphasis in the literature, I have not dwelt on the element of "risk" and very little on that of "consent." Discussion of the first is beyond the layman's competence; the emphasis on the second has been lessened because of its equivocal character. It is a truism to say that one should strive to minimize the risk and to maximize the consent. The more demanding concept of "identification," which I have used, includes "consent" in its maximal or authentic form, and the assumption of risk is its privilege.

XV. No Experiments on Patients Unrelated to Their Own Disease

Although my ponderings have, on the whole, yielded points of view rather than definite prescriptions, premises rather than conclusions, they have led me to a few unequivocal yeses and noes. The first is the emphatic rule that patients should be experimented upon, if at all, *only* with reference to *their disease.* Never should there be added to the gratuitousness of the experiment as such the gratuitousness of service to an unrelated cause. This follows simply from what we have found to be the *only* excuse for infracting the special exemption of the sick at all—namely, that the scientific war on disease cannot accomplish its goal without drawing the sufferers from disease into the investigative process. If under this excuse they become subjects of experiment, they do so *because,* and only because, of *their* disease.

This is the fundamental and self-sufficient consideration. That the patient cannot possibly benefit from the unrelated experiment therapeutically, while he might from experiment related to his condition, is also true, but lies beyond the problem area of pure experiment. I am in any case discussing nontherapeutic experimentation only, where *ex hypothesi* the patient does not benefit. Experiment as part of therapy–that is, directed toward helping the subject himself–is a different matter altogether and raises its own problems, but hardly philosophical ones. As long as a doctor can say, even if only in his own thought: "There is no known cure for your condition (or: you have responded to none); but there is promise in a new treatment still under investigation, not quite tested yet as to effectiveness and safety; you will be taking a chance, but all things considered, I judge it in your best interest to let me try it on you"–as long as he can speak thus, he speaks as the patient's physician and may err, but does not transform the patient into a subject of experimentation. Introduction of an untried therapy into the treatment where the tried ones have failed is not "experimentation on the patient."

Generally, and almost needless to say, with all the rules of the book, there is something "experimental" (because tentative) about every individual treatment, beginning with the diagnosis itself; and he would be a poor doctor who would not learn from every case for the benefit of future cases, and a poor member of the profession who would not make any new insights gained from his treatments available to the profession at large. Thus, knowledge may be advanced in the treatment of any patient, and the interest of the medical art and sufferers from the same affliction as well as the patient himself may be served if something happens to be learned from his case. But this gain to knowledge and future therapy is incidental to the bona fide service to the present patient. He has the right to expect that the doctor does nothing to him just in order to learn.

In that case, the doctor's imaginary speech would run, for instance, like this: "There is nothing more I can do for you. But you can do something for me. Speaking no longer as your physician but on behalf of medical science, we could learn a great deal about future cases of this kind if you would permit me to perform certain experiments on you. It is understood that you yourself would not benefit from any knowledge we might gain; but future patients would." This statement would express the purely experimental situation, assumedly here with the subject's concurrence and with all cards on the table. In Alexander Bickel's words: "It is a different situation when the doctor is no longer trying to make [the patient] well, but is trying to find out how to make others well in the future.[10]"

[10] *Proceedings,* p. 33. To spell out the difference between the two cases: In the first case, the patient himself is meant to be the beneficiary of the experiment, and directly so; the "subject" of the experiment is at the same time its object, its end. It is performed not for gaining knowledge, but for helping him–and helping him in the *act* of performing it, even if by its results it also contributes to a broader testing process currently under way. It is in fact part of the treatment itself and an "experiment" only in the loose sense of being untried and highly tentative. But whatever the degree of uncertainty, the motivating

But even in the second case, that of the nontherapeutic experiment where the patient does not benefit, at least the patient's own disease is enlisted in the cause of fighting that disease, even if only in others. It is yet another thing to say or think: "Since you are here–in the hospital with its facilities–anyway, under our care and observation anyway, away from your job (or, perhaps, doomed) anyway, we wish to profit from your being available for some other research of great interest we are presently engaged in." From the standpoint of merely medical ethics, which has only to consider risk, consent, and the worth of the objective, there may be no cardinal difference between this case and the last one. I hope that the medical reader will not think I am making too fine a point when I say that from the standpoint of the subject and his dignity there is a cardinal difference that crosses the line between the permissible and the impermissible, and this by the same principle of "identification" I have been invoking all along. Whatever the rights and wrongs of any experimentation on any patient–in the one case, at least that residue of identification is left him that it is his own affliction by which he can contribute to the conquest of the affliction, his own kind of suffering which he helps to alleviate in others; and so in a sense it is his own cause. It is totally indefensible to rob the unfortunate of this intimacy with the purpose and make his misfortune a convenience for the furtherance of alien concerns. The observance of this rule is essential, I think, to at least attenuate the wrong that nontherapeutic experimenting on patients commits in any case.

XVI. On the Redefinition of Death

My other emphatic verdict concerns the question of the redefinition of death–that is, acknowledging "irreversible coma as a new definition for death."[11] I wish not to be misunderstood. As long as it is merely a question of when it is permitted to cease the artificial prolongation of certain functions (like heartbeat) traditionally regarded as signs of life, I do not see anything ominous in the notion of "brain death." Indeed, a new definition of death is not even

---

anticipation (the wager, if you like) is for success, and success here means the subject's own good. To a pure experiment, by contrast, undertaken to gain knowledge, the difference of success and failure is not germane, only that of conclusiveness and inconclusiveness. The "negative" result has as much to teach as the "positive." Also, the true experiment is an act distinct from the uses later made of the findings. And, most important, the subject experimented on is distinct from the eventual beneficiaries of those findings: He lets himself be used as a means toward an end external to himself (even if he should at some later time happen to be among the beneficiaries himself). With respect to his own present needs and his own good, the act is gratuitous.

[11] "A Definition of Irreversible Coma," Report of the *Ad Hoc* Committee of the Harvard Medical School to Examine the Definition of Brain Death, *Journal of the American Mental Association* 205, no. 6 (August 5, 1968), pp. 337–40.

necessary to legitimize the same result if one adopts the position of the Roman Catholic Church, which here at least is eminently reasonable—namely that "when deep unconsciousness is judged to be permanent, extraordinary means to maintain life are not obligatory. They can be terminated and the patient allowed to die."[12] Given a clearly defined negative condition of the brain, the physician is allowed to allow the patient to die his own death by *any* definition, which of itself will lead through the gamut of all possible definitions. But a disquietingly contradictory purpose is combined with this purpose in the quest for a new definition of death—that is, in the will to *advance* the moment of declaring him dead: Permission not to turn off the respirator, but, on the contrary, to keep it on and thereby maintain the body in a state of what would have been "life" by the older definition (but is only a "simulacrum" of life by the new)—so as to get at his organs and tissues under the ideal conditions of what would previously have been "vivisection."[13]

Now this, whether done for research or transplant purposes, seems to me to overstep what the definition can warrant. Surely it is one thing when to cease delaying death, another when to start doing violence to the body; one thing when to desist from protracting the process of dying, another when to regard that process as complete and thereby the body as a cadaver free for inflicting on it what would be torture and death to any living body. For the first purpose, we need not know the exact borderline between life and death—we leave it to nature to cross it wherever it is, or to traverse the whole spectrum if there is not just one line. All we need to know is that coma is irreversible. For the second purpose we must know the borderline with absolute certainty; and to use any definition short of the maximal for perpetrating on a *possibly* penultimate state what only the ultimate state can permit is to arrogate a knowledge which, I think, we cannot possibly have. *Since we do not know the exact borderline between life and death,* nothing less than the maximum definition of death will do—brain death plus heart death plus any other indication that may be pertinent—before final violence is allowed to be done.

[12] As rendered by Dr. Beecher in *Proceedings,* p. 50.

[13] The Report of the *Ad Hoc* Committee no more than indicates this possibility with the second of the "two reasons why there is need for a definition"; "(2) Obsolete criteria for the definition of death can lead to controversy in obtaining organs for transplantation." The first reason is relief from the burden of indefinitely drawn out coma. The report wisely confines its recommendations on application to what falls under this first reason—namely, turning off the respirator—and remains silent on the possible use of the definition under the second reason. But when "the patient is declared dead on the basis of these criteria," the road to the other use has theoretically been opened and will be taken (if I remember rightly, it has even been taken once, in a much debated case in England), unless it is blocked by a special barrier in good time. The above is my feeble attempt to help in doing so.

It would follow then, for this layman at least, that the use of the definition should itself be defined, and this in a restrictive sense. When only permanent coma can be gained with the artificial sustaining of functions, by all means turn off the respirator, the stimulator, any sustaining artifice, and let the patient die; but let him die all the way. Do not, instead, arrest the process and start using him as a mine while, with your own help and cunning, he is still kept this side of what may in truth be the final line. Who is to say that a shock, a final trauma, is not administered to a sensitivity diffusely situated elsewhere than in the brain and still vulnerable to suffering, a sensitivity that we ourselves have been keeping alive. No fiat of definition can settle this question.[14] But I wish to emphasize that the question of possible suffering (easily brushed aside by a sufficient show of reassuring expert consensus) is merely a subsidiary and not the real point of my argument; this, to reiterate, turns on the indeterminacy of the boundaries between *life and death,* not between sensitivity and insensitivity, and bids us to lean toward a maximal rather than a minimal determination of death in an area of basic uncertainty.

There is also this to consider: The patient must be absolutely sure that his doctor does not become his executioner, and that no definition authorizes him ever to become one. His right to this certainty is absolute, and so is his right to his own body with all its organs. Absolute respect for these rights violates no one else's right, for no one has a right to another's body. Speaking in still another, religious vein: The expiring moments should be watched over with piety and be safe from exploitation.

I strongly feel, therefore, that it should be made quite clear that the proposed new definition of death is to authorize *only* the one and *not* the other of the two opposing things: only to break off a sustaining intervention and let things take their course, not to keep up the sustaining intervention for a final intervention of the most destructive kind.

XVII. Conclusion

There would now have to be said something about nonmedical experiments on human subjects, notably psychological and genetic, of which I have not lost sight. But I must leave this for another occasion. I wish only to say in conclusion that if some of the practical implications of my reasonings are felt to work out toward a slower rate of progress, this should not cause too great dismay. Let us

[14] Only a Cartesian view of the "animal machine," which I somehow see lingering here, could set the mind at rest, as in historical fact it did at its time in the matter of vivisection. But its truth is surely not established by definition.

not forget that progress is an optional goal, not an unconditional commitment, and that its tempo in particular, compulsive as it may become, has nothing sacred about it. Let us also remember that a slower progress in the conquest of disease would not threaten society, grievous as it is to those who have to deplore that their particular disease be not yet conquered, but that society would indeed be threatened by the erosion of those moral values whose loss, possibly caused by too ruthless a pursuit of scientific progress, would make its most dazzling triumphs not worth having. Let us finally remember that it cannot be the aim of progress to abolish the lot of mortality. Of some ill or other, each of us will die. Our mortal condition is upon us with its harshness but also its wisdom—because without it there would not be the eternally renewed promise of the freshness, immediacy, and eagerness of youth; nor would there be for any of us the incentive to number our days and make them count. With all our striving to wrest from our mortality what we can, we should bear its burden with patience and dignity.

THE MORAL JUSTIFICATION FOR RESEARCH USING HUMAN SUBJECTS

ARTHUR J. DYCK
AND HERBERT W. RICHARDSON

Introduction

The use of human subjects in research that is of some risk to them is not only morally justifiable but, in certain instances, morally required. Two kinds of considerations lead to this conclusion: the limitations of research with animals and the stringent moral demand to alleviate or eradicate human suffering.

With respect to the limits of research with animals, there is surely no need here to rehearse the abundant evidence for the species-specific nature of various chemical reactions.[1] Until, therefore, much more about species differences and similarities is known, clinical research that establishes sound medical practice will have to rely upon the knowledge gleaned from the reactions of human subjects. In addition, whenever the use of certain food additives and environmental toxins is considered to be morally desirable, it will be necessary to establish safety levels for their use. In certain cases, these safety levels can only be established by using human subjects under carefully controlled test conditions.

Medical research affords abundant evidence that we sometimes require, on moral grounds, that humans be subjected to risk. The development and use of vaccines is but one case in point. Vaccines involve statistically minimal, though very real, risks to their recipients. Yet we have undertaken the necessary research with humans, once animals are tested, and now make compulsory the use of some of these vaccines despite our knowledge that persons can and do die from

From A. J. Dyck and H. W. Richardson, "The Moral Justification for Research Using Human Subjects," in *Use of Human Subjects in Safety Evaluation of Food Chemicals–Proceedings of a Conference.* National Academy of Sciences and National Research Council, Pub. 1491 (Wash., D.C., 1967), pp. 217–227. Reproduced with permission of the authors and the National Academy of Sciences.

[1] For a thorough discussion of the issues at stake here, see other papers in *Use of Human Subjects in Safety Evaluation of Food Chemicals–Proceedings of a Conference:* "Limitations of Animal Data for Predicting Safety for Man" by Karl Beyer (page 43) and "Limitations in the Value of Studies in Human Subjects" by Alastair Frazer (page 63).

them. The very great benefits derived from vaccines, and the very small risks associated with their use, serve as the moral justification for what we do.

The benefits obtained by the use of vaccines and a vast array of other medical products are not luxury items; they are designed to prevent existing, highly undesirable threats to our existence. The struggle against existing threats to human life has a much higher moral urgency for us than the possibility to create benefits that add to human pleasure. In instances where great harm is to be prevented, the use of human beings in experiments involving risk is much more easily justified and more apt to be viewed as morally required. However, even risk-experiments for the sake of ascertaining safety levels for the use of food additives and pesticides, which are primarily for the sake of increasing human pleasure, can also be morally justified under certain circumstances.

The question, then, is not whether we shall at all use human subjects in scientific research or whether such use can ever be justified. The real question, the question we wish to deal with in this paper, has to do rather with the conditions under which we would consider the use of human subjects to be justifiable in those instances where the research in which they are to be used may, in some way, harm them. We shall take the position that the use of humans in research involving risk to them can only be justified if the following two requirements are satisfied: such research must anticipate specific benefits accruing from its results, and these benefits must more than offset all the potential, associated risks, and such research must be in accord with certain inviolable moral values that can only be secured by satisfying the requirement of *informed consent.* We shall distinguish three types of research, namely, medical research, research with food additives, and research with environmental toxins, and examine what it takes in each of these cases to comply with our two moral requirements for justifying research involving risk to humans.

It should be noted that nothing we have said thus far, and nothing we expect to say, should be construed as an attack upon the general value of sheer knowledge or of sheer research for the sake of research. No research project, however, is totally exempt from moral scrutiny. It is important that the evaluation of any contemplated research project consider the possible harm such research might inflict upon human beings, for in doing research for the sake of research, we could, at some point, seriously undermine the quality of our lives in our very efforts to enhance it. In short, the value of research, including its value as a contribution to knowledge, can be offset, in a given instance, by some harm it may cause.

Two Types of Value

Let us now characterize the two sets of moral criteria by which the use of humans in research involving risks to them is to be morally justified. These criteria are specified by two types of moral values: benefits and structural values. Benefits may be determined by the calculation of relative pleasures and pains (or benefits and harms); structural values are maintained by satisfying the requirement of informed consent.

Benefits presuppose specific social systems. Benefits are for the sake of increasing happiness by sustaining and improving the quality of human life. What things are considered to be benefits and the means thereto are matters decided upon and adjudicated by the members and institutions of a given social system. Social systems, to the extent that they are congruent with the aspirations of their members, seek to maximize these benefits.

Any social system not only values the benefits that secure the happiness of its members, but also values restraints upon actions that threaten harm to any of its members. Activities that carry with them a risk of harm to human welfare can, however, sometimes be morally sanctioned when they promise to bring about very great benefits or to prevent very great suffering. There is, then, a contextually calculated ratio of benefits and harms by means of which potentially harmful actions can be morally evaluated.

Although there are harms that can be balanced against benefits to help us decide whether a certain risk is morally justifiable, there is a certain type of harm that cannot be outweighed by any benefit, no matter how great that benefit may be. This "harm" is to the second kind of moral value we have mentioned above, that is, violation of a structural value. The loss of benefits is tolerable within a pleasure–pain calculus, but "harm" to, or violation of, structural values is not tolerable since these values maintain the social systems that are the presupposition of all benefits. To violate a structural value, therefore, is to violate the very condition that makes benefits possible and meaningful. With regard to experiments involving human beings, the underlying network of structural values is protected by the requirement that no person be an experimental subject without first giving his informed consent.

Among the inviolable structural values that the requirement of informed consent protects are veracity, freedom, and justice: veracity refers to relationships characterized by promise-keeping and truth-telling; freedom refers to a person's general right to pursue his own happiness and to decide what risks to his own body and his own welfare he will take; justice refers to the distribution of goods in accord with a standard of equity and to redress for undeserved harm. These are values people will die to preserve, values without which social systems and interpersonal relations break down and cannot function. Because they are essential to the very existence of society itself, these structural values may be distinguished from all those other "benefits" that society seeks to attain—benefits that are legitimately ascertained within a benefit–harm calculus. Structural values are not part of this benefit–harm calculus because the loss of such values is so harmful that benefits become meaningless. What meaning would "benefits" have for those who had no freedom, or for those who were victims of extreme injustice, or for those who lived in a society where no one could be trusted and where no contract was honored?

Three Types of Research

Having clarified the two sets of moral criteria to be satisfied if investigations that involve risk to their human subjects are to be morally justified, it is now necessary to see how these criteria can be satisfied. The calculation of benefit–harm ratios and the provisions for informed consent vary considerably from one situation to another. For this reason, it is impossible to assume that what is being said about these matters in medical ethics will apply to research with food additives and environmental toxins. Indeed, the discussion that now follows will distinguish between the situations created by medical research, research with food additives, and research with environmental toxins, and seek in turn to demonstrate the marked differences in what it takes, in each of these instances, to calculate benefit–harm ratios and to honor the principle of informed consent.

It should be recalled that all the research that concerns us in this paper involves risk to humans. For this reason, such research must contemplate some specific benefit that can justify the risk inherent in the research itself. Moreover, if the specific kind of research being undertaken is for the sake of establishing *safety levels* for the use of particular drugs, food additives, or pesticides, then we must assume that the particular benefit being contemplated by such research as its justification is the *use* of such drugs, food additives, or pesticides. For this reason, the decision to undertake research involving safety levels implies also a decision to use the materials under investigation should this investigation prove successful, and, of course, it is only undertaken on the supposition that it will prove successful. If, on the other hand, someone would wish to argue that the decision to undertake research involving safety levels for the use of drugs, food additives, or pesticides does not imply a decision to use such materials if that research is successful, then the obvious reply is, "Why expose people to the risks of such investigation when no benefits accruing from the use of such materials are contemplated?"

By noting that the decision to test for safety levels implies a decision to use a given material because of its contemplated benefits, we have made a distinction that helps us to discriminate among various kinds of research involving humans. In certain kinds of research, e.g., medical research, the "test conditions" and the "use conditions" closely resemble each other; in these cases, the contemplated benefits from use are usually not beyond the capacity of the individual investigator to determine. However, in other kinds of research, e.g., research with environmental toxins, the "test conditions" for safety levels and the "use conditions" may be very dissimilar. The contemplated benefit from use is not easily determined by the individual investigator since it involves not only possible harmful effects to the primary user, but also extended effects upon secondary users and even the entire ecological system.[2] Moreover, because test and use conditions vary from one kind of experiment to another, we must be wary of all arguments that attempt to justify one kind of experimentation involving humans from the accepted legitimacy of another kind of experimentation. We can better see why this is the case by considering the difference between test and use conditions in the three types of research under discussion in this paper.

In medical research the test conditions, and the conditions in which the results of the tests will be used, are quite comparable to one another. In both instances, the conditions are usually well controlled. A given drug, for example, is investigated under carefully controlled circumstances, and once the drug has been found to be acceptable, and a safe dosage has been established, the use of the drug will often be confined to circumstances that are quite similar to those under which the drug was accepted for use. Furthermore, the effects of the drug, whatever they may be in use, will generally be restricted to those who are given the drug, to those, in other words, who stand to benefit directly from its use. In medical research, therefore, we have to deal with a comparatively simple investigator–subject relationship.[3]

When, however, we turn to research involving food additives and environmental toxins, we are confronted with a much more complex nexus of relationships. Investigations in both of these instances encompass much more than simply risk to the subject under the controlled conditions of the laboratory. In the case of food additives, there are sometimes risks to the people who eat the food, risks that it is the purpose of the investigator to minimize by establishing safety levels.[4] However, it is not easy to determine the extent to which a given food will be eaten and the extent to which, therefore, the controlled conditions of the laboratory are or are not approximated in society. It is also difficult to

[2] The terms "primary" and "secondary" user need to be roughly clarified. Primary users are those who directly seek to benefit from the use or consumption of the chemical in question; secondary users are those who are inadvertently affected by the use or consumption of a given chemical. Thus the farmer who sprays his apple orchard is a primary user of a pesticide; those who eat apples from that orchard are secondary users of that pesticide.

[3] Drugs are not, of course, always used under carefully controlled conditions. Antiobiotics have been used as a growth food and medicine for animals and also to preserve freshness in raw poultry, fish, and shellfish. It is noteworthy that the FDA has found it necessary to rescind hitherto permitted residues in the latter case as of August 23, 1966, and to investigate the former use of antibiotics as growth food and medicine.

[4] One should not assume from this discussion that food additives generally are toxic substances and that safety levels must always be established for their use. Some food additives, for example, are nutritionally beneficial and widely used foods and need not, therefore, be tested for their safety.

know or to predict the long-term effects of ingesting food additives in dosages that are otherwise harmless. Research involving food additives differs from medical research in still another way. Whereas in medicine it is sometimes quite difficult or even impossible to develop alternatives to certain risky therapeutic procedures, in research with food additives it is usually easier to do so. We can think here, for example, of a certain flavoring that may be quite toxic, a flavoring, however, that may have a less toxic flavoring substituted for it, a flavoring that will be much less of a risk both to the subject of any investigation and hence to anyone who will eventually eat the food.

If we turn to investigations with environmental toxins (e.g., pesticides), we note that these not only involve risks to the subjects of such investigations and risks to many other people, but even risks to a considerable portion of the environment in which people live. Environmental toxins may very well have unpredictable or even irreversibly ill effects upon our environment, so that the use of these does not depend solely upon the establishment of safety levels for subjects under highly controlled situations. In such cases, this means that the individual investigator is not in a position to decide for himself whether a given study is to be undertaken. What food additives and environmental toxins will be used in research and subsequently in the environment is something that demands considerable knowledge of the larger environment within which these products of research will be employed. In connection with environmental toxins, as with food additives, we should bear in mind that alternatives exist and can be developed for achieving the same effects, alternatives that minimize risks and maximize benefits, alternatives that are subject to greater degrees of control in use.

The Moral Justification

To summarize briefly what we have been saying, let us note that as we move from medical research, to research with food additives, to research with environmental toxins, the *uses* of the products of such research are more widespread in their effects, more difficult to control, and more likely to involve unanticipated hazards. For these reasons, the benefits accruing from the use of food additives and pesticides cannot be calculated with the precision usually characteristic of the products of medical research. Hence, the benefit–harm ratio that justifies risk-experiments involving food additives and pesticides is generally less exact than in medical experimentation; for this reason the preponderance of contemplated benefits over possible harms required to justify testing and using these materials must be concomitantly greater. There are, however, two steps that both can and should be taken in order to increase precision in calculating these benefit–harm ratios. We should increase our research into the complex and widespread environmental effects of possible food additives and pesticides in order to be better able to calculate their potential benefits and harms, and we should take steps to bring the use of such products under more exact control. Let us turn, therefore, to a consideration of these two matters.

Controlling Benefit–Harm Ratios

In this paper, we have stressed that the decision to test toxins on human subjects for the sake of establishing safety levels implies a decision to use these toxins if the tests are successful. Whenever there is a significant discrepancy between test and use conditions, the researcher is not in a position to determine whether the possible benefits arising from his tests are sufficient to justify the potential risks to the subjects of his proposed experiment. In these cases, there must be extensive gathering of information about the short- and long-range effects of various materials on all living beings and their environment; there must be a consideration of various alternatives to attain the same goal; and there must be follow-up studies of materials already in use in order to discover possible unanticipated effects. Because the gathering of such information is for the sake of making a disinterested judgment regarding the effect of any given product upon the human environment, we suggest the necessity for a public data-pooling agency that could coordinate and promote research concerning the effects of different substances. The weighing of possible benefits and harms belongs more properly to some such disinterested data-pooling agency than it does to the representatives of particular vested interests.[5]

Although this data-pooling agency should carry out and promote research in ecology, especially in those areas where our knowledge of the effects of various toxins on our environment is still deficient, it should not replace agencies already contributing such research. One example of a public agency that has done valuable research on the effects of environmental toxins is the United States Fish and Wildlife Service. It is essential that agencies like this receive adequate funds for research. Industries that develop and distribute environmental toxins also need to concern themselves with studies of the environment, and they should either sponsor research in this direction themselves or take steps to see that research of this kind is abundantly supported.

Extensive research into the effects of toxins and other materials on various human and biological systems and the pooling of this information in one or more centralized agencies will help to provide a clearer specification of possible

[5] Such a data-pooling agency should perhaps be a public corporation that would be responsible to both industry and government. *Consumer Reports* of March 1963 carried an article suggesting just such an agency. It should be noted that the agency proposed there was more limited in scope, limited as it was to the evaluation of the safety of drugs.

benefits to be attained from the use of particular chemicals. With respect to food additives, prior to testing them on human beings, there should be extensive study of their potential benefits, possible long-range effects, and consumption patterns. On this basis, the final testing of all such products to determine safety levels for use can be more exactly calculated in terms of benefit–harm ratios. Moreover, because the benefit–harm ratio is essential to morally justifying all experimentation involving human subjects, it should be clear that our emphasis on the necessity for data-pooling to determine possible effects *before* testing, and upon continuous follow-up to discover unanticipated effects *after* testing, bears directly upon the legitimacy of such experiments. Only as we are striving to increase our knowledge and control over the use of various chemicals can we adequately calculate benefit–harm ratios and discriminate among various experimental proposals. The heart of moral discrimination in these matters is knowledge and control, and progress here is essential if the increasing use of humans in experimental projects is to be at all justified.

A second way to gain increased control over our calculation of benefit–harm ratios is through more careful study of, and control over, the actual handling of chemicals by the primary users themselves. We have already seen that there is generally less control over the primary users in instances involving food additives and pesticides than in instances involving drugs. Not only are the former available without the equivalent of prescriptions, even in unlimited quantities, but directions for their use are often less exact and detailed and are followed with less precision.[6] To gain greater control over the actual use of these products, there must be greater stress on accurate, sufficiently complete labeling and education regarding the necessity for use as directed.

Even with these efforts, however, there will remain a clear difference in degree with respect to both the possibility of misuse and the calculation of the effects of possible misuse as we move from drugs to food additives to pesticides. The restrictions governing the use of many drugs make it possible, where such restrictions obtain, to calculate with considerable accuracy the results that accrue from their misuse. Since food additives are much more readily available, the effects of their misuse are more difficult to ascertain. Nevertheless, the effects of their misuse, as in the case of drugs, are likely to be confined to primary users, and the possible harm they may cause is calculable. But the possibility for the misuse of pesticides, and hence the calculation of benefit–harm ratios for such substances, is presently so great as to be almost out of

[6] In July 1965 *Consumer Reports* reprinted a letter warning gardeners to read and heed the labels on pesticides by a man who died, soon after he wrote the letter, of chlordane poisoning–he had been dusting his roses. His letter implied that, although the label did provide the necessary cautions for users, the dangers of misuse were not emphasized; certainly he did not expect what happened to him. *Consumer Reports* draws the conclusion that the wording on labels is as unalarming as the law allows.

control.[7] Not only are pesticides easily available in concentrated forms, but they may irreversibly affect both primary and secondary users and even entire ecological systems. As long as their use is relatively unrestricted, there is no precise way of calculating whether their purported benefits significantly outweigh their harm—or even outweigh their harm at all.

For these reasons, both pesticide research and the use of pesticides require much more stringent control against their possible misuse—if such products are to be morally justified. For example, it would not seem inappropriate that farmers and other large users of pesticides in concentrated forms be licensed to obtain and use them on the basis of qualifying examinations testing requisite competence. Misuse, thereafter, would constitute grounds for a fine or for revocation of the license. And, on the other hand, the availability of these same pesticides to the general public might be curtailed or otherwise regulated. By bringing the use and possibilities for misuse of pesticides and other environmental toxins under greater control, their potential benefits can be calculated more exactly.

Satisfying the Requirement of Informed Consent

The central contention of our paper, as we noted at the outset, is that experiments involving risk to human subjects should be justified by meeting the requirement of informed consent and by promising the attainment of certain benefits that considerably outweigh all possible harms. In discussing the latter criterion, we have tried to show that in all experiments involving risk to human subjects, where these subjects are not themselves the immediate beneficiaries of the knowledge gained, the difficulty of calculating benefit–harm ratios varies in direct proportion to the degree of dissimilarity between test and use conditions. We have argued, therefore, that such calculations of the benefits and harms associated with the use of a given product, calculations that are needed for morally justifying experiments using humans to determine the safety level of that product, depend upon a thorough knowledge of the effects of that product and a stringent control over its use. Without such knowledge and control, unanticipated harms may arise from the use and/or misuse of a given product which may outweigh contemplated benefits.

As we have sought to demonstrate in the second section of our paper, however, a highly favorable balance of pleasure over pain (benefit over harm)

[7]We are not unaware of the existing power of the FDA to establish tolerances for pesticide residues in foods. Laudable and necessary as this is, it does not prevent residues in the Mississippi River lethal enough to kill numerous fish, residues now to be found in the drinking water of New Orleans. (See *Consumer Reports* of June 1964.)

will not by itself morally justify investigations that involve risks to humans. Such investigations must also satisfy the principle of informed consent. How, then, can the requirement of informed consent be met by the three types of research under discussion in this paper?

Satisfying the requirement of informed consent for medical research, though problematic, is certainly easier on the whole than it is for research with food additives and environmental toxins. In the medical setting, it is essential that the physician–patient and investigator–subject relations be kept distinct.[8] The physician as physician is pledged to the welfare of his patient and to the use of practices in accordance with the standards of his profession.[9] Striving to keep the patient as informed as possible, the physician will be called upon, nevertheless, to make difficult but necessary judgments as to what information, if any, might jeopardize the recovery of his patient.[10]

Informed consent should also be the general guide for the investigator in the clinical setting. It is not always technically possible, however, to obtain the direct consent of the subjects of a clinical investigation.[11] When this is the case, only those persons should be sought as subjects whose physicians are not participants in the research under consideration. This stipulation permits the physician in charge of a given patient to serve as that patient's "advocate," representing his interests and withholding him from serving as a subject whenever the risks to the patient seem excessive or unnecessary.[12] Additional

[8] The importance of this distinction has been recognized in the literature. See, for example, Henry K. Beecher, "Experimentation in Man," *J. Am. Med. Assoc.*, *169*, 461–478 (1959); O. E. Guttentag, "The Problem of Experimentation in Human Beings: II. The Physician's Point of View," *Science*, *117*, 207–210 (1953); and Irving Ladimer, "Human Experimentation: Medical Legal Aspects," *New Engl. J. Med.*, *257*, 18–24 (1957).

[9] These standards do, of course, constantly change as new discoveries are made. This means that some person does have to be the first patient to initiate what can reasonably be expected to become standard practice. Before a physician begins to initiate a new practice on his own patient, there must be some reasonable assurance, where the risk is great or relatively unknown, that nothing else will do or, in accord with what is known, that the new procedure has at least as good a chance of success as existing procedures. If the initiation of a procedure is part of an attempt to procure information, it must not be an experiment by a physician on his own patient.

[10] A physican should generally not withhold information from his patients nor *assume* that it is desirable to "protect" them from what he knows. Each case has to be judged on its own merits.

[11] Technical reasons include, for example, the "double-blind" technique. In every case, however, the investigator should have necessary and sufficient reasons for using research techniques that require uninformed subjects. It should be evident that highly competent scientific personnel are needed to design and carry out research that will minimize the risk of harm to humans.

[12] The public is justifiably indignant when physicians use their own patients to conduct research, especially when this is done without the patients' knowledge that they are involved in research. Drugs not cleared for sale, thalidomide, for example, were made

advocates have been, and indeed should be, provided at various institutions engaged in medical research in the form of review boards set up to scrutinize proposals for research thought to be potentially harmful to the participants.[13]

In testing the safety level of food additives and environmental toxins, the informed consent of the persons serving as subjects should also be secured. It is already the case that many investigators enter into written agreements with their subjects, i.e., contracts describing a particular study and also indicating to the subjects the possibility of removing themselves from the investigation should they, for any reason, deem it advisable or necessary to do so. In many cases, provisions are also made to care for persons in the event of injury to them. The basic rights, therefore, of individuals are by these means recognized and honored in practice. For this reason, these practices should be universalized. Moreover, those who find themselves in certain offices or professions also have rights and privileges and should get the same kind of legal protection being given to the subjects of these investigations. Both investigators and subjects should have some form of insurance to indemnify possible injuries as well as protection from liability suits.

There would seem to be no insurmountable barriers to obtaining informed consent as long as we are dealing with the subjects of a controlled experiment, who have advocates to represent their interests, medical supervision during the course of the investigation, the right to cease being subjects when they so choose, and the possibility of insurance or indemnification for injuries that are unanticipated, though possible.

The more serious problems arise when we consider the right to informed consent of those persons who are affected by the *use* of drugs, food additives, and environmental toxins such as pesticides. For, as we have maintained, unless the principle of informed consent can be satisfied *here,* the very research for the sake of establishing safety levels of such products is morally questionable. Procuring the informed consent of the subjects of a carefully controlled test to establish the safety level of a given pesticide is not enough to justify testing when it is known or even suspected that the pesticide will be widely and indiscriminately used. For example, techniques such as aerial spraying subject many people to the effects of such a pesticide without their explicit and informed consent and, in some instances, against their express wishes. Instances

available to consumers by physicians conducting clinical tests for the William S. Merrell Company. (See *Consumer Reports,* March 1963.) Conducting research in this manner is a violation of the patient's right to expect therapy and not research to be the basis on which his own physician dispenses medicine.

[13] This practice is not and should not be restricted to medical research. For the sake of impartiality, it is important that board members not all be employees of the same institution and not all come from the same discipline or department.

of this kind were reported by Rachel Carson in her book *Silent Spring.* If we are to understand the public protest evoked by the knowledge of such instances, and if we are to understand why the public was not dissuaded from these protests by assurances from pesticide manufacturers of the great benefits of their products, we need to recognize that such practices violate the principle of informed consent. If the principle of informed consent is to be satisfied, we must invoke the notion of society as an implicit social contract and understand various public agencies as advocates of the interests of all individuals against the risks and possible harms resulting from the use of various products. Hence, what we have said in our earlier discussion of benefit–harm ratios regarding institutions for ensuring and increasing our knowledge of and control over the use of potentially harmful chemicals applies here as well.

Our previous discussion, however, of existing agencies and of agencies that will need to be established (e.g., a data-pooling center) requires some supplementation and extension at this point. If the interests of the public are to be fairly and sufficiently represented *vis-à-vis* the vested interests of all specialized investigators, manufacturers, and primary users of food additives, pesticides, and other environmental toxins (including drugs), consumers and property owners should be protected and directly represented in specifically institutionalized ways. For one thing, the substances we are here concerned with should be licensed for use by some public agency, perhaps by the data-pooling agency we have proposed.[14] Moreover, such approval should be renewed periodically on the basis of a continuous assessment of effects. It is, perhaps, necessary to remind ourselves that there are no perfectly disinterested men and that the best way that such a public agency can function is by allowing all the special interests to plead their cases before a judicial board. Such a device is not foolproof, but it is nevertheless needed to help to prevent use of environmental toxins in ways that would violate the principle of informed consent. Moreover, only as we protect this principle with respect to the *use* of such products can we justify the increasing need for experiments with human subjects for the sake of ascertaining safety levels related to their use. Those who call for an increased right to experiment with and to use various toxins cannot, at the same time, oppose the increased need to regulate experiments with and use of these toxins.

Licensing boards designed to represent the "informed consent" of the public to the environmental use of various toxins are not enough. Just as we have argued that, even in more restricted experimental contexts, those injured by statistically possible though actually unexpected injuries should be compen-

[14] The FDA already has the power to license drugs. Nevertheless, the power to license pesticides and other environmental toxins might better be given to a public agency of the sort we are suggesting, beholden as it would be both to the government and to private industry.

sated, so we suggest that justice requires compensation to persons injured by food additives, pesticides, and other experimental toxins outside the confines of the controlled conditions of the laboratory. Such compensation should also be institutionalized independently of the licensing boards. It should be stressed, however, that compensation for injuries occurring from licensed food additives, pesticides, and other environmental toxins should not come in the form of a liability suit against the investigators and manufacturers or primary users of these products whenever they have been given permission to use these materials and whenever they have met the criteria established for such use. Rather, the compensation should come from public insurance or even from taxes upon the product itself–in recognition of the inevitable statistically probable risks associated with its use. If agencies were required to provide compensation for injury resulting from products licensed by these agencies, they would surely stimulate efforts to reduce the need for such compensation.

The Food and Drug Administration already provides considerable protection to consumers through its powers, among other things, to certify antibiotic drugs and insulin, review evidence of the safety of food additives, establish tolerances of safety for food and drug color additives, establish tolerances for pesticide residues, and administer the Federal Hazardous Substances Labeling Act. Its hand as an enforcement agency should be strengthened. At the same time, it is vital that there be a data-pooling agency devoted to research so that the FDA's work as an enforcement agency does not discourage research that such a data-pooling agency could show to be necessary or clearly beneficial. The interests of consumers are represented in the federal government by the office of the Special Advisor to the President on Consumer Affairs. Though much remains to be done to increase the direct representation of consumers' interests at the federal level, this is a step in the right direction. A great public service on behalf of consumers is rendered by the Consumers Union, a voluntary organization, through their monthly publication, *Consumer Reports*. This publication procures and furnishes essential information about the benefits and harms, strengths and weaknesses of various products, information that helps to provide the possibility for *informed* consent and for the rejection of harmful and hazardous products, as well as those of otherwise dubious merit.[15]

Our argument in this section parallels our argument regarding benefit–risk ratios, namely, that greater knowledge of and control over use helps to satisfy the principle of informed consent and of legitimate research with human subjects designed to test and establish safety levels for using various chemicals.

[15] Recently, the FDA announced the publication of a new monthly magazine, *FDA Papers*, that will provide a great deal of valuable information, not only about its regulations and activities, but also about current practices in the food and drug industries as they relate to the health and welfare of the public. The first issue was published early in 1967.

A Further Moral Consideration

All these considerations concerning legitimate experiments on human beings have, of course, a certain air of unreality about them. For, as we all know, some food additives and pesticides are already in widespread use without the prior controlled testing we are calling for. In these cases, the widespread use of these products without something equivalent to public licensing and control constitutes *de facto* experimentation on humans without their consent. The build-up of DDT in human bodies from their consumption of various foods, the ingestion of antibiotics from fish or fowl that have been treated with these "preservatives" (often without informative labeling regarding this fact), and certain aerial spraying programs expose persons to various incalculable risks to which they have not given their informed consent—either in person or through competent public agencies. This widespread use of potentially hazardous materials has evoked a concomitant demand from many scientific investigators for the immediate testing of these products on human beings in order to establish safety levels and to achieve some degree of control over their dispersal.

The extremity of the situation we are describing does, in fact, constitute a justifiable reason for some immediate research on humans, as long as the principle of informed consent is not violated, and *as long as serious efforts to control the unregulated use of these materials are intensified and extended.* In this situation, of course, it is necessary to proceed with experimentation without all the pretesting *ideally required* to justify such experimentation. Rather, the research with humans to establish safety levels for products already in use must be done immediately. The moral justification for the curtailment of this extensive pretesting arises from the fact that incalculable and uncontrollable potential and actual *harms* are already being inflicted upon persons and their environments by the relatively indiscriminate use of the products in question. The existence of these incalculable and uncontrollable harms and risks legitimates a procedure that is less than ideal. Given the situation, it is incumbent upon us to take the morally best action and, in this instance, that means conducting carefully controlled investigations of the chemicals now in use whose potential or actual harmfulness to humans has not been scientifically established. Materials that are thus proven to be hazardous can, with just cause, be withdrawn from the market, and for other materials, whose benefits outweigh their negligible risks, more accurate safety levels can be determined to guide usage.

Of course, the situation we are here describing is not just the exception; rather, it is the rule. In actual life, we are always confronted with existing evils, and our ethical decisions never can accord perfectly with ideal norms. But this acknowledgment only makes clearer the status of our considerations in this paper, namely, that we have been concerned with moral norms as *factors* in determining particular procedures. By themselves, moral norms do not decide how we shall proceed. Rather, procedural decisions must take account of existing circumstances, moral norms, and technical capacities. All these limit the options open to us and the choice we can make.

Now, the recognition of the inevitability of falling short of what is morally required (in the fullest sense of this term) sometimes gives rise to despair or to cynicism regarding the possibility of any moral action in certain spheres, for example, in relation to war, or to the use of environmental toxins, or to the application of technology. Some of this despair is generated by the recognition that none of our decisions in these matters can perfectly accord with all moral requirements, since every choice involves a selection from among a number of competing claims. To this we reply that morally responsible action has always consisted in doing what is possible and what is best in the situation confronting us. All our arguments in this paper are direct expressions of this point: that we must do all we can to determine the differences among values and situations and do all that we can to maximize our discrimination of and control over these differences in order to discern those options that appear to be the best possible in the light of all moral claims, existing circumstances, and technical capacities.

If we acknowledge that actual moral behavior is never perfect, taking its cue as to what is right from among competing claims and choosing among the options actually available at the time, then we shall also recognize the *moral* requirement of foresight and planning. For by the maximal anticipation of the consequences of particular actions and the interrelationships of particular actions, we help to prevent certain harms from occurring that create a situation that is less than ideal and that close off the possibility of relatively more ideal actions. If, for example, there had previously been greater anticipation of the pervasive ill effects of certain pollutants on the atmosphere, the kind of radical government control over the dissemination of these substances which now seems imminent might not have been required. It is the occurrence of actual harms to society that justifies and calls forth regulatory legislation and enforcement agencies. The maximization of foresight and the anticipation of the harmful consequences of certain courses of action is, therefore, not only morally required, but also is the best technique for maintaining optimum freedom consistent with justice and for preserving a plurality of types of social control. For these reasons, advances in science and technology, upgrading the training and skills of scientific personnel, and rendering more precise our methods of understanding society and larger ecological systems are indispensable aids to moral discrimination and ethical insight.

Summary

The use of humans in research that is of some risk to them is required because of the limitations of research with animals and the moral demand to alleviate human suffering. Though research for the sake of overcoming pain has a greater moral urgency than research for the sake of conferring additional pleasures or benefits, even this second type of research may use human subjects so long as two conditions are met: the anticipated benefits arising from use of a given material must be sufficiently greater than all concomitant harms also associated with its use and all possible harms to experimental subjects, and no person must be exposed to any of these potential harms unless he has first given his informed consent.

The observance of these two conditions in research to establish safety levels for the use of drugs, food additives, and pesticides implies that all such research presupposes the use of the chemicals under consideration (if the research is successful). Since such use involves potential harms, these potential harms must be calculated before testing such products on human subjects in order to determine benefit–harm ratios required to justify such research. The difficulty of calculating these ratios increases with the degree of dissimilarity between "test" and "use" conditions of various products. In general, therefore, benefit–harm ratios are most easily calculated in medical research and least easily calculated in research involving pesticides and other environmental toxins.

To the extent that the incalculability of potential harms arising from the use of certain materials originates in our ignorance of their total effects within ecological systems and from our lack of control over their use, we can remedy this situation by promoting systematic research into the effects of all materials on living systems and by controlling the use of all chemicals through, for example, clearer labeling, licensing, and packaging. It should be noted that whenever there is a significant difference between test and use conditions of a given product, the individual investigator is usually not in a position to calculate the complex benefit–harm ratios necessary to justify experiments using human subjects. Rather, such calculations must represent the interests of various groups and the employment of data-pooling mechanisms.

Whenever test and use conditions are similar, informed consent should and can be secured directly from the subject or, where technically necessary, from his immediate advocate (e.g., his attending physician). However, to the extent that test and use conditions are not similar, procuring the direct consent of subjects, though necessary, is not sufficient to satisfy the principle of informed consent. In such instances, information must be obtained concerning the effects of using the product being tested, and controls over its use must be implemented. Indemnity for those injured by chemicals, whether as subjects of experiments or as users of the harmful chemicals in question, should be provided. Similarly, investigators who exercise professional care in executing approved research should be protected from liability suits. Procedures for such insurance and protection were suggested.

Finally, since many products are now in use even though their actual harms on living systems have never been adequately calculated, immediate steps should be taken both to curtail their indiscriminate use and to ascertain their total effects by immediate research using human subjects. Although this procedure does not allow the pretesting *ideally* required to determine the benefit–harm ratios that justify risk-experiments using human subjects, it is here justified by the fact that incalculable and incontrollable harms are being done by products already in use.

Our argument throughout this paper stresses the moral necessity of maximizing knowledge of and control over procedures and products and increasing both feedback and planning in order to anticipate the widest range of consequences. Such knowledge, control, and planning are necessary if experiments with human subjects are to be morally justified. Moreover, such control and planning are the best techniques for maintaining the optimum freedom consistent with justice within a pluralistic society.

REALITIES OF PATIENT CONSENT TO MEDICAL RESEARCH

JOHN FLETCHER

I

The theme of coercion and freedom is at the center of moral concern about the ethics of medical research in human beings. Intense efforts by groups in government, medical societies, and related professions have produced interpretations, codes, and regulations in the conduct of research.[1] Among the many valued objectives clustered around the discussion of morality in medical research, obtaining informed consent from the subject in an experiment is emphasized most. The literature on the principles of informed consent is enormous. Ninety-nine percent of it considers what ought to be; only a small fraction contains reports of what *happens* when consent is given by patient to investigator. A handful of studies by Renée Fox, Henry K. Beecher, and Fellner and Marshall expose the fragility of the consent contract and lay bare many "myths" about the supposed freedom and rationality in informed consent.[2] Due to the considerable doubts about the possibility of obtaining informed consent in many

Reprinted from John Fletcher, "Realities of Patient Consent to Medical Research," *Hastings Center Studies, 1,* No. 1 (1973), pp. 39–49 by permission of the author and the Institute of Society, Ethics and the Life Sciences, Hastings-on-Hudson, New York.

[1] John Fletcher, "Human Experimentation: Ethics in the Consent Situation," *Law and Contemporary Problems 34:*620–649, 1967; and "A Study of Ethics of Medical Research," Th.D. thesis, Union Theological Seminary, New York, 1969.

[2] Renée C. Fox, *Experiment Perilous* (Glencoe, Illinois: The Free Press, 1959); Henry K. Beecher, "Ethics and Clinical Research," *New England Journal of Medicine 274:*1354–1360, 1966, and "Some Guiding Principles for Clinical Investigation," *Journal of the American Medical Association 195:* 1135–1136, 1966; W. J. Curran and Henry K. Beecher, "Experimentation in Children," *Journal of the American Medical Association 210:* 71–73, 1969; C. H. Fellner and J. R. Marshall, "Kidney Donors–The Myth of Informed Consent," *American Journal of Psychiatry 126:*1245–1251, 1970.

circumstances, a fresh treatment of the actualities of giving consent in medical experiments is in order.

The most rewarding inquiry in ethics takes place between the "ought" and the "is." The dangers in the debate on informed consent are a too legalistic approach from the side of those who uphold the law and a too secretive approach from those who practice. H. Richard Niebuhr[3] noted that the two purposes of ethics are self-understanding and guidance in the concrete problems of using our freedom. Researchers need to know more about the ways they and their subjects actually make decisions, and those who are charged with regulating research need to keep rules within the reach of obedience.

In order to inquire into the dynamics of the consent situation in practice, I first interviewed twenty clinical investigators at the Clinical Center, National Institutes of Health, Bethesda, Maryland, to discern the major moral problems confronting them in their work. A majority named the difficulty of obtaining a truly informed consent from a sick person to do research for nontherapeutic purposes. Those interviewed felt morally compelled to seek a highly qualitative consent for studies which carried no benefit to the patient, but felt equally that in practice high standards could not be attained. A study of the relationship of subject and investigator was designed which would allow me to interview patient and doctor both before and after visits for consent to nontherapeutic studies were made. I participated in the conduct of four such studies and interviewed eighteen patient-volunteers extensively after they entered the studies.[4] Each had signed a consent document and had been given an explanation of the study.

The Clinical Center is a unique medical research institution controlled by the government. Investigators often contact patients in different parts of the country, or have patients referred to them, who suffer from the diseases the Institutes exist to study. Thus, the "consent situation" begins in most cases before the patient enters the hospital. Fellner and Marshall discovered that kidney donors actually made a decision to donate long before they were ever interviewed by physicians and "informed" about the risks of donating. My findings were similar in that the decision to enter research was made prior to entering the hospital, though decisions about particular studies and their related demands were not made until after admission. Patients made decisions agreeing to the importance of human studies before they arrived. Consenting must be defined dynamically to include all of those encounters between investigator and subject in which the expectations of the research agreement are being built, maintained, or repaired.

[3] H. Richard Niebuhr, *The Responsible Self* (New York: Harper and Row, 1963), p. 48.
[4] Fletcher, "A Study of Ethics of Medical Research."

II

When consent is being sought for medical research, a human encounter involving decisions to take risks occurs between an investigator and a subject. I assume that in the majority of these encounters little discussion of whether the investigator is *really* justified in pursuing his study takes place. Usually, he has already had to justify his plan with a group of his peers. Further, I assume that it would be rare, if ever, that a question would be raised by either party as to whether the subject were *really* free to choose to participate, or whether there is an authentic condition called "freedom" anyway. As Henry D. Aiken, a moral philosopher, pointed out, these sorts of questions usually occur some distance from practice and are occasioned by changing social conditions or a severe conflict of rules.[5] I am not claiming that the raw materials for such questions are not present in the consent encounter, for wherever one self puts a claim on another and risk is a factor, ethical and theological questions are implicit. I am claiming that one observing the consent process would seldom hear self-conscious discussions of the ethics of consent.

I did assume, however, in approaching a field study of the consent process, that patient-volunteers in medical research would be able to reflect upon their impressions of the possibilities and limits of their freedom of choice. Since individualism or the sense of autonomy is the primary value orientation within which these patients might be expected to speak of their sense of freedom,[6] one might expect that when the patient reflects on his sense of being a "person" or being denied such a status, the issue of his freedom would be to the forefront. Thus, as an ethicist studying the issue of freedom in research, my research was designed to test the hypothesis that the conduct of the consent situation was decisive for the patient's sense of being treated as a "person."

Reading and interviews had convinced me that many investigators and administrators maintained a ritualistic attitude toward informed consent. That is, working from the assumption that a sick person will do almost anything a physician suggests, and feeling that the signing of a consent form entailed a ritual which covered over the impossibility of informed consent, many investigators went about their practice having despaired of attaining genuine informed consent from their patients. Several early interviews had opened with an investigator remarking that patients are so dependent on physicians that they would not be seriously affected by anything the physician did in his explanation of a study. These interviews were also spiced with a heavy antipathy to the legal profession

[5] Henry D. Aiken, *Reason and Conduct* (New York: Knopf, 1962), p. 75.

[6] C. Kluckholm, H. A. Murray, and D. M. Schneider, *Personality in Nature, Society, and Culture* (New York: Knopf, 1964), p. 352.

and administrators who, according to several physicians intereviewed, were interested only in protecting the reputation of the institution.

In short, it appeared to me that many people supposed that patients who entered the research situation left their sense of personhood behind, having surrendered their autonomy to the "white coat" world. From a deep professional interest in inquiring into dependency relations between "laymen" and experts who control valuable yet risky techniques, I expected that these patients would give signs, gestures, and verbal expression to show how their experience in a risk-taking study reflected on their sense of having the status of a "person." I assumed, with those who have labored on the concept of informed consent, that the real meaning of the rule was to protect the status of personhood which is enshrined in the traditions of law, morality, and religion surrounding the concept of consent. Arthur Dyck and Herbert Richardson have expressed better than anyone the social-structural values which nourish the need for consent: freedom, justice, truth-telling, and equality.[7] They affirmed that unless these values were responded to in the actual practice of medical research no one would interpret the progress of medical science in terms of social "benefits." If being socialized in institutions which are maintained by such values creates the capacity (or status) of perceiving that one is a "person," then would he divest himself of this status, lower himself, when he came within the orbit of medical investigators? Here was an opportunity to test the relation between a theory of the ethics of informed consent and the practice of investigator and subject in the process of consent-giving.

What does the "sense of being treated as a person" mean? Three norms were accepted which nourish the basis of the symbolic status of "person."

1. The patient might perceive himself as a being who was addressable as "never merely as means . . . but at the same time as ends in themselves."[8] To use more religious terms to describe this sense, one may perceive himself as a "thou and not an it"; as Buber stated: "Without *It* man cannot live. But he who lives with It alone is not a man."[9]

2. He might give signs of perceiving himself as a responsible being, capable of choice, exercising some control over his body and general welfare. To quote Tillich: "As a centered self and individual, man can respond in knowledge and action to the stimuli that reach him from the world to which he belongs; but because he also *confronts* his world, and in this sense is free from it, he can

[7] Arthur J. Dyck and Herbert W. Richardson, "The Moral Justification for Research Using Human Subjects," *Uses of Human Subjects in Safety Evaluation of Food Chemicals* (National Academy of Sciences, 1967), p. 231.

[8] I. Kant, "The Metaphysics of Morals," *Great Books of the Western World,* edited by R. M. Hutchins, Vol. 42 (Chicago: Encyclopedia Britannica, Inc., 1952), p. 274.

[9] Martin Buber, *I and Thou* (New York: Scribner's Sons, 1958), p. 34.

respond "responsibly," namely, after deliberation and decision rather than through a determined compulsion."[10]

3. The patient might give signs of including himself as a member of a community in which the transcendence of self-interest is an ever-present possibility. Both psychologist Jean Piaget and theologian Reinhold Niebuhr agree that social relations of cooperation, involving the transcendence of one's own interests, are the hallmark of personal life.[11]

These three norms can be shown to be embodied in the law and morality surrounding human experimentation. Moreover, they can be shown to arise, at the deepest level, from a confidence that we are related to reality in a way that calls forth actions based on these norms which can be trusted to survive. In short, as a theologian, I am interested in pressing to uncover the sources of morality in human encounter of great risk. Persons disagree on the final source of morality. A theological perspective is one of several ways of exploring the depths of human encounters. Because medical research is an encounter of persons with persons, and not I's and It's, one final court of appeal to establish the limits of that encounter is a moral imperative derived from an unconditional source establishing our personhood beyond any threat of destruction.

When one examines all of the possible perspectives within which to answer the question of why investigators should be moral at all in their consent relations with subjects, he finds that an answer can only arise from the actual commitments in decisions made by those who practice. Persons are ultimately accountable only in the light of their decisions and that to which they are committed beneath their decisions. The reason why a theologian should be interested in inquiring into the ethics of informed consent is that when pushed to its final justification, the morality of any action rests upon some form of commitment and loyalty. The task of a theologian in all times and places is to clarify the forms of commitment, for upon them rests the substance of culture and science.

A description of two studies and their chief investigators will illustrate my conclusion that patient-volunteers in this setting maintained their relation to personhood, but that several gave signs of having to defend themselves against their consent being "engineered" rather than "coerced."

Study of Dyslipoproteinemia

The physician-investigator in this study was a twenty-eight-year-old Clinical Associate interested in discovering the basic causes of dyslipoproteinemias.

[10] Paul Tillich, *Morality and Beyond* (New York: Harper and Row, 1963), p. 19.

[11] Jean Piaget, *The Moral Judgment of the Child* (Glencoe, Ill.: The Free Press, 1965), p. 395; Reinhold Niebuhr, *Man's Nature and His Communities* (New York: Scribner's Sons, 1965), p. 107.

These genetically inherited conditions result in a radically high blood cholesterol level for those who suffer from one form of this disease. Dr. A. had prepared the design of his study for one year, doing preliminary studies in animals. His work was designed to investigate the metabolism of labeled lipoproteins in these disease states. He explained that if the production and destruction rate of the lipoproteins could be determined, as these proteins are essential to the transportation of cholesterol in the body, physicians could better attack the control of this problem through drugs as well as understand the basic mechanisms of these diseases.

The method of implementation was as follows: blood would be collected from each subject, plasma separated and lipoproteins isolated by ultracentrifugation. The lipoproteins would then be labeled with a radioactive iodine tracer, sterilized, and tested extensively for pyrogenicity. This preparation would be injected intravenously into each subject and small samples of blood collected ten minutes later and then daily or every other day for 14–21 days. Complete urine collections would be made. The blood and urine collections would enable a complete study of the rate at which the labeled lipoproteins were being destroyed. A special safety measure included doses of potassium iodide to protect the thyroid. Dr. A described the risks of radiation in these cases as less than that received from a standard X ray. Other than the discomfort of an injection and the regimen of diet and collections, there would be no real risks in this study. The risk of hepatitis had been greatly reduced by using the patient's own blood for isolating lipoproteins.

Dr. A.'s study had been approved, after careful scrutiny, by (1) group consideration of his peers, (2) a radiation committee of the Center, and (3) the clinical research committee of his Institute.

Dr. A. had visited each patient-volunteer several times prior to his call to get consent forms signed. During these visits he had given a complete explanation of the study and had tried to answer questions. He showed that he knew a great deal about each subject personally, as well as physiologically. Dr. A. described this study as "investigative," one in which the possibility of therapeutic value might be indirect. All of his patients had children who had inherited the same disease, and each had been involved in research prior to this study.

It is important to note here that Dr. A.'s study was carried out in the context of his patient's participation in another drug study which carried possible therapeutic benefits. A double-blind drug study, designed to test a particular preparation and its effect on plasma cholesterol levels, was accepted by the patient volunteers Dr. A. also intended to study. Each patient was told *before* and *after* entry to the Center that their hospitalization would be prolonged to allow Dr. A.'s study to be integrated with the drug study.

Study of Biogenic Amines

Dr. C. is a thirty-five-year-old neurologist. He works on several projects with patients who have diseases of the brain or disorders which affect the central nervous system. His interest lay in studying a group of chemicals, the biogenic amines, believed to play a central role in synaptic transmission in the brain. Dopamine is an example of one main subgroup of these chemicals, the catecholamines. He had designed a study, after a year of animal testing, to investigate the difference in the metabolism of dopamine in the brain as contrasted to its metabolism in the body. He explained that neurologists did not know if there were a difference; if there were, they will know more about dopamine's nature and the uses drug therapy can have in treating brain disease.

The following experiment was designed: patients with indwelling intraventricular catheters would be chosen as subjects. Some patients with brain tumors have such catheters to permit safer and easier treatment with chemotherapeutic agents. Radioactively labeled dopamine would be injected intravenously *or* intraventricularly in these patients, in a very small dose. By testing urine collections for four days following the injection, Dr. C. hoped to study the difference in metabolism.

He explained that the study carried no direct benefit to the patient-volunteer, although the potential information to be gained was within proportion to the risks. Animal tests showed no pharmacologic effect. The major risk of any injection, he stated, was infection, but every known precaution had been taken. Ninety-eight percent of injected radioactivity would be passed in the urine of each patient. His study has been approved by a group of his peers and by a radiation committee.

Dr. C. mentioned more than once the possibility that what transpires in the consent situation can be a "charade." He defined this term as the disappointment of society's expectation that the investigator will always obtain an informed consent, and that the patient will fully understand the risks and benefits. He frankly admitted that some patients are, through serious illness, unable to measure up to these expectations, even after serious attempts to communicate. In such cases, he stressed, he always sought third-party consent.

Dr. C. had chosen two female patients, both of whom had had nursing educations, as participants in dopamine infusion. Both had come to the Center expecting research with treatment for brain tumor. Since they were roommates, he had met with them together several times prior to obtaining signed consent, six hours in all, and he showed he was very aware of their limitations. He had gradually unfolded his explanations for them, giving them ample time to discuss it. Dr. C. had noticed that Mrs. S., who was somewhat aphasic as a result of her brain tumor and had difficulty understanding others and expressing herself, had become very dependent upon Mrs. N. for interpretation and cues. Following the session in which both had signed forms, Dr. C. said that in his opinion Mrs. N. had given informed consent, but if strict standards were applied, Mrs. S. had not. "Her illness is so pronounced." He observed that "Mrs. N. really understands what is going on," but that Mrs. S. "kept looking at the other lady to find out what the right thing to do was." Under the circumstances, he concluded that Mrs. N. had given an authentic third-party consent for Mrs. S., and that the previous sessions had indicated to him that this was an acceptable route to follow.

It is seriously debatable as to whether one should choose a third party who is participating in the same experiment to authenticate the consent of another. In principle, such should never be the case. As the following interview shows, Mrs. S. gives dramatic signs that she felt treated humanely and as a person. Yet several hard questions should be raised here about the conclusion of Dr. C. that she could be included in the study on the basis of Mrs. N.'s consent. The two patient-volunteers made the following comments after each had already received an injection:

Q.: *You had met with the doctor several times before yesterday?*

Mrs. N.: *Yes. He saw us four or five times, no . . . every day last week. Then one day he just sat down, for about an hour, and talked and talked. He told us all about the project. He explained everything. He told us that this chemical he was studying was in the body anyway, so it wouldn't put anything new into us.*

Q.: *How do you feel about Dr. C.?*

Mrs. S.: *He talks to you like you are a human being and not a glass. He has a lot of warmth. He took our anxieties away. The doctor is the most important thing in something like this. If he talks to you then you know what is happening.*

Q.: *Then you liked that part of it?*

Mrs. S.: *Oh, yes. I like being told what is going on. It isn't that often a doctor will talk to you that long. Dr. C. took his time and answered all of our questions.*

Q.: *What do you understand about the purpose of this study?*

Mrs. N.: *Well, they are going to study how this chemical acts . . . maybe they will be able to use it on others.*

Q.: *Will your participating in this study help you?*

Mrs. N.: *It won't help the illness I have. Maybe it will help someone else someday, or the world of medicine.*

Mrs. S.: *At least we have contributed something. So much of the time you are just like a robot; now you can do something, you can give. Everyone is doing something for us all the time, now we can repay them in this way.*

Q.: *Did any questions occur to you after we met yesterday?*

Mrs. N.: *No, no questions. We had them all answered before. Mainly, we were concerned about how long it would last and if it would do any harm. Since it was just a teeny bit of radioactivity, and they would get that later, it didn't seem that it would do any harm.*

When the three norms of the presence of personhood discussed above are applied, each appears clearly confirmed here. Being treated as an end and not a means only is affirmed by Mrs. S. (the aphasic patient) as being a "human being and not a glass." Her metaphor is possibly a more apt description of personhood in this age than the "thou–it" terminology. Mrs. N. showed that she understood the meaning of a nontherapeutic study, that she had had her questions answered, and had exercised her capacity to choose. Mrs. N. also answered all of the "technical" questions, showing the special role Dr. C. had used with her. Mrs. S. perceived that she was a member of a community in which the consuming self-interest of illness can be transcended; her use of the term *robot* is especially compelling to describe the dehumanization of being ill in a hospital. When contrasted with her "*now* you can give," a clear picture of membership in a special community emerges.

When one contrasts Mrs. S.'s warm statements with the probability that there is no firm basis for assuming that Dr. C. got consent in this case, he can understand how complex a moral judgment there is to render in this case. First, there is Dr. C.'s own estimate that Mrs. S. was too ill to speak for herself. Secondly, Mrs. N. was never told that she was acting as a third party to consent for her roommate. Thirdly, even if Mrs. N. had been told, there is the likelihood that she was a poor choice, since she was a fellow-participant in the same study. In my opinion, a more independent third-party consent should have been obtained by Dr. C. To the extent that I played that role unknowingly in discussion with Dr. C. about Mrs. S., I would have said yes. Yet I must conclude that the wrong third party was used and that his choice did not constitute a sound basis for including Mrs. S. in the study. Dr. C.'s conduct of the consent situation was decisive for Mrs. S.'s sense of being treated as a person, but it was not sufficient to provide a firm moral basis for consent when studied by an independent party. Thus, the sense of "personhood" on the part of the patient is not sufficient evidence that all is well. A physician can be compassionate and extraordinarily informative, as was Dr. C., and his patient can feel quite free about his choices; yet each be found in serious legal and moral question. This case is an excellent prism for the problems of informed consent and a good illustration of the "finite freedom" of men.

Two additional matters can be pointed out about these two patients. The guilt of illness, and especially of the very ill, as commented on by Otto Guttentag and Beecher,[12] may be present in Mrs. S.'s great need to "give." At any rate, the guilt of illness and the need to sacrifice may be exploited by insensitive investigators. Secondly, I asked no questions of these two patients about their understanding of their right to withdraw, since each had already received an injection and one had felt somewhat dizzy as a result.

Dr. A's patient-volunteers also showed signs of maintaining their status as persons during participation in research. Mr. J., a Navy veteran with an eighth-grade education:

Q.: *How do you feel about being here in such a study?*

Mr. J.: *Well, it makes me feel good. Everybody is here for a purpose. I just happen to be part of discovery. Like I say, I have two children who have this, and I feel obligated to them to do something. If my life had been different, and I hadn't had any children maybe I wouldn't feel this way.*

Q.: *I can see you feel strongly about it.*

Mr. J.: *You've go to. Even if nothing comes of it, you are still part of something.*

Mr. J. put special emphasis on being part of a community of "discovery," where everyone is there for a "purpose." To have purposes is a particularly human activity, and he joins nicely the first and third norms.

Mrs. B. in her early sixties, whose daughter had died as a result of the same inherited condition, complained in her interview with me that she lacked knowledge about the purpose of Dr. A.'s study, and that she had difficulty understanding the technical parts. When I reminded her that she did have ample opportunity in the consent process to ask such questions of Dr. A., she said, "Well, I didn't want him to think I was stupid!" She had pretended to understand some things and did not question Dr. A. as much as she wanted. This action might be taken as a negative confirmation of the hypothesis. That is, Mrs. B. was *pretending* to be a person who understood and chose rationally. She in fact withheld her questions and capacity to control for fear of alienating the doctor. It is debatable whether a pretense at personhood is better, in the moral sense, than no attempt at all; I tend to believe so. At least, Mrs. B. gives evidence that she was aware of the possibility of exercising more autonomy.

Another example of negative confirmation came from Mr. C., who after saying that he had been in the Navy for twenty-one years and had learned to take orders from officers (whom he compared to doctors), stated: "My philos-

[12] Otto Guttentag, "The Problem of Experimentation on Human Beings: The Physician's Point of View," *Science 117:*209, 1953; Beecher, "Some Guiding Principles," p. 1135.

ophy in the Navy was 'yours not to reason why, yours but to do or die.' You will find that people who have been in the service for a long time think my way. *That may not be a good outlook,* but it is the way I have been raised." [Emphasis added.] In the process of "lowering" himself from the status of a responsible person who asks questions, Mr. C. acknowledged his wrongdoing, appealing in a negative way to values which might restore him to a relation with his own best sense of himself.

The term *guinea pig* when applied to human subjects represents the ultimate lowering of humans from personal to animal status. Mrs. F., a forty-year-old patient-volunteer, showed how she defends against it:

Q.: *Had you met with Dr. X. (principal investigator) before?*

Mrs. F.: *No, that was my first time. I have confidence in him and had it the first time I saw him. I have trust in him, and I have been around enough doctors to know something about it. He knows his business, and he takes his time.* (She mentions another doctor who appears to her to be always in a hurry.) *Dr. X. sat and talked with me. My family does not understand my illness at all, and they call me a "guinea pig." I wish I could change their minds about that, but they would have to go through what I have to understand, and be a part of what I know to feel different.*

In short, one would have to be a member of the community of Mrs. F. to maintain the sense of being a person. Persons are creatures who, when they perceive that their stature in that role is being limited, threatened, or coerced, will respond to values which promise to restore them to their sense of occupying a special and unique capacity in their relationships.

One could not conclude, from these examples, that these patients in research leave their personhood behind. Ample signs were shown that the sense of freedom in a patient is related to being part of a community of healing in which his own contribution might be ultimately valuable. At the same time, research patients will require assistance in keeping them open to ask questions of physicians and themselves, for there are many diverse forces at work in the consent process.

III

Many factors impinge on the patient-volunteer to limit or reduce his autonomy. First, serious illness is threatening and limiting. Patients tend to believe that the slightest change in the arrangements, such as the introduction of a tape recorder, will make them better. Almost everything that happens is filtered through illness, including consent to research. Bennett, a social worker in an experimental therapeutics ward of the Clinical Center, corroborated my findings in her study of twenty-three patients, and observed that "patients do not conceptualize the principal investigator as a scientist and equate treatment and research as one and the same."[13] Park and others reported on the same phenomenon in psychiatric patients.[14]

Secondly, the arrangements at an imposing institution like the Clinical Center tend to diminish the patient's willingness to complain or question. Surely the fact that treatment is free, and that each has a serious disease for which no cure has yet been found, limits a patient's freedom to question. Savard, also a social worker, reports for the whole social work staff: "One of the functions the social work staff sees for itself is to help the patient work through this conflict (guilt over being ungrateful when the wish to gripe crosses their minds). This could include second thoughts about continued participation in a research project."[15]

Thirdly, the expectations of the investigator for the subject are a strong force which may operate often to limit the subject's freedom. Dr. A. actually believed Mrs. B. to be very well informed and highly curious about the technical side of medicine. He spent much time on his explanation, but at the same time she was embarrassed to ask more questions. Patients are also aware from the attitudes shown by investigators that they have much of their own prestige invested in studies. They are eager not to disappoint. The investigator's knowledge of human behavior has become as important for his work as his skill in carrying out his studies.

None of the three restrictions on freedom of the patient to question or withdraw mentioned above are comparable to outright examples of coercion. No investigator I observed ever used force, threat, or his authority to make a patient submit to research. Patients come to the Center knowing that they will be in studies, even though they tend to confuse research and treatment. The particular question of coercion and freedom in human studies revolves in part around the right of the patient to withdraw at any time. Mr. H., a thirty-eight-year-old patient in Dr. A.'s study, declared that he had a right to withdraw but that the subject had not come up. "It is something I knew about. I never signed any agreement to be held to, and they can't hold anyone against his will. . . . The choice was really made when they called me on the telephone and asked me to come."

[13] C. M. Bennett, "Motivation, Expectations and Adjustment of Patients on an Experimental Therapeutics Service," 1970. Available from Social Work Department, Clinical Center, NIH.

[14] L. C. Park, *et al.*, "The Subjective Experience of the Research Patient," *Journal of Nervous and Mental Disease 143:*199–206, 1966.

[15] R. J. Savard, "Serving Investigator, Patient and Community in Research Studies," *Annals of the New York Academy of Science 169:*429–434, 1970.

Coercion can be normally defined as the act of influencing others against their will. It has been said that the ultimate in coercion is manipulating another's emotions, "forcing him to will that which you will." If an entire will had to be created *ex nihilo* for the subject in research, this definition might hold; however, patients come prepared for research by cultural, medical, and institutional conditions. It is more true to say that a specific will to do a particular study has to be developed in the patient; he must be persuaded. A more capacious concept than "coercion" must be sought to describe the *pressure,* bordering on coercion, which some patients felt to accept research. I observed that some patients found an opportunity to defend themselves against their consent to research being "engineered," a term used by Cahn to describe the control of the conditions of consent by experts controlling information and technique.[16]

The participants in Dr. A.'s study were asked to discuss their right of withdrawal, should they so choose. Mr. J. said that he wanted to participate and was "wide open for suggestions." He discussed the question of his freedom of choice, and his right of withdrawal, without distinguishing between Dr. A.'s study and the double-blind drug study. He made it clear that if he were inclined to leave the nontherapeutic study he would risk continuing in treatment. Smiling wryly, he commented, "that is pretty strong pressure." Mrs. B. also discussed her freedom to say yes or no in association with the risk of continuing in treatment. "Since I came here for treatment," she said, "if I didn't cooperate, it would mean ending treatment." She strongly wanted treatment, since she felt the drug might be helping her for the first time. The apprehension of the patients was that the condition of the right of withdrawal had been severely weakened by the risk of losing treatment.

When I reported these interpretations to Dr. A. in writing, he immediately responded, along with his colleague who was in charge of the drug study. They assured me and the patients that there was no intention of tying research to treatment. Under the arrangements which had been made for the two studies, if these patients had chosen to withdraw, they would not have been excluded from treatment, even though their hospitalization would have been shorter, excluding the time necessary to complete Dr. A.'s study. In reviewing the process of this feedback, Dr. A. stated: "These patients were aware that they were brought in to do two studies. They were aware that the 'turnover' study was superimposed on the drug study. What wasn't made clear to them was that they didn't have to do the second study."

He also affirmed that the most important finding, for him, in participating in our study of consent, was the effect of not informing the patient-volunteer as to his right of withdrawal from nontherapeutic studies. He reported that he had

[16] E. Cahn, "The Lawyer as Scientist and Scoundrel: Reflections on Francis Bacon's Quadricentennial," *New York University Law Review 36*:8, 1961.

not encountered the same sense of pressure in patients who had been informed of the right to withdraw.

I would attribute this sense of pressure to accept research "tied" to treatment, in these patients, as a result of institutional arrangements plus the unfinished business of a consent process. The patients entered an impressive research center for therapeutic and nontherapeutic studies. They knew that they had a right to withdraw, but they had not discussed it with their physician. One of the most important findings of the study, in my estimation, was the opportunity the presence of my inquiry gave to the physician and subject to maintain, build and repair their agreement. Savard and Bennett also report such effects. Although I did not present myself as an "ombudsman" or advocate for the patient to physicians, the implication of my position was similar. An intermediary person who has access to the consent process might function to enhance the freedom of patient and investigator to change the terms of an unfinished agreement. The range of options open to an individual constitutes his sphere of freedom. Whenever a third party enters, there may be new options available.

What is needed to remedy the possibility of coercion is not more exhortation, but more practical action within the consent process to insure a maximum number of options for improving qualitative consent between physician and patient. The addition of institutional review committees in grantee institutions funded by the government, obligated to assure the rights of subjects in research carried out by their own institutions, seems a creative step. The following table indicates a reduction in the number of "problem projects," deferrable on scientific or ethical grounds, proposed through the NIH system.

Table I is concrete evidence that unpromising studies, or studies involving unethical risks, are being screened at a much more elementary level so far as those studies funded through public funds are concerned. Further evaluation of the actual effectiveness of institutional review committees, and their composition, should be eagerly anticipated.

In view of the increasing demand for medical research, it is past the time when research institutions should act concretely on Guttentag's proposal of the provision of a "physician-friend" to supplement the physician-investigator for purposes of advocacy for the patient.[17] In my experience, a busy investigator cannot sort out the many signals he or she receives from needy patients in research, and he or she requires assistance from one who is not invested in the project. In some institutions this role may be played by social workers or psychiatrists who are delegated to represent the total welfare of the patient. The part psychiatrists have played in screening organ donors is an important pre-

[17] Guttentag, "The Problem of Experimentation on Human Beings," p. 210.

Table I. Two-Year Study of Relation of Problem Projects to Applications

Council	Total applications	"At Risk" human subject applications	Problem projects No.	% total applica.	% "at risk"
June '66	4,100	1,230	93	2.24	7.4
June '67	3,931	1,180	38	0.96	3.2
Nov. '67	3,677	1,100	38	1.03	3.4
Mar. '68	4,001	1,200	27	0.67	2.2
June '68	4,078	1,250	21	0.52	1.7

Source: Office of Associate Director for Extramural Programs, National Institutes of Health. "Status Report of Experience with PPO #129," May 31, 1968 (Memorandum).

cursor of the emerging role of medical "ombudsman."[18] In other institutions where chaplains are charged with such duties their roles might be enlarged to include special relations with investigators and patients attendant to conflicts over research demands. Special training and sensitivity to the complexities of medical research are definitely required of such a figure, whether he be a medical or paramedical person. However, the time has arrived for all medical institutions which engage in human research to consider the step of designating one or more persons to bear the responsibility of advocating *for* the patient, to assure him and the public that consent is within the reach of obedience. Such steps definitely need to be taken wherever especially necessitous groups are involved in research such as prisoners, children and the mentally ill. The principal investigator in each human study is charged with the responsibility to be the final judge of the quality of consent obtained. My proposal would not shift that responsibility. It would help assure him that every step had been taken by providing maximum feedback between himself and the subject. The issue of informed consent has become too socially charged to relegate to the realm of goals alone. New moral initiatives should be taken by those responsible for regulating medical research to invent a flexible role for one working between what ought to be and what is in each research institution.

[18] Fellner and Marshall, "Kidney Donors"; J. P. Kemph, "Renal Failure, Artificial Kidney and Kidney Transplants," *American Journal of Psychiatry 122:*1270–1274, 1966, and "Psychotherapy with Patients Receiving Kidney Transplant," *American Journal of Psychiatry 124:*623–629, 1967.

A proposal for a special representative of the subject in research may find its most productive work in the study done with poor or otherwise deprived persons who possess the least defenses for maintaining their moral status. As the history of human experimentation shows, the poor have been a "captive group" for medical experiments. A review of the policy of municipal hospitals in New York City cited evidence that research exploitation of the poor was a common occurrence known to city health officials.[19] A familiar method of exploitation is to make the receiving of continued treatment in outpatient clinics conditional upon participation in experimental drug trials.

Every moral resource in the religious traditions urges special attention to the needs of the sick, the defenseless, and the poor. Lying behind this moral concern is an ethic of universal responsibility, inhering in the double-love commandment. Yet, love has not yet become visible until it is embodied in concrete human relations which establish the "weighty matters" of justice and mercy. Now that the period of concern about the principle of informed consent has crested, and lest this concern be found to be mere sentimentality, more thorough steps to institutionalize and embody the value of personhood in medical research are required.

[19] R. Burlage, *New York City's Municipal Hospitals* (Washington, D.C.: Institute for Policy Studies, 1967), p. 329.

NIH GUIDELINES ON RESEARCH WITH HUMAN SUBJECTS

ORIGINAL AND REVISED NIH GUIDELINES

Background Information

Certain classes of persons have limited capacity to consent to their involvement in experimental activities. Recognizing this, the Department of Health, Education, and Welfare, through the National Institutes of Health (NIH), appointed a special study group to review and recommend policies for the protection, in biomedical and behavioral research, of children, prenatal beings, prisoners, and the institutionalized mentally disabled. On November 16, 1973, the NIH published a document entitled "Protection of Human Subjects: Policies and Procedures" (hereafter called NIH original guidelines). This was a draft working document upon which the public was invited to comment. After comments were received and considered, a revised version of the original document was published, with public comment again being solicited. Once comments upon this document had been received and evaluated, a final set of recommendations would have been drafted and sent to the Secretary of Health, Education, and Welfare for promulgation as guidelines governing human research in the United States.

Although both the original and revised guidelines were published and commented upon, a final set of recommendations has not yet been formulated. There is a good reason for this. In July 1974, Congress passed a National Research Act, establishing an eleven-member National Commission for the Protection of Human Subjects in Biomedical and Behavioral Research. The Commission was sworn in on December 2, 1974, with a life of two years. Among other things, the Commission is expected to develop guidelines for research involving human subjects, and to recommend them to the Secretary of Health, Education, and Welfare for promulgation as national policy. In its deliberations, the National Commission will give some attention to the original and revised guidelines formulated by NIH, as well as to public comments received in reaction to these working documents.

Original Guidelines are from the *Federal Register,* Vol. 38, No. 221 (November 16, 1973).
Revised Guidelines are from the *Federal Register,* Vol. 39, No. 165 (August 23, 1974).

Both the original and revised NIH guidelines recommend procedures for protecting different classes of persons having limited capacity to consent to their participation in research activities. In what follows, we have included only those portions of the documents which deal with prenatal beings, for the majority of public comments were directed to these sections of the guidelines.

There are several respects in which the revised guidelines may be said to weaken protections afforded prenatal beings in the earlier document; and it will be an interesting exercise for the student to see if he can locate those portions of the revised recommendations which relax restrictions imposed in the original guidelines.

NIH GUIDELINES
ORIGINAL RECOMMENDATIONS

Protection of Human Subjects

Introduction

The mission of the Department of Health, Education, and Welfare includes the improvement of the health of the Nation's people through biomedical research. This mission requires the establishment of policy and procedures for the protection of subjects on whose participation that research depends. In DHEW policy, as well as in ethical codes pertaining to research in human subjects, the keystone of protection is informed consent.

An uncoerced person of adult years and sound mind may consent to the application of standard medical procedures in the case of illness, and when fully and properly informed, may legally and ethically consent to accept the risks of participating in research activities. Parents and legal guardians have authority to consent on behalf of their child or ward to established therapeutic procedures when the child is suffering from an illness, even though the treatment might involve some risk.

There is no firm legal basis, however, for parental or guardian consent to participation in research on behalf of subjects who are incompetent, by virtue of age or mental state, to understand the information provided and to formulate the judgments on which valid consent must depend. In addition, current policies for clinical research afford such subjects inadequate protection. Nevertheless, to proscribe research on all such subjects simply because existing protections are inadequate, would be to deny them potential benefits, and is, therefore, inequitable. Knowledge of some diseases and therapies can be obtained only from those subjects (such as children) who suffer from the disease, or who will be receiving the therapy. Their participation in research is necessary to progress in those fields of medicine. When such subjects participate in research, they need more protection than is provided by present policy . . .

This addition to existing policy is offered as a means of providing adequate protection to subjects who, for one reason or another, have a limited ability to give truly informed and fully autonomous consent to participate in research. The aim is to set standards which are both comprehensive and equitable, in order to provide protection and, to the extent consistent with such protection, maintain an environment in which clinical research may continue to thrive.

I. *Definitions.* For purposes of this policy:

A. *Subject at risk* means any individual who might be exposed to the possibility of harm (physical, psychological, sociological, or other) as a consequence of participation as a subject in any research, development, or demonstration activity (hereinafter called "activity") which goes beyond the application of established and accepted methods necessary to meet his needs.

B. *Clinical research* means an investigation involving the biological, behavioral, or psychological study of a person, his body, or his surroundings. This includes but is not limited to any medical or surgical procedure, any withdrawal or removal of body tissue or fluid, any administration of a chemical substance, any deviation from normal diet or daily regimen, and any manipulation or observation of bodily processes, behavior, or environment. Clinical research comprises four categories of activity:

1. Studies which conform to established and accepted medical practice with respect to diagnosis or treatment of an illness.
2. Studies which represent a deviation from accepted practice, but which are specifically aimed at improved diagnosis, prevention, or treatment of a specific illness in a patient.
3. Studies which are related to a patient's disease but from which he or she will not necessarily receive any direct benefit.
4. Investigative, non-therapeutic research in which there is no intent or expectation of treating an illness from which the patient is suffering, or in which the subject is a "normal control" who is not suffering from an illness but who volunteers to participate for the potential benefit of others.

It is important to emphasize that "non-therapeutic" is not to be understood as meaning "harmful." Understanding of normal processes is essential; it is the prerequisite, in many instances, to recognition of those deviations from normal which define disease. Important knowledge can be gained through such studies of normal processes. Although such research might not in any way benefit the subjects from whom the data are obtained, neither does it necessarily harm them.

Patients participating in studies identified in paragraph B-1, above, are not considered to be at special risk by virtue of participating in research activities, and this policy statement offers no special protection to them. When patients or subjects are involved in procedures identified in paragraphs B2, B3, and B4, they are considered to be "at risk," and the special policy and procedures set forth in this document pertain. Excluded from this definition are studies in which the risk is negligible, such as research requiring only, for example, the recording of height and weight, collecting excreta, or analyzing hair, deciduous teeth, or nail clippings. Some studies which appear to involve negligible physical risk might, however, have psychological, sociological, or legal implications which are significant. In that event, the subjects are in fact "at risk," and appropriate procedures described in this document shall be applied . . .

D. *Pregnancy* encompasses the period of time from implantation until delivery. All women during the child-bearing years should be considered at risk of pregnancy; hence, prudence requires definitive exclusion of pregnancy when women in this period of life are subjects for experimentation which might affect the fetus.

E. *Fetus* means the product of conception from the time of implantation to the time of delivery from the uterus.

F. *Abortus* means a fetus when it is expelled whole, whether spontaneously or as a result of medical or surgical intervention undertaken with the intention of terminating a pregnancy, prior to viability. This definition, for the purpose of this policy, excludes the placenta, fetal material which is macerated at the time of expulsion, a dead fetus, and isolated fetal tissue or organs excised from a dead fetus.

G. *Viability of the fetus* means the ability of the fetus, after either a spontaneous delivery or an abortion, to survive to the point of independently maintaining vital functions, such a "viable" fetus is a premature infant. Determination of viability entails a subjective and objective judgment by the physician attending labor or examining the product of conception, and must be made by a physician other than the investigator wishing to use fetal tissue in research. In general, and all other circumstances notwithstanding, a beating heart is not sufficient evidence of viability. At least one additional necessary condition is the possibility that the lungs can be inflated. Without this precondition, no currently available mechanisms to initiate or maintain respiration can sustain life; and in this case, though the heart is beating, the fetus or abortus is in fact non-viable.

H. *In vitro fertilization* is any fertilization of human ova which occurs outside the body of the female, either through admixture of donor sperm and ova or by any other means . . .

K. *Informed consent* has two elements; comprehension of adequate information and autonomy of consent. Consent is a continuing process. The person giving consent must be informed fully of the nature and purpose of the research and of the procedures to be used, including identification of those procedures which are experimental, the possible attendant, short or long term risks and discomforts, the anticipated benefits to himself and/or others, any alternative methods of treatment, expected duration of the study, and of his or her freedom to ask any questions and to withdraw at any time, should the person wish to do so. There must also be written evidence of the process used for obtaining informed consent, including grounds for belief that the subject has understood the information given and has sufficient maturity and mental capacity to make such choices and formulate the requisite judgment to consent. In addition, the person must have sufficient autonomy to choose, without duress, whether or not to participate. Both the comprehension of information and the autonomy of consent are necessary elements; to the extent that either of these is in doubt, the adequacy of informed consent may be in doubt.

L. *Supplementary judgment* is the judgment made by others to assent, or to refuse to assent, to procedures for which the subject cannot give adequate consent on his or her own behalf. For the purposes of this document, supplementary judgment will refer to judgments made by local committees in addition to the subject's consent (when possible) and that of the parents or legal guardian (where applicable), as to whether or not a subject may participate in clinical research. This supplementary judgment is to be confirmed by the signature of the Chairman of the Protection Committee on the consent form. In accordance with the procedures approved by the agency for the Protection Committee, the Chairman's signature may be affixed on a standard-consent form, or may need to be withheld until the Committee approves the participation of the individual subject.

II. *General policy considerations.* In general, clinical research, like medical practice, entails some risk to the subjects. When the potential subject is unable fully to comprehend the risks which might be involved, or to make the judgment essential to consent regarding the assumption of those risks, current guidelines suggest obtaining the consent of the parents or legal representative.

Whereas it is clear by law that consent of a parent or legal representative is valid for established and generally accepted therapeutic procedures performed on a child or an incompetent adult, it far from clear that it is adequate for research procedures. In practice, parental or guardian consent generally has been accepted as adequate for therapeutic research, although the issue has not been definitively resolved in the courts. When research might expose a subject to risk without defined therapeutic benefit or other positive effect on that subject's well-being, parental or guardian consent appears to be insufficient . . .

The law is not clear on these issues. Even if the law were clear, however, ethical questions would remain; specifically, whether, and under what conditions research involving these subject groups may proceed. Resolution of these ethical questions requires judgments concerning both the ethics of conducting a particular research project, and the adequacy of procedures for protecting the individual subjects who will be asked to participate. The intention of this policy is to broaden the scope of review, preclude or resolve conflicts of interest, and invoke social as well as scientific judgments to protect potential subjects who might have diminished capacity to consent . . .

III. *The fetus.* Respect for the dignity of human life must not be compromised whatever the age, circumstance, or expectation of life of the individual. Therefore, all appropriate procedures providing protection for children as subjects in biomedical research must be applied with equal rigor and with additional safeguards to the fetus.

The recent decision of the Supreme Court on abortion does not nullify the ethical obligation to protect the developing fetus from avoidable harm. This obligation, along with the right of every woman to change her decision regarding abortion, requires that no experimental procedures entailing risk to the fetus be undertaken in anticipation of abortion. Further, since the fetus might be at risk in research involving pregnant women, all research involving pregnant women must be reviewed by the Ethical Review Board, unless the Primary Review Committee determines that the research involves no risk to the fetus. Recruitment of pregnant subjects for research reviewed by the Board must involve the institution's Protection Committee in a manner approved by the Board, to provide supplementary judgment.

The consent of both parents must be obtained for any research involving the fetus, any statutes to the contrary on consent for abortion notwithstanding. Both the mother and the father have an interest in the fetus, and legal responsibility for it, if it is born. Therefore, the father's consent must be obtained for experimental procedures involving the fetus; consent of the father may be waived if his identity or whereabouts cannot be ascertained, or if he has been judged mentally incompetent.

IV. *Special categories.*

A. *The abortus.* Prematurity is the major cause of infant death in this country; thus, research aimed at developing techniques to further viability is of utmost importance. Such research has already contributed significantly to improvement in the care of the pregnant woman and of her fetus. In addition, knowledge of fetal drug metabolism, enzyme activity, and the development of organs is essential to progress in preventing or offsetting certain congenital defects. After thorough research in animal models, it often eventually becomes essential to undertake studies in the non-viable human fetus.

The decision of the Supreme Court on abortion does not eliminate the ethical issues involved in research on the non-viable human fetus. No procedures should be undertaken on the non-viable fetus which clearly affront societal values. Nevertheless, certain research is essential to improve both the chance of survival and the health status of premature infants. Such research must meet ethical standards as well as show a clear relation either to the expectation of saving the life of premature infants through the development of rescue techniques, or to the furthering of our knowledge of human development and thereby our capacity to offset the disabilities associated with prematurity. It is imperative, however, that the investigator first demonstrate that appropriate studies on animals have in fact been exhausted and that therefore the research in question requires that the work be done on the non-viable human fetus. Specific reasons for this necessity must be identified. A thorough review of the ethical issues in proposed research involving the non-viable fetus is of utmost importance.

It must be recognized that consent for abortion does not necessarily entail disinterest on the part of the pregnant woman in what happens to the product of conception. Some women feel strongly about what may, or may not, be done to the aborted fetus; others do not. In order to give every woman the opportunity to declare her wishes, consent of the pregnant woman for application of any research procedures to the aborted fetus must be secured at the time of admission to the hospital for the abortion.

Because research on the abortus involves ethical as well as scientific issues, all projects involving the abortus must be reviewed by the Ethical Review Board and recruitment of individual pregnant women for such research must involve the institution's Protection Committee in a manner approved by the Board to provide supplementary judgment. In addition to the requirement for maternal consent, both the Ethical Review Board and the Protection Committee shall, in their deliberations, consider the ethical and social issues surrounding research on the non-viable fetus. The Protection Committee must be satisfied that maternal consent is freely given and based on full disclosure, each time approved research is conducted on an abortus.

In order to insure that research considerations do not influence decisions as to timing, method, or extent of a procedure to terminate a pregnancy, no investigator engaged in the research on the abortus may take part in these decisions. These are decisions to be made by the woman and her physician.

The attending physician, not the investigator, must determine the viability of the abortus at the termination of pregnancy. If there is a reasonable possibility that the life of the fetus might be saved, experimental and established methods may be used to achieve that goal. Artificial life-support techniques may be employed only if the physician of record determines that the fetus might be viable. If the physician determines that the fetus is not viable, it is not acceptable to maintain heart beat or respiration artificially in the abortus for the purpose of research. Experimental procedures which of themselves will terminate respiration and heart beat may not be undertaken.

This policy and these protections apply with equal force to the products of spontaneous abortions.

B. *The products of in vitro fertilization.* In the interest of improving human health and development, the biology of human fertilization and the early events surrounding this phenomenon, including implantation, should be studied. To the extent that in vitro studies of human fertilization might further this aim, they are permissible at the present time within the limits outlined below.

Current technology limits the in vitro development of the human fertilized ovum to a period of several days. This is a rapidly advancing field of biomedical research, however, and the time might come when it is possible to extend in vitro development beyond the stage of early cell division and possibly even to viability.

It is contrary to the interests of society to set permanent restrictions on research which are based on the successes and limitations of current technology. Still, it is necessary to impose restraints prospectively in order to provide reasonable protections, while at the same time permitting scientific advancements which might well benefit society. A mechanism is required to weigh, at any given time, the state of the art, a specific proposal, legal issues, community standards, and the availability of guidelines to govern the research situation. This mechanism is provided by the Ethical Review Board. Ultimately, the Board will determine the acceptability of a project involving in vitro fertilization, and by recognizing the state of the art, as well as societal concerns, propose appropriate research policy.

Care must be taken not to bring human ova fertilized in vitro to viability—whether in the laboratory or implanted in the uterus—until the safety of the technique has been demonstrated as far as possible in subhuman primates. To this end:

1. All proposals for research involving human in vitro fertilization must be reviewed by the Ethical Review Board.
2. No research involving the implantation of human ova fertilized in the laboratory into recipient women should be supported until the appropriate scientific review boards are satisfied that there has been sufficient work in animals (including subhuman primates) to demonstrate the safety of the technique. It is recommended that this determination of safety include studies of natural born offspring of the products of in vitro fertilization.
3. No implantation of human ova fertilized in the laboratory should be attempted until guidelines are developed governing the responsibilities of the donor and recipient "parents" and of research institutions and personnel.

Subpart C—Additional Protection for Certain Classes of DHEW Activities

Section 46.31 *Applicability.*

(a) The regulations in this subpart are applicable to all Department of Health, Education, and Welfare research, development, or demonstration activities: (1) Involving pregnant women, unless there is a finding by DHEW that the activity will have no adverse effect on the fetus, or is clearly therapeutic with respect to the fetus involved, (2) involving the abortus or the non-viable fetus, or (3) involving in vitro fertilization of human ova.

(b) Nothing in this subpart shall be construed as indicating that compliance with the procedures set forth herein will in any way render inapplicable pertinent State or local laws bearing upon activities covered by this subpart.

(c) To the extent the requirements of subpart A of this part are applicable to activities also covered by this subpart, the requirements of this subpart are in addition to those imposed under subpart A.

Section 46.32 *Purpose.* It is the purpose of this subpart to provide additional safeguards in reviewing activities to which this subpart is applicable to assure that they conform to appropriate ethical standards and relate to important societal needs.

Section 46.33 *Definitions.* As used in this subpart:

(a) "DHEW" means the Department of Health, Education, and Welfare.

(b) "DHEW activity" means: (1) the conduct or support (through grants, contracts, or other awards) of biomedical or behavioral research involving human subjects; or (2) research, development, or demonstration activities regulated by any DHEW agency.

(c) "Board" means the Board established under § 46.25.

(d) "Protection Committee" means a committee referred to in § 46.26.

(e) "Pregnancy" means the period of time from implantation of a fertilized ovum until delivery.

(f) "Fetus" means the product of conception from implantation until delivery.

(g) "Abortus" means the fetus when it has been expelled whole, whether spontaneously or as a result of medical or surgical intervention to terminate a pregnancy, prior to viability. This definition, for the purpose of this policy, excludes the placenta, fetal material which is macerated at the time of expulsion, a dead fetus, and isolated fetal tissue or organs excised from a dead fetus.

(h) "Viability of a fetus" means capability given the benefit of available therapy, or independently maintaining heart beat and respiration.

(i) "In vitro fertilization" means any fertilization of human ova which occurs outside the body of a female, through admixture of human sperm and such ova.

Section 46.34 *Duties of the Ethical Review Board.*

(a) It shall be the function of the Board to review each activity to which this subpart applies and advise the agency concerning the acceptability of such activities from the standpoint of societal need and ethical considerations, taking into account the assessment of the appropriate Primary Review Committees as to: (1) the potential benefit of the proposed activity, (2) scientific merit and experimental design, (3) the sufficiency of studies involving animals demonstrating the clear potential benefit of the proposed procedures, and (4) whether the information to be gained may be obtained from further animal or adult human studies.

(b) The Board may recommend the establishment by the sponsoring institution of a Protection Committee to carry out such functions as the Board deems necessary.

Section 46.35 *Maternal consent to activities involving the abortus.*

(a) No activity to which this subpart is applicable may involve an abortus or a non-viable fetus unless maternal consent has been obtained.

(b) No activity to which this subpart is applicable may involve an abortus or a non-viable fetus unless (1) individuals involved in the activity will have no part in the decision as to timing, method, or extent of the procedure used to terminate the pregnancy, or in determining viability of the fetus at the termination of the pregnancy; (2) vital functions of the abortus will not be maintained artificially for purposes of research; and (3) experimental procedures which would terminate heart beat or respiration in the abortus will not be employed.

Section 46.37 *Prohibition on certain activities involving pregnant women where the fetus may be adversely affected.* The Board shall review all research, development, and demonstration activities involving pregnant women. No activity to which this subpart is applicable may involve pregnant women unless all the requirements of this subpart are satisfied.

Section 46.38 *Parental consent to activities which might affect the fetus.* No activity involving a pregnant woman which might affect the fetus but which nevertheless is permissible under § 46.37 shall be conducted unless maternal consent has been obtained, as well as the consent of the father if he is available and capable of participating in the consent process.

Section 46.39 *Activities to be performed outside the United States.* In addition to satisfying all other applicable requirements in this subpart, activities to which this subpart is applicable, which are to be conducted outside the United States, must include written documentation satisfactory to DHEW that the proposed activity is acceptable under the legal, social, and ethical standards of the locale in which it is to be performed.

NIH GUIDELINES
REVISED RECOMMENDATIONS

Protection of Human Subjects

Definitions

C. Comments on "Informed Consent" suggested the addition of language concerning (i) full and complete disclosure, (ii) the likelihood of success or failure of the experiment, (iii) the use of placebos or other control procedures, (iv) provision of information as to the progress of the research, (v) publication of names of all persons, institutions, and review committees involved in approval of consent procedures, (vi) provision of legal counsel and technical advice, and (vii) assurance that the subject comprehends the disclosure.

The Department, having considered these comments, notes that "Informed Consent" is presently defined in 45 CFR 46.3(c) and not in the present proposed rulemaking. With respect to the specific suggestions the Department notes that: as far as (i) is concerned, the regulations already call for a "fair explanation" of the procedures and a description of risks and benefits reasonably to be expected; (ii) reflects a basic misunderstanding of the experimental process which begins, essentially, with the comparison of two or more methods, procedures, or modalities on the *a priori* hypothesis that there will be no difference; (iii) is implicit in the existing regulations and is better emphasized in interpretive materials; (iv) would not be an element of informed consent unless interim findings affected the risk or benefit involved; and (v) touches on the subject of a possible future proposed rulemaking and the Department reserves its options for the present. The suggestion in (vi) is met in part by the proposals in the present proposed rulemaking to employ consent committees to advise potential subjects. The last suggestion (vii) goes beyond requirements for informed consent as they have generally been articulated by the courts . . .

E. Several commentators criticized provisions of the draft policy that would have required that activities to be conducted outside the United States satisfy all requirements of the Departments regulations including those based on ethical concepts peculiar to the Judeo-Christian moral heritage or to English common law. It was noted that this would create substantial problems for United States investigators working overseas since these concepts are often inconsistent if not in conflict with normal, ethical, and legal concepts in certain foreign countries. For the same reasons, it was argued that these provisions would create problems for United States citizens assigned, detailed, seconded, or acting as consultants to international organizations or to foreign governmental or private institutions.

Having considered these objections, the Department proposes to retain the basic concept that activities supported by Departmental funds should, in general, be subject to a uniform ethical policy wherever they are conducted, but to permit the Secretary to modify consent procedures if it can be demonstrated to his satisfaction that such procedures, as modified, are acceptable under the legal, social, and ethical standards of the locale in which the activities are to be performed.

Fetuses, Abortuses, and Pregnant Women

Since comments on the draft provisions in 38 CFR 31738 providing additional protections for fetuses, abortuses, *in vitro* fertilization, and pregnant women were integrated with those on children, it is difficult to identify the communications specifically concerned with these subjects. However, it is estimated that the majority of the more than 400 letters received on research with children, born and unborn, touched on one or more aspects of research with fetuses, abortuses, and pregnant women.

A. A large number of respondents disagreed entirely with the idea of permitting research with the fetus, with the abortus (whether living or dead), or with the pregnant woman if the research might conceivably endanger the fetus.

The Department, having carefully considered these comments and similar proposals reflected in general correspondence and in articles in the public media, notes that their adoption would seriously hamper the development of needed improvements in the health care of the pregnant woman, the fetus, and the newborn. The opposition to research involvement of the fetus and abortus appears to be based in part on the assumption that the needed information can be obtained through research with animal species or with adults. Unfortunately, these assumptions are not valid. While much useful research can be conducted in animals, differences in species are nevertheless so great that any research finding in nonhuman species must ultimately be repeated in man before its general application in human medicine. In addition, the fetus and the newborn are not small adults. They suffer from some diseases not encountered in the adult. They may react differently to the diseases commonly affecting both adult and young, and they may have a different response to the same treatment, both with regard to its effectiveness and to its safety. The Department therefore proposes that (i) the ethical probity of any application or proposal for the support of any activity covered by subpart C be reviewed by an Ethical Advisory Board as described in § 46.304, and (ii) the conduct of any such activity supported by the Department be subject to oversight and monitoring by a consent committee as described in § 46.305.

B. Opinion was divided as to the need for an Ethical Advisory Board. Many respondents called it a welcome addition in the review process. Others felt that it would duplicate the function of the local organizational review committee and that its existence would encourage the organizational review committee to be less critical and would impose an additional roadblock that would delay or prohibit important research while needlessly consuming time, energy, and money, and posing potential danger to a patient waiting for treatment. Complaints were voiced that such decisions should be made locally, not in Washington, and that the investigator should be able to present his case in person. Numerous comments suggested that the Board's function should be limited to advising on policy, guidelines, or procedures, and not be concerned with the review of individual projects. This would avoid duplicating the function of the organizational review committee. Others suggested that the Ethical Advisory Board should serve as an appeal body from the organizational review committee.

There were also numerous comments to the effect that it is unwise and impossible to totally separate ethical and scientific review. Approval based only on ethics would be unethical if the science were bad. Both should be reviewed jointly.

The Department, having reviewed these comments, concludes that Ethical Advisory Board remains, in concept, a useful addition to the review process. It does not duplicate the functions of the local organizational review committee since the latter is primarily concerned with matters of organizational regulations, local standards of professional practice, applicable law within its jurisdiction, and local community attitudes. The Ethical Advisory Board will be primarily concerned with similar issues at the national level. Applications and proposals should be capable of passing scrutiny at both levels. It is therefore proposed that the Ethical Advisory Board be retained as part of the additional protection mechanism.

Specific comments regarding the establishment of an Ethical Advisory Board touched principally on (i) the possibility that appointment of members at an agency level might lead to "loaded" Boards, while appointment at a higher level, i.e., by a joint Congressional committee or by independent outside bodies, might produce a more objective group, and (ii) disagreement as to the proper balance between scientist and nonscientist members, with a majority of the commentators suggesting that more than one-third of the members should have the scientific expertise necessary to identify risks and their possible consequences. It was specifically suggested that different sizes, compositions, and administrative locations of the Board be tried before selecting a final mechanism. In addition, it was suggested (iii) that a fifteen-member Board was too large, (iv) that all members be human geneticists, (v) that at least one member be a psychologist, if behavioral issues were to be considered, (vi) that there be an absolute ban on departmental agency employees, (vii) that all proceedings be confidential, (viii) that all meetings be open to the public, and (ix) that an appeal mechanism be established.

The Department, having considered these views, proposes that while an Ethical Advisory Board to deal with biomedical research involving fetuses, abortuses, pregnant women, and in vitro fertilization might logically be established at the National Institutes of Health, (i) the power of appointment should be reserved to the Secretary, (ii) while the membership should include research scientists, physicians, lawyers, clergy or ethicists, and representatives of the general public, and balance between callings should rest with the Secretary as should also (iii) the number of members, so that the membership (iv, v) can be adjusted to the needs of the Board as the work-load and the issues before it dictate. The specific suggestion (see vi) that departmental agency employees be excluded is adopted and expanded to include all full-time employees of the Federal Government. The decisions with regard to suggestions (vii) and (viii) will be governed by the provisions of the Federal Advisory Committee Act which generally require that meetings of similar advisory groups be open to the public for the purposes of policy discussion, but closed and confidential for the purpose of review of specific applications and proposals. Since the Board will be advisory to funding agencies, the final action will be that of existing awarding authorities, and appeal mechanisms (ix) will be provided only to the extent available under other existing departmental regulations and policies. These proposals are incorporated into § 46.304.

C. A number of respondents recommended that the policy governing in vitro fertilization be strengthened, on the one hand, or liberalized, on the other. The Department has considered these recommendations, and has provisionally chosen not to stipulate at this time protections for the product of in vitro fertilization which is not implanted, but rather to leave that series of issues to the Ethical Advisory Board established under § 46.304(a). The Board will be required to weigh, with respect to specific research proposals, the state of the art, legal issues, community standards, and the availability of guidelines to govern each research situation.

Because biomedical research is not yet near the point of being able to maintain for a substantial period the non-implanted product of in vitro fertilization, no clear and present danger arises from not stipulating in these regulations the protections for it. Given the state of the research, we believe that such stipulation would be premature.

It is the Department's intent that the definition of the term *fetus* (§ 46.303 (d) be construed to encompass both the product of in vivo conception and the product of in vitro fertilization which is subsequently implanted in the donor of the ovum. Whatever the nature of the conception process, it is intended that upon implantation the protections of subpart C apply to all fetuses. It is only with respect to the protections available to the non-implanted product of in vitro fertilization that the regulations are silent.

With respect to the fertilization of human ova in vitro, it is expected that the Board will consider the extent to which current technology permits the continued development of such ova as well as the legal and ethical issues surrounding the initiation and disposition of the products of such research.

With respect to implantation of fertilized human ova, it is expected that the Board will consider such factors as the safety of the technique (with respect to offspring) as demonstrated in animal studies, and clarification of the legal responsibilities of the donor and recipient parent(s) as well as the research personnel.

Since the Department does reserve the option of later specifying such protections by regulation, we invite comment on the question of appropriate regulations in the future.

D. The draft proposals included a suggestion for the establishment of a protection committee which elicited numerous comments that the use of the term *protection committee* implies that the Department recognizes a clear, present need for protection against the investigator, the uncertain relation of this committee to the organizational review committee, and the uniform need for and desirability of such protection.

Having reviewed these comments the Department proposes an extensive revision in this innovative concept. Initially it acknowledges that the term *protection committee* is pejorative and proposes the term *consent committee* as more appropriate and consistent with the primary purpose of such bodies. Further, it proposes to eliminate specific requirements for the size and composition of such committees. Instead, applicants and offerors are to propose the establishment of such a committee, specifying its size, composition, and rules of procedure. In addition, where the applicant or offeror believes that the activity involves only negligible risks, it may ask the Secretary to waive or modify the requirement for a consent committee. All proposals for the establishment, modification, or waiver of a consent committee shall be subject to review and approval at the local level by the organizational review committee and at the departmental level by the Ethical Advisory Board. The Ethical Advisory Board may prescribe additional duties for the consent committee. These changes are incorporated in § 46.405. In view of this drastic change in concept of the committee, detailed discussion of the many excellent and often thought-provoking comments concerned with details of the original draft seems inappropriate.

E. Many critical comments were addressed to the definitions used in this subpart, specifically:

1. "Pregnancy." It was suggested that pregnancy should be defined (i) conceptually to begin at the time of fertilization of the ovum, and (ii) opera-

tionally by actual test unless the woman has been surgically rendered incapable of pregnancy.

While the Department has no argument with the conceptual definition as proposed above, it sees no way of basing regulations on the concept. Rather, the order to provide an administerable policy, the definition must be based on existing medical technology which permits confirmation of pregnancy. This approach is reflected by § 46.303(c).

2. "Viability of the Fetus." Many recommendations were received concerning viability of the fetus after premature delivery or abortion. Some respondents urged that presence of fetal heart beat be definitive (whether or not there is respiration) while others urged that identifiable cortical activity be specified as an alternative sign of viability. The Department has concluded that the issue of viability is a function of technological advance, and therefore must be decided with reference to the medical realities of the present time. We reserve the option of redefining the parameters as conditions warrant.

Only upon the basis of a definition which is both precise and consistent with current medical capability can a regulation realistically be interpreted and enforced. Current technology is such that a fetus, given the benefit of available medical therapy, cannot survive unless the lungs can be inflated so that respiration can take place. Without this capability, even if the heart is beating, the fetus is non-viable. In the future, if technology has advanced to the point of sustaining a fetus with non-inflatable lungs, the definition can and should be modified.

The Department has therefore chosen to specify, in the definition of viability of the fetus (§ 46.303(e)), that heart beat and respiration are, jointly, to be the indicator of viability.

3. "Abortus." Various comments noted that this definition is more restrictive than the usual medical definition of the abortus as a "non-viable fetus," and suggested substitution of the broader definition.

The Department proposes to retain the original definition for the purposes of these regulations. There is general agreement that there are distinct ethical problems involved in decisions concerning research use of the intact fetus, or use of organs or tissues obtained from a fetus that has died in utero or from an abortus at autopsy. The definition recurs with minor editorial changes in § 46.303(f).

F. Several comments were critical of the draft regulation's provisions limiting activities involving pregnant women to those not adversely affecting the fetus, except where the primary purpose of the activity was to benefit the fetus. It was suggested that the regulations (i) should contain language permitting exceptions for research necessary to meet the health needs of the mother, and (ii) should grant the right to participate in research aimed at improvement of methods of abortion, birth control, and genetic intervention.

The Department concurs with the first suggestion (i), and proposes that the regulations permit research whose primary interest is to benefit the particular fetus or to respond to the health needs of the pregnant woman. It does not fully accept the second suggestion, (ii), and proposes that the regulations permit fetal research concerned with diagnosis and prevention of perinatal disease, and to offset the effects of genetic abnormality or congenital injury, but only when such research is done as part of a procedure properly performed to terminate a pregnancy. These changes are incorporated into § 46.306(a). The Department has tentatively concluded that consideration of risk vs. benefit with respect to fetal research does not seem to be appropriate.

G. Draft regulation provisions required maternal consent and the consent of the father if he were available and capable of participating in the consent process. This provision was strongly criticized on the grounds that it could permit the father of the fetus to deny needed health care to the woman or to the fetus even though he had no marital obligations, and that it might result in undue delay in the delivery of health care. It was also pointed out that the regulations did not touch on the question of the validity of consent by a pregnant minor.

The Department agrees. It is now proposed that paternal consent be sought only if the activity is not responding to the health needs of the pregnant woman and the father is reasonably available. These changes are reflected by § 46.306(b).

H. The Department has provisionally chosen, in § 46.306(a), to permit research to be undertaken from which there will be risk of harm to the fetus if such research is conducted as part of the abortion procedure. This decision, upon which we invite comment, has been made in the expectation that such research may produce new technology which will enable countless premature infants to live who now cannot.

It is not intended that this provision be construed to permit fetal research in anticipation of abortion prior to the commencement of the termination procedure itself.

While it is true that the class of fetuses for whom abortion is contemplated will be placed at greater research risk than all fetuses in general, such risk can arise only after implementation of the double safeguard of parental consent to the contemplated abortion, and second parental consent to the research procedure itself.

I. Comments regarding activities involving the abortus were concerned with the issue of maintaining vital functions and signs. It was argued that maintaining vital functions at the level of the organ, tissue, or cell is essential to studies and involves no prolongation of the dying of the abortus. At the same time, it was argued that termination of the heart beat should not be prohibited since temporary cardiac arrest has proved essential in the development of surgical techniques necessary to correct congenital heart defects.

Neither of these objections appear valid and no significant changes in § 46.307 are proposed. However, in order to emphasize again the distinction between research with the whole fetus or abortus, functioning as an organism with detectable vital signs, and with the dead fetus or abortus, the Department has added § 46.308, concerning activities involving a dead fetus or abortus, and § 46.309, concerning the abortus as an organ or tissue donor. Also § 46.307(d) has been expanded to permit the artificial maintenance of vital functions of an abortus where the purpose is to develop new methods for enabling the abortus to survive to the point of viability.

The Department feels that there is evident distinction between "termination" and "arrest" of the clinical signs as applied to the fetus or premature infant, but that no such distinction is valid or applicable where the abortus is concerned.

Subpart C—Additional Protections Pertaining to Biomedical Research, Development, and Related Activities Involving Fetuses, Abortuses, Pregnant Women, and in Vitro Fertilization

§ 46.301 *Applicability.*

(a) The regulations in this subpart are applicable to all Department of Health, Education, and Welfare grants and contracts supporting biomedical research, development, and related activities involving: (1) the fetus in utero, (2) the abortus, as that term is defined in § 46.303, (3) pregnant women, and (4) in vitro fertilization. In addition, these regulations are applicable to all such activities involving women who could become pregnant, except where the applicant or offeror shows to the satisfaction of the Secretary that adequate steps will be taken in the conduct of the activity to avoid involvement of women who are pregnant.

(b) Nothing in this subpart shall be construed as indicating that compliance with the procedures set forth herein will in any way render inapplicable pertinent State or local laws bearing upon activities covered by this subpart.

(c) The requirements of this subpart are in addition to those imposed under the other subparts of this part.

§ 46.302 *Purpose.*

It is the purpose of this subpart to provide additional safeguards in reviewing activities to which this subpart is applicable to assure that they conform to appropriate ethical standards and relate to important societal needs.

§ 46.303 *Definitions.* As used in this subpart:

(a) "Secretary" means the Secretary of Health, Education, and Welfare or any other officer or employee of the Department of Health, Education, and Welfare to whom authority has been delegated.

(b) "Biomedical research, development, and related activities" means research, development, or related activities involving biological study (including but not limited to medical or surgical procedures, withdrawal or removal of body tissue or fluid, administration of chemical substances or input of energy, deviation from normal diet or hygiene, and manipulation or observation of bodily processes).

(c) "Pregnancy" encompasses the period of time from confirmation of implantation until delivery.

(d) "Fetus" means the product of conception from the time of implantation to the time of delivery.

(e) "Viability of the fetus" means the ability of the fetus, after either spontaneous or induced delivery, to survive (given the benefit of available medical therapy) to the point of independently maintaining heart beat and respiration. If the fetus has this ability, it is viable and therefore a premature infant.

(f) "Abortus" means a fetus when it is expelled whole, prior to viability, whether spontaneously or as a result of medical or surgical intervention. The term does not apply to the placenta; fetal material which is macerated at the time of expulsion; or cells, tissue, or organs excised from a dead fetus.

(g) "In vitro fertilization" means any fertilization of human ova which occurs outside the body of a female, either through admixture of donor sperm and ova or by any other means.

§ 46.304 *Ethical Advisory Board.*

(a) All applications or proposals for the support of activities covered by this subpart shall be reviewed by an Ethical Advisory Board, established by the Secretary within the National Institutes of Health, which shall advise the funding agency concerning the acceptability of such activities from an ethical standpoint.

(b) Members of the Board shall be so selected that the Board will be competent to deal with medical, legal, social, and ethical issues and shall include, for example, research scientists, physicians, lawyers, and clergy and/or ethicists, as well as representatives of the general public. No Board member may be a regular, full-time employee of the Federal Government.

§ 46.305 *Establishment of a consent committee.*

(a) Except as provided in paragraph (c) of this section, no activity covered by this subpart may be supported unless the applicant or offeror has provided an assurance acceptable to the Secretary that it will establish a consent committee (as provided for in the application or offer and approved by the Secretary) for each such activity, to oversee the actual process by which individual consents required by this subpart are secured, to monitor the progress of the activity and intervene as necessary, and to carry out such other duties as the Secretary (with the advice of the Ethical Advisory Board) may prescribe. The duties of the consent committee may include: (1) participation in the actual selection process and securing of consents to assure that all elements of a legally effective informed consent, as outlined in § 46.3, are satisfied. Depending on what may be prescribed in the application or offer approved by the Secretary, this might require approval by the committee of individual participation in the activity or it might simply call for verification (e.g., through sampling) that procedures prescribed in the approved application or offer are being followed. (2) Monitoring the progress of the activity. Depending on what may be prescribed in the application or offer approved by the Secretary, this might include: visits to the activity site, identification of one or more committee members who would be available for consultation with those involved in the consent procedure (i.e., participants) at the participant's request, continuing evaluation to determine if any unanticipated risks have arisen and that any such risks are communicated to the participants, periodic contact with the participants to ascertain whether they remain willing to continue in the activity, providing for the withdrawal of any participants who wish to do so, and authority to terminate participation of one or more participants with or without their consent where conditions warrant.

(b) The size and composition of the consent committee must be approved by the Secretary, taking into account such factors as: (1) the scope and nature of the activity; (2) the particular subject groups involved; (3) whether the membership has been so selected as to be competent to deal with the medical, legal, social, and ethical issues involved in the activity; (4) whether the committee includes sufficient members who are unaffiliated with the applicant or offeror apart from membership on the committee; and (5) whether the committee includes sufficient members who are not engaged in research, development, or related activities involving human subjects. The committee shall establish rules of procedure for carrying out its functions and shall conduct its business at convened meetings, with one of the members designated as chairperson.

(c) Where a particular activity, involving fetuses in utero or pregnant women, presents negligible risk to the fetus, an applicant or offeror may request the Secretary to modify or waive the requirement in paragraph (a) of this section. If the Secretary finds that the risk is indeed negligible and other adequate controls are provided, he may (with the advice of the Ethical Advisory Board) grant the request in *while* (*sic*) or in part.

(d) The requirements of this section and § 46.304 do not obviate the need for review and approval of the application or offer by the organizational review committee, to the extent required under Subpart A of this part.

§ 46.306 *Activities involving fetuses in utero or pregnant women.*

(a) No activity to which this subpart is applicable, involving fetuses in utero or pregnant women, may be undertaken unless: (1) the purpose of the activity is to benefit the particular fetus or to respond to the health needs of the mother, or (2) the activity conducted is part of (but not prior to the commencement of) a procedure to terminate the pregnancy and is for the purpose of evaluating or improving methods of prenatal diagnosis, methods of prevention of premature birth, or methods of intervention to offset the effects of genetic abnormality or congenital injury.

(b) Activities covered by this subpart which are permissible under paragraph (a) of this section may be conducted only if the mother and father are legally competent and have given their consent, except that the father's consent need not be secured if: (1) the purpose of the activity is to respond to the health needs of the mother or (2) his identity or whereabouts cannot reasonably be ascertained.

(c) Activities covered by this subpart which are permissible under paragraph (a) (2) of this section may not be undertaken unless individuals engaged in the research will have no part in: (1) any decisions as to the timing, method, or procedures used to terminate the pregnancy, and (2) determining the viability of the fetus at the termination of the pregnancy.

§ 46.307 *Activities involving abortuses.*

No activity to which this subpart is applicable, involving an abortus, may be undertaken unless:

(a) appropriate studies on animals have been completed;

(b) the mother and father are legally competent and have given their consent, except that the father's consent need not be secured if his identity or whereabouts cannot reasonably be ascertained;

(c) individuals engaged in the research will have no part in: (1) any decisions as to the timing, method, or procedures used to terminate the pregnancy, and (2) determining the viability of the fetus at the termination of the pregnancy;

(d) vital functions of an abortus will not be artificially maintained except where the purpose of the activity is to develop new methods for enabling the abortus to survive to the point of viability; and

(e) experimental procedures which would terminate the heart beat or respiration of the abortus will not be employed.

§ 46.308 *Activities involving a dead fetus or abortus.*

Activities involving a dead fetus or abortus shall be conducted in accordance with any applicable State or local laws governing autopsy.

§ 46.309 *Activities involving the abortus as an organ or tissue donor.*

Activities involving the abortus as an organ or tissue donor shall be conducted in accordance with any applicable State or local laws governing transplantation or anatomical gifts.

§ 46.310 *Activities to be performed outside the United States.*

Activities to which this subpart is applicable, to be conducted outside the United States, are subject to the requirements of this subpart, except that the consent procedures specified herein may be modified if it is shown to the satisfaction of the Secretary that such procedures, as modified, are acceptable under the laws and regulations of the country in which the activities are to be performed and that they comply with the requirements of Subpart A of this part.

FURTHER READINGS

For lists of further sources see the National Library of Medicine, Literature Search: #5-69 "Heart Transplantation," and #73-4 "Human Experimentation," and #6-69 "Kidney Transplantation in Man," and #70-35 "Psychological Aspects of Transplantation," and #70-32 "Psychological Response to Hemodialysis," and #71-3 "Transplantation in the Therapy of Hearing Disorders," and #73-6 "Transplantation-Ethical, Legal and Religious." All of these lists are obtainable from the Literature Search Program, National Library of Medicine, Bethesda, Maryland 20014.

Books

George J. Annas, *The Rights of Hospital Patients: The Basic ACLU Guide to a Hospital Patient's Rights* (New York: Avon, 1975); Bernard Barber *et al., Research on Human Subjects: Problems of Social Control in Medical Experimentation* (New York: Russell Sage Foundation, 1973); Henry K. Beecher, *Experimentation in Man* (Springfield, Ill.: Charles C Thomas, 1959), and also by this author: *Research and the Individual: Human Studies* (Boston: Little, Brown and Co., 1970); Claude Bernard, *An Introduction to the Study of Experimental Medicine,* Tr. Henry Copley Green (New York: Dover Publications, 1957); I. S. Copper, *The Victim is Always the Same* (New York: Harper & Row, 1973); G. R. Dunstan, *The Artifice of Ethics* (London: S.C.M. Press, 1974); R. T. Eastwood, *Cardiac Replacement: Medical, Ethical, Psychological and Economic Implications,* a report by the Ad Hoc Task Force on Cardiac Replacement, National Heart and Lung Institute, National Institutes of Health (Washington, D.C.: Government Printing Office, 1969); V. Fattorusso, ed., *Biomedical Science and the Dilemma of Human Experimentation* (Paris: Council for International Organization of Medical Sciences, UNESCO House, 1967); Paul A. Freund, ed., *Experimentation with Human Subjects* (New York: Braziller, 1970); Renée Fox and Judith P. Swazey, *The Courage to Fail: A Social View of Organ Transplants and Dialysis* (Chicago: University of Chicago Press, 1974); Bernard Haring, *Medical Ethics* (Notre Dame, Indiana: Fides Publications, 1972); Jay Katz, with Alexander M. Capron and Eleanor Swift Glass, *Experimentation with Human Beings* (New York: Russell Sage Foundation, 1972); Herbert Kelman, *A Time to Speak: On Human Values and Social Research* (San Francisco: Jossey-Bass, 1968); Jack Kevorkian, *Capital Punishment Or Capital Gain* (New York: Philosophical Library, 1962), and also by this same author: *Medical Research and The Death Penalty* (New York: Vantage Press, 1960); Irving Ladimer, and

Roger Mewman, eds., *Clinical Investigation in Medicine: Legal, Ethical and Moral Aspects* (Boston: Law-Medicine Research Institute, Boston University, 1963); Donald Longmore, *Spare-Part Surgery: The Surgical Practice of the Future* (Garden City, N.Y.: Doubleday and Co., 1968); Catherine Lyons, *Organ Transplants: The Moral Issues* (Philadelphia: Westminster Press, 1970); Sidney Merlis, ed., *Non-Scientific Constraints on Medical Research* (New York: Raven Press, 1970); David W. Meyers, *The Human Body and the Law* (Chicago: Aldine Press, 1970); Stanley Milgram, *Obedience to Authority* (New York: Harper & Row, 1973); George W. Miller, *Moral and Ethical Implications of Human Organ Transplants* (Springfield, Ill.: Charles C Thomas, 1971); Alexander Mitscherlich and Fred Mielke, *Doctors of Infamy: The Story of the Nazi Medical Crimes* (New York: Henry Schuman, 1949); New York University School of Medicine, The Student Council, *Ethical Issues in Human Experimentation: The Case of Willowbrook State Hospital Research* (New York: New York University Medical Center, The Urban Affairs Program, 1973); M. H. Pappworth, *Human Guinea Pigs: Experimentation on Man* (Boston: Beacon Press, 1968); Paul Ramsey, *The Ethics of Fetal Research* (New Haven: Yale University Press, 1975), and also by this author: *The Patient as a Person* (New Haven: Yale University Press, 1970); Felix T. Rapaport, ed., *A Second Look at Life* (New York: Grune and Stratton, 1973); C. Reagan, *Ethics for Scientific Researchers,* second edition (Springfield, Ill.: Charles C Thomas, 1971); Alfred M. Sadler and Blair L. Sadler, *Organ Transplantation: Current Medical and Medical-Legal Status: The Problems of an Opportunity* (Washington, D.C.: U.S. Government Printing Office, 1970); Bernard Shaw, *The Doctor's Dilemma: A Tragedy* (Baltimore: Penguin Books, 1965); Harmon L. Smith, *Ethics and the New Medicine* (Nashville: Abingdon Press, 1970); Richard Titmuss, *The Gift Relationship: From Human Blood to Social Policy* (New York: Pantheon, 1971); Hans-Reudi Weber, ed., *Experiments with Man: Report of an Ecumenical Consultation,* World Council of Churches Studies, No. 6 (New York: Friendship Press, 1969); G. E. W. Wolstenholme and Maeve O'Connor, *Ethics in Medical Progress: With Special Reference to Transplantation* (Boston: Little, Brown and Co., 1966).

Articles

Leo Alexander, "Medical Science Under Dictatorship," *New England Journal of Medicine 241* (July 14, 1949), pp. 39–47; Ralph J. Alfidi, "Informed Consent: A Study of Patient Reaction," *Journal of the American Medical Association 216* (1971), pp. 1325–29; Lawrence K. Altman, "Auto-Experimentation: An Unappreciated Tradition in Medical Science," *New England Journal of Medicine 286* (1972), pp. 346–52; American College of Surgeons/National Institutes of Health Organ Transplant Registry, Advisory Committee to the Registry, "Third Scientific Report," *Journal of the American Medical Association 226* (Dec. 3, 1973), pp. 1211–16; American Medical Association House of Delegates, "Statement on Heart Transplantation," *Journal of the American Medical Association 207* (Mar. 3, 1969), pp. 1704–05; American Medical Association Judicial Council, "Ethical Guidelines for Organ Transplantation," *Journal of the American Medical Association 205* (Aug. 5, 1968), pp. 341–42; James A. Baker, "Court Ordered Non-Emergency Medical Care for Infants," *Cleveland Marshall Law Review 18* (1969), pp. 296–307; B. H. Beard, "Fear of Death and Fear of Life: The Dilemma in Chronic Renal Failures, Hemodialysis, and Kidney Transplantation," *Archives of General Psychiatry 21* (1969), pp. 373–80; Henry K. Beecher, "Experimentation in Man," *Journal of the American Medical Association 169* (Jan. 31, 1959), pp. 461–78; and also by this author: "Scarce Resources and Medical Advancement," *Experimentation with Human Subjects* ed. Paul Freund (New York: George Braziller, 1970), pp. 66–104, and "Some Guiding Principles for Clinical Investigation," *Journal of the American Medical Association 195* (Mar. 28, 1966), pp. 1135–36; Emile Z. Berman, "The Legal Problems of Organ Transplantation," *Villanova Law Review 13* (1968), pp. 751–58; Arthur H. Bernstein, "Consent to Operate, to Live, or to Die," *Hospitals: J.A.H.A.* (Oct. 1, 1972), pp. 124–28; Board on Medicine of the National Academy of Sciences, "Cardiac Transplantation in Man," *Journal of the American Medical Association 204* (May 27, 1968), pp. 805–06; British Medical Association, "New Horizons in Medical Ethics: Research Investigations in Adults," *British Medical Journal* (Apr. 28, 1973), pp. 220–24; British Medical Association, "Report of the Special Committee on Organ Transplantation," *British Medical Journal* (Mar. 21, 1970), pp. 750–51; Alexander Capron, "Legal Consideration Affecting Clinical Pharmacological Studies in Children," *Clinical Research 21* (Feb., 1973), pp. 141–50; Thomas Chalmers *et al.*, "Controlled Studies in Clinical Cancer Research," *New England Journal of Medicine 287* (July 13, 1972), pp. 75–78; James F. Childress, "Who Shall Live When Not All Can Live?" *Readings on Ethical and Social Issues in Biomedicine* ed. Richard W. Wertz (Englewood Cliffs, N.J.: Prentice-Hall, 1973); André Cournard *et al.*, "Symposium on Organ Transplantion in Man," *Proceedings of the National Academy of Sciences 63* (Aug., 1969), pp. 1018–38; W. J. Curran, "A Problem of Consent: Kidney Transplantation in Minors," *New York University Law Review 34* (1959), pp. 891 ff, and also by this author: "Kidney Transplantation in Identical Twin Minors," *New England Journal of Medicine 287* (July 6, 1972), pp. 26–27, and "The Law and Human Experimentation," *New England Journal of Medicine 275* (Aug. 11, 1966), pp. 323–25, and with Henry K. Beecher, "Experimentation in Children," *Journal of the American Medical Association 210* (Oct. 6, 1969); Fred Cutter, "Transplants and Psychological Survival in the Treatment of Kidney Disease," *Omega 3* (Feb., 1973), pp. 57–65; J. Dukeminier, Jr., "Supplying Organs for Transplantation," *Michigan Law Review 68* (1970), pp. 811–66, and also with D. Sanders, "Organ Transplantation: A Proposal for Routine Salvaging of Cadaver Organs," *New England Journal of Medicine 279* (Aug. 22, 1968), pp. 413–419; A. J. Dyck and H. W. Richardson, "The Moral Justification for Research Using Human Subjects," *Use of Human Subjects on Safety Evaluation of Food Chemicals,* Proceedings of a Conference of the National Academy of Sciences and National Research Council (1967), pp. 229–47; Gary S. Dyer, "Kidney Transplant–Mentally Incompetent Donor," *Missouri Law Review 85* (Fall, 1970), pp. 538–44; J. R. Elkinton, "Moral Problems in the Use of Borrowed Organs, Artificial and Transplanted," *Annals of Internal Medicine 60* (1964), pp. 309–13; Carl H. Fellner, "Altruism in Disrepute," *New England Journal of Medicine 284* (Mar. 18, 1971), pp. 582–85, and other articles by this author are: "Kidney Donors–the Myth of Informed Consent," *American Journal of Psychiatry 126* (Mar., 1970), p. 9, and "Selection of Living Kidney Donors and the Problem of Informed Consent," *Seminars in Psychiatry3* (Feb., 1971), pp. 79–85, and "Twelve Kidney Donors," *Journal of the American Medical Association 206* (Dec. 16, 1968), p. 2703; John C. Fletcher, "Dialogue between Medicine and Theology: Death and Transplantation," in *Should Doctors Play God?* ed. Claude A. Frazier (Nashville: Broadman Press, 1971), pp. 150–63, and "Human Experimentation: Ethics in the Consent Situation," *Law and Contemporary Problems 32* (1967), pp. 620–49, and "Our Shameful Waste of Human Tissue," *Updating Life and Death* ed. Donald Cutler (Boston: Beacon Press, 1969), chap. 1, and "Realities of Patient Consent to Medical Research," *Hastings Center Studies 1* (No. 1, 1973), pp. 39–49; Paul A. Freund, "Ethical Problems in Human Experimentation," *New England Journal of Medicine 273* (1965), pp. 687–92; Iago Galdston, "The History of Research, with Particular Regard to Medical Research," *Ciba Symposium 8* (1946), pp. 338–72; Roger Greenberg *et al.,* "The Psychological Evaluation of Patients for a Kidney Transplant and Hemodialysis Program," *American Journal of Psychiatry 130* (March, 1973), pp. 274–77; Germain G. Grisez, "Rational Ethics Says 'No,'" *Commonweal 86* (April 14, 1967), pp. 112–25; J. Grundel, "Ethics of Organ Transplantation," *Organ Transplantation Today* eds. N. A. Mitchison, J. M. Greep, and J. C. M. Hattinga Verschure (Amsterdam: Excerpta Medica Foundation, 1969); O. E. Guttentag, "Ethical Problems in Human Experimentation," *Ethical Issues in Medicine* ed. E. Fuller Torrey (Boston: Little, Brown and Company, 1968), pp. 195–226; J. Hamburger and J. Crosnier, "Moral and Ethical Problems in Transplantation," *Human Transplantation* eds. Felix T. Rapaport and Jean Dausset (New York: Grune and Stratton, 1968); C. C. Havighurst, "Compensating Persons Injured in Human Experimentation," *Science 169* (July 10, 1969), p. 154; Adele Hofmann and Harriet F. Pilpel, "The Legal Rights of Minors," *Pediatric Clinics of North America 20* (Nov., 1973), pp. 989–1004; Angela R. Holder, "Transplant Problems," *Journal of the American Medical Association 223* (March 12, 1973), pp. 1315–16; "Human Experimentation," *Medical World News* (June 8, 1973), pp. 37–51; Hans Jonas,

"Philosophical Reflections on Experimenting with Human Subjects," in *Experimentation with Human Subjects* ed. Paul A. Freund (New York: Braziller, 1970), pp. 1–31; Albert R. Jonsen, "The Totally Implantable Artificial Heart," *Hastings Center Report 3* (Nov., 1973), pp. 1–4; Alexander M. Kidd, "Limits of the Right of a Person to Consent to Experimentation on Himself," *Science 117* (Feb. 27, 1953), pp. 211–12; A. L. Knutson, "Body Transplants and Ethical Values," *Social Science and Medicine 2* (1968–69), pp. 393–414; Irving Ladimer, ed., "New Dimensions in Legal and Ethical Concepts for Human Research," *Annals of the New York Academy of Sciences 169* (1970), pp. 293–593; Louis Lasagna, "Some Ethical Problems in Clinical Investigation," *Human Aspects of Biomedical Innovation* eds. E. Mendelsohn, J. P. Swazey, and I. Taviss (Cambridge: Harvard University Press, 1971), pp. 98–111; Michael T. Malloy, "A 'No' to Research on Aborted, Live Fetuses," *The National Observer,* Vol. 38 (1973); Robert Q. Marston, "Medical Science, the Clinical Trail and Society," *Hastings Center Report 3* (April, 1973), pp. 1–4; Daniel Martin *et al.,* "Human Subjects in Clinical Research–A Report of Three Studies," *New England Journal of Medicine 279* (1968), pp. 1426–31; Curtis R. Morris *et al.,* "Guidelines for Accepting Volunteers: Consent, Ethical Implications, and the Function of Peer Review," *Clinical Pharmacology and Therapeutics 13* (Sep., 1972), pp. 782–802; Howard N. Morse, "Legal Implications of Clinical Investigation," *Vanderbilt Law Review 20* (1967), p. 747; Robert C. Neville and Peter Steinfels, commentators, "Case Studies in Bioethics: Blood Money: Should a Rich Nation Buy Plasma from the Poor?" *Hastings Center Report 2* (Dec., 1972), pp. 8–10; Harriet F. Pilpel, "Minors' Rights to Medical Care," *Albany Law Review 36* (1972), pp. 462–87; Paul Ramsey, "The Ethics of a Cottage Industry in an Age of Community and Research Medicine," *New England Journal of Medicine 284* (April 1, 1971), pp. 700–06; Gary Reback, "Fetal Experimentation: Moral, Legal and Medical Implications," *Stanford Law Review 26* (May, 1974), pp. 1191–07; Nicholas Rescher, "The Allocation of Exotic Medical Lifesaving Therapy," *Ethics 79* (April, 1969), pp. 173–86; Alfred M. Sadler, Jr. and Blair L. Sadler, "Transplantation and the Law: The Need for Organized Sensitivity," *Georgetown Law Review 57* (1968), p. 5, and other articles by these same authors are: "Transplantation and the Law: Progress Toward Uniformity," *New England Journal of Medicine 282* (March 26, 1970), pp. 717–23, and "Recent Developments in the Legal Aspects of Transplantation in the United States," *Transplantation Proceedings 3* (March, 1971), pp. 293–97; Alfred Sadler, Jr., Blair L. Sadler, and E. B. Stason, "The Uniform Anatomical Gift Act: A Model for Reform," *Journal of the American Medical Association 206* (December 9, 1968), pp. 2501–06; "Scarce Medical Resources," *Columbia Law Review 69* (April, 1969), pp. 620–92; Herbert A. Schwartz, "Children's Concepts of Research Hospitalization," *New England Journal of Medicine 287* (Sep. 21, 1972), pp. 589–92; Joseph E. Simonaitis, "Recent Decisions on Informed Consent,"

Journal of the American Medical Association 221 (July 24, 1972), pp. 441–42, and also by this author, "More About Informed Consent," *Journal of the American Medical Association 224* (June 25, 1973), pp. 1831–32; Strunk v. Strunk, Ky., 445 S. W. 2d. 145, (1969); D. Surgenor *et al.,* "Blood Services: Prices and Public Policy," *Science 180* (April 27, 1973), pp. 384–89; Robert M. Veatch and Sharmon Sollitto, "Human Experimentation–the Ethical Questions Persist," *Hastings Center Report 3* (June, 1973), pp. 1–3; Leroy Walters, "Ethical Issues in Experimentation on the Human Fetus," *Journal of Religious Ethics* (Spring, 1974); Wolf Wolfenberger, "Ethical Issues in Research with Human Subjects," *Science 155* (1967), pp. 47–51.

PART IV: HUMAN GENETICS

INTRODUCTION

This section deals with the unique moral and legal problems attending the acquisition and possible uses of genetic knowledge.

Tracy Sonneborne's opening essay, "Ethical Issues Arising from the Possible Uses of Genetic Knowledge," presents a panoramic view of the more pressing moral problems emerging from the uses of genetic information and raises very basic questions pertinent to the possibility of achieving moral agreement.

Probing the ethical issues of genetic counseling and prenatal diagnosis, Leon Kass's "Implications of Prenatal Diagnosis for the Human Right to Life" focuses on the moral problems centering on the abortion of the genetically defective fetus. Although sympathetic to the practice of abortion for reasons of severe genetic defect, the author argues that there are no reasons which can morally justify such a procedure without simultaneously justifying the killing of "defective" infants, children, and adults. Robert Morison, however, argues against this latter view in his "Implications of Prenatal Diagnosis for the Quality of, and Right to Human Life: Society as a Standard."

Turning from the more specific questions on abortion for reasons of severe genetic defect, Kurt Hirschhorn's paper, "Practical and Ethical Problems in Human Genetics," urges that the current methods of positive and negative eugenics can not significantly improve the gene pool of the population and allow for adequate evolutionary improvement of the race.

In "Reproductive Rights and Genetic Disease" Lawrence Ulrich treats the question of whether or not a person's moral right to reproduce is subject to encroachment or repeal on the part of society. He argues that reproductive rights are not absolute and that those who are a high risk for passing on clearly identifiable severely dileterious genes and debilitating genetic disease should not be allowed to exercise their reproductive prerogative. The general reasoning behind this latter view is rejected by Thomas Beauchamp in his "On Justification For Coercive Genetic Control."

The last two essays in the section deal extensively with the legal problems generated by current practice in genetic counseling and improvement. In "Legal Rights and Moral Rights" Alexander Capron argues in part that the genetic counselor has a clear legal duty to give competent advice so as to place parents into the position of informed decision-makers. If a counselor's negligence results in the birth of a defective child, parents and child alike have a valid claim for damages against him. The essay describes in detail not only current legal attitude and practice, but also the problems the law will need to solve in order to have an effective legal code dealing with the rights of individuals in the face of current techniques and practice employed for the genetic improvement of the species. Pursuing the legal dimension further, Herbert Lub's closing paper, "Privacy and Genetic Information," reexamines the concept of privacy and seeks to provoke thought about what test cases might be most suitable to clarify the issues of medical ethics in relation to human genetics. He also discusses what new laws may be desirable.

ETHICAL ISSUES ARISING FROM THE POSSIBLE USES OF GENETIC KNOWLEDGE

TRACY M. SONNEBORN

I realize that I am probably a fool to rush into the ethical domain where angels—the philosphers, ethicists, theologians, and lawyers—do not fear to tread. But these ethical problems concern us all, and not the least among us, the geneticists. I submit that after we listen carefully to what the professional theologians, ethicists, and philosophers have to say, as many of us have, then we, too, may speak up and tell about our own attempts to see our way through the difficult problems that beset us. This dialogue has now been going on with increasing frequency during the last eight years, and some of my fellow scientists have written and spoken very thoughtfully on the subject. They are not in complete agreement, but neither are the theologians, ethicists, and philosophers. I assume that the purpose of this conference is to encourage further communication between physicians and geneticists, on the one hand, and philosophers, theologians, ethicists, and lawyers on the other hand. Perhaps it is significant that a geneticist has been given the opening spot at the conference, but that lawyers and a historian will have the last words.

The present and potential uses of genetic knowledge and technology are, in a general way, widely known. They have been frequently presented to the public by the popular press and other mass communication media. So, I think it is not necessary for me to do more than recall them briefly. Most of the major ethical issues arise directly or indirectly from the genetic knowledge and correlated technologies that are concerned with human procreation. Genetic knowledge permits a degree of counseling to prospective parents in regard to the probabilities for the occurrence of certain traits among their future children. Perhaps the most spectacular technological advances along these lines have been those that

---

This paper originally appeared in *Ethical Issues in Human Genetics*, Bruce Hilton *et al.*, eds. (New York: Plenum Press, 1973). Reprinted with permission of the author and Plenum Publishing Corporation.

are useful in predicting characteristics of a developing baby a considerable time before it is due to be born. Certain abnormal genes and chromosomal conditions can be detected by this technology with virtually 100 percent accuracy. The number of conditions that can be predicted prenatally and the accuracy with which these predictions can be made is increasing every year. Many ethical issues arise in connection with the interrelations among the genetic counselor, the physician, the pregnant woman, and her spouse. These will doubtless be brought out in some of the later papers and discussions. But the major ethical issue is whether abortion is justified when the child is found to have defective or abnormal genes or chromosomes. In a beautiful and eminently humane paper known to many of us, Dr. Lejeune—who has been a pioneer and remains a leader in the field of human chromosomal abnormalities—has presented many reasons for doubting the justification of such induced abortions. Later I shall come back to this problem.

A second and very different set of ethical issues arises from the possibility of using for genetic purposes forms of procreation other than the normal one of sexual intercourse between husband and wife. Among these forms of procreation, only one is at present feasible; artificial insemination of the woman with sperm from a donor other than her husband. The other forms of procreation have been carried out with higher animals; although not yet possible in man, research toward this objective is in progress. One of them is fertilization outside of the body, using eggs from any female and sperm from any male; the fertilized egg, after proceeding to a very early developmental stage, is then implanted into the uterus of any properly prepared female. As Arno Motulsky vividly phrased it, this raises among other ethical problems the problem of "wombs for rent." The other form of procreation is to remove the nucleus from any female's egg and replace it by a nucleus obtained from a body cell of the same or any other individual. This, theoretically, should result in the development of an individual whose heredity is identical with that of the individual who provided the donor nucleus. As you are well aware, this is called cloning and can be used to produce as many genetically identical individuals as desired—not merely twins or triplets, but multiplets. Cloning has been subjected to searching ethical analysis by Kass, Ramsey, and others.

Finally, a third set of ethical issues is raised by genetic surgery, which lies still further in the future. Genetic surgery refers to anticipated possibilities of changing the genetic constitution of the reproductive cells in a person, or by changing it in the initial or very early stage of development of a new individual. Doubtless Dr. Sinsheimer will have much to say about this.

All of the possibilities mentioned, along with simple differential amounts of normal procreation by people with different genetic endowments, are components of a vision held in some quarters that foresees man consciously and purposefully guiding his own future evolution or, as some like to characterize it, trying to play God. Clearly the ethical issues that arise from these possibilities are among the most important that could be envisaged. They all deal with problems of life and death, with the character or quality of life, with the active interference of human beings in deciding who shall live and who shall die, and with what kinds of people shall live or die. These decisions could affect not only those now living and their children but our successors many generations hence.

These, then, are the major questions, problems, and ethical issues that have been and will continue to be discussed. Yet there is, I believe, a deeper and more encompassing question. The way we answer this question largely determines how we will answer all the others. It is a touchy and highly sensitive question, which is bound to annoy, anger, or infuriate many of us. Perhaps that is why it is so seldom put explicitly and clearly at the center of discussion where it really belongs, although answers to it are tacitly implied by our actions and the principles of action to which we adhere. It is, therefore, with fear and trembling that I put the question before you: Who or what decides what is right or good? By what authority? What do we really mean when we ask about anything—"Is it ethical?" In spite of not being a professional ethicist, I cannot avoid coming always to that central question. Do we assume that there are eternal verities—universal, self-evident, absolute truths about right and good human conduct? Do we assume that they have been transmitted to us from a supernatural divine authority via his mortal servants? Do we assume that knowledge of them is an inherent characteristic of man, knowledge that he can obtain by turning to his conscience? Or does ethics take a different form and have a different authority? If so, what? Or is ethics a chaotic no-man's land without authority of any kind?

The answers given by human beings to these central questions are by no means uniform. Some of us, though profoundly awed by the universe as we apprehend it, do not believe in a supernatural God and we reject divine authority for an ethical code. Nor can we accept the idea of an eternal, universal, absolute ethics imprinted in the conscience of man. We find in comparative religion, in anthropology, and in history the record of diverse ethical codes. We see them as man-made and variable from time to time, from culture to culture. We see them as codes of conduct authorized by common consent, or imposed to regulate particular social orders. Although, as Waddington and others have argued, man is an ethical animal and even an authority acceptor (as well as challenger), the particulars of his conceptions of right and good are varied and changeable. Even within one overall culture, different groups profess different ethical codes—for example, physicians, corporation executives, lawyers, congressmen, and presidents. It seems to me, therefore, that the authority for ethical decisions, for decisions as to what is right and good, comes from man himself, from his own choices, individually and in groups. The function then of conferences such as this is to debate what is right and what is good for man as part of the process of crystallizing individual and group choices which will become the authority for ethical decisions.

Viewing the general situation in this way, I should like now to apply this view to the ethical problems of life and death, of human procreation, that arises from present and potential uses of genetic knowledge. The first problem is abortion. Two opposed solutions to the problem are indicated by news reports. Superior Court Judge Jack G. Marks of Tucson, Arizona, is reported to have appointed a guardian of a nine-week old fetus on the ground that "the fetus has the rights of equal protection of this court." The suit was reported to have been filed by the Planned Parenthood Center of Tucson and by ten physicians in a challenge of the state's abortion laws. According to the papers, the suit claims that the mother of the fetus will probably not die if she gives birth to the child, but that she will be permanently injured. I have seen no mention of whether abnormality of the child is involved; presumably it is not. The issue that appears to be drawn is simply conformity to existing laws, based on the sanctity of life versus modification of the law to conform to changing public opinion about the range of applicability of the adversary principle.

The adversary principle has been recognized as valid even by the Judaeo-Christian tradition. Abortion is right if the choice is death of the fetus or death of the mother. Killing has also been justified—albeit regrettably—in self-defense against a life-threatening adversary, both at the level of the individual and, in the case of war, at the level of a nation. The ethical question at issue now is whether the adversary principle should be extended in the case of a fetus, to situations in which the fetus is deemed to be an adversary against not necessarily the life, but merely the well-being of the parents or society. In the past, man's ethical judgments have changed on the basis of discussion and experience as new situations have arisen—as, for example, in the case of contraception when new methods were developed and new social conditions came into being. At first, argument centered about the right to prevent the initiation of a new life by interposing mechanical blocks between sperm and egg. After long and bitter battles agains the law and previous ethical judgments, the issue was in effect decided by widespread practice. More recently the pill, working on a different principle, has been widely accepted. The intrauterine device, which may operate on what amounts to very early abortion, is also accepted by many. Extension to somewhat later abortion is now at issue, as well as the question of how late.

The great numbers and heterogeneity of mankind, as well as the existence of many adventurous, nonconformist or simply desperate people, make it highly unlikely that new possibilities will be tried by some people regardless of how they stand in the light of current civil, moral, or sacred law. Some of these trials will fail to win general acceptance; others will succeed. Acceptance may at first be limited to special cases and later become more general and comprehensive.

Submission to the test of public opinion and practice is sometimes a slow method of change. But a slow pace, permitting time for testing and weighing, affords some protection against precipitous unwise choices. Individuals and society thus have ways of sanctioning uses of new knowledge and technology and, in fact, they do so even if these uses initially conflict with current legal, moral, and religious codes. If the new ethics are eventually judged to operate contrary to the good of man, readjustments can be made. We went through that reversal, for example, in the adoption and then repeal of Prohibition. The touchstone of man's choices, of his ethical choices, is simply his judgment of whether it is right and good for man. Man is the measure of all things.

This I believe to be the way that all of the ethical issues mentioned earlier—abortion, methods of procreation, guiding human evolution—will be decided. I doubt whether man's present choices will be guided by long-range considerations of human evolution, at least not by enough people to have an appreciable effect. Actually, however, that doesn't greatly matter. What does matter are the choices made in each generation with regard only to the procreation of the next generation. People greatly desire not to have defective or abnormal children. Because this hits home hard, I believe man will eventually decide that it is right and good to use for this purpose means offered by genetic knowledge and procreative technology, including the techniques of genetic surgery if and when they become available. I believe that man will, in short, adopt as ethical Bentley Glass' dictum that every child has the right to be free of genetic defect and abnormality insofar as this can be achieved. If people come to desire to have above-average or outstanding children, however defined, with anything like the strength of their desire to avoid having defective and abnormal children, they will find it as easy, or as difficult, to resolve the ethical problem in the one case as in the other, for the problems have much in common. The conscious guiding of human evolution would then be in progress simply by active concern for one's children, without looking further ahead; the guidance could continually adjust to changing conditions and new knowledge.

My thesis, then, is that man develops his ethics by the method of public discussion, by individual decisions and actions, by public acceptance of what appears to be right and good for man, and by rejection of what appears to be wrong or bad. We agree that it is right and good to reduce misery and improve the quality of life for all those who live, by using environmental and social means. We now debate whether it is right and good to use genetic means. Our conceptions of what is ethical, right, and good change in the light of new knowledge and new conditions. What we lack is neither flexibility of mind nor adventurous spirits, but knowledge and experience. If the future can be judged by the present and the past, we shall get that knowledge and experience and eventually authorize the ethics that permits doing what is believed to be right and good for man.

If I have glossed over the pitfalls and difficulties in the process of arriving at particular ethical judgments about specific details of the uses of genetic knowledge and technology, it is not because I am unaware of them. But this, I am happy to say, is not my assignment. These tough problems will occupy us during the next four days—and beyond.* As we approach that task, we have no basis for being cocky. We are still full of ignorance in spite of the spectacular increase of knowledge. It would be both unwise and inhumane to proceed without the utmost humility and compassion.

*All papers in this section, taken from *Ethical Issues in Human Genetics*, were part of a symposium sponsored by The John E. Fogarty International Center For Advanced Study in The Health Sciences and The Institute of Society, Ethics and the Life Sciences, October 10–14, 1971.

IMPLICATIONS OF PRENATAL DIAGNOSIS FOR THE HUMAN RIGHT TO LIFE

LEON R. KASS

It is especially fitting on this occasion to begin by acknowledging how pleased I am to be a participant in this symposium. I suspect that I am not alone among the assembled in considering myself fortunate to be here. For I was conceived after antibiotics yet before amniocentesis, late enough to have benefited from medicine's ability to prevent and control fatal infectious diseases, yet early enough to have escaped from medicine's ability to prevent me from living to suffer from my genetic diseases. To be sure, my genetic vices are, as far as I know them, rather modest, taken individually—myopia, asthma and other allergies, bilateral forefoot adduction, bowleggedness, loquaciousness, and pessimism, plus some four to eight as yet undiagnosed recessive lethal genes in the heterozygous condition—but, taken together, and if diagnosable prenatally, I might never have made it.

Just as I am happy to be here, so am I unhappy with what I shall have to say. Little did I realize when I first conceived the topic, "Implications of Prenatal Diagnosis for the Human Right to Life," what a painful and difficult labor it would lead to. More than once while this paper was gestating, I considered obtaining permission to abort it, on the grounds that, by prenatal diagnosis, I knew it to be defective. My lawyer told me that I was legally in the clear, but my conscience reminded me that I had made a commitment to deliver myself to this paper, flawed or not. Next time, I shall practice better contraception.

Any discussion of the ethical issues of genetic counseling and prenatal diagnosis is unavoidably haunted by a ghost called the morality of abortion. This ghost I shall not vex. More precisely, I shall not vex the reader by telling ghost stories. However, I would be neither surprised nor disappointed if my discussion

---

This paper originally appeared in *Ethical Issues in Human Genetics,* Bruce Hilton *et al.,* eds. (New York: Plenum Press, 1973). Reprinted with permission of the author and Plenum Publishing Corporation.

of an admittedly related matter, the ethics of aborting the genetically defective, summons that hovering spirit to the reader's mind. For the morality of abortion is a matter not easily laid to rest, recent efforts to do so notwithstanding. A vote by the legislature of the State of New York can indeed legitimatize the disposal of fetuses, but not of the moral questions. But though the questions remain, there is likely to be little new that can be said about them, and certainly not by me.

Yet before leaving the general question of abortion, let me pause to drop some anchors for the discussion that follows. Despite great differences of opinion both as to what to think and how to reason about abortion, nearly everyone agrees that abortion is a moral issue.[1] What does this mean? Formally it means that a woman seeking or refusing an abortion can expect to be asked to justify her action. And we can expect that she should be able to give reasons for her choice other than "I like it" or "I don't like it." Substantively, it means that, in the absence of good reasons for intervention, there is some presumption in favor of allowing the pregnancy to continue once it has begun. A common way of expressing this presumption is to say that "the fetus has a right to continued life."[2] In this context, disagreement concerning the moral permissibility of abortion concerns what rights (or interests or needs), and whose, override (take precedence over, or outweigh) this fetal "right." Even most of the "opponents" of abortion agree that the mother's right to live takes precedence, and that abortion to save her life is permissible, perhaps obligatory. Some believe that a woman's right to determine the number and spacing of her children takes precedence, while yet others argue that the need to curb population growth is, at least at this time overriding.

Hopefully, this brief analysis of what it means to say that abortion is a moral issue is sufficient to establish two points. First, that the fetus is a living

[1] This strikes me as by far the most important inference to be drawn from the fact that men in different times and cultures have answered the abortion question differently. Seen in this light, the differing and changing answers themselves suggest that it is a question not easily put under, at least not for very long.

[2] Other ways include: one should not do violence to living or growing things; life is sacred; respect nature; fetal life has value; refrain from taking innocent life; protect and preserve life. As some have pointed out, the terms chosen are of different weight, and would require reasons of different weight to tip the balance in favor of abortion. My choice of the "rights" terminlogy is not meant to beg the questions of whether such rights really exist, or of where they come from. However, the notion of a "fetal right to life" presents only a little more difficulty in this regard than does the notion of a "human right to life," since the former does not depend on a claim that the human fetus is already "human." In my sense of terms "right" and "life," we might even say that a dog or a fetal dog has a "right to life," and that it would be cruel and immoral for a man to go around performing abortions even on dogs for no good reason.

thing with some moral claim on us not to do it violence, and therefore, second, that justification must be given for destroying it.

Turning now from the general questions of the ethics of abortion, I wish to focus on the special ethical issues raised by the abortion of "defective" fetuses (so-called "abortion for fetal indications"). I shall consider only the cleanest cases, those cases where well-characterized genetic diseases are diagnosed with a high degree of certainty by means of amniocentesis, in order to sidestep the added moral dilemmas posed when the diagnosis is suspected or possible, but unconfirmed. However, many of the questions I shall discuss could also be raised about cases where genetic analysis gives only a statistical prediction about the genotype of the fetus, and also about cases where the defect has an infectious or chemical rather than a genetic cause (e.g., rubella, thalidomide).

My first and possibly most difficult task is to show that there is anything left to discuss once we have agreed not to discuss the morality of abortion in general. There is a sense in which abortion for genetic defect is, after abortion to save the life of the mother, perhaps the most defensible kind of abortion. Certainly, it is a serious and not a frivolous reason for abortion, defended by its proponents in sober and rational speech—unlike justifications based upon the false notion that a fetus is a mere part of a woman's body, to be used and abused at her pleasure. Standing behind genetic abortion are serious and well-intentioned people, with reasonable ends in view: the prevention of genetic diseases, the elimination of suffering in families, the preservation of precious financial and medical resources, the protection of our genetic heritage. No profiteers, no sex-ploiters, no racists. No arguments about the connection of abortion with promiscuity and licentiousness, no perjured testimony about the mental health of the mother, no arguments about the seriousness of the population problem. In short, clear objective data, a worthy cause, decent men and women. If abortion, what better reason for it?

Yet if genetic abortion is but a happily wagging tail on the dog of abortion, it is simultaneously the nose of a camel protruding under a rather different tent. Precisely because the quality of the fetus is central to the decision to abort, the practice of genetic abortion has implications which go beyond those raised by abortion in general. What may be at stake here is the belief in the radical moral equality of all human beings, the belief that all human beings possess equally and independent of merit certain fundamental rights, one among which is, of course, the right to life.

To be sure, the belief that fundamental human rights belong equally to all human beings has been but an ideal, never realized, often ignored, sometimes shamelessly. Yet is has been perhaps the most powerful moral idea at work in the world for at least two centuries. It is this idea and ideal that animates most of the current political and social criticism around the globe. It is ironic that we should acquire the power to detect and eliminate the genetically unequal at a time when we have finally succeeded in removing much of the stigma and disgrace previously attached to victims of congenital illness, in providing them with improved care and support, and in preventing, by means of education, feelings of guilt on the part of their parents. One might even wonder whether the development of aminocentesis and prenatal diagnosis may represent a backlash against these same humanitarian and egalitarian tendencies in the practice of medicine, which by helping to sustain to the age of reproduction persons with genetic disease has itself contributed to the increasing incidence of genetic disease, and with it, to increased pressures for genetic screening, genetic counseling, and genetic abortion.

No doubt our humanitarian and egalitarian principles and practices have caused us some new difficulties, but if we mean to weaken or turn our backs on them, we should do so consciously and thoughtfully. If, as I believe, the idea and practice of genetic abortion points in that direction, we should make ourselves aware of it. And if, as I believe, the way in which genetic abortion is described, discussed, and justified is perhaps of even greater consequence than its practice for our notions of human rights and of their equal possession by all human beings, we should pay special attention to questions of language and, in particular, to the question of justification. Before turning full attention to these matters, two points should be clarified.

First, my question "What decision, and why?" is to be distinguished from the question "Who decides, and why?" There is a tendency to blur this distinction and to discuss only the latter, and with it, the underlying question of private freedom versus public good. I will say nothing about this, since I am more interested in exploring what constitutes "good," both public and private. Accordingly, I would emphasize that the moral question—What decision and why?—does not disappear simply because the decision is left in the hands of each pregnant woman. It is the moral question she faces. I would add that the moral health of the community and of each of its members is as likely to be affected by the aggregate of purely private and voluntary decisions on genetic abortion as by a uniform policy imposed by statute. We physicians and scientists especially should refuse to finesse the moral question of genetic abortion and its implications and to take refuge behind the issue, "Who decides?" For it is we who are responsible for choosing to develop the technology of prenatal diagnosis, for informing and promoting this technology among the public, and for the actual counseling of patients.

Second, I wish to distinguish my discussion of what ought to be done from a descriptive account of what in fact is being done, and especially from a consideration of what I myself might do, faced with the difficult decision. I cannot know with certainty what I would think, feel, do, or want done, faced with the knowledge that my wife was carrying a child branded with Down's syndrome or Tay-Sachs disease. But an understanding of the issues is not advanced by personal anecdote or confession. We all know that what we and others actually do is often done out of weakness, rather than conviction. It is all too human to make an exception in one's own case (consider, e.g., the extra car, the "extra" child, income tax, the draft, the flight from the cities). For what it is worth, I confess to feeling more than a little sympathy with parents who choose abortions for severe genetic defect. Nevertheless, as I shall indicate later, in seeking for reasons to justify this practice, I can find none that are in themselves fully satisfactory and none that do not simultaneously justify the killing of "defective" infants, children, and adults. I am mindful that my arguments will fall far from the middle of the stream, yet I hope that the oarsmen of the flagship will pause and row more slowly, while we all consider whither we are going.

Genetic Abortion and the Living Defective

The practice of abortion of the genetically defective will no doubt affect our view of and our behavior toward those abnormals who escape the net of detection and abortion. A child with Down's syndrome or with hemophilia or with muscular dystrophy born at a time when most of his (potential) fellow sufferers were destroyed prenatally is liable to be looked upon by the community as one unfit to be alive, as a second-class (or even lower) human type. He may be seen as a person who need not have been, and who would not have been, if only someone had gotten to him in time.

The parents of such children are also likely to treat them differently, especially if the mother would have wished but failed to get an amniocentesis because of ignorance, poverty, or distance from the testing station, or if the prenatal diagnosis was in error. In such cases, parents are especially likely to resent the child. They may be disinclined to give it the kind of care they might have before the advent of amniocentesis and genetic abortion, rationalizing that a second-class specimen is not entitled to first-class treatment. If pressed to do so, say by physicians, the parents might refuse, and the courts may become involved. This has already begun to happen.

In Maryland, parents of a child with Down's syndrome refused permission to have the child operated on for an intestinal obstruction present at birth. The physicians and the hospital sought an injunction to require the parents to allow surgery. The judge ruled in favor of the parents, despite what I understand to be the weight of precedent to the contrary, on the grounds that the child was Mongoloid; that is, had the child been "normal," the decision would have gone the other way. Although the decision was not appealed to and hence not affirmed by a higher court, we can see through the prism of this case the possibility that the new powers of human genetics will strip the blindfold from the lady of justice and will make official the dangerous doctrine that some men are more equal than others.

The abnormal child may also feel resentful. A child with Down's syndrome or Tay-Sachs disease will probably never know or care, but what about a child with hemophilia or with Turner's syndrome? In the past decade, with medical knowledge and power over the prenatal child increasing and with parental authority over the postnatal child decreasing, we have seen the appearance of a new type of legal action, suits for wrongful life. Children have brought suit against their parents (and others) seeking to recover damages for physical and social handicaps inextricably tied to their birth (e.g., congenital deformities, congenital syphilis, illegitimacy). In some of the American cases, the courts have recognized the justice of the child's claim (that he was injured due to parental negligence), although they have so far refused to award damages, due to policy considerations. In other countries, e.g., in Germany, judgments with compensation have gone for the plaintiffs. With the spread of amniocentesis and genetic abortion, we can only expect such cases to increase. And here it will be the soft-heated rather than the hard-hearted judges who will establish the doctrine of second-class human beings, out of compassion for the mutants who escaped the traps set out for them.

It may be argued that I am dealing with a problem which, even if it is real, will affect very few people. It may be suggested that very few will escape the traps once we have set them properly and widely, once people are informed about aminocentesis, once the power to detect prenatally grows to its full capacity, and once our "superstitious" opposition to abortion dies out or is extirpated. But in order even to come close to this vision of success, amniocentesis will have to become part of every pregnancy—either by making it mandatory, like the test for syphilis, or by making it "routine medical practice," like the Pap smear. Leaving aside the other problems with universal amniocentesis, we could expect that the problem for the few who escape is likely to be even worse precisely because they will be few.

The point, however, should be generalized. How will we come to view and act toward the many "abnormals" that will remain among us—the retarded, the crippled, the senile, the deformed, and the true mutants—once we embark on a program to root out genetic abnormality? For it must be remembered that we shall always have abnormals—some who escape detection or whose disease is undetectable *in utero,* others as a result of new mutations, birth injuries, accidents, maltreatment, or disease—who will require our care and protection. The existence of "defectives" cannot be fully prevented, not even by totalitarian breeding and weeding programs. Is it not likely that our principle with respect to these people will change from "We try harder" to "Why accept second best?" The idea of "the unwanted because abnormal child" may become a self-fulfilling prophecy, whose consequences may be worse than those of the abnormality itself.

Genetic and Other Defectives

The mention of other abnormals points to a second danger of the practice of genetic abortion. Genetic abortion may come to be seen not so much as the prevention of genetic disease, but as the prevention of birth of defective or abnormal children—and, in a way, understandably so. For in the case of what other diseases does preventive medicine consist in the elimination of the patient-at-risk? Moreover, the very language used to discuss genetic disease leads us to the easy but wrong conclusion that the afflicted fetus or person *is* rather than *has* a disease. True, one is partly defined by his genotype, but only partly. A person is more than his disease. And yet we slide easily from the language of possession to the language of identity, from "He has hemophilia" to "He is a hemophiliac," from "She has diabetes" through "She is diabetic" to "She is a diabetic," from "The fetus had Down's syndrome" to "The fetus is a Down's." This way of speaking supports the belief that it is defective persons (or potential persons) that are being eliminated, rather than diseases.

If this is so, then it becomes simply accidental that the defect has a genetic cause. Surely, it is only because of the high regard for medicine and science, and for the accuracy of genetic diagnosis, that genotypic defectives are likely to be the first to go. But once the principle, "Defectives should not be born," is established, grounds other than cytological and biochemical may very well be sought. Even ignoring racialists and others equally misguided—of course, they cannot be ignored—we should know that there are social scientists, for example, who believe that one can predict with a high degree of accuracy how a child will turn out from a careful, systematic study of the socio-economic and psychodynamic environment into which he is born and in which he grows up. They might press for the prevention of socio-psychological disease, even of "criminality," by means of prenatal environmental diagnosis and abortion. I have heard rumor that a crude, unscientific form of eliminating potential "phenotypic defectives" is already being practiced in some cities, in that submission to abortion is allegedly being made a condition for the receipt of welfare payments. "Defectives should not be born" is a principle without limits. We can ill afford to have it established.

Up to this point, I have been discussing the possible implications of the practices of genetic abortion for our belief in and adherence to the idea that, at least in fundamental human matters such as life and liberty, all men are to be considered as equals, that for these matters we should ignore as irrelevant the real qualitative differences amongst men, however important these differences may be for other purposes. Those who are concerned about abortion fear that the permissible time of eliminating the unwanted will be moved forward along the time continuum, against newborns, infants, and children. Similarly, I suggest that we should be concerned lest the attack on gross genetic inequality in fetuses be advanced along the continuum of quality and into the later stages of life.

I am not engaged in predicting the future; I am not saying that amniocentesis and genetic abortion will lead down the road to Nazi Germany. Rather, I am suggesting that the principles underlying genetic abortion simultaneously justify many further steps down that road. The point was very well made by Abraham Lincoln (1854)[3]:

> If A can prove, however conclusively, that he may, of right, enslave B–Why may not B snatch the same argument and prove equally, that he may enslave A?
>
> You say A is white, and B is black. It is color, then; the lighter having the right to enslave the darker? Take care. By this rule, you are to be slave to the first man you meet with a fairer skin than your own.
>
> You do not mean color exactly? You mean the whites are intellectually the superiors of the blacks, and, therefore have the right to enslave them? Take care again. By this rule, you are to be slave to the first man you meet with an intellect superior to your own.
>
> But, say you, it is a question of interest; and, if you can make it your interest, you have the right to enslave another. Very well. And if he can make it his interest, he has the right to enslave you.

Perhaps I have exaggerated the dangers; perhaps we will not abandon our inexplicable preference for generous humanitarianism over consistency. But we should indeed be cautious and move slowly as we give serious consideration to the question "What price the perfect baby?"[4]

Standards for Justifying Genetic Abortion

The rest of this paper deals with the problem of justification. What would constitute an adequate justification of the decision to abort a genetically defective fetus? Let me suggest the following formal characteristics, each of which still begs many questions. (1) The reasons given should be logically consistent, and should lead to relatively unambiguous guidelines–note that I do not say "rules"–for action in most cases. (2) The justification should make

[3] Lincoln, A. (1854). In *The Collected Works of Abraham Lincoln,* R. P. Basler, editor. New Brunswick, New Jersey, Rutgers University Press, Vol. II, p. 222.

[4] For a discussion of the possible biological rather than moral price of attempts to prevent the birth of defective children see Motulsky, A. G., G. R. Fraser, and J. Felsenstein (1971). In Symposium on Intrauterine Diagnosis, D. Bergsma, editor. *Birth Defects: Original Article Series,* Vol. 7, No. 5; Neel, J. (1972). In Early Diagnosis of Human Genetic Defects: *Scientific and Ethical Considerations,* M. Harris, editor. Washington, D.C. U.S. Government Printing Office, pp. 366–380.

evident to a reasonable person that the interest or need or right being served by abortion is sufficient to override the otherwise presumptive claim on us to protect and preserve the life of the fetus. (3) Hopefully, the justification would be such as to help provide intellectual support for drawing distinctions between acceptable and unacceptable kinds of genetic abortion and between genetic abortion itself and the further practices we would all find abhorrent. (4) The justification ought to be capable of generalization to all persons in identical circumstances. (5) The justification should not lead to different actions from month to month or from year to year. (6) The justification should be grounded on standards that can, both in principle and in fact, sustain and support our actions in the case of genetic abortion and our notions of human rights in general.

Though I would ask the reader to consider all these criteria, I shall focus primarily on the last. According to what standards can and should we judge a fetus with genetic abnormalities unfit to live, i.e., abortable? It seems to me that there are at least three dominant standards to which we are likely to repair.

The first is societal good. The needs and interests of society are often invoked to justify the practices of prenatal diagnosis and abortion of the genetically abnormal. The argument, full blown, runs something like this. Society has an interest in the genetic fitness of its members. It is foolish for society to squander its precious resources ministering to and caring for the unfit, especially for those who will never become "productive," or who will never in any way "benefit" society. Therefore, the interests of society are best served by the elimination of the genetically defective prior to their birth.

The societal standard is all too often reduced to its lowest common denominator: money. Thus one physician, claiming that he has "made a cost-benefit analysis of Tay-Sachs disease," notes that "the total cost of carrier detection, prenatal diagnosis and termination of at-risk pregnancies for all Jewish individuals in the United States under 30 who will marry is $5,730,281. If the program is set up to screen only one married partner, the cost is $3,122,695. The hospital costs for the 990 cases of Tay-Sachs disease these individuals would produce over a thirty-year period in the United States is $34,650,000.[5] Another physician, apparently less interested or able to make such a precise audit has written: "Cost-benefit analyses have been made for the total prospective detection and monitoring of Tay-Sachs disease, cystic fibrosis (when prenatal detection becomes available for cystric fibrosis) and other disorders, and in most cases, the expenditures for hospitalization and medical care far exceed the cost of prenatal detection in properly selected risk populations, followed by selective abortion." Yet a third physician has calculated that the costs to the state of caring for children with Down's syndrome is more than three times that of detecting and aborting them. (These authors all acknowledge the additional

[5]I assume this calculation ignores the possibilities of inflation, devaluation, and revolution.

non-societal "costs" of personal suffering, but insofar as they consider society, the costs are purely economic.)

There are many questions that can be raised about this approach. First, there are questions about the accuracy of the calculations. Not all the costs have been reckoned. The aborted defective child will be "replaced" by a "normal" child. In keeping the ledger, the "costs" to society of his care and maintenance cannot be ignored–costs of educating him, or removing his wastes and pollutions, not to mention the "costs" in non-replaceable natural resources that he consumes. Who is a greater drain on society's precious resources, the average inmate of a home for the retarded or the average graduate of Harvard College? I am not sure we know or can even find out. Then there are the costs of training the physician, and genetic counselors, equipping their laboratories, supporting their research, and sending them and us to conferences to worry about what they are doing. An accurate economic analysis seems to me to be impossible, even in principle. And even if it were possible, one could fall back on the words of that ordinary language philosopher, Andy Capp, who, when his wife said that she was getting really worried about the cost of living, replied: "Sweet 'eart, name me one person who wants t'stop livin' on account of the cost."

A second defect of the economic analysis is that there are matters of social importance that are not reducible to financial costs, and others that may not be quantifiable at all. How does one quantitate the costs of real and potential social conflict, either between children and parents, or between the community and the "deviants" who refuse amniocentesis and continue to bear abnormal children? Can one measure the effect of racial tensions of attempting to screen for and prevent the birth of children homozygous (or heterozygous) for sickle cell anemia? What numbers does one attach to any decreased willingness or ability to take care of the less fortunate, or to cope with difficult problems? And what about the "costs" of rising expectations? Will we become increasingly dissatisfied with anything short of the "optimum baby?" How does one quantify anxiety? humiliation? guilt? Finally, might not the medical profession pay an unmeasurable price if genetic abortion and other revoluntionary activities bring about changes in medical ethics and medical practice that lead to the further erosion of trust in the physician?

An appeal to social worthiness or usefulness is a less vulgar form of the standard of societal good. It is true that great social contributions are unlikely to be forthcoming from persons who suffer from most serious genetic diseases, especially since many of them die in childhood. Yet consider the following remarks of Pearl Buck[6] on the subject of being a mother of a child retarded from phenylketonuria:

[6]Buck, P. S. (1968). Foreward to *The Terrible Choice: The Abortion Dilemma*, New York, Bantam Books, pp. ix–xi.

My child's life has not been meaningless. She has indeed brought comfort and practical help to many people who are parents of retarded children or are themselves handicapped. True, she has done it through me, yet without her I would not have had the means of learning how to accept the inevitable sorrow, and how to make that acceptance useful to others. Would I be so heartless as to say that it has been worthwhile for my child to be born retarded? Certainly not, but I am saying that even though gravely retarded it has been worthwhile for her to have lived.

It can be summed up, perhaps, by saying that in this world, where cruelty prevails so many aspects of our life, I would not add the weight of choice to kill rather than to let live. A retarded child, a handicapped person, brings its own gift to life, even to the life of normal human beings. That gift is comprehended in the lessons of patience, understanding, and mercy, lessons which we all need to receive and to practice with one another, whatever we are.

The standard of potential social worthiness is little better in deciding about abortion in particular cases than is the standard of economic cost. To drive the point home, each of us might consider retrospectively whether he would have been willing to stand trial for his life while a fetus, pleading only his worth to society as he now can evaluate it. How many of us are not socially "defective" and with none of the excuses possible for a child with phenylketonuria? If there is to be human life at all, potential social worthiness cannot be its entitlement.

Finally, we should take note of the ambiguities in the very notion of societal good. Some use the term "society" to mean their own particular political community, others to mean the whole human race, and still others speak as if they mean both simultaneously, following that all-too-human belief that what is good for me and mine is good for mankind. Who knows what is genetically best for mankind, even with respect to Down's syndrome? I would submit that the genetic heritage of the human species is largely in the care of persons who do not live along the aminocentesis frontier. If we in the industrialized West wish to be really serious about the genetic future of the species, we would concentrate our attack on mutagenesis, and especially on our large contribution to the pool of environmental mutagens.

But even the more narrow use of society is ambiguous. Do we mean our "society" as it is today? Or do we mean our "society" as it ought to be? If the former, our standards will be ephemeral, for ours is a faddish "society." (By far the most worrisome feature of the changing attitudes on abortion is the suddenness with which they change.) Any such socially determined standards are likely to provide too precarious a foundation for decisions about genetic abortion, let alone for our notions of human rights. It we mean the latter, then we have transcended the societal standard, since the "good society" is not to be found in "society" itself, nor is it likely to be discovered by taking a vote. In sum, societal good as a standard for justifying genetic abortion seems to be unsatisfactory. It is hard to define in general, difficult to apply clearly to particular cases, susceptible to overreaching and abuse (hence, very dangerous), and not sufficient unto itself if considerations of the good community are held to be automatically implied.

A second major alternative is the standard of parental or familial good. Here the argument of justification might run as follows. Parents have a right to determine, according to their own wishes and based upon their own notions of what is good for them, the qualitative as well as the quantitative character of their families. If they believe that the birth of a seriously deformed child will be the cause of great sorrow and suffering to themselves and to their other children and a drain on their time and resources, then they may ethically decide to prevent the birth of such a child, even by abortion.

This argument I would expect to be more attractive to most people than the argument appealing to the good of society. For one thing, we are more likely to trust a person's conception of what is good for him than his notion of what is good for society. Also, the number of persons involved is small, making it seem less impossible to weigh all the relevant factors in determining the good of the family. Most powerfully, one can see and appreciate the possible harm done to healthy children if the parents are obliged to devote most of their energies to caring for the afflicted child.

Yet there are ambiguities and difficulties perhaps as great as with the standard of societal good. In the first place, it is not entirely clear what would be good for the other children. In a strong family, the experience with a suffering and dying child might help the healthy siblings learn to face and cope with adversity. Some have even speculated that the lack of experience with death and serious illness in our affluent young people is an important element in their difficulty in trying to find a way of life and in responding patiently yet steadily to the serious problems of our society[7]. I suspect that one cannot generalize. In some children and in some families, experience with suffering may be strengthening, and in others, disabling. My point here is that the matter is uncertain, and that parents deciding on this basis are as likely as not to be mistaken.

The family or parental standard, like the societal standard, is unavoidably elastic because "suffering" does not come in discontinuous units, and because parental wishes and desires know no limits. Both are utterly subjective, relative, and notoriously subject to change. Some parents claim that they could not tolerate having to raise a child of the undesired sex; I know of one case where the woman in the delivery room, on being informed that her child was a son, told the physician that she did not even wish to see it and that he should get rid of it. We may judge her attitude to be pathological, but even pathological suffering is suffering. Would such suffering justify aborting her normal male fetus?

[7]Cassell, E. (1969). *Death and the Physician, Commentary,* (June) pp. 73–79.

Or take the converse case of two parents, who for their own very peculiar reasons, wish to have an abnormal child, say a child who will suffer from the same disease as grandfather or a child whose arrested development would preclude the threat of adolescent rebellion and separation. Are these acceptable grounds for the abortion of "normals"?

Granted, such cases will be rare. But they serve to show the dangers inherent in talking about the parental right to determine, according to their wishes, the quality of their children. Indeed, the whole idea of parental rights with respect to children strikes me as problematic. It suggests that children are like property, that they exist for the parents. One need only look around to see some of the results of this notion of parenthood. The language of duties to children would be more in keeping with the heavy responsibility we bear in affirming the continuity of life with life and in trying to transmit what wisdom we have acquired to the next generation. Our children are not our children. Hopefully, reflection on these matters could lead to a greater appreciation of why it is people do and should have children. No better consequence can be hoped for from the advent of amniocentesis and other technologies for controlling human reproduction.

If one speaks of familial good in terms of parental duty, one could argue that parents have an obligation to do what they can to insure that their children are born healthy and sound. But this formulation transcends the limitation of parental wishes and desires. As in the case of the good society, the idea of "healthy and sound" requires an objective standard, a standard in reality. Hard as it may be to uncover it, this is what we are seeking. Nature as a standard is the third alternative.

The justification according to the natural standard might run like this. As a result of our knowledge of genetic diseases, we know that persons afflicted with certain diseases will never be capable of living the full life of a human being. Just as a non-necked giraffe could never live a giraffe's life, or a needle-less porcupine would not attain true "porcupine-hood," so a child or fetus with Tay-Sachs disease or Down's syndrome, for example, will never be truly human. They will never be able to care for themselves, nor have they even the potential for developing the distinctively human capacities for thought or self-consciousness. Nature herself has aborted many similar cases, and has provided for the early death of many who happen to get born. There is no reason to keep them alive; instead, we should prevent their birth by contraception or sterilization if possible, and abortion if necessary.

The advantages of this approach are clear. The standards are objective and in the fetus itself, thus avoiding the relativity and ambiguity in societal and parental good. The standard can be easily generalized to cover all such cases and will be resistant to the shifting sands of public opinion.

This standard, I would suggest, is the one which most physicians and genetic counselors appeal to in their heart of hearts, no matter what they say or do about letting the parents choose. Why else would they have developed genetic counseling and amniocentesis? Indeed, the notions of disease, of abnormal, of defective, make no sense at all in the absence of a natural norm of health. This norm is the foundation of the art of the physician and of the inquiry of the health scientist. Yet, as Motulsky and others have pointed out, the standard is elusive. Ironically, we are gaining increasing power to manipulate and control our own nature at a time in which we are increasingly confused about what is normal, healthy, and fit.

Although possibly acceptable in principle, the natural standard runs into problems in application when attempts are made to fix the boundary between potentially human and potentially not human. Professor Lejeune[8] has clearly demonstrated the difficulty, if not the impossibility, of setting clear molecular, cytological, or developmental signposts for this boundary. Attempts to induce signposts by considering the phenotypes of the worst cases is equally difficult. Which features would we take to be the most relevant in, say, Tay-Sachs disease, Lesch-Nyhan syndrome, Cri du chat, Down's syndrome? Certainly, severe mental retardation. But how "severe" is "severe"? As Abraham Lincoln and I argued earlier, mental retardation admits of degree. It too is relative. Moreover it is not clear that certain other defects and deformities might not equally foreclose the possibility of a truly or fully human life. What about blindness or deafness? Quadriplegia Asphasia? Several of these in combination? Not only does each kind of defect admit of a continuous scale of severity, but it also merges with other defects on a continuous scale of defectiveness. Where on this scale is the line to be drawn: after mental retardation? blindness? muscular dystrophy? cystic fibrosis? hemophilia diabetes? galactosemia? Turner's syndrome? XYY? club foot? Moreover, the identical two continuous scales—kind and severity—are found also among the living. In fact, it is the natural standard which may be the most dangerous one in that it leads most directly to the idea that there are second-class human beings and subhuman beings.

But the story is not complete. The very idea of nature is ambiguous. According to one view, the one I have been using, nature points to or implies a peak, a perfection. According to this view, human rights depend upon attaining the status of humanness. The fetus is only potential; it has no rights, according to this view. But all kinds of people fall short of the norm: children, idiots, some adults. This understanding of nature has been used to justify not only abortion and infanticide, but also slavery.

There is another notion of nature, less splendid, more humane and, though less able to sustain a notion of health, more acceptable to the findings of modern science. Animal nature is characterized by impulses of self-preservation and by

[8] Lejeune, J. (1970). *American Journal of Human Genetics, 22,* 121.

the capacity to feel pleasure and to suffer pain. Man and other animals are alike on this understanding of nature. And the right to life is ascribed to all such self-preserving and suffering creatures. Yet on this understanding of nature, the fetus—even a defective fetus—is not potential, but actual. The right to life belongs to him. But for this reason, this understanding of nature does not provide and may even deny what it is we are seeking, namely, a justification for genetic abortion, adequate unto itself, which does not simultaneously justify infanticide, homicide, and enslavement of the genetically abnormal.

There is a third understanding of nature, akin to the second, nature as sacrosanct, nature as created by a Creator. Indeed, to speak about this reminds us that there is a fourth possible standard of judgments about genetic abortion: the religious standard. I shall leave the discussion of this standard to those who are able to speak of it in better faith.

Now that I am at the end, the reader can better share my sense of frustration. I have failed to provide myself with a satisfactory intellectual and moral justification for the practice of genetic abortion. Perhaps others more able than I can supply one. Perhaps the pragmatists can persuade me that we should abandon the search for principled justification, that if we just trust people's situational decisions or their gut reactions, everything will turn out fine. Maybe they are right. But we should not forget the sage observation of Bertrand Russell: "Pragmatism is like a warm bath that heats up so imperceptibly that you don't know when to scream." I would add that before we submerge ourselves irrevocably in amniotic fluid, we take note of the connection to our own baths, into which we have started the hot water running.

IMPLICATIONS OF PRENATAL DIAGNOSIS FOR THE QUALITY OF, AND RIGHT TO, HUMAN LIFE

SOCIETY AS A STANDARD

ROBERT S. MORISON

My framework will be one, as Dr. Kass has already predicted, of a rather crass pragmatist. I have no expertise such as is represented by the other contributors, but I once had a fairly close association with public health. And since I was given the topic of talking about amniocentesis from the standpoint of the social standard, I have drawn a little bit on my past public health experience.

Actually, I want to ally myself with the ethical attitude which relies principally on a careful weighing of the probable practical consequences of a given choice of alternatives.

In doing this, I withdraw in horror from the tendency exhibited, I am sorry to say, by Dr. Kass and others to reduce this utilitarian procedure to a cheap matter of dollars, cents, and simple linear equations. It clearly is a much more complicated matter than that and involves judgment of all kinds of values, as Dr. Kass hinted as he went through this. But he would like you to remember that it is really dollars that we use, and I don't believe that is really the case.

I am also relatively unimpressed with the camel's head argument. I don't think that looking at an ethical viewpoint from the standpoint of "You do this and inevitably this terrible thing is going to happen later on" is an acceptable way of going about it.

I have noticed that principles, whatever their logical base, are valid only over a certain limited range; when pushed toward their limit, they tend to conflict with one another or become absurd. Even Oliver Wendell Holmes could not

This paper originally appeared in *Ethical Issues in Human Genetics,* Bruce Hilton *et al.*, eds. (New York: Plenum Press, 1973). Reprinted with permission of the author and Plenum Publishing Corporation.

extend the principle of free speech to those who falsely cried "fire" in a crowded theatre. Even the most concerned believers in the sanctity of life do not always equate the life of a four-day-old, yet-to-be-implanted egg with the life of a Mahatma Ghandi.

Again, I follow the view that modern ethical decision-making should be guided by moral policies rather than by principles. Policies are—and indeed usually should be—developed in relation to principles, but they need not be completely bound by them. Policies do not determine decisions by themselves. There must always be a decision-maker in the background, or a group of decision-makers.

Fortunately, the decision-makers in the abortion case are already in the process of being identified for us, and we need not spend time on this problem. A steadily increasing proportion of the developed world is deciding that the principal decision-maker should be the woman carrying the fetus—assisted, perhaps, by her physician. This primacy of the prospective mother cannot be based only or even primarily on the feminist assertion of the woman's right to do what she wants with her own body. As many people have pointed out here, the biological evidence is conclusively that a fetus is something quite different from a right hand or even an eye, which even the Bible urges as to extirpate or pluck out if they are offensive. I realize the Bible is equivocal on this point, and one can quote other passages against self-mutilation. The woman in the case, however much she would like to, cannot help participating as a trustee or guardian of the developing human being. The fact that the fetus is so primitive and so helpless, far from justifying a ruthlessly destructive attitude, should be added reason for compassionate concern.

It seems to me, therefore, that society, in its capacity as protector of the weak and moderator of differences between individuals, should not become indifferent to the welfare of the fetus or to the effect that the continuance or discontinuance of its life will have on the fabric of society as a whole.

The recent worldwide move towards liberalizing the abortion laws should not be construed as reflecting the belief that such laws have no constitutional standing. The change in attitude is so sweeping and so recent, especially in this country, that it is difficult to assess all the reasons that have influenced so many different people to change their minds—including me. But it appears that the majority finally concluded that the old rules were simply not working as acceptable policies.

One way of describing what we have done with our recent legislation is to say we have converted a matter of principle into a series of individual problems in situational ethics. This by no means implies, however, that society should relinquish all interest in the question. In place of the previous formal and legal restraints, it must find informal ways of impressing the decision-makers with the gravity of the situation in which they find themselves. And—here again, I think I agree with Kass—pointing out the numerous issues which must be weighed before arriving at a decision.

In reviewing the legislative history of the new arrangements, it appears that principal attention has centered on allowing the mother to base her decision on the effect of another child on her own health, on the welfare of the other children, and on the standard of living of the entire family unit. The particular qualities of the particular fetus to be destroyed were not at issue, in part because they seemed destined to remain unknown.

It may be worthwhile to pause and note that the presumptive unknownness of the individual characteristics of the fetus may have contributed to the lack of concern that some people have felt for the right of the fetus as an individual. As long as it remained in the darkness of the womb, it could be safely dismissed as in limbo, a subhuman entity, a thing rather than a person.

What amniocentesis may have done for us, among other things, is to make it clear that the fetus is an individual with definite, identifiable characteristics. In considering whether or not to abort a fetus because of certain such characteristics we don't happen to like, we recognize her or his individuality in a way we could avoid when we talked simply about the addition of any new member, or an essentially unknown "X," to the family circle.

However much considerations of personal identity will heighten our awareness of what we are doing and increase our sense of what I have called the gravity of the situation, I believe these considerations will not keep us from weighing the finding of amniocentesis in the balance when it comes to deciding on particular abortions.

This conference is based on the assumption that we will indeed use such findings, and I understand my task to be to discuss if and how the interests of society as a whole are to be taken into account.

I will, in fact, argue that society does have substantial interest in both the quantity and quality of the children to be born into its midst, and that it is already expressing an interest in the results of amniocentesis and in the way these findings are used in arriving at individual abortion decisions.

I should explain that I am using the word *society* in a flexible and—you may think—ambiguous way. Part of the time I am thinking of society as what the mathematicians refer to as a "set" or collection of individuals with needs, interests, and rights. More often I include with this notion the formal and informal organizations and institutions through which people express, satisfy, and protect their collective interests and rights.

Any suggestion that society may have some sort of stake in the proceedings may be greeted with the prediction that such ideas lead directly and immediately to the worst abuses of Nazi Germany. Let me say at once, then, that in trying to analyze the interest of society in unborn babies, I am not for a moment suggesting that legislation be passed forcing people to have abortions or to be surgically sterilized. If it turns out that society does have defensible interests in the number and condition of the infants to be born, it can express these interests in ways that fall far short of totalitarian compulsion.

It can, for example, facilitate or ban abortion. It can develop or not develop genetic counseling and contraceptive services. It may offer bounties or impose taxes on parents for having or not having children. Purists may object to such procedures as invasions of personal liberty, as perhaps they are, but they are the kind of invasions which all societies find necessary for survival.

No modern society could long endure without making it as easy as possible for parents to send children to school. But expressions of interest in individual education do not stop there. Most societies require that children be sent to school for some years under penality of the law. In most countries the courts have the right to protect children from physical abuse, to order ordinary medical care in opposition to the parent's wishes, and even to remove them entirely from a family environment that appears to be inadequate.

Such protection and fostering of infants and children by the state can be explained wholly in terms of an obligation to protect the child's right to life, liberty, and the pursuit of happiness. But in point of fact, this is not the only consideration. Society is also protecting its own safety and welfare by ensuring that there will be maximal numbers of people capable of protecting the goods and services and composing the songs necessary for a good society. Note that "composing songs." It is not purely a matter of dollars and cents. We want people who are able to contribute at every level. At the same time, society is minimizing the number likely to become public charges or public enemies.

Similarly, the provision of maternal and child clinics and the requiring of vaccination against disease can be explained in part as an expression of society's interest in controlling the spread of disease to third parties, and in part by the desire to ensure that each individual citizen be as healthy as possible. Part of this concern is altruistic, but public health officials can also argue that such preventive measures help society avoid the cost of caring for the incapacitated and handicapped later on.

These social concerns about individual health are not limited to persons already born. They extend back to the period of pregnancy and fetal life. Perhaps one of the earliest legal expressions of society's concern for the unborn infant was the dramatic requirement by the ancient Romans that Cesarean section be performed on all woman who died while carrying a viable fetus. Less dramatic perhaps, but far more important quantitatively, are our present procedures for encouraging prospective mothers to eat the right foods, have their blood tested for syphilis and Rh antibodies, and even to give up smoking.

This concern for the unborn extends back into time before the new individual is even conceived. Mothers are urged to check their health and to undergo such procedures as vaccination against German measles before becoming pregnant. Vaccination of young women against German measles is now so fully accepted that we may overlook how sophisticated a procedure it really is. It is not proposed as a protection of the woman, herself, against what is usually a scarcely noticable illness; its purpose is purely and simply that of protecting an unborn, unknown individual against a small but identifiable probability that it may be born with a congenital defect.

Perhaps the foregoing examples are sufficient to establish the following points:

1. Society has an interest in the welfare of children and fetuses.
2. This interest is recognized by the general public as legitimate, and various expressions of it have the sanction of tradition and current practice.
3. In several important cases, the interest is thought to justify not only the use of educational persuasion, but even the force of law to control or shape the behavior of mothers and prospective mothers.

It is highly probable that different people have different reasons for supporting these policies, and of course there is a minority which has reasons for opposing them. Some will talk of the right of the infant to be born healthy as a part of a more general right to health, which some enthusiasts regard as having been established by the World Health Organization shortly after the close of World War II. Other more utilitarian types will simply point out that the results for both mother and child are usually "better"—and that the cost to society is less.

Still others maintain that the Christian commandment to love everyone requires society to analyze all the angles in each particular case and that, by and large, love will be maximized if society takes an intelligent interest in the welfare of infants and fetuses.

The important thing to note is not the variety of reasons, but the general agreement on the conclusions. In a pluralistic society one must never lose sight of the fact that it is much more important that people agree on a policy rather than that they agree on the reasons for agreeing.

We must now move from this all too brief discussion of how society is currently expressing its interest in the yet-unborn, to how we can build on this experience to formulate policies to guide mothers and their counselors in adding the facts revealed by amniocentesis to all the other matters in the minds and hearts as they decide whether to continue or interrupt a pregnancy.

In the first place, I will assume that a prospective mother has a right to consider the specific characteristics of a specific fetus as she weighs the probable results of carrying a given pregnancy to term. If so, it follows that a society like ours has an obligation to provide, either through public or private means, facilities for amniocentesis in the same way as it customarily provides diagnostic facilities in other areas of medicine and public health.

So far, everything seems to be in accord with normal public health practice. But as we look a little further we encounter perplexities. The major difficulty is that the facts revealed by amniocentesis do not tell us what to do as clearly as many other laboratory tests do.

When a laboratory test reveals the presence of a streptococcus or a treponema, there is not much doubt in anyone's mind that the thing to do is to get rid of it.

When amniocentesis reveals a hereditary or a congenital defect, the situation is rather different. In the first place, the object to be got rid of is not a bacterium but an incipient human being. In the second place, it is not always immediately clear what the results will be if the defective fetus is allowed to go to term. Who is to say whether a child with Down's syndrome is more or less happy than the child with an IQ of 150 who spends the first 16 years of his life competing with his peers for admission to an ivy league college? Who is to assess the complex effects on parents of coping with a handicapped child?

Most troublesome is the recognition that the interest of the individuals directly concerned may be more often at variance with those of society than is the case in most public health procedures. For example, most public health officials have little difficulty in urging or even requiring people to be vaccinated, since the risk is very small and the benefits appear to accrue more or less equally to society and to the individual concerned.

The interests of parents and the interests of society in the birth of children, however, may often be quite at odds with one another. In some rapidly growing societies, for example, the birth of additional children of any description may be looked upon with dismay by society as a whole, however much the new citizens may be welcomed by their parents. In societies where the overall rate of growth is under better control, society may properly still have an interest in maximizing the proportion of new members who can at least take care of themselves, and if possible return some net benefit. Some parents, however, may long for any child to love and to cherish. They may regard any talk of social costs and benefits as shockingly utilitarian.

Even more difficult situations could arise when the prospective child may, himself, be quite normal, a joy to his parents, and a net asset to society—except for the fact that he is a heterozygous carrier of a serious defect who would, as a matter of fact, not have been started at all had the way not been made clear for him by aborting one or more homozygous siblings.

The more one thinks of these problems in abstract, philosophical terms, the more appalled he becomes, and this is one of the reasons for meetings like this. I am beginning to suspect, however, that the situation will turn out to be less serious in practice than it is in theory.

In the first place, it seems highly improbable that the wishes of parents and the interests of society will really be at odds in cases of severe defects like Down's syndrome or Tay-Sachs disease. In other words, if the parents' views are such that abortion is an acceptable option on general grounds, they are likely to exercise it for their own reasons in the kinds of cases in which carrying the fetus to term would also be most costly and least beneficial to society. In cases where the option is unacceptable to the individual for religious or similar reasons, society, under existing and generally agreed upon policies, has no business interfering anyway.

At the other end of the spectrum, relatively mild defects which may alter only slightly the subject's chances of becoming a useful citizen, like six fingers, may be safely left to the parents to decide without comment from outside since the social costs of either outcome are likely to be negligible. There may be a number of areas where family attitudes may be quite variable. It would seem unwise to emphasize the possible conflicts between individuals and society in these gray areas until we know a lot more about them.

Particularly troublesome, of course, are those genetic conditions in which the homozygotes suffer a severe disorder while the heterozygotes show little or no deficit. There may come a day when the question of what to do about such heterozygous fetuses may indeed involve ethical problems between the long-run interest of society in its gene pool and the short-run interest of the individual in family life. At present, however, the technical difficulties of identifying heterozygotes in many conditions and especially of evaluating the possibly unusual "fitness" of the latter may excuse us from taking any firm position which might breathe life into a latent conflict between individuals and society.

To point out that the number of potential conflicts is probably not so large as some people have feared is not to say that there is no problem. But the realization may be of some comfort to prudent men as they start to formulate policies.

It is also true that the above analysis depends very largely on the assumption that the general public knows as much about the disorders in question as the representatives of society do and that decisions will be arrived at more or less autonomously. This, of course, is not the case, as we are thus brought to the heart of the problem that worries many moralists and men of good will.

In most instances the facts revealed by amniocentesis must be interpreted to the prospective parents. In advanced countries these interpretative services are likely to be at least partially supported by society. How are we—and perhaps equally important, how is the counselor—to be sure that he is providing the interpretation in the best interests of the patient? Alternatively, how far is he justified in presenting the interests of society, especially if these may appear from some angles to be in conflict with those of the parents? These are indeed difficult questions. But I am not persuaded that they are so difficult as some people think they are.

Some commentators appear to feel that society is making an unjustifiable invasion of personal liberties if it undertakes to persuade an individual to make any decision in the reproductive sphere other than that suggested by his uninstructed instincts. The criticism is particularly severe if the proposed learning experience is coupled with rewards and penalties. It is hard to understand this squeamishness, since societies have been rewarding people from time immemorial for doing things which benefited society and even more enthusiastically penalized its citizens for actions against the public interest.

Even without this background, it is hard to see how offering a man a transistor radio if he refrains from having a child is interfering with his personal freedom. Indeed, one would ordinarily suppose that by increasing the number of options open to him, we increase rather than restrict his freedom. Furthermore, transistor radios may be looked upon primarily as educational devices, a way of making real to a not very sophisticated mind that children impose costs on society—and that society can in turn use its resources to benefit existing individuals if they cooperate in reducing the potential competition.

Again, this increase in understanding should be considered liberating rather than confining.

If society has a definable and legitimate interest in the number of children to be born, then it is clear that it also may have an interest in their quality. This is especially true in a period of sharply reduced birth and death rates. In an earlier day when both were high, normal attrition worked differentially to eliminate the unfit. A much smaller proportion of society's resources was devoted to keeping the unfit alive. Now, when a defective child may cost the society many thousands of dollars a year for a whole lifetime without returning any benefit, it would appear inevitable that society should do what it reasonably can to assure that those children who are born can lead normal and reasonably independent lives. It goes without saying that if the model couple is to be restricted to 2.1 children, it is also more important to them that all their children be normal than it was when an abnormal child was, in effect, diluted by a large number of normal siblings. And the over-all point here again is that the interests of society and the interests of the family become coincident at this stage, when they both realize that there are a limited number of slots to be filled.

That is really the situation that we are approaching in many countries. There is a limited number of slots that human beings can occupy, and there do seem to be both social and family reasons to see that those increasingly rare slots are occupied by people with the greatest possible potential for themselves and for others.

What then can society reasonably do to work with potential parents to insure as high quality a product as possible? It certainly can encourage and support the development of research in genetics and the physiology of fetal life. It can certainly support, in the same way, work on the technique of amniocentesis. It can make abortion available under appropriate auspices and regulations. Many societies are doing some or all of these things right now.

Our problem, if there is one, centers on what society can and should do in helping the individual mother to use the available information not only in her own best interests, but also in those of the society of which she is part. And the whole assumption here is that we are not going to require any behavior. We seem to be working toward a position which gives to the mother's physician or genetic counselor the greatest possible influence. Sometimes the two functions may be discharged by the same individual. In either case, the relationship is a professional one, with the individuals giving the advice owing primary responsibility to their patient or client. I think all of us with medical training feel our first obligation in any such situation is to the patient before us.

Just what the amniocentesis counselor will say to a given client is likely to remain a professional matter, biased in favor of the client when her interests and those of society conflict. On the other hand, there seems no good reason to discourage the advisor from pointing out to the client what the interests of society are. Indeed, there seem to be several good reasons for encouraging him to do so.

In the first place it seems proper, even if it is not always precisely true, to assume that the client wishes to be a good citizen as well as a good parent. Thus, it is in some sense an insult to her intelligence and good will to withhold information about social repercussions on the grounds that such advice might curtail her Anglo-Saxon freedom to act ignorantly in her own selfish interest.

It may also be remembered that every professional man is an officer of society—although I realize not everybody agrees with this—as well as an individual counselor. True, most lawyers, clergymen, and physicians like to think of themselves as owing a paramount obligation to their patients, clients, and parishioners. But the clergyman has at least a nominal interest in the deity, the lawyer in justice, and the physician in the health of the community.

Fortunately, in medicine at least, the welfare of the patient usually coincides reasonably closely with the welfare of society. But there are well-known cases of conflicts in which the physician is supposed to act in the best interest of society even at some cost to his patients, and sometime he does.

Thus, the conscientious physician reports his cases of veneral infection and T.B. in order to protect society, even at the risk of embarrassment and inconvenience to the patient. In a similar vein he urges his epileptic patients and those with seriously defective vision not to drive automobiles. These and other examples provide ample precedent for injecting the interests of society into a counseling situation.

I hope it is obvious that when he does so, the counselor will wish to make clear exactly what he is doing and why. It would of course be totally inadmissible for a professional counselor of any kind to deceive his patient or client, in order to gain what he thinks of as a socially desirable end.

We have heard of cases where it may be necessary to deceive a patient for his own good (or what is thought to be his own good), but I don't think any of us would countenance deceiving a patient for society's good.

On the other hand, it would be almost equally reprehensible to conceal the long-term effects on the gene pool of substituting a heterozygous carrier for a homozygote of limited life expectancy, or to fail to mention social dislocations which might follow from a decision to use abortion as a method of sex determination.

In summary, the position I have described is a simple and straightforward one. Many people are likely to feel it is too simple. It is based on the fact that society has placed or is in the process of placing the burden of decision as to whether or not to have an abortion in the hands of the woman carrying the fetus.

It accepts the proposition that in arriving at that decision, she will be allowed to add to a number of already stated considerations, evidence on the probable characteristics of the child.

It does not accept the inference that society no longer has interest in the matter simply because it has released the formal decision-making power.

It suggests that the best way of expressing its interest is through the counselor-physician, who in effect has a dual responsibility to the individual whom he serves and to the society of which he and she are parts.

Finally, it holds out the hope that instances in which the interest of the individual woman are clearly seen to conflict with the interests of society will be less numerous than might be feared.

In review, I find that I must add one point. Just as I believe there are no simple principles which take precedence over all others, I believe that everything one does affects everything else. We are now all so interdependent that there are no such things as purely private acts.

There is danger that the present effort to make abortion a crime without a victim will obscure the seriousness of all such decisions.

As Dr. Kass has pointed out, we will all certainly be diminished as human beings, if not in great moral peril, if we allow ourselves to accept abortion for what are essentially trivial reasons. On the other hand we will, I fear, be in equal danger if we don't accept abortion as one means of ensuring that both the quantity and quality of the human race are kept within reasonable limits.

PRACTICAL AND ETHICAL PROBLEMS IN HUMAN GENETICS

KURT HIRSCHHORN

The past 20 years, and more particularly, the past five or ten years have seen an exponential growth of scientific technology. The chemical structure of the hereditary material, as well as its language, have essentially been resolved. Cells can be routinely grown in test tubes by tissue culture technics. The exact biochemical mechanisms of many hereditary disorders have become clarified. Computer programs for genetic analysis are in common use. All of these advances and many others have inevitably led to discussions and suggestions for the modification of human heredity, both in individuals and in populations. This has been called genetic engineering. Among the many recent articles on this subject three of the most challenging are by the Nobel Prize winning geneticist Joshua Lederberg,(1) by the world's leading experts on evolution, Theodosius Dobzhansky,(2) and by Bernard Davis.(3)

One of the principal concerns of the pioneers in the field is the problem of the human genetic load, that is, the frequency of disadvantageous genes in the population. It is well established that each of us carries between three and eight genes which, if present in double dose in the offspring of two carriers of identical genes, would lead to severe genetic abnormality or even to death of the affected individual before or after birth. In view of the rapid medical advances in the treatment of such diseases, it is likely that affected individuals will be able to reproduce more frequently than in the past. Therefore, instead of a loss of genes due to death or sterility of the abnormal, the mutant gene will be transmitted to future generations in a slowly but steadily increasing frequency. This is leading the pessimists to predict that we will become a race of genetic cripples requiring a host of therapeutic crutches.

The optimists, on the other hand, have great faith that the forces of natural evolution will continue to select favorably those individuals who are best

---

This paper originally appeared in *Advances in Human Genetics and Their Impact on Society*, Birth Defects Original Article Series, D. Bergsma (ed.) Vol. VIII, No. 4, White Plains: The National Foundation, 1972, pp. 17–30. Reprinted with permission of the author and The National Foundation.

adapted to the then current environment. It is important to remember in this context that the "natural" environment necessarily includes man-made changes in the medical, technical, and social spheres. Since it appears that at least some of the aspects of evolution and a great deal of genetic planning will be in human and specifically in scientific hands, it is crucial at this relatively early stage to consider the ethical implications of these proposed maneuvers. Few scientists today doubt the feasibility of at least some forms of genetic engineering, and there is considerable danger that common use of this practice will be upon us before its ethical applications are defined.

While examining the various possibilities for genetic engineering, it will be useful to keep in mind the concepts of freedom and coercion as they may apply to the various technics. For the purpose of this paper, freedom is defined primarily as the right to make a decision with full knowledge of all available data. This definition of freedom must, however, be somehow limited so that the decisions do not impinge upon the immediate needs of or benefits to other individuals and the general population. Coercion is defined as a direct, indirect, or subliminal attempt to influence the decision of one or more individuals for an assumed, but unproved, benefit of other individuals or the general population, or secondary to a conscious bias of the counselor.

It must be recognized that these concepts apply to three groups of individuals, each quite different in their objectives and interests. The first, consisting of individuals and families, is primarily concerned with their own and their children's welfare. How open-ended is their freedom of decision to be, when the result may or will have harmful social consequences, such as the birth of a child who will remain totally dependent on society? Should society or how could society control organized voluntary practices by groups of individuals, practices known or believed to be detrimental to the genetic well-being of society, such as widespread inbreeding?

The second group, consisting of research scientists and genetic counselors, poses perhaps even greater problems of definition of limitation of freedom and coercion. How free should the investigator be in pursuing certain lines of investigation? If widespread use of some technics, eg., the utilization of frozen sperm banks, is assumed to have long-term detrimental effects, should a group be allowed to implement these on a more limited basis? Is governmental regulation required or should scientists regulate themselves? How can one (or should one) limit scientific investigation which could result in potentially harmful technics? Does the genetic counselor have the right to persuade families, to direct (or coerce) their decision, in even an obviously beneficial manner, or should he be restricted in simply presenting all the data and alternatives?

The third group is society itself or its governmental representatives. Do public authorities have the freedom to intervene, either by establishing laws or through regulatory agencies in the field of genetics, by limiting the scope of decisions by individuals and by regulating the limits of research or its applications?

I do not propose to answer these questions, but simply raise them as a background for consideration while examining the various technics suggested for genetic intervention. It is obvious that conflicts arise in the answers to these questions, when considering limitations imposed on individuals as against society.

A number of individually quite different methods have been proposed for the control and modification of human hereditary material. Some of these methods are meant to work on the population level, some on the family level and others directly on the affected individual. Interest in the alteration of the genetic pool of human populations originated shortly after the time of Mendel and Darwin in the latter part of the 19th century. The leaders were the English group of eugenicists headed by Galton. *Eugenics* is nothing more than planned breeding designed to alter the genetic makeup of future generations. This technic, of course, has been successfully used by the agricultural community in the development of hybrid breeds of cattle, corn, and numerous other food products.

Human eugenics can be divided into *positive eugenics* and *negative eugenics*. *Positive eugenics* is the preferential breeding of so-called superior individuals in order to improve the genetic stock of the human race. The most famous of the many proponents of positive eugenics was the late Nobel Prize winner Herman J. Muller. It was his suggestion that sperm banks be established from a relatively small number of donors, chosen by some appropriate panel, and that this frozen sperm remain in storage until some future panel had decided that the chosen donors truly represented desirable genetic studs. If the decision is favorable, a relatively large number of women would be inseminated with these samples of sperm, and it is hoped by the proponents of this method that a better world would result. The qualifications for such a donor would include high intellectual achievement and a socially desirable personality, qualities assumed to be affected by the genetic makeup of the individual, as well as an absence of obvious genetically determined physical anomalies.

A much more common effort is in the application of *negative eugenics*. This is defined as the discouragement of the legal prohibition of reproduction by individuals carrying genes leading to disease or disability. This can be achieved by genetic counseling or by sterilization, either on a voluntary or enforced basis. There are, however, quite divergent opinions as to which genetic traits are to be considered sufficiently disadvantageous to warrant the application of negative eugenics.

A diametrically opposite solution is that of *euthenics*, which is a modification of the environment in such a way as to allow the genetically abnormal individual to develop normally and to live a relatively normal life. Euthenics can be applied both medically and socially. The prescription of glasses for near-sighted individuals so that they may live normally in our automotive world is an example of medical euthenics. The provision of special schools for the deaf, a great proportion of whom are genetically abnormal, is an example of social euthenics. The humanitarianism of such efforts is obvious, but it is exactly these types of activities that have led to the concern of the pessimists who assume that evolution has selected for the best of possible variations in man and that further accumulations of genes considered abnormal can only lead to decline.

One of the most talked-about advances for the future is the possibility of altering an individual's genetic complement. Since we are well on the way to understanding the genetic code, as well as to deciphering it, it is suggested that we can alter it. This code is written in a language of 64 letters, each being determined by a special arrangement of three out of four possible nucleotide bases. A chain of these bases is called deoxyribonucleic acid or DNA and makes up the genetic material of the chromosomes. If the altered letter responsible for an abnormal gene can be located, for example in a fertilized egg, and the appropriate nucleotide base substituted, the correct message would again produce its normal product, which would be either a structurally or enzymologically functional protein. Another method of providing a proper gene, or code word, to an individual having a defect has been suggested from an analysis of viral behavior in bacteria. It has long been known that certain types of viruses can carry genetic information from one bacterium to another or instruct a bacterium carrying it to produce what is essentially a viral product. Viruses are functional only when they live in a host cell. They use the host's genetic machinery to translate their own genetic codes. Viruses living parasitically in human cells can cause diseases such a poliomyelitis and have been implicated in the causation of tumors. Other viruses have been shown to live in cells and be reproduced along with the cells without causing damage either to the cell or to the organism. If such a harmless virus either produces a protein that will serve the function of one lacking in an affected individual or if it can be made to carry the genetic material required for such functions into the cells of the affected individual, it could permanently cure the disease without additional therapy. If carried on to the next generation, it could even modify the inheritance of the disease.

An even more radical approach has been outlined by Lederberg.(1) It has become possible to transplant whole nuclei, the structures which carry the DNA, from one cell to another. It has become easy to grow cells from various tissues of any individual in tissue culture. Such tissue cultures can be examined for a variety of genetic markers and thereby screened for evidence of new mutations. Lederberg suggests that it would be possible to use nuclei from such cells derived from known human individuals, again with favorable genetic traits, for the asexual human reproduction of replicas of the individuals whose nuclei are being used. For example, a nucleus from a cell of the chosen individual could be transplanted into a human egg whose own nucleus has been removed. This egg, implanted into a womb, could then divide just like a normal fertilized egg to produce an individual genetically identical to the one whose nucleus was used. One of the proposed advantages of such a method would be that, as in positive eugenics, one could choose the traits that appear to be favorable, and one could do this with greater efficiency since one eliminates the somewhat randomly chosen female parent necessary for the sperm bank approach. Another advantage is that one can mimic what has developed in plants as a system for the preservation of genetic stability over limited periods of time. Many plants reproduce intermittently by such parthenogenetic mechanisms, always followed by periods of sexual reproduction which results in elimination of disadvantageous mutants and increase in variability.

Another possibility derives from two other technologic advances. Tissue typing, similar to blood typing, and some immunologic tricks have made it possible to transplant cells, tissues, and organs from one individual to another with reasonably long-term success. Over the past few years scientists have also succeeded in producing hybrid cells containing some of the genetic material from each of two cell types either from two different species or two different individuals from the same species. Recently Weiss and Green(4) at New York University have succeeded in hybridizing normal human cultured cells with cells from a long-established mouse tissue culture line. Different products from such fusions contain varying numbers of human chromosomes and, therefore, varying amounts of human genes. If such hybrids can be produced which carry primarily that genetic information which is lacking or abnormal in an affected individual, transplantation of these cultured cells into the individual may produce a correction of his defect.

These are the proposed methods. It is now fair to address oneself to the question of feasibility. Feasibility must be considered not only from a technical point of view; of equal importance are the effects of each of these methods on the evolution of the human population and the effect of evolution upon the efficacy of the method. It should also be remembered that adverse effects on evolution raise the question whether freedom alone is sufficient reason for attempting some of these technics. In general, it can be stated that most of the proposed methods either are now or will in the not too distant future be technically possible. We are, therefore, not dealing with hypothesis or science fiction but with scientific reality. Let us consider each of the propositions independently.

Positive eugenics by means of artificial insemination from sperm banks has been practiced successfully in cattle for many years. Artificial insemination in man is an everyday occurrence. But what are some of its effects? There is now ample evidence in many species, including man, of the advantages for the population in terms of individual genetic variation, mainly flexibility of adaptation to a changing environment. Changes in environment can produce drastic effects on some individuals, but a population which contains many genetic variations of that set of genes affected by the particular environmental change, will contain numerous individuals who can adapt. There is also good evidence that individuals who carry two different forms of the same gene, that is, are heterozygous, appear to have an advantage. This is even true if that gene in double dose, that is, in the homozygous state, produces a severe disease. For example, individuals homozygous for the gene coding for sickle-cell hemoglobin invariably develop sickle-cell anemia which is generally fatal before the reproductive years. Heterozygotes for the gene are, however, protected more than normals from the effects of the most malignant form of malaria. It has been shown that women who carry the gene in single dose have a higher fertility in malarial areas than do normals. This effect is well known to agricultural geneticists and is referred to as hybrid vigor. Fertilization of many women by sperm from few men will have an adverse effect on both of these advantages of genetic variability since the population will tend to be more and more alike in their genetic characteristics. Also, selection for a few genetically advantageous factors will carry with it selection for a host of other genes present in high numbers in the population. Therefore, the interaction between positive eugenics and evolution makes this method not desirable on its own, and raises the question of the need for limitation upon individual, or even collective freedom to engage in such efforts.

Negative eugenics is, of course, currently practiced by most human geneticists. It is possible to detect carriers of many genes, which when inherited from both parents will produce abnormal offspring. Parents, both of whom carry such a gene, can be told that they have a one in four chance of producing such an abnormal child. Individuals who carry chromosomal translocations are informed that they have a high risk of producing offspring with congenital malformations and mental retardation. But how far can one carry out such a program? Some states have laws prescribing the sterilization of individuals who are mentally retarded to a certain degree. These laws are frequently based on false information regarding the heredity of the condition. The marriage of people with reduced intelligence is forbidden in some localities, again without adequate genetic information. While the effects of negative eugenics may be quite desirable in individual families with a high risk of known hereditary disease, it is important to examine its effects on the general population. These effects must be looked at individually for conditions determined by genes that express themselves in single dose (dominant), in double dose (recessive) and those which are due to an interaction of many genes (polygenic inheritance). With a few exceptions, dominant diseases are rare and interfere severely with reproductive ability. They are generally maintained in the population by new mutations.

Therefore, there is either no need or essentially no need for discouraging these individuals from reproduction. Any discouragement, if applicable, will be useful only within that family but not have any significance for the general population. One possible exception is the severe neurologic disorder, Huntington's chorea, which does not express itself until most of the patient's children are already born. In such a situation it may be useful to advise the child of an effected individual that he has a 50% chance of developing the disease and a 25% chance of any of his children being affected. Negative eugenics in such a case would at least keep the gene frequency at the level usually maintained by new mutations.

The story is quite different for recessive conditions. Although detection of the clinically normal carriers of these genes is currently possible only for a moderate number of diseases, the technics are rapidly developing whereby most of these conditions can be diagnosed even if the gene is present only in single dose and will not cause the disease. Again, with any particular married couple it would be possible to advise them that they are both carriers of the gene and that any child of theirs would have a 25% chance of being affected. However, any attempt to decrease the gene frequency of these common genetic disorders in the population by prevention of fertility of all carriers would be doomed to failure. First, we will carry between three and eight of these genes in a single dose. Secondly, for many of these conditions, the frequency of carriers in the population is about 1 in 50 or even greater. Prevention of fertility for even one of these disorders would stop a sizable proportion of the population from reproducing and for all of these disorders would prevent the passing on to future generations of a great number of favorable genes and would, therefore, interfere with the selective aspects of evolution which can only function to improve the population within a changing environment by selecting from a gene pool containing enormous variability. It has now been shown that in fact no two individuals, with the exception of identical twins, are likely to be genetically and biochemically identical, thereby allowing the greatest possible adaptation to changing environment and the most efficient selection of the fittest.

The most complex problem is that of negative eugenics for traits determined by polygenic inheritance. Characteristics inherited in this manner include many measurements that are distributed over a wide range throughout the population, such as height, birthweight, and possibly intelligence. The latter of these can serve as a good example of the problems encountered. Severe mental retardation in a child is not infrequently associated with perfectly normal intelligence or in some cases even superior intelligence in the parents. These cases can, a priori, be assumed to be due to the homozygous state, in the child, of a gene leading to mental retardation, the parents representing heterozygous carriers. On the other hand, borderline mental retardation shows a high association with subnormal intelligence in other family members. This type of deficiency is assumed by some to be due to polygenic factors, more of the pertinent genes in these families being of the variety that tends to lower intelligence. However, among the offspring of these families there is also a high proportion of individuals with normal intelligence and a sprinkling of individuals with superior intelligence. All of these comments are made with the realization that our current measurements of intelligence are very crude and cannot be compared between different population groups. It is estimated that, on the whole, people with superior intelligence have fewer offspring than do those of average or somewhat below average intelligence, and therefore, a lack of replacement of superior individuals from the offspring of less intelligent parents would lead to a general decline of intelligence in the population.

It can be seen therefore, that neither positive nor negative eugenics can ever significantly improve the gene pool of the population and simultaneously allow for adequate evolutionary improvement of the human race. The only useful aspect of negative eugenics is in individual counseling of specific families in order to prevent some of the births of abnormal individuals. One recent advance in this sphere has important implications from both a genetic and a social point of view. It is now possible to diagnose genetic and chromosomal abnormalities in an unborn child by obtaining cells from the amniotic fluid in which the child lives in the mother.

It is now known that some 24 biochemical inborn errors of metabolism are potentially diagnosable prenatally. The chromosomal disorders can also be detected by this method, and one group that it would be particularly useful to study in this manner would be mothers over 40 who have at least 1%, and probably closer to 3%, risk of having chromosomally abnormal offspring. One could enlarge upon the number of people to be studied in this way if we were to screen populations for heterozygotes for particular biochemical defects that can be detected in utero and therefore could be prewarned when two carriers happen to get married and the wife gets pregnant. At least 40 genes in the heterozygote can now be detected and presumably these 40 genes in the homozygous state will also be diagnosable by amniocentesis. An example is Tay-Sachs disease. This disease is recessively inherited, that is, it expresses itself in the homozygote (in double dose) and happens to be frequent in Ashkenazic Jews. In this population about one in 25 individuals is asymptomatic carrier of this particular disease. It is now quite easy to detect the carrier from a simple blood sample and it is also easy to detect the disease from an amniotic fluid sample so that one knows whether two carriers in fact have produced a defective child. This is a particularly nasty genetic disease in that the child is born apparently normal and then at about eight or nine months of age begins to deteriorate and then dies a rather miserable death by two or three years of age. If one were to go to the Ashkenazic Jewish population and do universal screening for heterozygotes one could conceivably within one generation stop the birth of affected individuals. One would, of course, not eradicate the gene unless one also were to diagnose for carriers by means of prenatal diagnosis and abort the clinically normal carriers that would be born to these individuals. This, of course, raises a number of legal, religious, and ethical questions. One has to balance, however, some of these questions with the knowledge that in a relatively inbred group such as the Ashkenazic Jews where the gene frequency is as high as it is for this particular gene, abortion of the affected would be compensated for by the birth of clinically normal offspring which will lead to a slow increase in the gene frequency for the particular gene. The reason for this is that two out of three of the normal offspring of such families will be carriers. The earlier you abort, the more quickly reproductive compensation is going to happen, leading to an additional rise in the rate of increase of the gene frequency. These types of problems, of course, need to be worked out and discussed very thoroughly. There is, however, one additional aspect which must be considered. The fact is that the inbreeding of Ashkenazic Jews is rapidly decreasing and they are beginning to outbreed to the extent that there is now an increasing frequency of religious intermarriage. This, of course, would avoid the concentration of this particular gene and its increase, and will lead to a sharp decrease in the number of couples at risk.

The future may bring further advances allowing one to start treatment on the unborn child and to produce a functionally normal infant. We already have at least five or six disease such as phenylketonuria in which treatment is possible. If one starts treatment in the newborn child, it is possible to produce an intellectually normal child, whereas without treatment these children in most cases would be severely mentally retarded. Here again questions would arise for a number of genes, because in order to do this efficiently one might have to screen the parents for the heterozygous state. When should one screen? Should one screen all newborns and then give them a score card as to what genes they are carrying? Should one screen when people get married? Should one screen all pregnant women when they arrive for the first prenatal examination, and if one finds a particular carrier state, should the husband then be examined to see whether he also carries that particular gene?

There are many other problems, and not the least of these with regard to universal screening is the question of invasion of privacy, and what one does with this information beyond transmitting it to the individual who is being screened. The only currently possible solution, however, for many of these problems is restricted to termination of particular pregnancies by therapeutic abortion. This is, of course, applied negative eugenics in its most extreme form. The future may bring further advances, allowing one, then, to start treatment on the unborn child and to produce a functionally normal infant.

Euthenics, the alteration of the environment to allow aberrant individuals to develop normally and to lead a normal life, is currently in wide use. Medical examples include special diets for children with a variety of inborn errors of metabolism who would, in the absence of such diets, either die or grow up mentally retarded. Such action, of course, requires very early diagnosis of these diseases, and programs are currently in effect to routinely examine newborns for such defects. For example, PKU screening laws are in existence which provide for newborn screening in most states. Other examples of euthenics include the treatment of diabetics with insulin and the provision of special devices for children with skeletal deformities. Social measures are of extreme importance in this regard. As has many times been pointed out by Dobzhansky,(2) it is useless to plan for any type of genetic improvement if we do not provide an environment within which an individual can best use his strong qualities and obtain support for his weak qualities. One need only mention the availability of an environment conducive to artistic endeavor for Toulouse-Lautrec, who was deformed by an inherited disease.

The problem of altering an individual's genes by direct chemical change of his DNA presents technically an enormously difficult task. Even if it became possible to do this, the chance of error would be high. Such an error, of course, would have the diametrically opposite effect to that desired and would be irreversible; in other words, the individual would become even more abnormal. The introduction of corrective genetic material by viruses or transplantation of appropriately hybridized cells is technically predictable, and since it would be performed only in a single affected individual, would have no direct effect on the population. If it became widespread enough, it could, like euthenics, increase the frequency in the population of so-called abnormal genes, but if this treatment became routine, this would not develop into an evolutionarily disadvantageous situation. It must also be constantly kept in mind that medical advances are occurring at a much more rapid rate than any conceivable deterioration of the genetic endowment of man. It is, therefore, very likely that such corrective procedures will become commonplace long before there is any noticeable increase in the load of disadvantageous genes in the population.

The growing of human beings from cultured cells, while again theoretically feasible, would, on the other hand, interfere with the action of evolutionary forces. There would be an increase, just as with positive eugenics, of a number of individuals who would be alike in their genetic complement with no opportunity for the degree of genetic recombination which occurs during the formation of sperm and eggs and which becomes manifest in the resultant progeny. This would diminish the adaptability of the population to changes in the environment and, if these genetic replicas were later permitted to return to sexual reproduction, would lead to a marked increase in homozygosity for a number of genes with the disadvantages pointed out before.

We see, therefore, that many of the proposed technics are feasible although not necessarily practical in producing their desired results. We may now ask the question, which of these are ethical from a humanistic point of view? Both positive and negative eugenics when applied to populations presume a judgment of what is genetically good and what is bad. Who will be the judges and where will be the separation between good and bad? We have had at least one example of a sad experience with eugenics in its application in Nazi Germany. This alone can serve as a lesson on the inability to separate science and politics. The most difficult decisions will come in defining the borderline cases. Will we breed against tallness because space requirements become more critical? Will we breed against nearsightedness because people with glasses may not make good astronauts? Will we forbid intellectually inferior individuals from procreating despite their proved ability to produce a number of superior individuals? Or should we rather provide an adequate environment for the offspring of such individuals to realize their full genetic potential?

C. C. Li(5) in his presidential address to the American Society of Human Genetics in 1960 pointed out the real fallacy in eugenic arguments. As he points out, man has continuously improved his environment to allow so-called inferior individuals to survive and reproduce. The movement into the cave and the putting on of clothes have protected the individual unable to survive the stress of the elements. Should we then consider that we have reached the peak of man's progress, largely determined by environmental improvements designed to increase fertility and longevity, and that any future improvements designed to permit anyone to live a normal life will only lead to deterioration? Nineteenth century scientists, including such eminent biologists as Galton, firmly believed that this peak was reached in their time. As analyzed by Li, this obviously fallacious reasoning must not allow a lapse in ethical considerations by the individual and by humanity as a whole, just to placate the genetic pessimists. The tired axiom of democracy that all men are created equal must not be considered from the geneticist's point of view, since genetically all men are created unequal. Equality must be defined purely and simply as equality of opportunity to do what one is best equipped to do. When we achieve this, the forces of natural evolution will choose those individuals best adapted to this egalitarian environment. No matter how we change the genetic make-up of individuals, we cannot do away with natural selection. We must always remember that natural selection is determined by a combination of truly natural events and the artificial modifications which we are introducing into our environment at an exponentially increasing rate.

It is obvious that the possibility of coercion exists in each of the technics proposed. The false concepts of what is good or bad in the genetic makeup of an individual or population, or even the misapplication of true concepts, can lead well-meaning scientists or politicians (in the latter case occasionally not so well-meaning) to urge or decree a variety of maneuvers designed to fulfill the aims of their bias. This does not imply that, therefore, none of these methods is justified. It does mean that their application must be controlled by avoiding, as much as possible, any coercive aspects. The genetic counselor can easily fall into the trap of letting his biased opinion, good or bad, influence the decision of a family regarding procreation or abortion. Perhaps it will become necessary to submit such opinions to teams of experts in order to correct the bias, or to make clear the existence of divergent opinions. Lack of understanding that what is "bad" at this time may act quite differently in an altered environment, may lead well-meaning counselors to change the gene pool or to do away with clinical carriers for certain genes by means of selective abortion.

With these points in mind, we can try to decide what in all of these methods is both feasible and ethical. I believe that the only logical conclusion is that all maneuvers of genetic engineering must be judged for each individual and, in each case, must take primary consideration of the rights of the individual. Obviously each situation will differ as to whether we are considering the rights of present individuals or those not yet born or conceived. Consideration of individual rights is patently impossible in any attempt at positive eugenics. Negative eugenics on a population level suffers from similar problems, but applied in this form of intelligent genetic counseling is the answer for some. Our currently changing attitudes about practicing negative eugenics by means of intelligent selection for therapeutic abortion must be encouraged. Basic to this change is a more accurate definition of a living human being. Such restricted uses of negative eugenics will prevent individual tragedies. Correction of unprevented genetic disease, or that due to new mutation, by introduction of new genetic material may be one answer for the future; but until such a new world becomes universally feasible, we must on the whole restrict ourselves to environmental manipulations from both the points of view of allowing affected individuals to live normally and permitting each individual to realize his full genetic potential.

There is no question that genetic engineering in many forms, here described or not yet conceived, will come about. It is a general rule that whatever is scientifically feasible will be attempted. The application of these technics must, however, be examined from the point of view of ethics, individual freedom, and coercion. Both the scientists directly involved and, perhaps more important, the political and social leaders of our civilization must exercise utmost caution in order to prevent genetic, evolutionary, and social tragedies.

References

1. Lederberg, J.: Experimental genetics and human evolution. *American Naturalist* **100:** 519, 1966.
2. Dobzhansky, T.: Changing man. *Science* **155:**409, 1967.
3. Davis, B. D.: Prospects for genetic intervention in man. *Science* **170:**1279, 1970.
4. Weiss, M. C. and Green, H.: Human-mouse hybrid cell lines containing partial complements of human chromosomes and functioning human genes. *Proc. Nat. Acad. Sci. (Wash.)* **58:**1104, 1967.
5. Li, C. C.: The diminishing jaw of civilized people. *Amer. J. Hum. Genet.* **13:**1, 1960.

REPRODUCTIVE RIGHTS AND GENETIC DISEASE

LAWRENCE P. ULRICH

This paper attempts to examine each of the four issues contained in the title: (1) rights; (2) rights with regard to the reproductive function of human organisms; (3) disease in general; (4) disease which has a currently identifiable genetic etiology. The thesis which I wish to adopt is this: Reproductive rights are not absolute and those who are at high risk for passing on clearly identifiable, severely deleterious genes and debilitating genetic disease should not be allowed to exercise their reproductive prerogative.

I. Rights

Rights are frequently classified in two general ways: legal and human. In many ways the first are much easier to treat than the second, primarily because of their specificity and the fact that they are witten down and consciously agreed to by the society which is concerned enough about its citizens to work them out. While legal rights may be specific, they are not so specific that they do not admit of disputes in their application to particular situations. Hence the need for the legal profession.

Much less clear are those rights which some members of the human population claim under the guise of natural human rights—the most frequently quoted of which are life, (its means) liberty, and (its end) the pursuit of happiness. This is not to say that legal rights and human rights are two completely separate *kinds* of rights. They probably differ only as two parts of the same fabric of claims with the difference being that legal rights are written down while human rights are not. Because of this common bond between legal and human rights, the former are frequently taken as a reinforcement and guarantor of the latter. And in cases that are serious enough, the latter are taken as a substitute for the former.

The attempts at constitutional democracy in the last several hundred years have been largely attempts at "legalizing" fundamental human rights. This has

Research for this paper was funded by the University of Dayton's Research Council. Printed with permission of the author.

generally been done by guaranteeing that the first rights of the citizen are his rights as a person. The constitutions and legal codes were developed, among other reasons, to further guarantee what were claimed to be fundamental human rights. The Nuremburg trials exemplify the situation in which human rights are taken as a substitute for legal rights. The trials were predicated on the acknowledgement that there are such claims as fundamental human rights which every decent person and civilized nation acknowledges and that the violator of them is just as liable as the violator of a human right guaranteed by a legal code.

An important issue in this paper is whether the class of rights which we call human rights are indeed inalienable and non-negotiable. The natural rights theorist believes that certain rights are intrinsic to man and are his by reason of his nature. This position leads to a number of dubious claims for virtually any natural right. For the right that is claimed reflects the claimant's cultural heritage and specifically what he means by being a person and what he thinks will insure a better quality of life for the persons in the human condition. Can we say, then, that there are no such rights which we customarily speak of as natural human rights? Attempts at falsification here would get us no further than attempts at proof. In neither case do we have anything identifiable to use as a criterion for adjudication. Because of these stalemate I should like to advance what I take to be a reasonable account of human rights which are a great concern behind many of the medical ethical problems which we face today. The following considerations allow us to account for the addition of rights to the class of so-called human rights and the variation that we find in these rights-claims from population to population.

I suggest that the rights which we call human rights are nothing more than claims which are made with regard to the performance or prohibition of certain actions and which are acknowledged by other members of the population. Now merely to lay claim to a right does not authorize the right. The factor of social acknowledgement is absolutely central. Part of the problem here lies in the historical perspective from which we view the right. While the 18th-century liberal may have viewed liberty as a natural right, it would not have occurred to the 13th-century serf to claim such a right, much less would it have occured to the 13th-century lord to acknowledge such a right. Rights as recognized claims seem to be correlated with the state of awareness–political, economic, moral, etc.–of the claimants and their social environment. Similarly, it probably never occurred to early man to claim the right to life as a natural right. We do not know when the right to life was claimed and acknowledged but from the earliest days of civilization as we know it there seems to be agreement that the wanton killing of another human organism is a wrongful act.

What can be asserted about rights which are directly claimed as one's own is also true about rights claimed for another. This would be true of rights claimed for the infant, the fetus, the demented, the unconscious. We frequently assign to them human rights which we wish to claim for ourselves, with the hope that society at large will recognize the rights for them just as society recognizes the rights for us. But it is immediately obvious that the rights which articulate claimants assert for themselves are not always claimed for others. We can reflect on those societies where the old are cast out of the tribe to die or where the deformed infant is immediately executed. It would be useless to say that in such cases the basic human rights of such individuals have been violated. For there is no criterion for making such a judgment other than the acknowledged claims of the population as they have developed through practice and tradition. Thus it may be that a society may develop which does indeed hold the "right to life" for its citizens but does not hold it as absolute for its fetuses.

In conclusion, there is indeed a class of rights that can be called "human" but that these are to be understood as unwritten claims acknowledged by society rather than as intrinsic properties of "human nature."

II. Reproductive Rights

I would like now to turn to the matter of the rights of individual humans in the reproductive process. The right to conceive and bear offspring is seemingly implied in the right to life, liberty, and the pursuit of happiness articulated, for example, in the Declaration of Independence of the United States. And the implied claim is that such activity is a natural human right. However, the legal codes, in the United States at least, do not respect this right to reproduce as absolute. We can cite two cases: (1) marriages of close kinship and (2) marriages of parties who have not passed the venereal disease screening test. (Marriage in these cases is considered the normal vehicle for reproduction.) Society imposes this abridgment of reproductive rights for the promotion of the public good in the area of public health.

There is no intrinsic reason for prohibiting marriages of close kinship. The high risk of diseased offspring, because of the combination of recessive genes, is the only reason. The purpose of the law, then, is to protect society and the individuals involved from a situation that can easily lead to undesirable consequences. For one who would believe that reproductive rights are absolute either the law would have to be considered wrong in this case or the concession would have to be made that reproductive rights are not absolute. Of course there is the option, not advanced by society that two closely related individuals can marry but may not bear offspring. This procedure would admit the negotiability of reproductive rights.

The case is similar with the venereal disease test. Legally, society claims the right to prohibit or delay marriage until such time as a cure is verified or at least until reasonable treatment is undertaken and the individual had indicated his or her intention to continue the treatment to its successful conclusion.

There is implied in these two situations more than the right of society to guard its public health even at the sacrifice of individual liberties. Concomitant with this concern is the obligation of individuals to respect and promote public health. In this connection I would like to turn briefly to the matter of obligation.

Obligations, like rights, are not immediately deducible from actions or a series of actions. Just because one can perform an action does not mean that he has a right to perform the action. In natural biological processes, e.g., reproduction, the ability to perform the action does not make the action good nor does it confer upon the actors the right to perform the action. The action itself, the action as good, and the performance of the action as a right are three distinctly discernable features. In the case of obligations we find similar distinctions. The performance of, or ability to perform, an action does not entail an obligation to respect the carrying out of that action. Even if that action is evaluated as good, obligation to respect its performance is not attached. Only when the action is construed as a right does there follow an obligation to respect the performance of the action by either providing the means to facilitate the performance of the action or by prohibiting circumstances which might obstruct it. Thus, the acknowledgement and discharge of obligations are directly related to rights and are accredited and negotiated in a way similar to the accreditation and negotiation of rights. Just as rights are not directly deducible from actions but are mediated by acknowledged claims, so also obligations are not directly deducible from actions but are mediated by rights. Actions, then, and states of affairs only provide clues for the ultimate accrediting of rights and obligations. This accrediting is done with an eye toward actions, actions as good, and actions that fall within the range of priorities of goods. While obligations in general cannot be directly deduced from actions or states of affairs, nevertheless it would be reckless to impose or accept obligations without taking into account actions or states of affairs. Thus, there are clues to obligations which must be considered in any human context.

We can now turn to the question of species survival. In the past 100 years we have come to know that the actions of a species and the interactions of a species with other species or with its environment frequently lead to extinction. The species always "attempts" to survive and is frequently temporarily successful. Its actions, then, are survival oriented, but the end result of those actions over time is generally extinction. If rights were directly deducible from actions then the answer to the question "Does a species have the right to survive?" would be both "yes" and "no" and we would get nowhere. A reasonable case could be made for the affirmative position in view of some natural selective tendencies which select for the survival of the fit. A reasonable case could be made for the negative position because of the general end result of natural selective tendencies toward extinction. But if survival is accredited as a good as opposed to extinction, then we might be in a position to claim survival as a right and attempt to gain acknowledgement of that right. We humans are egocentric enough to think that our continued presence on the planet is important and that it is to be construed as a good to be pursued. Moreover, we humans seem to be the only species on this planet, that we know of, that can pursue its own survival as a *good.* While other animals survive instinctively, we engage in long-range actions to insure (somewhat) our survival as a good. (Many conservation groups accredit the survival of other species in the same way.) I suggest that in the past few generations, we humans have at least tacitly accepted our survival as a species as a good and even laid claim to it as a right. If we have tacitly accepted our species survival as a good and indirectly claimed it as a right, it seems appropriate to articulate the right more strongly, attempt to gain explicit acknowledgement of it, and examine the obligations that this right entails. The area of species survival and thus of species obligation with which I am particularly concerned is that of reproductive practices.

This matter is complicated by the tendency in the history of western civilization to attain individual liberty in an acceptable social context. If this social and political tendency is true, then the advent of our understanding of ourselves as a species with concomitant obligations must certainly temper our pursuit of individual liberties. However, this species awareness does not have to temper all of our actions and claimed rights. Many or our actions could be unaffected by species awareness. But reproductive activity is directly affected because, in the evolutionary context, the pivotal point upon which turns the survival of a species is its reproductive pattern. We can mention the production of more offspring than is needed for species survival, an inverse proportion between the number of offspring and the availability of parental care and the length of time of parental care. Two features that are of particular concern in this paper are the quick demise of some offspring that are genetically defective and, if the demise is not so quick, the demise of some of the genetically defective before they have the ability themselves to reproduce. The early death (or death before reproductive age) of those with serious genetic deficiencies occurs frequently enough in the natural selective process to control the passing on of deleterious genes and the burden which the genetically defective imposes on the population as a whole. With regard to the human population, modern medicine has modified these natural selective factors considerably, primarily in the area of maintenance. The result being that many of those with serious genetic defects are kept alive, even into reproductive age, with the accompanying strain on the population as a whole and the passing on of deleterious genes. We can now turn directly to this matter of genetic disease.

III. Disease

Disease, like health, is an extremely complex matter to analyze conceptually. Philosophers of medicine are only now turning to it. I shall make only a few preliminary remarks before going into the more specific matter of genetic disease. It is important at the outset to distinguish between disease and a trait. Baldness, for example, is a trait, not a disease. In the following section I am discussing genetic diseases, not genetic traits.

To discuss disease as a deviation from the "normal" is a particularly unsatisfying, but common, approach. For the norm, when it comes to many bodily functions, will vary from individual to individual. Some bodies may have a high tolerance for a particular substance while others may have a relatively low tolerance. The variances in diabetes would be a good example. Currently physicians acknowledge this phenomenon by dealing in ranges that constitute the normal. This tolerance of variation within a range retains some notion of normalcy and so we are still left to decide if the "normal" is a statistical average, the expectation of a group of experts, a medical stipulation of the optimal, or whatever. This variability is further complicated by the fact that within a given organism, changes occur which may seem to be disease but as possibly temporary or defensive in character, are ambiguous in their diagnosis. Hypertension is a case in point. It could be that an individual is undergoing a specific stressful situation of fairly long duration and his blood pressure rises to protect the organism from other forms of harm or to provide the resources to carry out its operation with optimal efficiency. Some quite compelling examples might be employed here in the area of mental disease.

At the very minimum I suggest that we call an organism "diseased" if the relative independence of its existence is jeopardized and/or its well-being needs supervised health care, e.g., chemical or mechanical support (when the latter is a part of treatment) for its restoration or maintenance. The jeopardy results from some metabolic (in the case of physical disease) malfunction whether caused by an agent external to the organism such as a virus or by a breakdown in the chemistry of the organism. Some mental disease might result in a similar malfunction but of an emotional sort. There are other cases where the relative independence of an organism is jeopardized, such as a broken limb, and in such cases the organism is said to be "ailing."

We have observed that living organisms generally enjoy a good measure of independence. They ingest food on their own, they eliminate waste products on their own, they ward off minor infections on their own, etc. This independence is designated as relative because we do rely on our natural and social environments in many ways which would not lead to designating the organism as "diseased," e.g., the need for warmth, shelter, companionship, etc.

Supervised health care is generally executed by the use of chemical means—drugs—or mechanical means—repirators, transfusion equipment, etc. Some supervised health care may be dispensed by a regimen of rest and/or exercise or some monitoring system to watch a given condition. Mechanical support may be of two kinds. (1) When the support is needed for actual treatment such as a dialysis machine we can readily identify the organism as diseased. (2) When it is needed after treatment, e.g., in the case of amputees and victims of paralysis, such support is not an indication of disease. One need not actually be treated to call the organism "diseased." Some conditions may admit of no treatment or the diseased individual may not pursue treatment.

Perhaps, in the end, there is no general characterization for disease and each kind, e.g., mental, genetic, etc. will have some characterization peculiar to it. With the acknowledgement of this open question we can now turn to the notion of genetic disease.

IV. Genetic Disease

Every disease has a genetic component whether we are referring to gross genetic defects of early manifestation, a later breakdown in the DNA coding or the quality of the response of organism to external invaders. Thus, a further specification of genetic disease must be made for present purposes. At the outset I wish to indicate that I do not take a very rigid stance against reproduction on the part of those who suffer from genetic disease in general or of those who are carriers of any deleterious genes. (It is estimated that each of us carries approximately five deleterious, but recessive, genes in our organism that will never be expressed.) My focus of concern is genetic defects which, from birth (or within a relatively short time thereafter), manifest themselves as disease conditions and can only be treated, not corrected, as great economic expense. Therefore, I am not including such genetic defects as some carcinomas, hypertension, myopia, etc. These diseases manifest themselves late in individual development and a decent quality of life can be enjoyed before the disease is expressed. Nor does it include diseases which are easily treatable even though in an untreated condition they are quite lethal, e.g., diabetes and asthma.

Genetic diseases that fall within the scope of my position are conditions such as gross chromosomal translocations, some of the trisomies, Tay Sachs, hemophilia, and possibly cystic fibrosis. These diseases are clearly grossly debilitating and cause severe economic strain on the part of those associated with the afflicted. The severe debilitation is apparent almost from birth. The individual affected can only be treated and maintained but not cured and his dependence is virtually absolute.

I do not concern myself here with the emotional strain caused by such defects. What is a matter of species concern and is so frequently obscured is the economic strain involved in this type of disease. We are becoming more and more aware that we are living with very limited resources. The limitation of our resources even overrides cases where parents could affort to pay high medical costs. Beyond the costs of drugs, blood, or sophisticated machinery is the cost of medical care personnel who are devoting much of their time and efforts to cases that could have been avoided instead of devoting their time to many other medical care problems that may command a higher priority in terms of species survival or the quality of life of those affected. The time of medical care personnel can be diverted to those problems which admit of cure as well as treatment. This involves a priority of goods and it is here that we meet the issue of species obligation for the first time.

There are several ways we could deal with the problem of genetic disease in the area to which I have confined myself.

1. The classical libertarian would leave the choice to the parents, even in high risk situations, and medical resources would be available to those who can pay the price.

2. An amniocentesis could be performed in each pregnancy where the parents are at risk and parents could have the option of aborting identifiable abnormalities or laws could be passed requiring that identifiable abnormalities be aborted. The former would be an outgrowth of the libertarian position, the latter would be part of a social control package. There are some risks involved in amniocentesis and some errors in extraction of samples can occur. In the case of more than one fetus in the uterus, one of which is diseased and the other not, the appropriate action is unclear. This alternative, of course, involves a positive resolution to the abortion argument.

3. The classical libertarian might suggest genetic counseling so that parents could make an informed decision concerning their reproductive practices. I suggest that genetic counseling with the libertarian attitude is most appropriate in cases (a) where the risk of disease is not exceptionally high, (b) where the disease is easily treatable but nevertheless incurable, (c) where the disease does not manifest itself until well into the growth process. Parents in such cases can be given all the genetic information, the options can be laid open to them and support and aid given in the decision-making process.

4. Reproductive controls along much the same line as our current legislation regarding marriages of close kinship and venereal disease screening are the only approach that I find satisfactory in dealing with genetic disease of the high-risk, early-appearance type within the context of species obligation. Research should

be directed to clearly and economically identifying carriers of severely debilitating diseases of the kind to which I have narrowed my remarks. This is no simple task because a variety of tests may be necessary, chemical analysis, karyotyping, pedigree analysis, etc. Premarital screening should take place to compare disease-producing genetic conditions. For those couples identified at high risk, reproductive rights should be suspended. By "high risk" I mean a chance of one in four that the offspring would be affected and a chance of two in four that the offspring would be a carrier. In other words, if parents can produce a genetically healthy noncarrier for the clearly identifiable trait in only one chance in four, then the risk is too great. I suggest that there is both legal precedent and a species obligation to suspend reproductive rights under these conditions in line with my argument in the first two parts of my paper.

Finally, research should be redoubled on safe and effective means of contraception so that no danger accrues to those involved in this situation. Techniques should also be developed for reversible sterilization so that in the case of marital breakdowns and remarriage where the risk is no longer present, reproductive practices can then be resumed.

To summarize my argument briefly: My fundamental commitment is that the survival of the human species is a good and that it is a good of such importance and value that it can be accredited as a right. From this I deduce that individuals and social units have the concomitant obligation to pursue courses of action that will foster and protect the right of species survival. Among these acknowledged and traditional courses of action is general health care. One segment of that health care involves the protection of the population from the transmission of identifiable, seriously deleterious genes and from debilitating and costly (in terms of natural, economic, and human resources) genetic disease which can neither be cured nor treated with any preservation of the quality of life and relative independence of the afflicted. Because individual human rights are negotiable according to their historical context and because there is legal precedent for restriciting the exercise of reproductive rights, those who are at high risk for passing on clearly identifiable and severely deleterious genes and debilitating genetic disease should not be allowed to exercise their reproductive prerogative.

Bibliography

Brown, Stuart. "Inalienable Rights," *Philosophical Review,* Vol. 64, No. 2 (April, 1955), 192–211.

Callahan, Daniel and Murray, Robert. "Genetic Disease and Human Health," *The Hastings Center Report,* Vol. 4, No. 4 (September, 1974), 4–7.

Feinberg, Joel. *Social Philosophy.* Englewood Cliffs: Prentice-Hall, Inc. 1973.

Frankena, William. "Natural and Inalienable Rights," *Philosophical Review,* Vol. 64, No. 2 (April, 1955), 212–232.

Harris, Maureen. (ed.) *Early Diagnosis of Human Genetic Defects.* Fogarty Symposium, 1970. International Center for Advanced Study in the Health Sciences. National Institutes of Health, Bethesda, Maryland. Washington, D.C.: Government Printing Office, 1971.

Hart, H.L.A. "Are There Any Natural Rights?" *Philosophical Review,* Vol. 64, No. 2 (April, 1955), 175–191.

Hart, H.L.A. *The Concept of Law.* Oxford: The Clarendon Press, 1961.

Hilton, Bruce, *et al.* (ed.) *Ethical Issues in Human Genetics: Genetic Counseling and the Use of Genetic Knowledge.* Fogarty Symposium, 1971. International Center for Advanced Study in the Health Sciences and the Institute of Society, Ethics and the Life Sciences. New York: Plenum Press, 1973.

Lappe, Marc. "Allegiances of Human Geneticists," *The Hastings Center Studies,* Vol. 1, No. 2, (1973), 63–78.

Lappe, Marc. "Moral Obligations and the Fallacies of Genetic Control," *Theological Studies,* Vol. 33 (September, 1972), 411–427.

Melden, A. I. *Human Rights.* Belmont, California: Wadsworth Publishing Co., 1973.

Rawls, John. *A Theory of Justice.* Cambridge, Mass.: Harvard University Press, 1971.

Rensch, Bernhard. *Biophilosophy.* Translated by C.A.M. Sym. New York: Columbia University Press, 1971.

Watson, J.D. *Molecular Biology of the Gene.* 2nd edition. Menlo Park, California: W.A. Benjamin, Inc., 1970.

ON JUSTIFICATIONS FOR COERCIVE GENETIC CONTROL

THOMAS L. BEAUCHAMP

Good laws function to safeguard individual and societal rights and liberties. But the law has two sides. By ensuring liberty to one set of persons, the law may restrict the liberty of others. A law by its very function is coercive because it places a limit on what was formerly a free exercise or action. It is often said that we trade some liberties either for the insurance of other liberties or for some form of protection by the state. The acceptability of these liberty-limiting laws ultimately depends upon the adequacy of the justification offered for them; and when an adequate justification is not forthcoming, the law can easily become an instrument of oppression. In recent years a number of proposals by doctors, moral philosophers, and legislators have been offered which focus on what we ought to do legislatively in the near future to control various health and population problems by genetic engineering. In this paper I test the adequacy of some of the more sweeping of these endorsements of coercive genetic intervention by considering the justifications offered for them.

Two independent liberty-limiting principles are most frequently employed by the supporters of genetic intervention: the harm principle and the paternalistic principle. The harm principle says that coercive interference with a person's liberty is justified if through his actions he produces harm to *other* persons or perhaps to public institutions. It is sometimes argued, for example, that failure to enact this or that genetic program will produce serious and needless suffering to proximate future generations or will lead to avoidable and highly expensive costs to the state in the form of confinement and treatment. The moral force of such claims is generated by the harm principle. The paternalistic principle is slightly but significantly different. It says that coercive interference with a person's liberty is justified if through his actions he produces harm to *himself*. It is sometimes maintained, for example, that persons ill

This article is written especially for this volume. Two predecessors, on which the argument is partially based, are "Paternalism and Bio-behavioral Control," *The Monist*, Vol. 60 (January, 1976) and "On Justifying Genetic Intervention," an address delivered at the Symposium on Rights, Ethics, and Medicine at the University of Dayton, 1974.

equipped to care for themselves or potential offspring should be sterilized by the state in order to protect them from their own sexual activity. The moral force of such claims is generated by the paternalistic principle. While these two principles almost always are implicit rather than explicit in the arguments of the friends of genetic intervention, I cannot see that there are any other more basic moral principles underlying their arguments.

In this paper I argue (1) that paternalistic grounds should not be considered to provide good grounds for coercive genetic intervention, because the paternalistic principle itself is not an acceptable justifying principle, and (2) that while the harm principle is an acceptable justifying principle, those who invoke it for purposes of genetic intervention have not provided an acceptable moral justification, because they have not shown that the potential of harm to others and/or to public institutions is sufficient to warrant the loss of liberties which would accompany the adoption of coercive genetic laws. I begin with a discussion of the modern developments in biology which have created moral problems of genetic control. In the next two sections I explain further the nature of the harm and paternalistic principles, and provide examples of some of the actual proposals for genetic intervention which are at least mixed paternalistic and harm principle justifications. In the final section I argue that no such justification thus far offered ought to be accepted as satisfactory, despite the exciting and hopeful promise extended by some enthusiasts. I do not, however, argue the bolder thesis that eugenic laws which coerce persons cannot by any principles and under any conditions be justified.

Ethics and the New Biology

What are the actual and potential ethical problems which motivate us to be concerned about the possibilities for engineering people by biological means? Some problems have long been obvious: By lowering infant mortality rates and devising means of increasing fecundity, while also extending life spans, we increase population; something must be done to control the increase. Which among the possible means are we to select and/or prohibit? And, far more importantly for the purpose of this paper, should we under any circumstances allow the state to establish requirements such as mandatory sterilization for retarded persons or for those carriers of deleterious genes which might be passed on to future generations?

While these more familiar ethical problems are certainly important, they are only the beginning. Once the so-called genetic code was broken, and we began to learn the genetic processes which control human development, an area rife with possibilities for human engineering emerged. Recently, inheritable alterations in human cells (developing in tissue culture) have been produced by means such as the initiation of viral infection. This produces the possibility that both "favorable" and "unfavorable" alterations may be required by the state, or at least introduced on a massive scale. Further, medical diagnostic procedures now allow not only prenatal determination of the sex of infants, but also limited detection of crucial genetic deficiencies. This capacity will in all likelihood enormously increase. When perfected, genetic engineering will also enable us to alter and create new genetic capacities which will be transmitted to subsequent generations. We will be thus enabled to control both the sex and the genetic character of the population.

Although these projected biological possibilities stimulate great scientific interest, they are in themselves of little ethical interest. It is on the engineering side of "genetic engineering" that the specter of control emerges. Direct alteration of genes themselves is the least practicable but in some ways most exciting possibility, for it might result in the direct alteration of genotypes and ultimately in DNA artistry where new types of biological organisms are concocted. Various forms of genetic surgery on actual living persons may soon be available, and even deliberately initiated mutations may be planned. With adequate knowledge it will be tempting to require doctors to treat either infants or fetuses when diseases such as cystic fibrosis, Tay-Sachs, phenylketonuria (PKU), or sickle cell anemia have been detected. Other programs might be directed not so much at the actual living person but rather at his or her future progeny. One could in theory have women superovulated and their eggs frozen at a prime age, in order to minimize the threat of mongolism and other defects; and one could even alter the genetic structure of human ova prior to its being united with sperm. Theoretically less remote and perhaps more likely of actualization are eugenic programs where the genes themselves are not altered but where breeding ideals established criteria for marriage and reproduction. Here persons could for genetic reasons be prevented from marrying or from having offspring, perhaps by sterilization or forced abortion. Some eugenicists believe the latter program should be effected immediately. Indeed, as we shall see, some on the vanguard of the aforementioned biological contingencies contend that we can and should "improve" as a whole the human stock we breed.

In the mass of literature on genetic control an important distinction between "negative eugenics" and "positive eugenics" has been introduced. Eugenics is the technological science which promotes the improvement of hereditary qualities by preventing the transmission of genetically inferior conditions. The distinction centers around the nature of the "improvement" to be made. Negative eugenics supposedly has only the medical objective of eliminating or otherwise treating inherited genetic diseases. Since genetic diseases are both prevalent and flourishing, and since they are universally regarded as worthy of elimination, negative eugenics is generally praised as an advance in biomedical technology, though it certainly does raise a number of ethical issues. Positive eugenics raises still other issues, since its purported objective is the positive betterment of the human hereditary condition. Hermann Muller is now well known for the following sketch of the objectives he thinks could be and ought to be eugenically promoted: "a genuine warmth of fellow feeling and a cooperative disposition, a depth and breadth of intellectual capacity, moral courage and integrity, an appreciation of nature and of art, and an aptness of expression and of communication."[1] However well intentioned, comments such as these have quite naturally raised fears of gross abuse by the state.

Harm and Paternalism as Justifying Principles

On what grounds might the state be justified in adopting a coercive program of either positive or negative eugenics? As previously mentioned, most proponents finally rest their case on an appeal either to the harm principle or to the paternalistic principle. Actual examples of such arguments will be surveyed momentarily, but first it will be useful to reach a somewhat more precise understanding of the nature of these two justifying principles.

The Harm Principle

The harm principle, as I construe it,[2] says that when specific kinds of harm are caused to a person or a group of persons the state is justified in coercively intervening for the purpose of protection. In some cases the liberty of those causing the harm may be limited and in others the liberty of those harmed may be limited (at least temporarily). Harm might be produced in many ways, but certainly the fact that the harm was produced by negligence or other inadvertent means would not render coercive interference unjustified. There are also cases where an individual or group of individuals has been or will be physically or mentally harmed not directly by another party but rather by some cause or condition which is to the first party not known or not within its control or both. An earthquake or an epileptic seizure would be examples of the latter. In such cases the state is normally thought to be justified in interfering with the liberty of the persons threatened, at least in order to inform them of an impending

[1] "Should We Weaken or Strengthen Our Genetic Heritage?," *Daedalus*, Vol. 90 (Summer, 1961), p. 445.

[2] My construal of both principles is more elaborately explicated and defended in "Paternalism and Bio-behavioral Control," *The Monist*, Vol. 60 (January, 1976).

danger. Some think the state is justified in treating or protecting threatened persons whether or not they consent, though I would not agree to this construal of the harm principle, as the principle then incorporates elements of paternalism and no longer remains an independent liberty-limiting principle.

The role one grants to consent is a critical factor in one's interpretation of the harm principle. One may consent to actions and still be harmed (as a boxer who consents to a match and winds up in the hospital is harmed), though we might want to say that he is not *wrongfully* harmed because he consented. In my view, the state's only proper role is the prevention either of wrongful harm to persons or of conditions productive of harm which are unknown or uncontrollable by the affected persons. The state should not be in the business of preventing harm when that harm or its possibility have been consented to in an informed manner. I construe the harm principle, then, to allow the state to intervene only in the former cases and not in cases where informed consent is present.

The Paternalistic Principle

In recent years we have come to take with increasing seriousness laws which coerce human persons against their will on grounds that the coercive actions protect those persons against consequences of their own conduct. Such justifying grounds are said to be paternalistic, and statutes allowing or requiring such coercion are said to be paternalistic laws. The best known example of such practices is the nonvoluntary institutionalization of persons capable of free choice but allegedly ill mentally. However, as H. L. A. Hart has pointed out, such laws are found throughout our civil and criminal statutes.

Gerald Dworkin has provided a reasonable definition of paternalism, which I accept with only two innocuous modifications (in barackets): Paternalism is "the [coercive] interference with a person's liberty of action justified by [protective or beneficent] reasons referring exclusively to the welfare, good, happiness, needs, interests or values of the person being coerced."[3] Any supporter of the paternalistic principle will specify with care precisely which goods, needs, interests, etc. are acceptable. In most recent formulations, it has been said that the state is justified in coercively interfering with a person's liberty if by its interference it protects the person against his own actions where those actions are extremely and unreasonably risky (waterfall-rafting, e.g.), or are genuinely not in the person's own best interest when his best interest is knowable by the state (as some believe in the case of suicide), or are potentially dangerous and

[3] Gerald Dworkin, "Paternalism," *The Monist*, Vol. 56 (January, 1972), p. 65.

irreversible in effect (as some drugs are). Hart's characterization, though loose, is representative:

> Paternalism—the protection of people against themselves—is a perfectly coherent policy. . . . No doubt if we no longer sympathise with [Mill's] criticism this is due, in part, to a general decline in the belief that individuals know their own interests best, and to an increased awareness of a great range of factors which diminish the significance to be attached to an apparently free choice or to consent. . . . Harming others is something we may still seek to prevent by use of the criminal law, even when the victims consent to or assist in the acts which are harmful to them.[4]

Various qualifications are added by Hart and others—e.g., that the general presumption against coercion must be outweighed by the significance and real possibility of the danger involved. However, for our purposes the details of these qualifications are largely irrelevant.

Although controversial, I restrict use of the term *paternalism* to cases where the state coercively protects or benefits a person when his contrary choices are informed and voluntary. Intervention in cases of nonvoluntary or of uninformed conduct is not paternalism in any interesting sense, because it is not based on a liberty-limiting principle independent of the harm principle. It is important to be clear about this distinction since some interventions are both coercive and justified on what might deceptively appear to be paternalistic grounds. John Stuart Mill believed that a person ignorant of a potential danger which might befall him could justifiably be restrained, so long as the coercion was temporary and only for the purpose of rendering the person informed, in which case he would be free to choose whatever course he wished. Mill regarded this—correctly, I think—as temporary but justified coercion which is not "real infringement" of liberty: "If either a public officer or anyone else saw a person attempting to cross a bridge which had been ascertained to be unsafe, and there were no time to warn him of his danger, they might seize him and turn him back, without any real infringement of his liberty; for liberty consists in doing what one desires, and he does not desire to fall into the river."[5]

It is not a question of protecting a man *against himself* or of interfering with his liberty of action. He is not acting at all in regard to this danger. He needs protection from something which is precisely not himself, not his intended action, not in any remote sense of his own making. While I am here embellishing Mill, this seems to me clearly the direction of his argument. Mill goes on to say that once the man has been fully informed and understands the dangers of the bridge, then he should be free to traverse it, if he wishes. I shall be arguing in support of Mill's conclusion, though not by appeal to his utilitarian grounds.

[4] H. L. A. Hart, *Law, Liberty, and Morality* (Stanford: Stanford University Press, 1963), pp. 31–33.

[5] Mill, *On Liberty* (Indianapolis: Liberal Arts Press, 1956), p. 117.

Contemporary Justifications for Eugenic Methods

At various points in American history, eugenics has been put to some rather disconcerting purposes, only recently documented by Kenneth Ludmerer.[6] Sterilization laws and other forms of reproductive control have been the most prominent means to eugenic ends, but to contemporary eugenicists these pioneering efforts recede in significance when compared with the broader cultural controls on selective breeding which they propose. I shall discuss, as representative samples, three recent eugenic proposals. An instance of a justification for an immediately effective eugenic sterilization law will first be cited, and then two more speculatively bold programs will be discussed.

Eugenic Sterilization. Eugenic sterilization laws are still in effect in over half the states in the United States. The retarded have been a special target, since they along with criminals, epileptics, alcoholics, and other vulnerable groups have been alleged to have genetically rooted mental and physical disabilities. Since the retarded are often childlike and relatively inattentive to their responsibilities, the rearing of children is frequently a heavy burden. For such reasons it has been considered in their own best interest that they be sterilized, even if they do not agree or fail to comprehend the decision. Irvin B. Hill, writing about the sterilization of mentally deficient persons in prison, argues as follows:

> A mentally deficient person is not a suitable parent for either a normal or a subnormal child, and children would be an added burden to an already handicapped individual, who does well to support himself. It would be *unfair to the state, to the individual, and particularly to his potential children,* to permit his release without the protection of sterilization.... It has been the policy of the State of Oregon to sterilize mentally deficient persons before releasing them from its institution and ... this program has been of benefit from *economic, social, and eugenic* standpoints.... It *assists the individual* in his transition to a non-institutional life; and it relieves the state of the financial burden....[7]

This justification appeals to both the paternalistic and the harm principles. This mixture is common to most such justifications. Pure paternalistic justifications are especially rare, but once canonized into law, they can easily become purely paternalistic in coercive environments. Free and informed consent is unlikely in the context of penal institutions, especially when one is dealing with mentally deficient persons. They can be bribed with offers of freedom and intimidated by threats that their confinement will be extended. Although we now know both that many retarded persons are born from parents of normal intelligence and that the retarded often have children of normal intelligence, prison and other

[6] *Genetics and American Society: A Historical Appraisal* (Baltimore: The Johns Hopkins University Press, 1972).

[7] Irvin B. Hill, "Sterilizations in Oregon," *American Journal of Mental Deficiency,* Vol. 54 (1950), p. 403. Italics added.

custodial environments continue to give rise to the sort of paternalistically motivated interventions suggested by Hill.

Negative Eugenics

In a recent book-length study, *The Ethics of Genetic Control,* Joseph Fletcher trumpets the virtues of negative eugenics and throws a few kisses in the direction of positive eugenics. His arguments are tenaciously and somewhat dogmatically based on his version of the harm principle, with implicit support obliquely squeezed from the paternalistic principle. Fletcher is especially concerned about carriers of "genetic faults" who transmit genetic deficiencies and thereby "victimize" their children, their marital partners, and the general public. The following is a typical Fletcher argument:

> Each of us has a genetic "load" of three to eight defects. This could double in two hundred years if we go on spreading genetic disorders through random sexual reproduction, multiplying the illnesses and costs that result from bad genes. . . . Terrible and uncorrectable fetuses will have to be aborted or, after birth, let go; for those that *are* preserved and are able to live to reproductive maturity sterilization can prevent the spread of the bad genes and obviate the dysgenic side effect.
>
> What, then, of the "right to reproduce"? . . . Humanistic or personistic moralists will say, "A right depends on human well-being, and if the parents are both carriers of a recessive gene causing lifelong pain and misery for the child they would have, then they should not conceive–the right is null and void." The right to be parents ceases to run at the point of victimizing the offspring or society.[8]

Fletcher often seems most concerned about what present laws do not *permit.* He complains, for example, that laws force obstetricians and pediatricians to "stretch the truth" in the case of defective fetuses by "telling a woman her baby was 'stillborn' when it wasn't, after having simply not respirated the delivered fetus–out of mercy for her and her family."[9] The latter argument appears to be a harm principle justification for more permissive laws. The argument is hard to understand, however, since Fletcher approves the doctors' paternalistic treatment of the patient and the family, and presumably would continue to approve such merciful paternalism if the law were changed.

At other points Fletcher seems more concerned about what eugenic laws should *require* rather than simply about what they should permit. He advocates, for example, "preconceptive and uterine control" in the production of children in order to avoid congenital infirmities. He refers to such measures as "compassionate control" and contends that it is immoral to avoid the use of controls

[8] *The Ethics of Genetic Control* (Garden City: Doubleday, 1974), pp. 29f, 125f.
[9] *Ibid.,* p. 153.

when the medical knowledge is available.[10] He explicitly endorses control by the state:

> If people could be relied upon to be compassionate we would have no reason to even consider mandatory controls. But there are too many who do not control their lives out of moral concern; . . . the common welfare often has to be safeguarded by compulsory control or what Garrett Hardin calls "mutual coercion mutually agreed upon." . . . Ideally it is better to do the moral thing freely, but sometimes it is more compassionate to force it to be done than to sacrifice the well-being of the many to the "rights" of the few. This obviously is the ethics of a sane society. Compulsory controls on reproduction would not, of course, fit present interpretations of due process in the fifth and fourteenth amendments to the Constitution. Here, as in so many other ways, the law lags behind the ethics of modern medicine and public health knowledge.[11]

Positive Eugenics

Perhaps the best known advocate of positive eugenics in recent years has been Hermann Muller. He is, like Fletcher, mostly worried about "an increasing accumulation of deterimental mutations that occur at random" and which "adversely affect health, intellect, powers of appreciation and expression, and even the genetic bases of our cooperative disposition itself."[12] How the state rightly ought to promote eugenic reform is never extensively explored by Muller; generally he is content to allow enlightened individual decisions and social pressures to determine the course of genetic improvement. Still, he thinks the state should record the genetic "pedigrees" of its citizens, favors extensive use of AID and sperm banks for cross-family usage, and suggests that genetic selection is not to be feared if "conducted in a democratic way." If the majority seeks to "promote the diverse abilities and proclivities of specific types" he finds this control both natural and permissible.[13] Muller is also interested in more than what is merely permissible. As a eugenicist who believes that a clear and present danger exists, he thinks we cannot avoid making some hard decisions which will have to be enforced. It follows that if a democratic majority decides, for paternalistic or for harm principle reasons, to improve the genetic health of its deficient members, then this can only be regarded as a means for improving their lives (Fletcher and Hardin's "mutual coercion"?). It is to their advantage and to ours that some controls be initiated.

To be sure I have been putting some words in Muller's mouth, and it should be explicitly acknowledged that he rejects governmental controls "at this stage

[10] *Ibid.*, p. 158f.
[11] *Ibid.*, p. 180.
[12] Muller, *op. cit.*, p. 433.
[13] *Ibid.*, p. 447.

of world affairs."[14] He insists on voluntary action even in programs of negative eugenics. Nonetheless, I think any realistic appraisal of his proposals inescapably leads to the conclusion—as both Martin Golding and Bernard Davis have argued—that such eugenic controls could be achieved only by some rigorous form of state regulation.[15]

A Critical Estimate

In conclusion I shall briefly outline some arguments against the legislative enactment of the aforementioned genetic policies, whether based on the harm principle or on the paternalistic principle.

Use of the Harm Principle

It is universally acknowledged that the harm principle justifiably permits coercive state interventions. But proper application of the harm principle is always contingent upon evidence both that the quanity and quality of harm forecast will be produced unless the state intervenes and that the balance of harm will be minimized by the state intervention. Any assessment of the eugenic point of view, then, must consider the extensiveness and character of the genetic threat. It is in precisely these respects that eugenicists' arguments are lacking.

As regards positive eugenics, any proposal must *select* a set of traits which are allegedly of great positive value to the human species and therefore to be promoted. However, there are problems in regard to the wisdom of the choice of traits, in regard to the original source of the traits, and in regard to the alleged advantageousness of the traits. First, how are we to determine which traits ought to be promoted? Paul Ramsey has complained that when positive eugenicists "describe those human qualities to be selected and bred into the race of men, they write remarkably as if they were describing the attributes of mind and of character that make a good geneticist, or at least a good community of scientists."[16] While no doubt intentionally exaggerated, Ramsey's comment presses the important point that the choice of traits is a value-laden one. It is by no means obvious which traits are the most valuable to mankind, "bad" traits

[14] Muller, "What Genetic Course Will Man Steer?", in James F. Crow & James V. Neel, eds., *Proceedings of the Third International Congress of Human Genetics* (Baltimore, 1967), p. 536.

[15] Martin Golding, "Ethical Issues in Biological Engineering," 15 *UCLA Law Review* (1968), pp. 470f, reprinted in T. Beauchamp, ed., *Ethics and Public Policy* (Englewood Cliffs: Prentice-Hall, 1975). Bernard Davis, "Threat and Promise in Genetic Engineering," in *Ethical Issues in Biology and Medicine* (Cambridge, Mass.: Schenkman Publishing Co., 1973), p. 26.

being infinitely easier to ascertainment than "good" ones. And even if one were to select traits which virtually everyone admires (e.g., intellect and beauty), it would not follow that society would be improved if these traits were widely enhanced. Indeed, social problems would be enhanced if there were no employment for persons with similar talents or if we inadvertently enhanced the wrong traits.

There are many incompatible but strongly felt visions of man's future, and even those who are given to visions of social control often alter those visions in the light of new facts and theories.[17] Eugenicists generally tend to assume that we all agree or will come to agree on these traits, rather than present the required argument to this effect. I very much doubt that this argument is forthcoming, as their views seem to rest on false empirical assumptions about the extent of evaluative agreement and also appear to rest on ideological and individual-preferential bases.

Perhaps the most serious problem with positive eugenics is the assumption that the traits to be promoted in fact have a genetic rather than an environmental basis. The majority of the traits they mention involve higher cognitive and emotional capacities which are nurtured by education and which seem more subject to psychological control than genetic control.[18] This is not the place to mediate the age-old nativist–environmentalist debate, but eugenicists may fairly be accused, I think, of a simple fallacy of reasoning. They move from the perhaps acceptable premise that every human trait has *some* genetically controlled basis to the unacceptable conclusion that every trait has an *entirely* genetically controllable basis.[19] This claim not only downgrades causal consideration of environmental influence, but begs the critical question of whether the actual features of the desirable traits they promote are describable in purely genetic terms. It is not implausible to suppose that the most desirable features of their desirable traits are entirely or largely environmentally controlled, even if the traits are in some respects subject to genetic intervention. But, in any case, eugenicists have never, to my knowledge, made any serious experimental attempt to prove that the desired trait, under their description of it, can be engineered genetically.

[16] Paul Ramsey, *Fabricated Man* New Haven: Yale University Press, 1970), p. 22.

[17] Muller himself once had Marx, Lenin, and Sun Yat Sen on his list of highly valuable sperm donors, whose genetic capacities might be saved for purposes of artificial insemination. These names are absent from similar later lists which he published, while he added Einstein and Lincoln. It is hard to account for this change except (at least partially) in terms of a political evaluation of desirable traits. (I owe this comparison of lists to Martin Golding and Kenneth D. Eberhard.)

[18] Some persuasive examples of social and psychological influence are given in a commentary on Muller's *Daedalus* paper by J. P. Scott, esp. pp. 456f. Cf. also the comments by Ernst Mayr, p. 461. Both in the same volume.

[19] Muller's equivocal use of "have genetic bases" is interesting in *Daedalus, op. cit.*, p. 438.

Furthermore, most if not all of their allegedly desirable traits are subject to different and conflicting descriptions. Consider "intellectual capacity," "moral courage," and "warmth of fellow feeling"–three typical traits endorsed for enhancement by Muller. Even if such terms were not relative to cultural and individual preference, as I believe they are, they are so systematically vague that large numbers of philosophers skilled in conceptual analysis could spend careers on each without making significant headway. There is in addition an equally pressing need for an analysis of the notion of a "defective gene." L. K. Frank elliptically raises this question with an interesting example he adduces while commenting on one of Muller's papers. He asks whether the human female's loss of "the gene for heat" ought to be classified as a "genetic deterioration, or an emancipation from a strict biological control of sex."[20] Many parallel and probably unresolvable conceptual and normative problems would arise.

As regards negative eugenics, the major problem with all such proposals has been well stated by Martin Golding:

> It is not clear that a social program of negative eugenics wouldn't do more harm than good. This, of course, might depend upon the kind of defects that it would aim to eliminate. Nevertheless it is plain that "good" genes would be reduced in the process. A conflict could arise between the aims of positive and negative eugenics. Sweden has had a ban on the marriage of endogenous epileptics since 1757. This has been attacked by a Swedish geneticist not only on the grounds that the chances for an epileptic to have epileptic children are not very high, but also on the grounds "that many epileptics are highly intelligent and socially valuable members of their community, well fitted to bring up children, and that they may carry valuable genes whose transmission will be prevented by the existing law."[21]

In addition to this problem, for largely theoretical reasons, it is not clear that any program of negative eugenics, given current knowledge, is likely to be effective. The genes to be controlled often result from mutation rather than inheritance. Hence, one would need scientific knowledge of the causal laws governing mutation, as well as of the inheritance process, and science is not at this time capable of providing such knowledge. Moreover, many of the genes one would most want to monitor appear only infrequently and unpredictably. And even if this latter problem were eradicated, substantial problems of choice–paralleling the aforementioned problems in the choice of traits–would remain. According to modern genetic theory, each human person carries several recessive genetic diseases, and hence one would have to provide an order of priorities specifying which are to be combatted first. It is likely neither that this could be a value-free ordering nor that it could be linked in an unbiased way to the seriousness of the genetic threat posed by some but not by other diseases.

[20] *Daedalus, op. cit.*, p. 459.
[21] Golding, *op. cit.*, p. 475.

Use of the Paternalistic Principle

Finally, why ought paternalism to be judged an unacceptable justifying principle? The dominant reason is that paternalistic principles are intrinsically too broad and hence serve to justify too much. Robert Harris has correctly pointed out that H. L. A. Hart's description of paternalism, as above, would in principle "justify the imposition of a Spartan-like regimen requiring rigorous physical exercise and abstention from smoking, drinking, and hazardous pastimes."[22] Even the most thoughtful restrictions on paternalism known to me still leave unacceptable latitude, especially in contexts where behavioral controls are most likely to be abused. Prison environments and therapeutic agencies notoriously have thrived on the use of paternalistic justifications to confine and treat. Paternalism potentially gives prison wardens, psychosurgeons, and state officials a good reason for coercively using most any means in order to achieve ends they believe in subjects' or their offsprings' best interest. It is demonstrable that allowing this latitude of judgment is dangerous and acutely uncontrollable. Paternalistic justifications will become increasingly dangerous as genetic knowledge advances.

Most of us would surely be reluctant to allow state officials or even personal friends to judge the reasonableness of another's merely risk-running actions. The difficulty of disentangling judgments of unreasonableness and abnormality from judgments of moral attitude and evaluative outlook is notoriously difficult. We do not allow doctors coercively to give blood transfusions even in life and death situations without consent (except in certain special cases, as in an emergency, when the patient's wishes are unknown). Yet if ever there were, in the eyes of most, an extreme and unreasonable risk, surely this one qualifies.

Paternalism, then, leaves us with unresolved problems concerning the scope of the principle. Suppose, for example, that a man risks his life for the advance of medicine by submitting to an unreasonably risky genetic experiment, an act which most would think not in his own interest. Are we to commend him or coercively restrain him? Paternalism strongly suggests that it would be permissible to coercively restrain such a person. Yet if that is so, then the state is permitted to restrain coercively its morally heroic citizens, not to mention its martyrs, if they act—as such people do—in a manner "harmful" to themselves. I do not see how paternalism can be patched up by adding further conditions about the actions of heroes and martyrs. It would increasingly come to bear the marks of an ad hoc and gratuitious principle which is not genuinely independent of the harm principle, the deficiencies of which, in regard to genetic intervention, we have already surveyed.

[22] "Private Consensual Adult Behavior: The Requirement of Harm to Others in the Enforcement of Morality," 14 *UCLA Law Review* (1967), p. 585n.

LEGAL RIGHTS AND MORAL RIGHTS

ALEXANDER M. CAPRON

The relationship of a genetic counselor and his patients is a delicate, complex and important one. Treating as it does subjects of great moment—the prevention of crippling diseases, even life and death themselves—it commands growing public interest and scrutiny, especially as the counselor's predictive skills increase. It involves not only parents but geneticists, physicians, ministers, and others, in more lengthy and careful contemplation of the conception and birth of a child than occurs in any other type of "planned parenthood." Its highly charged subject matter and deeply involved participants open it to the internal and external pressures which encumber all significant decisions. Yet when we look at the moral and legal rights of the participants in genetic counseling, the picture before us begins to grow less distinct, and we inevitably see only the sharp features, outlined in black and white and not the interesting shadings of gray.

While this difficulty is inherent in any attempt to state general rules about complicated and varied relationships, it will be particularly pronounced in what follows. There are any number of rights which could be asserted on behalf of those who play a role in the relationship—genetic counselor, parents, unborn child, other family members, and even professional and public institutions. But I have chosen, not too arbitrarily I hope you will agree, to focus on the two sets of immediate participants, counselors and parents, and on only the one right which I believe ought to be of central concern to us.

Before we get to that right, perhaps we should pause to ask: What do we mean when we say a person has a right? The term *right* is customarily used when a person has a claim on the way another person or group behaves, especially toward himself. The concept is often defined by its reciprocal: to say that A has a right is to say that B has a duty to do, or abstain from doing, something at A's prompting. Sometimes "right" and "privilege" are used interchangeably, but "right" is by far the stronger term, suggesting that B's duty is not subject to

This essay originally appeared in *Ethical Issues in Human Genetics*, Bruce Hilton *et al.*, eds. (New York: Plenum Press, 1973). Reprinted with permission of the author and Plenum Publishing Corporation.

recall without A's approval. If A has a "legal right," he can compel B to perform his duty by calling on the organized power of the state; if his right is a "moral" one, he has to rely on B's conscience (abetted perhaps by persons or groups possessing "moral authority") to compel B to fulfill his duty.

It is perhaps only a reflection of my lawyerly literalness, but I find the term *moral right* to be something of an anomaly. While we speak easily of a "moral duty," the legal unenforceability of that duty makes strained any reference to a corresponding "right." In this sense of the word, we usually say a person has a "moral right" precisely when we believe that he ought to be able to compel another's behavior (or whatever) but legally cannot. You may, for example, have no legal right to collect on a debt which you foolishly and unnecessarily forgave me, but—perhaps to make you feel better if not richer—we might say you had a "moral right" to collect, particularly if I used the loan to get rich and you are now destitute. A desire to reduce this category of "moral right" (and the perceived injustice with which it may be associated), partly explains the development of courts of equity in the English (and by transplantation, the American) legal system; such courts sat only where the plaintiff was without "legal remedy" (legal in this case being distinguished from equitable).

My premise is that in genetic counseling the parents[1] have a *legal right* to be *fully informed* decision-makers about whether to have a child; or in the terms just discussed, the genetic counselor has the duty to convey to those he advises as clear and comprehensible a picture of the options open to them, and the relative risks, benefits, and foreseeable consequences of each option, as he can. This formulation is basically a legal one, and it suggests that the parents have recourse to the authority of the state, should the counselor breach his duty by negligently or intentionally withholding options or misdescribing their risks, benefits, etc. This right can also be seen from a "moral" vantage point, however. A dominant ethic in Western culture is the importance and inviolability of each individual human being; from this derives a right to make decisions about one's life, to be "master of one's fate."[2]

Despite the broad way in which this right to be an "informed decision-maker" can be stated, it is certainly narrower than many rights which might be (and have been) asserted to arise in genetic counseling. I would defend the primacy of this right for a number of reasons. First, it is as good a reflection of the underlying moral and legal principles (the sanctity of life; the protection of each member of the community) as any other, such as "the right of every child

[1] For the moment, I shall use the phrase "parents' rights" to include both those they assert on their own behalf and those they assert as the representatives of others, particularly their unborn child. This point is discussed *infra* in "Rights and Duties in Genetic Counseling."

[2] Berlin, I. (1969). "Two Concepts of Liberty," in *Four Essays on Liberty*. Oxford: Clarendon Press.

to be born with a sound physical and mental constitution, based on a sound genotype."[3] Second, it avoids, at least on the legal side, the ticklish problems which other such formulations of rights raise concerning "the quality of life." Third, it limits the issues under review to those which are comprehensible given the present state of genetic knowledge.

I anticipate that this limitation on the definition of the right which should concern us will meet with some objection, since our past discussions of genetics have ranged over a host of moral principles, which I would exclude from consideration. At the risk of stirring up the hornets' nest further, then, I will challenge a premise on which those discussions sometimes proceeded: that the moral conclusions reached were necessarily translatable into legal conclusions. My purpose in doing so is to try to demonstrate two things about the legal right set forth previously: (a) that it operates in the genetic counseling situation independently of any "moral" conclusions, and (b) that, for the time being at least, it is not only a necessary, but sufficient, legal right to regulate the participants in genetic counseling.

A Connection Between Law and Morality

I would like to endorse the hypothesis that "there is no necessary connection between law and morality" as the starting point for my discussion of "Moral Rights and Legal Rights." I embark on this course with some trepidation. First, not being a moral philosopher, there is danger in my saying anything about "moral rights," and I might be better advised simply to speak only of legal rights in genetic counseling. Second, by taking on the "no necessary connection" hypothesis, I run the risk of spawning a positivist–natural law debate. Moreover, there is a great deal of evidence that suggests this position is untenable. And finally, there is a part of me which wants to reject this premise and to assert rather that law not only must but should enact a set of morals—my set, of course. Nevertheless, I shall proceed on the premise of "no necessary connection," not merely to be provocative but because I believe it is useful to see the need for building separate, albeit related, lines of argument when we discourse in the moral and legal spheres, and because I hope to demonstrate that courts in disposing of the legal issues in genetic counseling cases need not venture into the "moral thicket," to paraphrase Justice Frankfurter.

[3]Glass, B. (1971), Science: Endless Horizons or Golden Age?, *Science, 171,* 38. See also *Smith v. Brennan,* 31 N.J. 353, 364 (1960), ("the right to begin life with a sound mind and body") quoted in *Glietman v. Cosgrove, 49* N.J. 22, 28 (1967), which is discussed extensively in "Rights and Duties in Genetic Counseling" in this paper.

Let us begin our discussion by exploring the different meanings which attach to the words "connection," "law," and "morality."

Connection

The term *connection* can imply a historical relationship, a chance overlap, or a strict cause-and-effect nexus. In a historical sense, there can be no denying that our system of laws has been closely related to morality, particularly that body of conventional morality known as religion. Even when a link is not provable as a matter of historical fact, the occurence of similar rules in the legal and moral domains suggest the existence of a connection, perhaps one with an anthropological explanation. Yet neither of these meanings, nor even that of cause-and-effect, is adequate once "connection" is modified by "necessary," for we are then referring to a "but for" relationship. In this sense, to speak of a necessary "connection" between a moral right and a legal right is to argue that the ultimate rationale for the existence and validity of the legal right is to be found in the moral one.[4] Any less rigorous meaning of "necessary connection" raises the danger of *post hoc* reasoning. For even without a knowledge of the moral views of a society, one can from a careful examination of its legal system derive a statement of that system's view of man, which could be cast in moral (or psychological or political) terms; but this falls far short of demonstrating the necessity of the connection, or even of showing which way it runs.

Morality

There are two ways in which "morals" might be said to be related to the law. The first—the one which people most often have in mind when they speak of law enacting morality—is illustrated by the set of criminal laws which punish abortion, prostitution, homosexual activities, and the like. As a historical matter, it is undeniable that the felt immorality of feticide and of variant sexual practices accounts for their prohibition by statute; however, since an examination of such relationships (whether perceived as what I term "codification" or "replacement") shows them to be merely historical, the definition of "connection" previously set out has not been satisfied.

Law as Codified Morals. Over the years, ethical thinkers have contributed to the growth of society's legal system as well as having guided its conscience.

[4] It is often assumed that the asserted "connection between law and morality" refers to law incorporating, or resting on, morality. The opposite meaning is equally plausible, but the interesting questions it raises go beyond the scope of this paper. What is at issue is whether the Old Testament view of the law as "a lamp unto the feet and a light unto the path" should be taken as a description of law's effect on morality or merely on conduct.

Most prominently, organized religion has had profound direct and indirect impact, through ecclesiastical law and general religious precepts, on legal rules not only on the Continent but also in the common law and in our own constitutional system.[5] This influence is even reflected in the language and customs of the law: for example, the "repentence" for which a judge looks in setting the sentence of a convicted man, or the oath on the Bible by which witnesses affirm that they will tell the truth.

During the period when the Anglo-American legal system was molded by the common law judges, legal rules represented what the courts took to be the community's view of the proper relationship of moral men. In these circumstances, "the law" was merely a formal endorsement by the community of views which its members (or most of them) already held. Once legislation came to play an important, or predominant, role in the legal system however, the more complex and less simply "moral" drives of law-makers became apparent.[6] It became more difficult simply to assert that since the people (directly or through their judges and legislators) create the law and the people (judges, legislators) hold moral views, therefore the law enacts morality.[7] This syllogism fails because under it we could as well say that law enacts biology, economics, psychology, astrology, or any other set of views which people hold and often believe in strongly.

[5] Under our constitution, the historical connection between a statute and that body of morality known as religion can present some intriguing questions of legislative intent and legal effect, but these need not detain us now. See generally John Hart Ely, "Legislative and Administrative Motivation in Constitutional Law," 79 *Yale Law Journal 1205* (1970).

[6] Even when the framers of a law speak in moral terms ("equality," "social justice"), the law-morality connection is not patent. Take, for example, Workmen's Compensation, which makes an employer liable for job-related injuries without a showing that he is at fault for their occurrence. There was much opposition to these statutes in the early years of this century, and for a while the courts held them unconstitutional on the grounds that they deprived the employer of his property without due process of law in violation of the fourteenth amendment. See, e.g., *Ives v. South Buffalo Railway Co.*, 201 N.Y. 271, 94 N.E. 431 (1911). (To do this, the courts found the doctrines of the fault system of liability, which along with numerous exceptions they had themselves created over the preceding centuries, to be an immutable part of the concept of "due process.") But while a "moral" rationale can be thought of for or against these statutes, the basis on which they were enacted by the legislatures and eventually accepted by the courts (as a valid exercise of the "police power" of the state) was, in the words of the draftsmen of the 1910 New York law, that the existing negligence system was "economically unwise . . . wasteful, uncertain and productive of antagonism between workmen and employers." The legislators concluded (1) that if employers, particularly in "dangerous trades," had to bear at least some of the cost of injuries, they would be more likely to improve the safety of working conditions, and (2) that both the costs of the accidents and of their prevention are costs of doing business, which ought to be reflected in the price of the goods purchased.

[7] Cahn, E. (1955). *The Moral Decision.* Bloomington and London: Indiana University Press.

Law in Place of Morals. There is another sense in which law and morals might be said to be connected—a sense in which a law-biology connection would not be even facetiously asserted. This relates to the role law has as a replacement for morality. Usually, in peaceful, smoothly functioning relationships and stable societies, there is little need for "the law" to play an active role in many areas, such as the family, education, or the like. The conduct of the individuals involved is guided by the commonly accepted norms for these relationships: e.g., children obey their parents (teachers, etc.), achieving greater independence with their increased maturity; parents are responsible for the well-being of their children; old people can rely on their position within the family to provide respect and, if need be, sustenance. These norms are not only provided by custom but explicitly sanctioned by moral codes, enforced either by internally assimilated standards or by outside forces operating with moral authority. Once this system breaks down, once the "commonly accepted" view of proper behavior is more widely questioned and less commonly accepted, however, people may turn instead to the law to restore relationships. In the process, law brings along new sanctions, new actors and new modes of action; in effect, it creates new relationships, although such relationships usually are presumed to be the heirs of the traditional ones and bear their names. While it is possible to see this as just another instance of law enforcing morality (akin to a common law judge creating the crime of "larceny by trick"),[8] in fact, the operant fact here is the failure of the moral system to operate, so that the legal system's intrusion might more accurately be seen as a desertion of morals for law (seen as a set of regulations fashioned on the basis of the sciences of human behavior rather than on *a priori* principles).

Justice and Fairness. In neither of these senses ("codification" nor "replacement"), then, does it seem accurate to say there is any necessary link between law and that notion of morals with which it is commonly associated. Yet one principle is not so easily disposed of—the principle of fairness. To hold that a law does not comport with a person's standards of sexual behavior is one thing, but to declare that it does not square with his standards of fairness or justice is quite another. The standard of fairness, like the standard of sexual behavior, can be seen as a moral one but unlike the latter, its connection to law seems almost indisputable. It provides the yardstick by which individual laws and decisions as well as whole legal systems are judged; its connection with "the very notion of proceeding by rule is obviously very close."[9] In deciding on the distribution of benefits and burdens or the compensation of injuries, it is the precept to which the law turns; as Professor Hart has formulated it: "Treat like cases alike, and treat different cases differently."[10]

[8] *King v. Pear,* 168 Eng. Rep. 208 (1779).
[9] Hart, H. L. A. (1961). *The Concept of Law.* Oxford: Clarendon Press.
[10] Id. at 155.

All this does not, however, make out a necessary connection between law and morality. First, the "Treat like cases alike" precept is functionally incomplete; a definition of likeness must also be given, and this need not be in "moral" terms. Some would argue that other standards of judgment (e.g., "rationality" or "reasonableness") are thereby simply made part of the concept of justice and that the overall concept remains a moral one. Under this view, to state that it is unreasonable to punish blue-eyed thiefs but not brown-eyed ones is to make a moral statement.[11] It can equally well be argued that the non-moral judgment (i.e., the "rational" definition of where lines are drawn for "like" classes) is the determinative one and the principle of fairness and justice only secondary. Moreover, even the fairness principle itself can have an amoral rationale: a legal system, or a judge deciding a case, might employ the principle because to do so decreases the costs of the system or increases subjects' allegiance to their sovereign, or whatever. These rationales might in turn reduce to a utilitarian calculus on closer inspection; while as such they would be part of a moral system this is a long way from what is usually taken to be the morality with which law is assertedly connected.

Law

In order to develop the distinction between two meanings of the third word–law–which concerns us, I should like to offer some definitions, albeit tentative ones, of morality and law. A moral system serves as a guide to conduct which is considered "good" in itself or which will lead to "the good" (variously defined as happiness, holiness, etc.). The system may be seen as a discoverable natural pattern or as a strictly human creation, or somewhere in between; that does not alter the basic definition. Law can be seen as the collection of enforceable rights and responsibilities through which the members of a society relate to one another and to their society as well as the system by which the society assigns these rights and responsibilities and resolves asserted conflicts among its members.[12]

Law and morality are similar in that both attempt to prescribe human behavior according to a set of rights and duties. If one adheres to a moral or legal

[11] See id. at 156. Professor Hart identifies the "moral outlook" of equality as the principle underlying justice and fairness, but he acknowledges that other bases for a "just" system are possible. Id. at 160–161.

[12] Some of these rules or arrangements are described as "customs" by anthropologists. Similarly, Prof. Hart argues that primitive societies which lack "secondary rules" (roughly, procedural rules) cannot be said to have law at all. (Unlike his fellow positivist John Austin, however, Hart holds that some customs count as law even before a law-making institution recognizes them.) I would argue further that those rules at the edge of the legal system, which enjoy their non-legal status primarily because the law has foreborne incorporating them for the present, ought to be regarded as part of our concept of law. What is said hereafter is not affected by these distinctions, however.

system, he does nothing more than it gives him the right to do and nothing less than it makes it his duty to do. But these definitions also suggest some differences between morals and law. First, the definition of morality includes on its face a normative element and that of law does not. This is more than a matter of chance or of arbitrary definition—the heart of a moral system is its judgments of right and wrong, in an outward-referring sense, while the heart of the legal system is the ordering of relationships, in which only consistency with the system's own rules need be sought.

A second difference arises from this matter of "reference." If you found yourself alone on the proverbial desert island, you might find cause to employ a moral system, but you would have no use in any reasonable sense for a legal system. If, however, I were then to be washed ashore on your island, a need for laws would arise. (We might decide not to have any "laws" as such but to talk through each matter on its own merits; that decision, however, would in itself create a legal system, one of negotiation without precedent.) If I decided that I did not like a law, you would then have to make reference to some "justification" which demonstrated my obligation to obey the law.[13]

To do this, you would make reference to some principle (that I acknowledged as binding) which either justified the particular disputed rule or established my general obligation to obey all valid rules, including the one in question. Law, in other words, has two meanings: that of individual rule and that of a system of rules. The Positivists argue, convincingly I believe, that law in the narrower sense can find its justification by reference to law in the broader sense; that is, a law need not rest on moral principles to have force. In the modern formulation of Professor Hart, binding legal duties (or, more generally, legal rules) are those which meet the criteria of what Hart calls "the rule of recognition" (e.g., "a law adopted by majority vote of the legislature and signed by the chief executive"), which in turn rests on its acceptance by members of the society.[14] Thus, neither individual rules nor the system as a whole need to be traced back to a "moral" wellspring.[15] The role of morality in this system is to supply a standard of judgment or criticism.

[13] The question of justification is what really lies behind most discussions of the connection between law and morality. The question underlying such discussions is whether one need obey a law for which the connection is missing.

[14] In addition, some rules—or customs, really—are binding in Hart's scheme because they too are "accepted" rather than achieving their validity by means of the master rule.

[15] Extra-legal (possilbly "moral") principles do have a place in Hart's system. When a "hard case" is not decided by an existing rule, or when an existing rule leads to a harsh result, the "rule of recognition" gives certain people (typically judges) discretion to fashion a new rule. In the exercise of their discretion, the judges may rely on extra-legal standards; consequently, some individual laws may (but need not) be grounded in moral principles. This explains the connection between morals and certain laws which was discussed under

In sum, the thesis of "no necessary connection between law and morality" is valid to a limited extent. If we adopt the positivist view, reference need not be made to moral rights or duties to justify either individual legal rights or duties or the aggregate legal system. We may sometimes invoke the moral judgment that "This law is unjust," but this is only a criticism of the law and does not reduce the law's binding force on its addressees, provided its valid pedigree is established. (Our discussion of this point would have to go into much greater detail if we were faced with deciding whether to punish a person who engages in civil disobedience against a law he believes is unjust, or conversely a person who committed what is now regarded as an unjust act in compliance with an apparently valid law which he believed was binding on him. The questions raised by our topic, while complex, are less knotty.)

Informed Decision-Making: A Legal Right?

The Legal Pedigree

Does the "informed decision-maker" rule I have suggested, then, state a legal right? There are two aspects to the right–information and decision–and both of them have been recognized by the law. The information component is of recent origin; it was developed out of recognition that knowledge is necessary to make meaningful the power to decide.[16] The doctrine of "informed consent" has been developed in cases involving malpractice claims arising out of the physician–patient relationship, which is basically the same as the relationship of genetic counselor and parents (if we assume that genetic counselors are either

---

the previous heading. Critics of the positivist model insist that the legal system is more than just a collection of individual laws. For example, Prof. Ronald Dworkin's contextualist approach attempts to supply a theory of legal obligation that squares with our social practices; particularly, he is concerned to avoid the positivists' use of "discretion" which leads to *ex post facto* results inconsistent with our social practice of blaming someone only for the breach of an existing social obligation. Unlike the positivist judge who may refer to standards outside the law as a guide to decision-making, a Dworkinian judge is bound by certain principles which as such are part of the law. In other words, Dworkin denies the dichotomy between law and morals and locates the notion of legal duty in the general practice of social obligation.

[16] See *Natanson v. Kline,* 186 Kan. 383, 350 P.2d 1093, clarified, 187 Kan. 186, 354 P.2d 670 (1960); *Salgo v. Leland Stanford,* etc., *Bd. of Trustees,* 154 Cal.App.2d 500, 317 P.2d 170 (1957). Cf. *Miranda v. Arizona,* 384 U.S. 436 (1966) (prescribing information about rights to remain silent and to confer with counsel which must be communicated to criminal suspect prior to interrogation); *Banzhaf v. F.C.C.,* 405 F.2d 1082 (D.C. Cir. 1968), *cert. denied* 396 U.S. 842 (1969) (broadcasters required to air anti-smoking information.)

physicians or fall into a special professional class of their own). The second right—the right to self-determination—has a fundamental place in Anglo-American law; its pedigree is indisputable, as reflected in the rules of consensual agreements in contracts law and consent as a defense to assault and battery in torts law. As previously suggested the two rights have come together in "the obligation of a physician to disclose and explain to the patient as simply as necessary the nature of the ailment, the nature of the proposed treatment, and probability of success or of alternatives, and perhaps the risks of unfortunate results and unforeseen conditions"[17] before obtaining the patient's consent.

Despite this rather clear rule, physicians often withhold information from their patients. Usually this reflects only an understandable desire to "keep things simple" and "move patients along" and if challenged in such a case, a physician would probably admit that he hadn't adhered to the standard of full disclosure but argue that no harm was done thereby. (If harm did occur, however, the physician would be in an unenviable position defending a malpractice or battery action.) In some instances, however, failure to disclose is based on intention rather than inadvertence. When he believes his patients' "best interests" would be served by ignorance, a physician may decide to withhold diagnosis, prognosis, or information about an impending medical intervention, as where a patient is found to have a malignant tumor but is told that the growth is benign so that "his final days can be happy ones." All the problems raised by this "therapeutic privilege," and few of its justifications, seem to be present in genetic counseling, although it is apparent that most counselors presently believe they have an unqualified right to withhold information from their patients. While the question is surely a ticklish one, I do not agree that a "therapeutic privilege" is wise or proper in this setting. I would rather that counselors proceeded on the premise, as the court said in a leading case, that "the law does not permit (the physician) to substitute his own judgment for that of the patient by any form of artifice or deception."[18]

The theoretical problems involved are highlighted by the statement that a patient's freedom is limited if knowledge (specifically, that he carries the sickle cell trait) is communicated without treatment being possible. Yet, as is so often the case, under the flag of "enhancing freedom," freedom (in the "positive" sense, of being able to make choices for oneself) has been severely limited—we deny the person before us freedom of choice (by depriving him of the knowledge which would inform, or even prompt, his choice) in order to increase his "true freedom" as we perceive it. Substituting our view of what should be done for that of the persons affected is always a dangerous course, and it is particularly so in the case of genetic counseling because of the great risk that counselors

[17] *Natanson v. Kline,* 186 Kan. at 417, 350 P.2d at 1106.
[18] Id. at 407, 350 P.2d at 1104.

will not choose as their patients would—or even in their patients' "best interests."

Briefly, my practical objections to the modus operandi adopted by genetic counselors are as follows. First, genetic counselors, unlike the family doctors of yore, are not intimately acquainted with their patients, their families, communities, etc. I have been impressed by the time and effort that we have heard the genetic counselors here devote to their clients, but I think this reflects qualities one would expect in men and women who are willing to come to such a conference and expose their practices to ethical and legal scrutiny. Similar sensitivity—and time—cannot be expected on the part of all counselors, particularly once the demands on counseling facilities increase substantially, as they are bound to. When counseling becomes much more routine, part of accepted practice should not be the routine withholding of information from the counselees on the spurious grounds that the counselors know what is best for patients they hardly know at all.

A second reason derives from the innumerable internal and external pressures operating on counselors which will interfere with an accurate assessment of their patient's "best interests." In efficiency/humanity terms, the really "efficient" course for most counselors is not full disclosure on a computer print-out, but the withholding of information which if disclosed would involve the counselor in a long and arduous process of truly "counseling" his patients. In short, it is more "efficient" (and certainly easier) for him to make the choices himself rather than to bring into open discussion facts (about carrier status, etc.) which are difficult to contemplate or discuss. This points toward another pressure which operates here: physicians' well-known tendency to overreact to disease. The phenomenon of regarding disease as an "enemy" to be "conquered" has its origins, I suspect, in the medical fraternity itself. As Professor Renee Fox has observed, this frame of mind may be quite necessary for physicians, particularly those working on the frontiers of medical science.[19] Whether it is necessary, or merely a reflection of doctors' training or the preexisting psychological makeup which brought them into the profession, this attitude is hardly conducive to a counselor's making a good choice for his patient. As Dr. Kaback commented yesterday, he is much more upset and distressed by the diagnosis of the Tay-Sachs trait than are the carriers to whom he communicates this fact.

Third, a physician's judgment may also be clouded by his own set of values, which will not necessarily correspond to his patient's. The potential for conflict is especially great in genetic counseling in which the options elected depend on one's opinions about such controversial matters as the importance of the traditional concept of family, the morality of divorce and of abortion, etc.

[19] Fox, R. C. (1959). *Experiment Perilous: Physicians and Patients Facing the Unknown.* Glencoe, Ill.: The Free Press.

Finally, the physician's map of social goals may differ markedly from the one held by the patient. A counselor who, for example, strongly believes in the elimination of a genetic disease for eugenic reasons, ought to convey his eugenic premises to any woman to whom he suggests an abortion for that disease, lest her choice be uninformed.[20]

To conclude this aside on "therapeutic privilege," not only do I think that withholding information is theoretically and practically unwise, but I find it unjustified in that the opposite course is perfectly acceptable. The terms in which the alternatives have been posed at this conference—revealing the "brutal truth" or keeping the patient in "benign ignorance"—give a false impression of the courses open to the counselor. They are reminiscent of the early discussions of how much dying patients should be told, discussions which also obscured much through the blinding dichotomies they employed. Only slowly did a few practicing physicians, and some enlightened sociologists, suggest that the question was not whether to tell but how to tell and how fast to tell.[21] Dr. Lejeune's description of his method of informing couples which of the marriage partners is a trait carrier illustrates the advantages of beginning with the assumption that the information should be conveyed and then applying one's creativity to devising a sensitive humane means of conveying it.[22]

Since its articulation by the courts establishes the "informed decision-maker" right as valid law (by reason of the "rule of recognition"), no moral justification for it need be given for its application to the participants before us. We may wish, however, to ask, "Is it just?" In legal terms, as we have already

[20] This formulation highlights the question of what "facts" must be disclosed. Ordinarily, this question may not be an easy one (how much of a diagnosis or description of risks, side-effects, etc., is "fact"?) but at least it is limited to "facts" about the patient (the results of his lab tests, etc.). But what of further information about the significance of the diagnosis, risks, etc., or about the physician's premises, etc. I believe that these too must be included in the information required to be disclosed, but while I may for economy's sake refer to "disclosing facts," I recognize that this phrase encompasses hard, factual data as well as opinions, beliefs and interpretations. Indeed, the very "non-factual" nature of the latter category suggests the great need that it be disclosed and that the disclosure include as full a statement of the competing opinions, beliefs, and interpretations as is possible.

[21] See, e.g., Glaser, B. G. and A. L. Strauss (1965), *Awareness of Dying,* Chicago: Aldine Publishing Co.; Weisman, A. D. (1967), "The Patient With a Fatal Illness: To Tell or Not to Tell," *J. Am. Med. Assn., 201,* 153.

[22] Another way of stating my thesis is that the "best interests" doctrine is acceptable to the extent it mirrors the physician's Hippocratic duty to "do no harm," but that it should be abandoned to the extent it would permit a physician to substitute his judgment for his patient's. Thus, this modified "best interests" would place a floor under the standard of acceptable conduct by physicians, by refusing to excuse intentional or reckless harm to patients, without allowing this protection against potential harm to swallow up the patient's whole right to information and consent.

noted, the right in question is an application of the same rule as that which is applied in torts and contracts law: that each person is free to govern his life as he chooses subject only to those constraints or interferences which have his voluntary assent. This principle also has its reciprocal: that each person is responsible for the consequences (sometimes limited to the foreseeable consequences) of his choices. A statement of rights in genetic counseling which denied parents "informed decision-maker" status would run afoul of the "like cases" precept on both points; the right is therefore necessary to a "just" treatment of counseling.

To conclude that the "informed decision-maker" right states a just principle of law in no wise denies that it also has moral equivalents.[23] Indeed, the very concept of man as a "moral being" is closely linked with this principle. In the view of many philosophers, as well as biologists and anthropologists, man's distinctive characteristics are his abilities to communicate, to reason, to imagine alternative possibilities so as to anticipate future events, and to act so as to alter them. Given his faculties, if man is to act morally he must take responsibility that each of his acts comports with moral rules (however conceived, i.e., "do no harm," "help thy neighbor," etc.). Giving each person power, as well as responsibility, for his own conduct also, in the view of some philosophers, assures that the good of the whole community is maximized.

The Question of Liberty

This raises a question which I have skirted. My primary interest in the "informed decision-maker" right is to identify its allocation of authority and responsibility between parents and counselor. Thus far, the state has entered the picture only as the enforcer of the right (through its courts). One implication of the right, however, is that the parents' decision takes precedence not only against the counselor's wishes but also against those of the state. I raise this point not to discuss it at any length,[24] but because it reflects one of the major forms of the law–morality debate. The view that one should have free choice

[23] One moralist's approach to informed consent is found in Prof. Paul Ramsey's description of the deontological dimension of consent: "The principle of an informed consent is a statement of fidelity between the man who performs medical procedures and the man on whom they are performed." Fidelity is thus an aspect of "the faithfulness that is normative for all the covenants or moral bonds of life with life." Ramsey, P. (1970), *The Patient as Person,* New Haven and London: Yale University Press.

[24] The issues involved would require lengthy treatment in a separate paper. The state's right to act against the parents' wishes would probably depend on such issues as: (a) in whose behalf is the state acting, that of the unborn child, the community, future generations, or science; (b) is its action premised on paternalism, the "common good," or the limitations which it places on the exercise of certain privileges it grants; and (c) how does it enforce its decisions, (by prohibiting abortions or commanding them, or by compulsory amniocentesis, contraception, sterilization, etc.).

about his own conduct is, of course, identified with John Stuart Mill. His critics argue that Mill's concept of liberty is not[25] and should not be accepted by society, for each member of a society owes the collectivity a duty to keep himself "physically, mentally and morally fit."[26] This argument provides another perspective on, or way into, the question of a necessary connection between law and morality. Rather than asking whether it is possible to view the legal system as a useful, justified entity, independent of morality, it asks whether one can conceive of a society operating successfully without imposing its moral views on its members. While this debate, between libertarians on the one hand and paternalists and collectivists on the other, is as fascinating as that between positivists and naturalists, I don't think we need go into it here. Suffice it to say that I defend the full implications of the "informed decision-maker" right, even against the state's authority.

Having concluded that genetic counselees do have a legal right to act on full information about the options open to them and their risks, I should like now to apply this analysis to a hypothetical situation presenting some of the potential conflicts among the rights and obligations of the participants in genetic counseling.

Rights and Duties in Genetic Counseling: A Case Analysis

Suppose that a couple who believe their potential offspring to be "at risk" for a genetically linked disorder consult a physician who specializes in genetics. The stage of the unborn child's development at the time of consultation, the type of advice, and the sorts of data used might vary greatly. Let us assume that the parents seek out the counselor after the child has been conceived but in time to terminate pregnancy safely; further assume that the data are restricted to family histories (the probabilities of the disease being calculable on that basis, but the disease not being amenable to diagnosis by amniocentesis, etc.); and finally assume that the advice given is a straightforward assertion that there is no known risk of the disease occurring. On this assumed set of facts, what consequences follow if the counselor intentionally withholds information or makes a negligent mistake in his advice and a child with the feared genetic disorder is born?

Although no such case has arisen to the best of my knowledge, analogies are available. Perhaps the closest of these is the decision of the Supreme Court of

[25] As Lord Devlin has observed: "Mill's doctrine has existed for over a century and no one has ever attempted to put it into practice." Devlin, P. (1965), *The Enforcement of Morals*, Oxford: Oxford University Press.

[26] Id. at 104.

New Jersey in *Gleitman v. Cosgrove,*[27] which provides us with a useful vehicle for analysis, not the least because Lord Kilbrandon takes a rather different view of the result reached from that which I take. I believe it will be worth our while to examine *Gleitman* in some detail, both because it is the leading case on this subject, and also because I believe both the New Jersey court and its critics have mistaken the rights and duties involved. Using the analysis we have developed, I believe we come inescapably to a different, and better, resolution of the contentions raised by the case.

The *Gleitman* court held that there was no cognizable claim against a physician who erroneously advised a woman that the German measles she suffered during the first month of pregnancy "would have no effect at all on her child."[28] Apparently Dr. Cosgrove knew the risk of rubella damage to be about 25 percent, but he withheld this information because he believed it unfair to abort three healthy fetuses to avoid one diseased one. By a divided vote, the justices ruled that neither parents nor child could sue for the child's substantial birth defects, because the mother had testified that if she had been properly warned of the risks she would have sought an abortion; the parents were foreclosed because at abortion even if legal (which the court assumed),[29] would have violated "the preciousness of human life," and the child was foreclosed because he would "not have been born at all" had his parents carried out the abortion. While it is not difficult to understand why the New Jersey court reached the conclusion it did, I believe its opinion rests on a misunderstanding of legal principles, a misapplication of precedent, and a misapprehension of the consequences of the result it reached as compared with the contrary result.[30]

Misunderstood Principles

The first question which arises is whether the legal rule established in this case is a just one. The reason it seems unjust is that the general rule—that (a) a person (b) who suffers injuries (c) will be made whole by (d) the person who

[27] 49 N.J. 22, 227 A.2d 689 (1967).

[28] *Gleitman v. Cosgrove,* 49 N.J. 22, 24 (1967).

[29] The court's assumption accords with one of the grounds for abortion proposed by American Law Institute (and now accepted in a dozen states): that a licensed physician believes there is substantial risk the child will be born with grave physical or mental defects. *Model Penal Code* Section 230.3.

[30] The discussion herein is limited to the child's right to recover, since the parents' claim is either derivative, dependent on the same theory, or governed by whether abortion is legal. The court assumed that the Gleitmans could theoretically have obtained an abortion, but cited "policy reasons" why recovery should be denied. If an abortion had been actually as well as theoretically, possible on a legal basis, the court would not have been able to rely on these policies.

caused the injuries—was not applied here. In consequence the court is open to criticism for not treating like cases alike.

As my statement of this tort rule suggests, in determining whether justice was done in the *Gleitman* case, we first inquire into the infant plaintiff's standing to sue. Is Jeffrey Gleitman, the defective child, "a person" in the eyes of the law? This question did not detain Justice Proctor of New Jersey for long. Looking to *Smith v. Brennan,*[31] in which the court upheld the right of a child to sue for injuries sustained *in utero,* he quoted that "justice requires that the principle be recognized that a child has a legal right to begin life with a sound mind and body."[32] In other words, the court relied on the principle of fairness to reach the conclusion that the protection of, and redress for, postnatal harm to "mind and body" should likewise be available to persons alleging prenatal injuries.

As to the second element of the rule—injuries—no question arose: the judges all agreed that Jeffrey suffered severe impairment. If we skip momentarily over the third element—compensation—the fourth element poses the question whether Dr. Cosgrove was "the person who caused (Jeffrey's) injuries." The physician did not cause the impairment in the sense of having given Mrs. Gleitman rubella; however, in torts parlance he was the proximate cause of the impairment because his mistaken advice prevented the Gleitmans from avoiding the manifestation of injuries. Had Jeffrey been a grown man who received negligently inaccurate advice from Dr. Cosgrove about a neurological disorder which thereafter, in absence of treatment, rendered him blind and deaf, the legal rules governing the doctor–patient relationship would require Dr. Cosgrove to compensate Jeffrey for his impairment.[33] The court's contrary conclusion that Dr. Cosgrove's conduct "was not the cause of infant plantiff's condition"[34] is nonsensical—as the court itself recognizes, the plaintiff would not have been in that condition had Dr. Cosgrove told the Gleitmans of the risk of impairment.

The reason why the New Jersey court felt constrained to deny Jeffrey a "just" application of the usual rule of recovery is not that Dr. Cosgrove did not cause the injuries, but its conclusion that there was no way to calculate how to make Jeffrey "whole" again. This aspect of *Gleitman* is very pertinent for us; given the present state of genetic counseling, the only "treatment" available in most cases is to abort the fetus.

[31] 31 N.J. 353 (1960).

[32] Id. at 364.

[33] Jeffrey's legal right to competent advice from an expert, and Dr. Cosgrove's duty to provide it, are paralleled by the moral rights and duties set forth in the *Principles of Medical Ethics* which state that physicians should render "to each (patient) a full measure of service and devotion."

[34] 49 N.J. at 28.

There are two grounds on which the court's holding is open to criticism. First, the conclusion that a court "cannot weigh the value of life with impairments against the nonexistence of life itself"[34] is contradicted by courts making similar subjective calculations (the value of lives cut short, of pain and suffering, and other intangibles) every day. Second, if the New Jersey court intended a broader point, that life with any handicap is per se better than no life at all, it cited no authority for this conclusion. What it did cite is Professor Tedeschi's argument that "no comparison is possible since were it not for the act of birth the infant would not exist."[35] But this adds nothing to the court's own *a priori* judgment in favor of impaired life versus abortion; it only serves to create confusion over the act for which plaintiff is suing. Jeffrey did not sue Dr. Cosgrove "for his life," although such a suit is not as illogical as Justice Proctor, relying on Professor Tedeschi, suggests.[36] Rather, Jeffrey sued the physician for his failure to give accurate advice on which a decision could be made by Jeffrey's parents, acting in their child's behalf, that for him not to be born would be preferable to being born deformed. If one objects to awarding damages for the violation of this right, it seems to me that the objection goes either to the policy of allowing abortions (the court assumed one could have been legally obtained) or to giving parents, who may have conflicting motivations, the authority to make this decision (as the law now does). The fact remains that the *Gleitman* court departed from the rule that the choice in this matter lies with the patient, not the physician,[37] and its action is no more defensible than that of a court which, faced with a patient who was ravaged by an untreated disease, were to dismiss the suit on the grounds that the standard treatment (about which the physician negligently failed to inform the patient) is highly dangerous and nearly always fatal.

[35] Tedeschi, G. (1966). "On Tort Liability for 'Wrongful Life'", *Israel Law Review, 1,* 529.

[36] Would a court throw out a suit brought by a patient who had contracted a disabling injury through a physician's negligence in administering transfusions, simply because the physician proved that but for the transfusions the patient would have died? Although the plaintiff would "owe his life" to the physician, he could, of course, still sue him.

[37] This choice may even extend to a patient's refusing "life-saving" therapy. See *In re Brooks Estate,* 32 Ill. 2d 361, 205 N.E.2d 435 (1965). The imposition of such therapy against the patient's wishes in *Application of the President and Directors of Georgetown College,* Inc., 331 F.2d 1010 (D.C. Cir.) cert. denied 377 U.S. 978 (1964), was defended by the judge there because the patient had shifted the "legal responsibility" for the choice over to the hospital. Moreover, cases involving adult patients turn on the applicability and interpretation of the policy against suicide which does not apply to cases involving fetuses under a "liberal" abortion law. And both the *Brooks Estate* and *Georgetown* lines of cases start from the position that if the patient's choice is overridden it can only be done by someone (usually a judge) officially enpowered to act as his guardian, and not by the physician alone, as in *Gleitman.*

Misapplied Precedent

If the New Jersey court's failure to heed prevailing legal doctrines led it into one sort of error, its application of prior cases led it into other errors, although certainly not all of its own making. The *Gleitman* court relied on "two cases from other states which have considered the theory of action for 'wrongful life',"[38] *Zepeda v. Zepeda,*[39] an Illinois case, and *Williams v. New York.*[40]

The New Jersey court's reliance on these cases is misplaced because, as the court observed, they "were brought by illegitimate children for damages caused by their birth out of wedlock, and in both cases policy reasons were found to deny recovery."[41] Policies relevant to illegitimacy clearly have limited, if any, application to a suit by a child made deaf and blind by rubella. Moreover, the opinions of the New York and Illinois courts are unsatisfactory on their own facts. In *Williams,* for example, the plaintiff was an infant who had been conceived when her mother, a mental defective in the custody of a state hospital, was raped by another patient. The child claimed that the state's negligence in protecting her mother had caused her (the child) to be deprived of a normal childhood and rearing and "to bear the stigma of illegitimacy."[42] The New York Court of Appeals recognized the "unfair burdens" the plaintiff would bear, as do "many other sons and daughters of shame and sorrow."[43] But, it concluded, "the law knows no cure or compensation for it, and the policy and social reasons against providing such compensation are at least as strong as those which might be thought to favor it."[43] If "the policy and social reasons" against making illegitimacy a "suable wrong" are of no assistance to the *Gleitman* court, perhaps it had in mind the arguments presented by Judge Keating's concurring opinion in *Williams.*[44] These concerned the "logico-legal" difficulty (derived by Judge Keating from Tedeschi's article) "of permitting recovery when the very act which caused the plaintiff's birth was the same one responsible for whatever damage she has suffered or will suffer."[43] We have already seen the error in the "same act" approach, which by characterizing the claim as one for "wrongful life" fails to distinguish between the act of conception and the circumstances under which it is done. Having intercourse is not a crime, but having it when unmarried is, and in this case that was the state's fault. Of course, "had the State

[38] 49 N.J. at 29.

[39] 41 Ill.App.2d 240, 190 N.E.2d 849 (App. Ct. 1963), cert. denied 379 U.S. 945 (1964).

[40] 18 N.Y.2d 481, 223 N.E.2d 343 (1966).

[41] 49 N.J. at 29.

[42] 18 N.Y.2d at 482.

[43] Id. at 484.

[44] The Appellate Division, whose decision was being reviewed, had based its decision in part on the Keating line of reasoning (that damages cannot be ascertained because they rest "upon the very fact of conception"). 25 App.Div.2d 907 (1966).

acted responsibly," as Judge Keating noted, the plaintiff "would not have been born at all."[45] Yet the state's failure to do so created not only the infant plaintiff but also her cause of action.

In *Williams* not only were the act (conception) and the tort (negligence in failing to protect the mother from men to whom she was not married) separable, but the latter was even partly remediable without abortion, since the illegitimacy could have been "cured" by subsequent marriage, adoption, etc. This is not so in the genetic counseling situation nor in the *Zepeda* case, where the defendant father was already married when he fraudulently induced the plaintiff's mother to have sexual relations with him by promising to marry her. Yet the *Zepeda* case is also of little comfort to the *Gleitman* court because the Illinois court agreed with plaintiff Zepeda "that the elements of a wilful tort are presented by the allegations of the complaint."[46] The *Zepeda* court saw no barrier to the suit in the tortious act or omission having occurred at, or even before, the plaintiff's conception; nor was the suit barred by the nature of the inquiry, which "is not as tangible as a physical defect but . . . is as real."[47] Yet the "radical" nature of the injury alleged—loosely, "bad" parentage—was the factor which led the court to deny recovery for the tort. If the *Gleitman* court relied at all on the "policy" set by *Zepeda*, it must be on that aspect of the opinion which held that recovery should be permitted only after the legislature had undertaken a "thorough study of the consequences."

Misapprehended Consequences

The Illinois court was not merely worried that entertaining Zepeda's suit would open the floodgates of litigation, leaving the courts inundated by the claims of the quarter million illegitimate children born each year in the United States, but also that damages would soon be sought "for being born of a certain color (or) race; . . . for being born with a hereditary disease, . . . for inheriting unfortunate family characteristics; (or) for being born into a large and destitute

[45] 18 N.Y.2d at 485.

[46] 41 Ill.App.2d at 259.

[47] The court built its theory of injury on a detailed review of the "lot of a child born out of wedlock." It contrasted the ignominy and hardships of illegitimacy in the past with the enlightened attitude of modern statutes, which do much to equalize the rights of bastards with those of legitimate offspring. It concluded, nonetheless, that:

> Praiseworthy as they are, they do not, and no law can, make these children whole. Children born illegitimate have suffered an injury. (Id. at 258.)

Earlier in the case the court had concluded that three more specific types of injury were not made out by the complaint—mental suffering was not properly averred; defamation requires communication to third persons, which was not alleged; and no child, legitimate or illegitimate, has a legal right to love or a happy home. Id. at 253–255.

family, (or to) a parent (who) has an unsavory reputation."[48] There is a surface appeal to the court's reasoning. Being born into a minority group of a "disadvantaged" family may subject a child to burdens similar to those of illegitimacy, and hereditary disease may cause greater suffering still.

But opening the court to the infant Zepeda would not necessarily open it to the others cited by the court, for poverty, race and genetic makeup do not constitute "moral wrongs(s) and . . . criminal act(s)"[49] which the court held Mr. Zepeda's sexual relations with the plaintiff's mother to be. Being poor or carrying an hereditary disease are not crimes; procreating in these circumstances violates no legal right of the child conceived.

Nevertheless, although nothing in the "policy reasons" of *Williams* or *Zepeda* is either convincing or applicable to *Gleitman,* we owe it to the New Jersey court to puzzle through the consequences of the result we believe it should have reached before criticizing as unjust the one that it did reach. Can it be said of the *Gleitman* decision "that, regrettable though it is, the demands of justice . . . must be overridden in order to preserve something held to be of greater value, which would be jeopardized if . . . discriminations (between Jeffrey Gleitman and other plaintiffs injured by a negligent failure to give complete medical advice) were not made"?[50]

To bring one possible countervailing value into view, let us alter the facts of the Gleitman case. Suppose the child alleged that the physician gave accurate advice to the parents but that the parents disregarded the risks and did not abort, resulting in his being born deformed. If the claim against Dr. Cosgrove is good, must not that against the parents also succeed? If we assume that there is no longer intrafamilial immunity in the jurisdiction,[51] there remains the simple

[48] Id. at 260.
[49] Id. at 253.
[50] Hart (1961), 158.
[51] English common law permitted tort actions as well as those involving property and contracts between children and parents. "But beginning in 1891 with *Hewlett v. George* (68 Miss. 703, 9 So. 885 (1891)), a Mississippi case of false imprisonment which cited no authorities, the American courts adopted a general rule refusing to allow actions between parent and minor child for personal torts, whether they are intentional or negligent in character." Prosser, W. L. (1971), *Handbook of The Law of Torts,* Section 122, 865. This result was justified as necessary to avoid introducing "discord and contention where the laws of nature have established peace and obedience." *Wick v. Wick,* 1972 Wis. 260, 262, 212 N.W., 787, 789 (1927); the danger of "fraud" has also been stressed. The "retreat" from this rule is now "under way," as Prof. Prosser notes, and parent-child immunity for personal torts may soon be a thing of the past. See, e.g., *Gibson v. Gibson,* 92 Cal. Rptr. 288 (1971); *Gelbman v. Gelbman,* 23 N.Y.2d 434, 245 N.E.2d 192 (1969): and *Coller v. White* 20 Wis.2d 402, 122 N.W.2d 193 (1963). The courts continue immunity, however, for matters subject to "parental discretion" over the care, etc. of children, and this would serve as a further bar to suits for inherited diseases, unless they involved wanton disregard of or intentional injury to the child's health.

fact that such suits are unlikely because the child's parents, as his guardians or "next friends," actually instigate suits on the child's behalf, and it is unlikely that they would, in effect, sue themselves. Yet even if the state routinely appointed special guardians for all defective children (or all children for that matter), with instructions to bring any necessary lawsuits, such suits would be of little practical value. Parents are already legally obliged to support their children, and most do so to the limits of their ability whether the child is "normal" or not. Consequently, unlike a recovery against an outside party like Dr. Cosgrove, a recovery against the parents would just shift family funds (less lawyers' fees and court costs) from one pocket to another.[52]

While there is at least some merit to these practical reasons why a suit against parents would be unlikely to follow had *Gleitman* been differently decided, the really persuasive argument denies that there is any claim against the parents at all. For there to be a recovery, the defendant must have breached a duty legally owed the plaintiff. Dr. Cosgrove violated such a legal (and moral) duty when he failed to give competent medical advice; by contrast, parents, in choosing not to abort, have exercised their legal right to make this choice. This right of the parents has two sources: (a) one derived from the child's own right, in which case the parents are considered to be making their decision on behalf of their offspring, in what they judge to be his "best interests"; and (b) one which focuses on the parents' own right to exercise control over an event which is of major importance to their lives (directly so in the mother's case, and indirectly in the father's). The second rationale is of more recent vintage and more narrow in scope, being applicable, so far as I know, only in the choice to have an abortion to safeguard the mother's life or health or, in a few jurisdictions, for any reason the parents may have prior to 24 weeks of gestation.[53] Since the decision not to have an abortion would probably be viewed by the courts as being based on both rationales, absent proof of intentional disregard of their child's interests or gross negligence in the exercise of their discretion, such an exercise of judgment would not subject the couple to liability. In the view of the law, it is up to them to weigh the probabilities and risks and to decide whether life with any defects is better than abortion or whether in some cases life with defects is "a fate worse than death."

Conclusion

From this discussion, I would conclude that our hypothetical genetic counselor has a legal duty to give competent advice so as to place the parents

[52] Additional funds would be injected only in the unlikely event that an insurance policy held by the parents covered this situation.
[53] Act 1, *Hawaii Sessions Laws of 1970;* N.Y. Penal Law Section 125.05 (McKinney 1970).

into the position of informed decision-makers,[54] and that if by his negligent or intentional breach of this duty a defective child is born, the child (and its parents) have a valid claim for damages against him. This is true whether the parents come to him for advice on whether to abort or on whether to conceive in the first place. (I take the latter situation to be an easier case to establish liability for medical advice and, consequently, have addressed myself only to the former).

Unless we accept as a valid legal rule the *Gleitman* court's dictum that every child has "a legal right to begin life with a sound mind and body,"[55] however, a child who suffers a genetic disease does not have a claim against its parents because they decided to give it birth despite the risks of the disease. The *Gleitman* court did not accept this principle at face value, and neither should we. As a moral precept it states an admirable guide for conduct and aspiration; as a legal rule it is too far-reaching. The legal rule which I have suggested should be applied protects courts from the nearly impossible task of reviewing the parents' good faith judgment about the "quality of life" which a child will experience; the child is protected against intentional harm by the parents, as he would be after birth; and the parents are protected in the prudent use of the capabilities with which nature endowed them. This comports with our moral sense that it is unjust to blame someone for something (such as his genetic makeup) which he cannot (presently) control. It would be cruel to add to the injury of a defective gene (and the undeserved self-blame which is felt when the disease manifests itself in an offspring) the insult of a suit by the offspring. On the other hand, parents who knowingly and recklessly took a drug with a substantial teratogenic risk would be liable if their offspring were deformed. Similarly, major manipulations of the birth process, done in the face of adverse or incalculable risks, would expose their creators to liability for injuries suffered.

None of these eventualities are pleasant to contemplate, and one can hope that they never pass from the hypothetical to the real. But if they do, I am confident that the courts, and in some instances the legislatures, will make clear

[54] Our discussion has focused solely on the rights and duties relating to liability for negligent advice. Time does not permit an exploration of the myriad other rights and duties which arise from the geneticist-patient relationship or of the limitations (and their remedies, if any) which are placed on the exercise of these rights and duties by internal and external constraints. Some exploration of these problems especially concerning informed consent, appears in my "Law of Genetic Therapy" in *The New Genetics and the Future of Man,* M. Hamilton, editor, Grand Rapids, Eerdmans (1972), and see Hans Jonas, "Philosophical Reflections on Experimenting with Human Subjects," 98 *Daedalus 219* (1969); Henry K. Beecher, "Consent in Clinical Experimentation: Myth and Reality," 195 *J. Am. Med. Assn. 124* (1966).

[55] 49 N.J. at 28, quoting 31 N.J. at 364.

the right of children to recover for their injuries. *Gleitman v. Cosgrove* neither will, nor should be, the final word on the subject.

Acknowledgment

The author is grateful to Dr. Jay Katz and Miss Barbara A. Brown for their comments on this paper, which also profited from work done on a related subject under grant HSM 110-69-213, Health Services and Mental Health Administration, DHEW.

PRIVACY AND GENETIC INFORMATION

HERBERT A. LUBS

Introduction

Privacy is a hot topic!

Invasion of privacy is decried on the streets and discussed in academic halls. The preservation of privacy is almost a holy cause in the United States and, indeed, it is somewhat un-American not to come to the vigorous defense of privacy. The other side of the coin, namely, the possible harmful effects of maintaining privacy, is less often displayed.

In this paper, I hope to provoke thought about what test cases might be most suitable to clarify the issues of medical ethics in relation to human genetics and what new laws might be desirable. However, one might ask whether we really would like to bring legal clarification to these issues. Perhaps instead a set of ethics for the medical geneticist would be more appropriate for handling this rapidly changing situation in human genetics.

One of the theses I submit is that over the next 10 to 20 years we must re-examine and possibly modify attitudes towards privacy. The geneticist must work in this sensitive area and it is critical for the future practice of medical genetics that questions of privacy be resolved. Stated succinctly, "How can optimal use of genetic data best be coupled with the maintenance of privacy?"

Genetic information may be used in three ways. A "good use" might be prevention of mental retardation by early treatment of a genetic disorder in a child known to be at risk. A "misuse" might be release of a report of an individual's abnormal chromosome complement in a way which would hinder his employment. "Nonuse," or complete privacy, would be a lost opportunity to use genetic information for the benefit of other family members.

The considerations must be made slightly more complex, however. Two levels of privacy are implicit even in these simple examples: (1) the patient's own

This paper originally appeared in *Ethical Issues in Human Genetics*, Bruce Hilton *et al.*, eds. (New York: Plenum Press, 1973). Reprinted with permission of the author and Plenum Publishing Corporation.

privacy and (2) that of other family members, including future offspring. There are also two general routes by which an individual's privacy may be invaded.

The first is through the proband (or propositus), who is the individual through whom a family comes to medical attention. This may occur in several ways. An individual with a genetic disorder may seek medical attention and the medical geneticist may subsequently seek information from or provide genetic counseling to other family members of the proband. Their privacy is therefore invaded. Similarly, geneticists interested in research may begin their clinical investigations with a group of individuals ascertained from clinics and hospital records and proceed to their families. If the investigation is on a larger scale, the genetics investigator might search health insurance records of major insuring agencies for patients with Tay-Sachs disease, for example. This would certainly be part of the "Big Brother" concern.

The second route is the identification of individuals with genetic disorders through surveys or screening programs. These may be either legally required, such as the PKU screening program, or for the purpose of medical research. The cytogenetic surveys of consecutive newborn infants are examples. In the future, medical screening of individuals in health plans will likely include certain genetic tests, such as tests for sickle and other abnormal hemoglobins. I emphasize this second route of invasion of privacy because here the individual does not come to the doctor with a medical problem. The relationship between the doctor and the patient is thus different, and he is an unusual proband. We are going to the patient and asking to help him.

Practical Problems

The majority of our work in the last five years in the Department of Pediatrics, University of Colorado Medical Center, has begun with cytogenetic studies of relatively large unselected populations of newborns or children, and we have become experts in how people react to having their privacy invaded. The range of reactions displayed by these children's mothers was enormous: from interest manifested by a letter every six months inquiring about the program of the study, to refusal to participate in the study because the mother was convinced my research nurse was really a commercial photographer. We have ventured out into several hundred of these thousands of families for more information and have experienced a wide range of reactions, often from the same person in a family. One hostile aunt of a child with a translocation, who initially resisted being studied, finally cooperated, but later again became hostile because more family studies had not been done sooner. Perhaps I oversold. Most people have been extremely cooperative and interested. Several examples will serve to illustrate certain of the problems we encountered. The names and precise pedigrees are hypothetical, but each situation has actually occurred.

Jack was found to have an XYY karyotype in a survey of newborn infants (Figure 1). Generally, the occurrence of an XYY karyotype is a sporadic event, and we are not greatly concerned with its transmission to offspring. We are concerned with the effect the XYY karyotype, as well as knowledge of this karyotype, on Jack's development and progress in life. The stereotype of the XYY "syndrome" has passed too quickly into the public domain, and the concern is what teachers, neighbors, and employers will think if they know that he is XYY. It is difficult to envision any benefit to his image, except in the eyes of an interested genetics investigator. Because of our knowledge of his karyotype, his development will be watched closely. Perhaps early referral for psychiatric help might help to prevent some of the potential psychological and social problems that appear to be associated with an XYY karyotype in some individuals. It is not known, however, that psychiatric intervention would be effective. Much has been written about the unproven association of the XYY karyotype with prisoner status, and I believe there is a real associated risk, since all surveys show a manyfold increase in frequency of the XYY karyotype in prisoners over that in newborn infants. Many XYY men, however, are normal, and the real problem is our ignorance. Jack may be the victim of this ignorance, and the question is how to safeguard him and protect society. My own temporary solution to this problem is to stall until we know more. The safest thing seems to be to withhold the information from everyone and to follow Jack closely. Ultimately, when perspective returns, society and physicians and the patient can handle the situation as it should be handled: by evaluating the particular person's performance and behavior. Privacy, ultimately, will not be so important in such cases.

What sort of informed consent should be obtained in studies such as this, which led to the detection of Jack's XYY karyotype? First, it should contain an

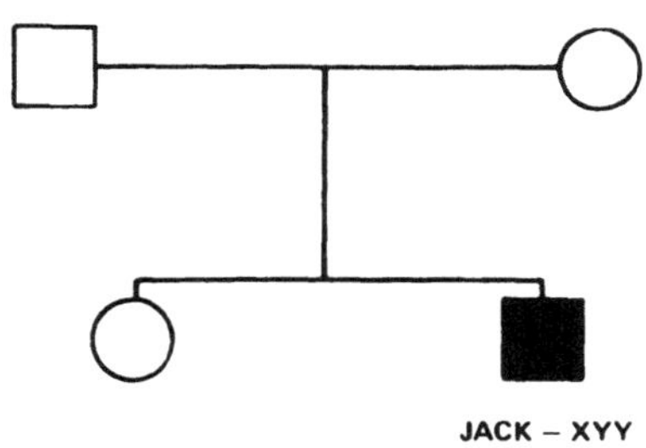

Figure 1. Ascertainment through screening programs: Problems for the individual.

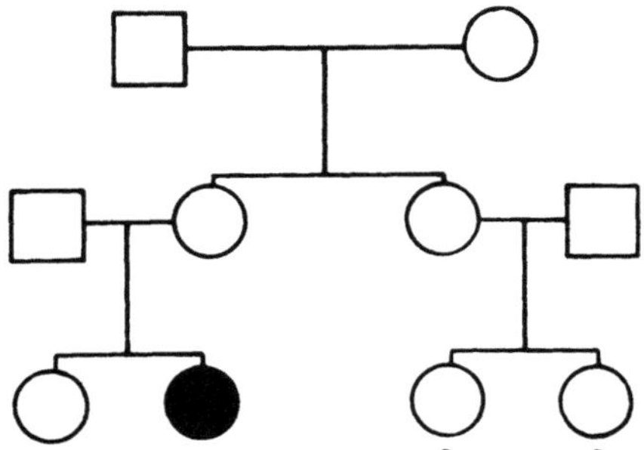

Figure 2. Ascertainment through screening programs: Problems for other family members.

assurance that the information will remain private. It should also include a statement that if important medical information is found, the investigator will inform the patient and his physician if it is felt to be helpful information. It should probably not include very much more. I do not see how we can discuss, when doing such surveys, all the possible disorders we may find and all of the implications. There are hundreds of abnormalities and we cannot possibly inform someone of each one.

Leslie was a normal appearing baby girl who showed evidence of a Y chromosome in a survey of amnions (Figure 2). This test was being done on each infant at the hospital where she was born. Further tests showed that she had an XY karyotype and the testicular feminization syndrome. She will develop as a girl and have normal intelligence, but she will have no uterus and a real risk that a tumor will develop in her abnormal gonads. Her older sister and two first cousins, all under 10 years of age, are at risk for the same problems. How are we to proceed? We do not wish to tell the whole truth (that Leslie is a chromosomal male), but we do wish to provide the best medical care for the family. Generally, we explain the risks and offer both prophylactic surgery and hormonal treatment, without being specific about the chromosomal information, and often we do not put the chromosome or pathology reports in the medical chart. The most difficult question arises when the family pursues the etiology of the problems and wants to know the results of the chromosome studies. Privacy may be essential to Leslie's psychological development and the geneticist for each family must decide how to proceed. It is a case of her privacy (on a very personal matter) versus her own and her relatives medical well being.

I believe we are obligated to offer similar chromosomal studies to others at risk in the family, and therefore to invade their privacy. Both real and imagined concerns will be introduced, but we hope we will do more good than harm. Once the issue is raised, it is critical to proceed with dispatch and return answers about the normality or abnormality of each person in question. An approach might be to wait until those at risk are age 5 or 10 and then carry out the studies. Americans are a mobile population, however, and we worry that the family will be lost to follow-up in the interval and that a tumor might develop in a family member. We do not have a nationwide medical data bank and it is likely that a relative will be lost to follow-up. Here, privacy is likely to be detrimental.

Who is the proper person to investigate, or at least to offer the possibilities of investigation, to the other family members? Is it the state, the geneticist, or the family? There is no current answer to these questions.

I believe that telling people the full truth is good and that this should be our goal, but I think it is too soon for this. People cannot yet cope with too much genetic knowledge about themselves.

The next case is presented as a complex situation involving an additional dimension, time, and should serve to caution us about rigid thinking and premature institution of laws. Hope was found to have 45 chromosomes and a D/G translocation in a survey of newborn infants five years ago (Figure 3). Her mother, grandmother, and aunt had the same translocation. No abnormal family members were found, yet it was felt at the time that a D/G translocation carrier had a significant risk of having a child with an unbalanced karyotype. The benefit from invading this family's privacy was the chance to have subsequent pregnancies monitored by amniocentesis and cytogenetic study, with therapeutic abortion of fetuses with an unbalanced translocation, as suggested by the question marks. The risk figure originally discussed with the parents was in fact a mean of several risk figures, since at least six possible combinations between D and G chromosomes were possible and at that time we could not determine

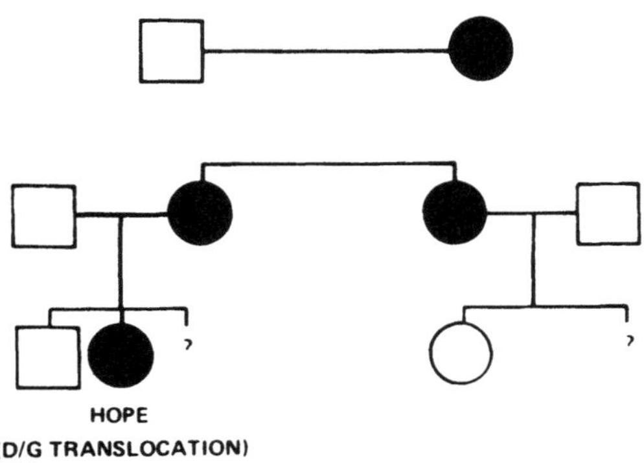

Figure 3. Ascertainment through screening programs: Problems created by imprecise methodology.

which combination was present. The family was restudied this fall with the more precise new techniques and the translocation was found to involve chromosomes 14 and 22. It is likely that there is little or no risk associated with this particular translocation and that we have caused five years of worry. We hope that by having offered amniocentesis to the involved parents that at least the worry was minimal. The point I want to illustrate here is the hazard of incomplete knowledge. The five-year interval between the initial ascertainment, nevertheless, provides an important perspective to the impact of such information on the involved families. Both the passage of time and precise information ultimately may produce a realistic acceptance and utilization of initially disturbing information. When first told that his daughter and wife had a translocation, the father responded half jokingly: "Is that grounds for divorce, Doctor?" Five years later he wrote a grateful letter, saying that they had given copies of my letters to their family physicians and were keeping a folder for each child with appropriate information to be given to them later. Lastly, he even referred to our "humane approach to research." Certain families, at least, are grateful for the invasion of privacy, even in the face of the uncertainties that may be raised in their minds.

Peter was brought to a hemophilia clinic at three months with severe bleeding and the diagnosis of hemophilia A was established (Figure 4). The family was shocked, disbelieving, and rejected the doctor's explanation of X-linked inheritance. No amount of persuasion would change their minds and they refused permission for the medical geneticist to contact relatives, even when it was explained that, by determining the sex of subsequent pregnancies in relatives at risk, it would be possible to prevent other cases of hemophilia in the family. The risks for having an affected male child are as follows (given only one affected male in the family):

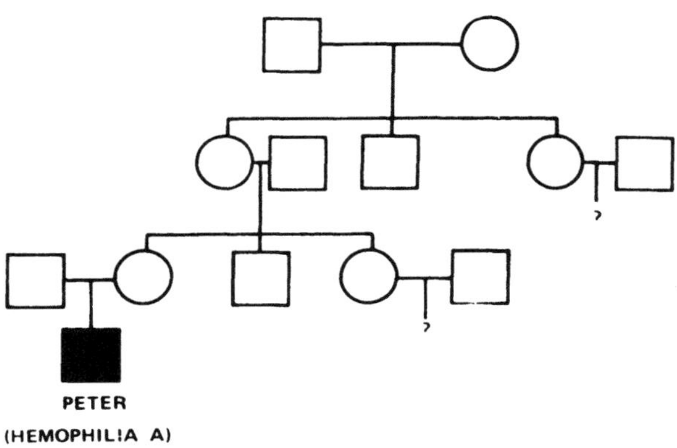

Figure 4. Ascertainment through a proband: Refusal of permission to see other family members.

Female Relatives	*Risk of Having an Affected Male Child*[1] %
Mother	25.0
Sister	12.5
Niece	6.3
Grand niece	3.2
Maternal grandmother	25.0
Maternal grand aunt	12.5
Maternal first cousin	6.3
Maternal first cousin once removed	3.2
Maternal second cousin	1.6

Many family members have a significant risk. How then are we to proceed? To my knowledge, this situation has not been tested in court and the usual procedure is to respect the privacy of Peter and his parents and hope that other family members at risk will hear of the problem and seek genetic counseling. They seldom do. This, then, is a case of nonuse. No study has been made to determine how many relatives are aware of their risks (or lack of risks) in similar situations. When such a case does come to court, the legal case would seem to rest on how far two laws are extended: the first being the 4th Amendment which covers invasion of privacy, and the second being the State's traditional right to control transmissable disease (e.g., infectious disease). We are on the horns of a dilemma. I can envision being sued by the mother if I went to other family members or by other family members if I did not.

The Data Bank and Big Brother

What has been discussed above are my concerns as a medical geneticist. Most people's concerns, I suspect, are about other possible misuses of computerized genetic data. Will "Big Brother" decide that certain parents cannot reproduce? How can we prevent computerized control of human behavior?

Are these real concerns? At the moment there is no prospect of a nation-wide genetic data bank, but a number of genetic registries are being developed. Dr. William Kimberling has a very large family registry in Oregon, and the aim is to record in the computer registry all the families in Oregon with genetic disorders. Dr. Marie-Louise Lubs in our laboratory is beginning a similar registry for Colorado. One of our specific aims in computerizing the data is to make it relatively easy to contact each family once a year, both to provide them with information about newly available treatments and to update our records. In addition, we hope it will be a resource to which the families can continue to turn

[1] New mutations not considered.

for help and information over the years as they move about the country. We proceed, I might add, only on a voluntary basis and one of our first projects will be to determine the families' reactions to this intrusion into their privacy.

There are a number of ways of maintaining the privacy of information on tapes and discs. The primary one is to code each person's data and identification, and to keep names and addresses on separate tapes. It is also possible to garble the coded information on the tape or disc in a number of ways, but this seems to me unnecessary. It would take a rather sophisticated intruder to find the number of the desired tape (there are thousands in the computer center), the format, and the coding system, hire a programmer, and have access to a computer. The hazard is not the computerized data, no matter how extensive, but the people around it. I don't see any safeguard from an underpaid programmer who might be bought, or a crusading geneticist who wished to prove how dangerous this approach was by exposing all data to the public. Basically, we have to depend on the integrity of scientists for the maintenance of privacy.

Let us assume, for the sake of discussion, that there is a nationwide data bank and that a request for names and addresses of all individuals with hemophilia is made by an insurance company who wished to offer special high risk policies to families with hemophilia. Should this information be released? Who should decide on the "goodness" of such a request? We can get some information about this by looking at other parts of the world. In Sweden such a data bank for certain diseases does exist and is workable. Information is released to bona fide investigators and no one seems troubled by it. I mention this simply to show that we should at least consider the benefits from such an approach as well as the hazards. Perhaps a board of medical geneticists, laymen, and social scientists would be helpful in the future if such a data bank even comes to pass in the United States. It is the history of science that, if something can be done, it will be done. Our role, I believe, is to see that something is done well, not to prevent its being done at all.

Summary and Conclusions

A simple balance sheet can serve to summarize what has been presented:

Potential Hazards of Loss of Privacy

Loss of feeling of "privacy," per se.

Prejudicial use of genetic information by other persons.

Disastrous effects on the patient of "loaded" information such as chromosomal sex.

Creation of unnecessary concerns in other family members.

Control of reproduction in a biased or unsound fashion by society.

Potential Benefits of Loss of Privacy

Opportunity of realistically dealing with a high reproductive risk.

Opportunity to alleviate unnecessary fears in relatives at complications of genetic disease in relatives.

Opportunity to offer selective reproduction via therapeutic abortion to relatives at risk.

Opportunity to offer benefits of research to affected families, without the usual 5–10-year delay.

How can these problems be resolved? Time, I believe, is the most important factor. The combination of better genetic information and better education of the public about genetics will resolve most of them. Examples of abnormal human karyotypes and biochemical disorders are now in many high school biology books, and sickle cell disease is becoming a household word, albeit for the wrong reasons. Privacy need not be such a hot topic. In the interim, however, we must proceed cautiously.

Discussion

SINGER: Dr. Lubs makes a point that is important for lawyers, physicians, and philosophers to remember–facts always seem to get in the way of our nice, theoretical analyses. They resist getting shoved comfortably into the pigeonholes we design for them. For example, one who is committed to full disclosure of information cannot but be made uncomfortable by the example of the XY female.

In analyzing issues of privacy we must ask to what extent and under what circumstances is it appropriate to invade the privacy of another? This question has two parts: the acquisition of information and the disclosure of information. It is one thing to acquire genetic data from A, about A; a different question is raised when we acquire data from A about B. On the disclosure side, should disclosure be made to the donor of the information or to some related party for whom the genetic information may be relevant? Perhaps the most difficult problem is disclosure of information to an unrelated person. Whose privacy is being invaded, by whom is it being invaded, and for what purposes is the invasion made?

We may also come to the point where we ask whether there are some types of information that we do not want to have acquired. For the research community, are there some types of experiments or information that you simply do not want to get into? One example is the relationship, if any, between race and intelligence. I am not anxious to acquire that information because I think it is irrelevant to current social issues. Thus for the issue of mass screening, we must balance the need to know and the right to maintain privacy among human beings who live very closely in one society.

CAPRON: The primary harm to a patient arises in the screening process when information which was never sought by the patient is found out. Giving this information to the patient is the second point at which he is harmed. If, however, a patient comes to you asking for information about himself, this information should be conveyed to him, although in a proper psychological setting. But it does not seem to me that there are any grounds, other than a misimpression of what is in the patient's best interest, which would lead to withholding information which is sought.

MURRAY: You have exaggerated the situation. It is the exceptional case in which one withholds information, and it is certainly not routine medical practice. The kinds of cases in which I would do that are those involving emotional stability. Lawyers ought to know that you can get some of the biggest verdicts in suits relating to emotional distress. You are trying to ignore the emotions in your rule that knowledge should always be transmitted, Also, you are guessing, just as the physician is guessing, that the patient will be able to deal with the information, no matter how it is conveyed. All you need is one person who takes an overdose of barbiturates or jumps off a bridge, and the lawyers have a case against the physician or counselor. You can get in trouble whether you give the information, and a patient kills himself, or whether you withhold the information. I prefer to commit a sin of omission rather than take the chance of committing a sin of commission.

CAPRON: I have stated a rule, which I defended on both theoretical and practical grounds, and yet I admit that you may present a case to me which I will find very difficult to insist should be guided by the principle of full disclosure. I don't pretend to say that these choices are easy or that they should always be decided according to that principle.

McLAREN: Mr. Capron has raised the problem of withholding information in the interests of the patient. This is really an example of the more general category of paternalism which is, of course, traditional to the medical profession. Medical practitioners should remember that if they do err, they are likely to err in that direction.

FRIED: Several non-physicians and I share Dr. McLaren's notion that the medical profession is indeed very paternalistic. Surely, if a physician has some findings, perhaps broad, inconclusive, or even incorrect, it would be very wrong to withhold this information on the mistakenly paternalistic notion that the people involved couldn't take it.

HIMSWORTH: Lord Kilbrandon made the point that a special relationship exists between a doctor and a patient, a relationship based on the expectation that the doctor will do the best for the patient that it is possible to do. The ultimate responsibility is on the physician to decide what information to give to the patient. If you get information that a fetus is abnormal, you have got to ask yourself what will be the effect of disclosing this, however tactfully and carefully, to the mother? Can she face bringing up this child? Or when she gets away from my persuasion and smoothness of talking, will she go take an overdose of barbiturates?

CAPRON: There seems to be no question that the relationship is a complex one, and there may be many doctors who believe, as Sir Harold does, that the doctor should make many choices for his patient. However, if doctors find themselves making choices not about the medical facts, but about future psychological factors of the child and the parents, then I think doctors have overstepped their role. This is not the role of the doctor; the doctor has the responsibility of conveying the medical import of the diagnosis.

MURRAY: Mr. Capron, in your analysis of one case you considered legal justification for performing abortion, and you supported the parents' right to decide to abort a fetus based on what they thought was in the best interests of the unborn child. I know a great deal more about a patient, even though I have only talked to him for a brief time, than those parents know about an unborn fetus. Therefore, I don't see how you can justify the rights of a parent to make a decision about a fetus, and say that the physician, who knows at least what the patient looks like, has no right to decide what is in the patient's best interests.

CAPRON: I am criticizing the substitution of judgment of best interests by one person for another. The person sitting across the table from you is able to make judgments for himself, whereas the fetus is not. Traditionally the role of making substitute judgment for children in our society has been given to the parents. If we were to conclude that the parent is no more able to make an accurate or an informed judgment for a fetus than a doctor is able to make for a patient, then we ought to find someone else to do it.

KILBRANDON: It may be a very serious question for the doctor whether he tells the truth or whether he gives the information which he judges to be most beneficial to his patient. If these two are in conflict, there is no doubt at all which he has to do. He has to give the information which is most beneficial to his patient. This may be a very difficult decision, and I really was quite horrified when Dr. Murray expressed the view that you can't win, because they will get you either way. If that were so, the law would be in a very shocking state. As I understand the law, a suit brought against a doctor raises the question of whether or not he has been negligent, not whether he has offered some treatment which another doctor would not have offered, or he has taken a view of the case which might not appeal to some of his other colleagues.

Whether you offer the truth or some modified part of the truth, or whether you offer beneficial information, is a very tricky question on which there is ample room for bona fide difference of opinion. If the law says that, having come down on one side in good faith, with adequate skill, and perhaps having discussed it with your colleagues, you are liable for damages, then the law ought to be changed.

HAVIGHURST: I pose the possibility that we litigate the matter. I will present some 20 expert witnesses drawn from this audience to indicate that the custom and practice in the profession of genetic counseling is to the effect that some information may be withheld. They would agree that in a particular case, it was professionally appropriate to withhold the information. Now, who would win, if that evidence were presented and not seriously challenged?

What I am asking is a clearer statement of what the law in fact is. I suggest that, in some states at least, the law on informed consent is that the doctor is obligated to inform his patient only to the extent that other practitioners in the area would inform the patient in similar circumstances.

CAPRON: Your premise is one which Lord Kilbrandon has already highlighted in that there is a great deal of "ought" as opposed to "is." Doctors have asked me exactly that question, and my answer has been that the prevailing law in judging the disclosure of information is, indeed, the standard which has always applied in malpractice, which is "the reasonable medical practitioner."

I suggest that some of the informed consent cases, for example, *Salgo v. Stanford University Board of Trustees* (154 Cal. App. 2d 560, 317 P 2d 170 (1957)) have conceived of the doctor's action not so much as a violation of the old standards of negligence, but under another tort standard whereby an intentional withholding of information or a deception would be judged by a reasonable man standard, so that the jury would be asked to decide whether it was reasonable to withhold information, and expert testimony as such would not be pertinent. Experts might persuade a jury that this was a reasonable course, but they would not be bound by that testimony under the instructions of the judge. An example of this approach in a therapeutic setting is *Berkey v. Anderson* (1 Cal. App. 3d 790, 805, 82 Cal. Rptr. 67, 78 (1969)) where the court stated:

> We cannot agree that the matter of informed consent must be determined on the basis of medical testimony . . . a physician's duty to disclose is not governed by the standard practice of the physician's community, but is a duty imposed by law which governs his conduct in the same manner as others in a similar fiduciary relationship. To hold otherwise would permit the medical profession to determine its own responsibilities. . . .

For all the reasons I gave that is the rule that should be adopted in these circumstances because the old idea of the family physician knowing so well his patient and his family and the conditions of the community is no longer an accurate one, particularly vis-à-vis consultants, such as genetic counselors, who see a family only briefly. For the theoretical reasons which prompt my "ought," the rule should be one of reasonableness and not of medical malpractice.

VEATCH: A dilemma arises when the norms of professional ethics conflict with the normative system operating more universally in society. Many medical professionals operate within the context of professional ethics which may be summarized as doing no harm or acting only for the benefit of the patient. However, it is not at all clear that that ethical principle is shared by the general public or by moral philosophers. Thus, a medical professional may be forced by the nature of his occupation to interact with a person who is not operating within the same ethical frame of reference. A very fundamental problem is created once one articulates moral responsibility in terms of a particular ethical system for the professional sphere.

MANGEL: There is a distinction between the information that you give a patient, depending on whether you are in a research situation or whether you are in a therapeutic situation. The test Mr. Capron enunciated is probably applicable to the pure clinical research situation, in which case it is difficult to justify the withholding of any relevant information. This position was held in the recent case of *Halushka* v. *University of Saskatchewan* (53 D. L. R. 2nd 436 (1965). I do not think that the case law in this country has yet taken the position that there must be full disclosure in a therapeutic situation. Courts always deal with the question of whether negligence existed in something done, and I haven't seen any case in which liability was based solely on a failure to disclose information in a therapeutic situation.

RAMSEY: Regardless of the law, the ethics of medical practice has to agree with the principle of full disclosure and must put upon itself the moral burden of justifying the withholding of information. This is not really out of accord with the ordinary man's understanding of truth-telling in his interpersonal relations.

FRIED: I want to remove any excessive sense of assurance that doctors may have about the state of the law, where it has been said that the standard is what do most reasonable practitioners do. There is clearly movement in the law in respect to all experts and special professional groups, not to let their relations to persons outside of their group be determined by the judgments of a majority of their own fraternity. There is a growing realization in the law that what doctors decide is subject ultimately to legal scrutiny. I suspect the only reasons that that movement has not gone further with doctors is that it might, one day, be applied to lawyers.

MacINTYRE: Dr. Lubs mentioned the testicular feminization syndrome. In my judgment, this case is one in which it is not only undesirable but highly dangerous to divulge complete information. Such individuals are genetically males, but externally they develop as females. In our society I can think of no psychological framework which is more emotionally important and sensitive than that associated with one's sexual identification.

Let's assume a situation in which a patient, married for some time and apparently female, comes to you asking why she has never menstruated or been able to become pregnant, and you ascertain that this individual is genetically male and is a case of the testicular feminization syndrome. If you divulge your complete findings and thereby destroy the patient's sexual identification as a female, I don't believe anyone could prevent emotional catastrophe in this patient and in her husband. I see no possible benefit, immediate or potential, to be derived from divulging complete information. I am familiar with two cases in which the total information was given carefully and with understanding and compassion. Nevertheless, the emotional impact was such that both previously happy marriages ended in divorce because of the inability of the members of each couple to look upon each other as they previously had.

CAPRON: However, isn't there a risk of cancer for these patients?

MacINTYRE: It is true that the testes probably should be removed because of the risk of malignant growth, but that does not mean they have to be specifically described as testes. They could be described by the general term *gonads,* and it should be pointed out that there is a risk of their becoming malignant and it is recommended that they be removed. It should be noted also that because of the developmental problem, the gonads would never be functional in a reproductive sense anyway. Counseling in this fashion protects the individual's identification with femininity which is all-important.

The question frequently will arise with respect to notifying the parents of a child with testicular feminization syndrome. Here, too, I think there is a potential danger. I have seen parents become terribly upset by such information to the extent that they are warped in their attitude toward the child thereafter. I believe that parents ought to know that the condition has a hereditary component but they don't have to find out that their "little girl" is really a "little boy."

It may be argued that by careful and lengthy counseling one could eliminate the dangers I have mentioned, but I don't think so. Regardless of how intelligent individuals may be, emotional stress and shock are a tremendous deterrent to clear understanding and full acceptance of a counselor's statements.

LEJEUNE: It is wrong to tell this woman she is a man, because she really is not. You may tell her that in general the chemical reactions which determine the male sex correspond to an XY chromosomal set, but that sometimes, as in her case, a special chemical change of the genes can produce a female with an XY chromosomal set. Thus there is no reason to conceal the truth. Do not tell her she is a male, which she is not, but that she is an exceptional female with an XY chromosome complement.

LUBS: The only long-term answer is education. If students learn more genetics and biology, it is possible that they can handle information such as this. But it is not appropriate right now with every patient.

MOTULSKY: The public is learning more and more about science, medicine, and genetics and often understands intellectually many of the processes involved. Intellectual understanding, however, does not mean that the emotional resources to accept bad news are available. Most people want a medical advisor or genetic counselor who knows science, but who is also a sympathetic human being who can decide what information will be in the best interest of the patient. I, as a patient, want that kind of physician. I would not want the "health technician" or the "genetics technician" who tells me the cold facts in an objective manner and then brings in a psychiatrist to make me feel better.

FURTHER READINGS

For lists of further sources see: *Current Literature in Family Planning* (Published Monthly by the Library, Planned Parenthood-World Population Information and Education Department, 810 Seventh Avenue, New York, N. Y. 10019); "Genetic Engineering: Evolution of a Technological Issue," (Science Policy Research Division, Congressional Research Service, Library of Congress, Serial W. November, 1972), pp. 109–119; Richard McCormick, "Genetic Medicine: Notes on the Moral Literature," *Theological Studies 33* (Sep., 1972), pp. 531–52; National Institute of Mental Health, National Clearinghouse for Mental Health Information, *Report on the XYY Chromosome Abnormality* (Public Health Service Publication No. 2103, October, 1970); National Library of Medicine, Literature search: #70-28 "Amniocentesis," and #70-42 "Genetics and Socially Deviant Behavior," and #71-2 "Genetic Counseling," (Literature Search Program, National Library of Medicine, Bethesda, Maryland 20014); James R. Sorenson, *Social and Psychological Aspects of Applied Human Genetics: A Bibliography* (Washington, D. C.: Fogarty International Center DHEW Publication No. NIH 73-412, 1973).

Books

Sir Macfarlane Burnet, *Genes, Dreams and Realities* (Aylesbury, Bucks: Medical and Technical Publishing Company, 1971); L. L. Cavalli-Sforza and W. F. Bodmer, *The Genetics of Human Populations* (San Francisco: Freeman, 1971); James F. Crowe and Motoo Kimura, *An Introduction to Population Genetics Theory* (New York: Harper & Row, 1970); T. Dobzhansky, *Genetics and the Evolutionary Process* (New York: Columbia University Press, 1970) and also by this author, *Mankind Evolving* (New Haven: Yale University Press, 1962); Albert Dorfman, ed., *Antenatal Diagnosis* (Chicago: University of Chicago Press, 1972); Lee Ehrman, Gilbert S. Omenn, and Ernst Caspari, eds., *Genetics, Environment and Behavior: Implications for Educational Policy* (New York: Academic Press, 1972); A. E. H. Emery, ed., *Antenatal Diagnosis of Genetic Disease* (Edinburgh: Churchill Livinstone, 1973); Amitai Etzioni, *Genetic Fix* (New York: Macmillan, 1973); Robert T. Francoeur, *Utopian Motherhood* (New York: Doubleday, 1970); Mark S. Frankel, *Genetic Technology: Promises and Problems* (Washington, D. C.: The George Washington University, Program of Policy Studies in Science and Technology, Monograph No. 15, March, 1973); Michael Hamilton, ed., *The New Genetics and the Future of Man* (Grand Rapids, Mich.: Eerdmans Publishing Company, 1972); Maureen Harris, ed., *Early Diagnosis of Human*

Genetic Defects: Scientific and Ethical Considerations (Fogarty International Center Proceedings, No. 6, 1972); Bruce Hilton et al., eds., *Ethical Issues in Human Genetics* (New York: Plenum Press, 1973); Arthur R. Jensen, *Educability and Group Differences* (New York: Harper & Row, 1973); Lawrence Lader, *Foolproof Birth Control* (Boston: Beacon Press, 1972); Gerald Leach, *The Biocrats* (New York: McGraw-Hill, 1970); Max Levitan and Ashley Montagu, *Textbook of Human Genetics* (New York: Oxford University Press, 1971); Kenneth M. Ludmerer, *Genetics and American Society* (Baltimore: Johns Hopkins Press, 1972); H. T. Lynch, *Dynamic Genetic Counseling for Clinicians* (Springfield, Ill.: Charles C Thomas, 1969); Victor McKusick, *Mendelian Inheritance in Man,* Third Edition (Baltimore: Johns Hopkins Press, 1971); Naccarato v. Grob, 162 *Northwestern Reporter* 2nd Series (1968), p. 305; Nuffield Provincial Hospitals Trust, *Screening in Medical Care* (Oxford: Oxford University Press, 1968); Planned Parenthood of Southeastern Pennsylvania, *Love, Sex and Birth Control for the Mentally Retarded–A Guide for Parents* (Planned Parenthood of Southeastern Pennsylvania and Family Planning and Population Center of Syracuse University, 1971); Robert A. Paoletti, ed., *Selected Readings: Genetic Engineering and Bioethics* (New York: MSS Information Corp., 1972); Ian H. Porter and Richard G. Skalko, eds., *Heredity and Society* (New York: Academic Press, 1973); Paul Ramsey, *Fabricated Man: The Ethics of Genetic Control* (New Haven: Yale University Press, 1970); L. E. Reisman and A. P. Matheny, Jr., *Genetics and Counseling in Medical Practise* (St. Louis: C. V. Mosby, 1969); J. A. Fraser Roberts, *An Introduction to Medical Genetics,* Fifth Edition (New York: Oxford University Press, 1970); Jonas Robotscher, ed., *Eugenic Sterilization* (Springfield, Ill.: Charles C Thomas, 1973); David Rorvik, *Brave New Baby: Promise and Peril of the Biological Revolution* (New York: Doubleday, 1971); Albert Rosenfeld, *The Second Genesis: The Coming Control of Life* (Englewood Cliffs, N. J.: Prentice-Hall, 1969); David Smith and Ann Asper Wilson, *The Child with Down's Syndrome* (Philadelphia: Saunders, 1973); Gordon R. Taylor, *The Biological Time Bomb* (Cleveland: World Publishing Company, 1968).

Articles

Margaret Adams, "Social Aspects of Medical Care for the Mentally Retarded," *New Journal of Medicine 286* (March 23, 1972), pp. 635–38; C. R. Austin, "Embryo Transfer and Sensitivity to Teratogenesis," *Nature 244* (August 10, 1973), pp. 333–34; B. Y. Barakat et al., "Fetal Quality Control in Pregnancies with High Risk for Genetic Disorders," *Fertility and Sterility 22* (July, 1971), pp. 409–15; Medora S. Bass, "Attitudes of Parents of Retarded Children toward Voluntary Sterilization," *Eugenics Quarterly 14* (March, 1967), pp. 45–53, and also by this author "Marriage, Parenthood, and the Prevention of Pregnancy for the Mentally Deficient," *American Journal of Mental Deficiency 55* (1963), pp. 318–33; Bruce Bennett and Oscar D. Ratmoff, "Detection of the Carrier State for Classic Hemophilia," *New England Journal of Medicine 288* (February 15, 1973), pp. 342–45; Dorothea Bennett and Edward A. Boyse, "Sex Ratio in Progeny of Mice Inseminated with Sperm Treated with H-Y Antiserum," *Nature 246* (Nov. 30, 1973), pp. 308–09; Samuel P. Bessman and Judith Swazey, "PKU: A Study of Biomedical Legislation," *Human Aspects of Biomedical Innovation,* eds. E. Mendelsohn, J. P. Swazey and Irene Taviss (Cambridge: Harvard University Press, 1971); Daniel Bergsma, ed. "Advances in Human Genetics and Their Impact on Society," *Birth Defects Original Article Series 8* (July, 1972), and the same author with others, "Contemporary Genetic Counseling," *Birth Defects Original Article Series 9* (April, 1973); Cecil Binney, "Legal and Social Implications of Artificial Insemination," *Eugenics Review 40* (April, 1948), pp. 199–204; C. P. Blacker, "Artificial Insemination: The Society's Position," *Eugenics Review 50* (1958), pp. 51–54; F. S. A. Brimblecombe and Jocelyn Chamberlain, "Screening for Cystic Fibrosis," *Lancet* (Dec. 22, 1973), pp. 1428–31; British Medical Association, Annual Report of the Council, "Appendix V: Report of Panel on Human Artificial Insemination," *British Medical Journal,* Supplement (April 7, 1973), pp. 3–5; Buck v. Bell 274 U.S. 200 (1927); Larry L. Bumpass and Harriet B. Presser, "Contraceptive Sterilization in the United States: 1965 and 1970," *Demography 9* (Nov., 1972), pp. 531–48; Daniel Callahan, "What Obligations Do We Have to Future Generations?" *American Ecclesiastical Review 164* (April, 1971), pp. 265–80; Patrick Carmody, "Tay-Sachs Disease: The Use of Tears for the Detection of Heterozygotes," *New England Journal of Medicine 289* (Nov. 15, 1973), pp. 1072–74; C. O. Carter et al., "Genetic Clinic: A Follow-Up," *Lancet* (Feb. 6, 1971), pp. 281–85; L. L. Cedarquist and F. Fuchs, "Antenatal Sex Determination: A Historical Review," *Clinical Obstetrics and Gynecology 13* (1970), pp. 159–77; C. A. Clarke, "Genetic Counseling," *British Medical Journal* (Mar. 4, 1972), pp. 606–609; Carol Clow et al., *"Management of Hereditary Metabolic Disease: The* Role of Allied Health Personnel," *New England Journal of Medicine 284* (June 10, 1971), pp. 1292–98; James F. Danielli, "Industry, Society, and Genetic Engineering," *Hastings Center Report 2* (Dec., 1972), pp. 5–7; Bernard D. Davis, "Prospects for Genetic Intervention in Man," *Science 170* (1970), pp. 1279–83; W. E. Dodson et al., "Cytogenetic Survey of XYY Males in Two Juvenile Court Populations, with a Case Report," *Journal of Medical Genetics* 9 (Sep., 1972), pp. 287–88; Editorial, "Genetic Engineering in Man: Ethical Considerations," *Journal of the American Medical Association 220* (May 1, 1972), p. 721; R. G. Edwards, "The Problem of Compensation for Antenatal Injuries," *Nature* (Nov. 9, 1973), pp. 54–55, and also by the same author: "The Problems of Artificial Fertilization," *Nature 233* (1971), pp. 23–25, and also with R. E. Fowler,

"Human Embryos in the Laboratory," *Scientific American 233* (1970), pp. 45–54, and also by the author with David J. Sharpe, "Social Values and Research in Human Embryology," *Nature 231* (May 14, 1971), pp. 87–91; A. E. H. Emery et al., "Social Effects of Genetic Counselling," *British Medical Journal* (March 24, 1973), pp. 724–26; C. J. Eptein et al., "Prenatal Detection of Genetic Disorders," *American Journal of Human Genetics 24* (March, 1972), pp. 214–26; R. J. Ericsson et al., "Isolation of Fractions Rich in Human Y Sperm," *Nature 246* (Dec. 14, 1973), pp. 421–24; Amitai Etzioni, "Doctors Know More Than They're Telling About Genetic Defects," *Psychology Today* (Nov., 1973), pp. 26 ff, and also by this author "Sex Control, Science and Society," *Science 161* (Sep. 13, 1968), pp. 1107–12; "Evaluating a Decade of PKU Screening," *Medical World News* (Nov. 19, 1971), p. 43; A. Falek, "Ethical Issues in Human Behavior Genetics: Civil Rights, Informed Consent and Ethics of Intervention", Position paper #6 presented at Chairman's Workshop on Developmental Human Behavior Genetics, Developmental Behavioral Science Study Section, Division of Research Grants, National Institute of Health, Bethesda, Md. April 16–17, 1974; E. J. Farris et al., "Emotional Impact of Successful Donor Insemination," *Obstetrics and Gynecology 3* (1954), pp. 19–20; Michael G. Farrow and Richard C. Juberg, "Genetics and Laws Prohibiting Marriage in the United States," *Journal of the American Medical Association 209* (July 28, 1969), pp. 534–38; George Felleenes, "Sterilization and the Law," *New Dimensions in Criminal Justice,* eds. H. K. Becker et al., (Metuchen, N.J.: Scarecrow Press, 1968); Joseph Fletcher, "Ethical Aspects of Genetic Controls," *New England Journal of Medicine 285* (Sep. 30, 1971), pp. 776–83, and also by this author: "Indicators of Humanhood: A Tentative Profile of Man," *Hastings Center Report 2* (Nov., 1972), pp. 104, and "Sterilization–Our Right to Foreclose Parenthood," *Morals and Medicine* (Boston: Beacon Press, 1960), chap. 1, and "Moral Problems in Genetic Counseling," *Pastoral Psychology 23* (April, 1972), pp. 47–60; A. T. Fort, "Counseling the Patient with Sickle Cell Disease About Reproduction: Pregnancy Outcome Does Not Justify the Maternal Risk!" *American Journal of Obstetrics and Gynecology 111* (1972), pp. 324–27; Clarke F. Fraser, "Genetic Counselling," *Hospital Practice* (Jan., 1971), pp. 49–56; G. R. Fraser, "The Short-Term Reduction in Birth Incidence of Recessive Diseases as a Result of Genetic Counseling after the Birth of an Affected Child," *Human Heredity 22* (1972), pp. 1–6; Theodore Friedmann, "Prenatal Diagnosis of Genetic Disease," *Scientific American 225* (Nov., 1971), pp. 34–42; Mark S. Frankel, "The Public Policy Dimensions of Artificial Insemination and Human Semen Cryobanking," Program of Policy Studies in Science and Technology (George Washington University, Monograph #18, Dec., 1973); Frank R. Freemon, "Pretesting for Huntington's Disease," *Hastings Center Report 3* (Sep., 1973), p. 13; Willard Gaylin, "Genetic Screening: The Ethics of Knowing," *New England Journal of Medicine 286* (June 22, 1972), pp, 1361–62, and also "We

Have the Awful Knowledge to Make Exact Copies of Human Beings," *New York Times Magazine* (March 6, 1972), pp. 10 ff; "Genetic Science and Man," *Theological Studies 33* (Sep., 1972); Albert Gerbie et al., "Amniocentesis in Genetic Counseling," *American Journal of Obstetrics and Gynecology 189* (March 3, 1971), pp. 765–70; Margaret Giannini and Lawrence Goodman, "Counseling Families During the Crisis Reaction to Mongolism," *American Journal of Mental Deficiency 67* (1963), pp. 740–47; Bentley Glass, "Human Heredity and Ethical Problems," *Perspectives in Biology and Medicine* (Winter, 1972), pp. 237–53; Martin Golding, "Ethical Issues in Biological Engineering," *U.C.L.A. Law Review 15* (Feb., 1968), pp. 443–79, and also by this author "Obligations to Future Generations," *The Monist 56* (Jan., 1972), pp. 85–99; H. Gordon, "Genetic Counseling: Considerations for Talking to Parents and Prospective Parents," *Journal of the American Medical Association 217* (Aug. 30, 1971), pp. 1215–25; I. I. Fottesman and L. Erienmeyer-Kimling, "Prologue: A Foundation for Informed Eugenics," *Social Biology 18,* Supplement: S1-S8 (Sep., 1971); Edward Grossman, "The Obsolescent Mother," *Atlantic Monthly* (May, 1971), pp. 39–50; James M. Gustafson, "Mongolism, Parental Desires, and the Right to Life," *Perspectives in Biology and Medicine 16* (Summer, 1973), pp. 529–57; Robert Guthrie, "Mass Screening for Genetic Disease," *Hospital Practice 7* (June, 1972), pp. 93–100; Michael Hemphill, "Pretesting for Huntington's Disease," *Hastings Center Report 3* (June, 1973), pp. 12–13; Bruce Hilton, "Will the Baby Be Normal? . . . and What is the Cost of Knowing?" *Hastings Center Report 2* (June, 1972), pp. 8–9; Kurt Hirschhorn, "On Re-Doing Man," *Annals of the New York Academy of Sciences 184* (June, 1971), pp. 103–12; A. R. Holder, "Voluntary Sterilization," *Journal of the American Medical Association 225* (Sep. 24, 1973), pp. 1743–44; Ernest B. Hook, "Behavioral Implications of the Human XYY Genotype," *Science 179* (Jan. 12, 1973), pp. 139–50; Lillian Y. F. Hsu et al., "Results in Pitfalls in Prenatal Cytogenetic Diagnosis," *Journal of Medical Genetics 10* (June, 1973), pp. 112–19; "Human Genetic Engineering," *Medical World News* (May 11, 1973), pp. 45–57; J. Huxley, "Eugenics in Evolutionary Perspective," *Perspectives in Biology and Medicine 6* (Winter, 1963), pp. 155–87; Institute of Society, Ethics and the Life Sciences, Research Group on Ethical, Social and Legal Issues in Genetic Counseling and Genetic Engineering, "Ethical and Social Issues in Screening for Genetic Disease," *New England Journal of Medicine 286* (May 25, 1972), pp. 1129–32; Patricia Jacobs, "Aggressive Behavior, Mental Sub-normality and the XYY Male," *Nature* (Dec. 25, 1965), pp. 1351–52, and the same author with others "Chromosome Surveys in Penal Institutions and Approved Schools," *Journal of Medical Genetics 8* (1971), pp. 49–53; Arthur R. Jensen et al., "Race, Intelligence and Genetics," *Psychology Today* (Dec., 1973), pp. 80–93; Michael M. Kaback. "Heterozygote Screening: A Social Challenge," *New England Journal of Medicine 289* (Nov. 15, 1973), pp. 1090–91; Leon R. Kass, "Babies by Means of

In Vitro Fertilization: Unethical Experiments on the Unborn?" *New England Journal of Medicine 285* (Nov. 18, 1971), pp. 1174–79, other articles by this author are: "The New Biology: What Price Relieving Men's Estate?" *Science 174* (Nov. 19, 1971), pp. 779–88, and "Making Babies: The New Biology and the Old Morality," *Public Interest 26* (Winter, 1972); Dale B. Kellon et al., "Physicians' Attitudes About Sickle Cell Disease and Sickle Cell Trait," *Journal of the American Medical Association 227* (Jan. 7, 1974), pp. 71–72; Harold Klawans et al., "Use of L-dopa in the Detection of Presymptomatic Huntington's Chorea," *New England Journal of Medicine 286* (June 22, 1972), pp. 1332–34; David Klein, "Genetic Manipulations," *Impact of Science on Society* (Jan.–Mar., 1973), pp, 21–27; F. I. D. Konotey-Ahulu, "Medical Considerations for Legalizing Voluntary Sterilization: Sickle Cell Disease as a Case in Point," *Law and Population Monograph Series* No. 13 (Fletcher School of Law, Tufts University, 1973); D. Kretzer et al., "Transfer of a Human Zygote," *Lancet* (Sep. 29, 1973), pp. 728–29; Marc Lappe, "Allegiances of Human Geneticists: A Preliminary Typology," *Hastings Center Studies* 1 (No. 1, 1973), pp. 63–78, other articles by this same author are: "The Genetic Counselor: Responsible to Who?" *Hastings Center Report 1* (Sep., 1971), pp. 6–11, and "Genetic Knowledge and the Concept of Health," *Hastings Center Report 3* (Sep., 1973), pp. 1–3, and "How Much Do We Want to Know About the Unborn?" *Hastings Center Report 3* (Feb., 1973), pp. 8–9, and "Human Genetics," *Annals of the New York Academy of Sciences 216* (May 18, 1973), pp. 152–59, "Risk-taking for the Unborn," *Hastings Center Report 2* (Feb., 1972), pp. 1–3, and with Peter Steinfels, "Choosing the Sex of Our Children," *Hastings Center Report 4* (Feb., 1974), pp. 1–4; J. Lederberg, "Experimental Genetics and Human Evolution," *Bulletin of the Atomic Scientists 23* (Oct., 1966), pp. 4–11, and also "Options for Genetic Therapy," *Medical Dimensions* (March, 1972), pp. 16 ff; J. LeJeune, "The William Allan Memorial Award Lecture–On the Nature of Man," *American Journal of Human Genetics 22* (March, 1970), pp. 121–28; Claire O. Leonard et al., "Genetic Counseling: A Consumer's View," *New England Journal of Medicine 287* (August 31, 1972), pp. 433–39; John Marshall, "Sterilization," *The Ethics of Medical Practice* (London: Darton, Longman and Todd, 1960), chap. 8; D. Mayo, "On the Effects of Genetic Counseling on Gene Frequencies," *Human Heredity 20* (1970), pp. 361–70; Charles McFadden, "Sterilization," *Medical Ethics* (Philadelphia: F. A. Davis, 1967), chap. 13; P. B. Medawar, "Do Advances in Medicine Lead to Genetic Deterioration?" *Mayo Clinic Proceedings 40* (1965), pp. 23–33; David W. Meyers, "Voluntary Sterilization" and "Compulsory Sterilization and Castration," *The Human Body and the Law* (Chicago: Aldine Press, 1970), pp. 1–47; A. Milunsky, "Prenatal Genetic Diagnosis," *New England Journal of Medicine* (Dec. 17, 1970), pp. 1370–81, (Dec. 24, 1970), pp. 1441–47, and (Dec. 31, 1970), pp. 1498–1504; Arno G. Motulsky, "Human and Medical Genetics: A Scientific Discipline and an Expanding Horizon," *American Journal of Human Genetics 23* (March, 1971), pp. 107–24, and also with others "Public Health and Long-term Genetic Implications of Intrauterine Diagnosis and Selective Abortion," *Birth Defects 7* (1971), pp. 22–32; H. J. Muller, "The Guidance of Human Evolution," *Perspective in Biology and Medicine 3* (1959), pp. 1–43; Charles W. Murdock, "Civil Rights of the Mentally Retarded: Some Critical Issues," *Notre Dame Lawyer* (1972), pp. 133 ff; John R. Murphy, "Sickle Cell Hemoglobin (Hb AS) in Black Football Players," *Journal of the American Medical Association 225* (August 20, 1973), pp. 981–82; Roberts F. Murray, Jr., "Problems behind the Promise: Ethical Issues in Mass Genetic Screening," *Hastings Center Report 2* (April, 1972), pp. 11–13; James V. Neel, "Thoughts on the Future of Human Genetics," *Medical Clinics of North America 531* (July, 1969), pp. 1001–11, and also "Lessons from a 'Primitive' People," *Science 170* (Nov. 20, 1970), pp. 815–22; John T. Noonan, Jr., "Sterilizing Operations," *Contraception* (Cambridge: Harvard University Press, 1966), pp. 451–60; Simon Olshansky, "Chronic Sorrow: A Response to Having a Mentally Defective Child," *Social Casework 43* (1962), pp. 190–93; Julius Paul, "Return of Punitive Sterilization Proposals," *Law and Society Review 3* (August, 1968), pp. 77–106; J. H. Pearn, "Patients' Subjective Interpretation of Risks Offered in Genetic Counseling," *Journal of Medical Genetics 10* (June, 1973), pp. 129–34; Nancy Phillips, "The Prevalence of Surgical Sterilization in Suburban Populations," *Demography 8* (May, 1971), pp. 261–70; P. Pinkerton, "Parental Acceptance of the Handicapped Child," *Developmental Medicine and Child Neurology 12* (April, 1970), pp. 207–12; Edward Pohlman and Daniel Callahan, commentators, "Case Studies in Bioethics: Good Incentives for Sterilization: Can They Be Just?" *Hastings Center Report 3* (Feb., 1973), pp. 11–12; Tabitha Powledge, "Laws in Question: Confusion over Sickle Cell Testing," *Hastings Center Report 2* (Dec., 1972), pp. 3–5, and also by this author: "New Trends in Genetic Legislation," *Hastings Center Report 3* (Dec., 1973), pp. 6–7, and "The New Ghetto Hustle," *Saturday Review of the Sciences* (Feb., 1973), pp. 38–47; Harriet B. Presser, "Voluntary Sterilization: A World View," *Reports on Population/Family Planning #5* (July, 1970), and also with Larry L. Bumpass, "The Acceptability of Contraceptive Sterilization amoung U. S. Couples: 1970," *Family Planning Perspectives 4* (Oct., 1972), pp. 18–26; Paul Ramsey, "Genetic Engineering," *Bulletin of the Atomic Scientists* (Dec., 1972), pp. 14–17, and also by this author: "Parenthood and the Future of Man by Artificial Donor Insemination, et cetera, et cetera," *Fabricated Man* (New Haven: Yale University Press, 1970), chap. 3, and "Shall We Clone a Man?" *Fabricated Man* (New Haven: Yale University Press, 1970), chap. 2, and "Shall We 'Reproduce'?" *Journal of the American Medical Association 220* (June 5, 1972), pp. 1346–50, and (June 12, 1972), pp. 1480–85; Warren T. Reich and Harmon Smith, commentators, "Case Studies in Bioethics: On the Birth of a Severely Handicapped Infant," *Hastings Center Report 3* (Sep., 1973), pp.

10–12; Philip Reilly, "Sickle Cell Anemia Legislation," *Journal of Legal Medicine 1* (Sep.–Oct., 1973), pp. 39–48, and (Nov.–Dec., 1973), pp. 36–40; Albert Rosenfeld, "Procreation Without Sex," *Physician's World* (May, 1973), pp. 61–63; Jeannie Rosoff, "Sterilization: The Montgomery Case," *Hastings Center Report 3* (Sep., 1973), p. 6; G. Sabagh and R. B. Edgerton, "Sterilized Mental Defectives Look at Eugenic Sterilization," *Eugenics Quarterly 9* (1962), pp. 213–22; Robert L. Sinsheimer, "Ambush or Opportunity?" *Hastings Center Report 2* (Sep., 1972), pp. 4–7; Keith D. Smith and Emil Steinberger, "Survival of Spermatazoa in a Human Sperm Bank: Effects of Long-Term Storage in Liquid Nitrogen," *Journal of the American Medical Association 223* (Feb. 12, 1973), pp. 774–77; James R. Sorenson, "Social Aspects of Applied Human Genetics," (The Russell Sage Foundation, 1971); Pierre Soupart and Larry L. Morgenstern, "Human Sperm Capacitation and In Vitro Fertilization," *Fertility and Sterility 24* (June, 1973), pp. 462–78; Zena Stein et al., "Screening Programme for Prevention of Down's Syndrome," *Lancet* (Feb. 10, 1973), pp. 305–09; Emil Steinberger and Keith D. Smith, "Artificial Insemination with Fresh or Frozen Sperm: A Comparative Study," *Journal of the American Medical Association 223* (Feb. 12, 1973), pp. 778–83; Norman St. John Stevas, "Human Sterilization," *Life, Death and the Law* (Bloomington: Indiana University Press, 1961), chap. 4; M. Stenchever, "An Abuse of Prenatal Diagnosis," *Journal of the American Medical Association 221* (July 24, 1972), p. 408; Robert L. Tips, "Impact of Genetic Counseling upon the Family Milieu," *Journal of the American Medical Association 184* (April 20, 1963), pp. 183–86; John R. G. Turner, "How Does Treating Congenital Diseases Affect the Genetic Load?" *Eugenics Quarterly 15* (Sep., 1968), p. 191; Carlo Valenti, "Antenatal Detection of Hemoglobinopathies," *American Journal of Obstetrics and Gynecology 115* (March 15, 1973), pp. 851–53, and also by this author with others "Prenatal Sex Determination," *American Journal of Obstetrics and Gynecology 112* (April 1, 1972), pp. 890–95; Robert M. Veatch, "The Unexpected Chromosome . . . A Counselor's Dilemma," *Hastings Center Report 2* (Feb., 1972), pp. 8–9; Angela M. Viana et al., "Searching for XYY Males through Electrocardiograms," *Journal of Medical Genetics 9* (June, 1972), pp. 165–67; Walter Wadlington, "Artificial Insemination: The Dangers of the Poorly Kept Secret," *Northwestern University Law Review 64* (1970), pp. 777–807; James D. Watson, "Moving Toward Clonal Man: Is This What We Want?" *Atlantic Monthly*, (May, 1971), pp. 50–53; World Health Organization, "Genetic Counseling: Third Report of the W.H.O. Expert Committee on Human Genetics," W.H.O. Technical Report Series, No. 416 (Geneva, 1969); G. H. Zuk, "The Religious Factor and the Role of Guilt in Parental Acceptance of the Retarded Child," *American Journal of Mental Deficiency 64* (July, 1959), pp. 139–47.

PART V: DYING

INTRODUCTION

Possessing doctor's degrees in both medicine and law, Professor Jonas Robitscher views the problems encountered in the phenomenon of death from a somewhat privileged vantage point. In his essay, "The Problems in Prolongation of Life," Robitscher introduces the reader to almost every issue of importance in this area of concern—the problems involved in defining "death," the distinction between active and passive euthanasia, the difficulties involved in securing legal recognition of "living wills," the distinction between ordinary and extraordinary means of preserving life, etc. The central theme of Robitscher's paper, however, is the right to die. Robitscher contends, albeit with some reservations, that this right is now both legally accepted and socially approved. Thus, the main issue as he sees it is whether or not we can devise ways of implementing this right. Robitscher's suggestion in this regard is that individuals make every effort to find physicians who share their own philosophy of life; for it is only in this way that a person can be assured that his wishes concerning his own death will be observed. But medical personnel are in short supply; and Robitscher acknowledges that his suggestion may not be practicable. If so—if such doctor–patient relationships are not feasible—then, Robitscher says, medicine has taken the wrong path in its search for greater technological proficiency.

One issue that Professor Robitscher does not discuss in detail is the question of how hospitals should go about allocating exotic medical lifesaving resources, when it is clear that demand for such resources outstrips supply. Professor Nicholas Rescher contends that the problem is not one that can be solved purely on the basis of medical considerations, and with this in mind he attempts to supply hospital administrators with a set of rational guidelines for making choices in life-and-death decisions.

The selection procedure ultimately defended by Rescher is a three-step affair. First, by utilizing general criteria of selection which Rescher calls "criteria of inclusion," a first phase selection group would be chosen from among all individuals available for treatment. This group would be substantially larger than the number of people actually able to be treated by exotic lifesaving techniques; hence, members of this group would be further culled by application of a second set of selection criteria involving individual comparisons. As conceived by Rescher, the resultant second phase selection group would only be slightly larger (e.g., by one-third or one-half) than the number of individuals able to be accomodated with exotic lifesaving therapy. From this group, then, final selection of patients would take place by random selection.

Rescher acknowledges that there is no one selection procedure which can lay claim to the title, "the best." At the same time, he defends his system, arguing that it is simple enough to be understood by the common man, reasonable, and most of all, fair.

Professor Henry Aiken argues that terms like *life* and *human* are ambiguous in that they may be used in more than one way. In a purely biological or biophysical sense, he says, a being may be said to be alive and human if he possesses certain specifiable characteristics. At the same time, a being may meet these criteria and be totally incapable of ever participating in forms of activity which yield significant experience. And where this is the case, Aiken contends, there is another sense in which it is improper to predicate either humanity or life of the organism. The question, then, is whether we see the right to life as being something possessed by beings who are "human" and "alive" in the purely descriptive, biological sense, or whether we see that right as somehow being tied to the potential for a full, *meaningful* human existence. Aiken acknowledges that there must be a right to life in the biological sense of the term, for if there were no such right, there would be no right to a meaningful life, and hence no right to life at all. Still, he insists that the right to biological survival is entirely contingent upon the ability of an individual to make, with the help of others, a meaningful life for himself.

If accepted, one of the logical implications of Aiken's position is that no doctor would be put under an obligation to keep alive a patient suffering from certain kinds of brain damage. Under some conditions, then, passive euthanasia (at least) would be condoned. This theme, along with several others, is taken up by Professor Robert Morison in his essay, "Dying."

In discussing the problems surrounding the various attempts to define "death," Morison notes that there are some patients in whom elementary vital signs have not failed although higher brain functions, e.g., thinking, consciousness, etc., have departed. Should patients of this sort be considered dead (or nonhuman), so that essential life-giving functions could be withdrawn? Morison does not favor such a "definitional" solution to the problem. Rather he argues, in essential agreement with Aiken, that "it seems preferable to face up to the fact that under these circumstances a patient may still be living in some sense, but that the obligation to treat the living is neither absolute nor inexorable." But whereas Aiken is never clear as to whether this position could (or should) be used to support active euthanasia, Morison rejects such a course of action on the basis of "simple human reactions."

Another major issue discussed by Morison concerns the patient's right to a death with dignity. He considers Robitscher's suggestion that patients seek out physician-friends, and rejects it as being somewhat "idyllic," at least under present conditions. In addition, he rejects attempts to solve the problem by means of legislation, or by assigning authority concerning treatment of terminally ill patients to next-of-kin, ombudsmen, or committees of social scientists, philosophers, and theologians. In actual practice, he says, "the conduct of a drawn-out terminal illness involves a series of small decisions, based on repeated evaluations of the physical and emotional condition of the patient. . . ." And given this as the situation, he does not see how an outsider can be fitted into the unobtrusive incremental process, nor does he see how the problem lends itself to tidy legislative resolution. Still, Morison is not wholly pessimistic. He notes that the problem is being dealt with in informal ways, and he holds out hope that these modes of attack may be productive.

The final two essays in the section on dying are concerned with the definition of "death." Professors Capron and Kass argue that the definition of "death" is not a purely medical matter, and that the public's interest would be best served if the standards for determining human death were set by legislative action. Given this belief, the authors take the following steps: First, they outline in some detail the principles which they feel should govern the formulation of any acceptable statutory definition of death. Second, they evaluate the recently enacted Kansas Death Statute in the light of these principles, and find it wanting. And finally, they offer an alternative to the Kansas Statute, expressing hope that their proposal will serve as a catalyst for future public debate.

Professor MacKinnon attacks the problem of the definition of "death" in a slightly different way. Her essay focuses on a debate between Professors Morison and Kass concerning the nature of death. Professor Morison's position is that death is typically a slow, continuous process, taking place over time. Thus, on his view, it is virtually impossible to specify any time T at which a patient can be said to have died. Professor Kass, on the other hand, argues that we need not abandon the concept of death as an event, nor need we cease in our efforts to find criteria for determining when death has occurred. Professor MacKinnon seeks a middle course. She argues that "death could be an event which is a definite happening causing a dramatic change and yet take some time." Further, she suggests that the debate over whether death is a process or event is one which rests upon different presuppositions concerning the nature of time and change. And if this is so, adequate resolution of the disagreement will not be possible until the alternate assumptions have been brought to the light of day.

THE PROBLEMS IN PROLONGATION OF LIFE

JONAS ROBITSCHER

The issue of the right to die was the focus of controversy a few years ago. Could someone who was ill and tired of the struggle elect not to fight for life anymore? Could we allow our fellow travelers in life to give up or were we forced to force them to continue a life that might be pain-ridden and without hope?

This was a new problem on the scene, a problem caused by such advances in medical care as blood transfusions, intravenous feeding, antibiotics, respirators, pacemakers, and other drugs, procedures, and techniques that enabled us to maintain life in the critically ill, the terminally ill, and even in those who met what had been the criteria for the declaration of death—the cessation of heartbeat and respiratory activity.

Before these new advances an occasional moribund or comatose patient had been maintained for years and even for decades, but this was a rarity. Doctors were not concerned about keeping people alive too long, and there had not been a demand for the clarification of criteria for the decision to end life support.

When we discovered that we had a problem on our hands—that life, on one hand, could sometimes be maintained embarrassingly long as one example, that doctors did not know about their liability for harvesting organs from patients who did not meet the traditional criteria for "dead people" as another—we began to work with this problem and we looked to court decisions as one way of determining a doctor's responsibility.

Some of the first court decisions on this subject only served to confuse us. The leading cases all dealt with Jehovah's Witnesses who refused blood transfusions, sometimes to their own detriment, sometimes additionally to the detriment of an unborn child. The legal argument might have been in these cases that everyone who is mature and competent has traditionally had a right to decide what medical treatment he wants and what medical treatment he does not want, and the right to die then could have been established with less confusion. Instead, these cases were argued on a narrower ground, that freedom of religion, guaranteed by the First Amendment to the Constitution, required that we let

This paper was originally presented at a Colloquium on Bioethics entitled "Man in Our Image" held at Princeton University, March 1, 1975.

these people refuse treatment. The fact that some people gained the right to die by virtue of their religious convictions left us with the implication that those who did not have the same religious conviction might not have the right to die, and so the law in this area was off on the wrong foot.

Doctors who perhaps needlessly feared criminal indictments if they let patients die a natural death without attempting to extend their lives maintained patients for what we would now see as unconscionably long periods of time; the relatives who had to make the decision which the doctors would follow did not feel society's support for a decision to terminate life maintenance. So further resort to the courts was needed to let us know what we probably should have known all along—that we did indeed have a right to die. This right is not as clearly established as some other of our legal rights. Not all doctors interpret this right as freely as others do, the word has not gotten around to all people involved with these decisions, and some families still fear legal liability or excessive guilt if they order the discontinuance of treatment. We have the great problem of protecting patients who because of their illness are no longer competent to make informed decisions. We have the further problem of close decisions in difficult cases. (A doctor who can clearly recommend that intravenous feeding be discontinued might find it more difficult to order oral feeding to be terminated.) Nevertheless, with a few exceptions and some qualms, we can say that the right to die is both legally accepted and socially approved.*

The cases which have recently given us some assurance that we do indeed have a right to die are not very plentiful, and they are not from courts sufficiently high so that they can be entirely relied on for legal precedent. We do not have anything comparable to *Roe vs. Wade* or *Doe vs. Bolton,* the Supreme Court's famous abortion decisions in 1973, to serve as expressions of judicial authority. But we have had enough cases from courts of inferior jurisdiction, enough physicians have publicly described their withdrawal of life support and

*This article was written before the Karen Ann Quinlan case in which a New Jersey judge declared that the decision to prolong the life of an incompetent (comatose) patient in spite of her persistent and apparently irreversible vegetative state was a "medical decision" in which the wishes of next-of-kin had no weight. The judge in this case relied on medical testimony that life prolongation was normal procedure for a patient in this condition, but public discussion of this case since the decision indicates most doctors would not prolong the life of this patient indefinitely; the judge was therefore placing his credence in medical testimony which represented a minority medical view and he was making the doctor responsible for the decision (although traditionally the patient or if the patient is incompetent someone close to him representing his best interest should make the decision after receiving the doctor's advice on medically acceptable alternatives). *In the Matter of Karen Quinlan,* Docket No. C-201-75, Superior Court of New Jersey, Chancery Division, Morris County, November 10, 1975. If other jurisdictions adopt a similar view, so-called Death With Dignity statutes holding the doctor free from harm for terminating prolongation procedures under specified conditions will be pushed in state legislatures.

have not been prosecuted or even condemned for this so that we can be satisfied that the right to die has a fairly firm foundation.

Now that we are reasonably satisfied of this, a whole other series of questions raises itself—who shall exercise this right? How do we satisfy ourselves of the competency of the person who is asserting this right? If the decision maker in this process is not the dying patient but someone who will substitute a judgment for the patient's judgment, how do we know that the substituted judgment approximates the judgment that the patient would have made if he were able to make his own decision?

We ask ourselves these questions, we get involved in these concerns, because we are worried not only about the duration of life but the circumstances of its ending. We wish for all patients what we wish for ourselves—a peaceful death when all possible has been done to restore health and function but before the maintenance of life has become an indignity at the least and possibly a torture. How can we attain this wished-for kind of death in a society where on the one hand an overzealous physician may want to put us out of our misery too soon to make our bed available for a more promising patient, or on the other hand an overzealous physician may want to continue our life beyond a point where that life should be continued in order to satisfy himself personally that he is stronger than death? I have just used the concept of a point beyond which life should or should not be continued, and that would be stricken down by any court as unconstitutionally vague. Still, if we have a right to die that means we can determine to some extent the timing and conditions of our death, and that in turn means that we can strive for a certain *quality of death,* a death with dignity, a death that saves us from a meaningless prolongation of a painful existence or the maintenance of some body processes in the absence of consciousness and the hope of regaining consciousness.

The law, which embodies and creates popular opinion, has always told us that although we do not have to submit to medical treatment we do not want—even though it may save our lives—we do not have a right to take our own lives in a more active way. And the law has often leaned over in the direction of the assumption of mental incompetence when it has seen people trying actively to end their lives or passively allowing their lives to be ended. The law has traditionally been opposed to suicide, although some commentators say that American law, in contrast to English law, is ambivalent on the criminality of suicide. Some states still follow the English Common Law, now repealed in England, that suicide is a crime and—what is more important to the individual bent on self-destruction—that attempted suicide is also a crime. The old English law had the property of the suicide reverting to the Crown, but American law has not been that punitive. Nevertheless, attempted suicide is a crime or a misdemeanor in many jurisdictions and although we do not actually prosecute the attempter we use the police power of the state to prevent the suicide and to take custody of the person. In states where suicide is not an offense, we act on the assumption that the attempter is insane and send him to a psychiatric hospital for evaluation. We are determined not to allow people to kill themselves if they do this in a violent way or if we are not sure they are competent to make decisions in their own best interests.

"So-called 'rational' suicide is a rarity," says Dr. George Murphy, Professor of Psychiatry at Washington University. "The descriptive facts are that most persons who commit suicide are suffering from clinically recognizable psychiatric illnesses often carrying an excellent prognosis; that the majority have sought help from physicians for their symptoms; and that few have received the indicated treatment."[1] He excepts from this description the "few persons who commit suicide" who are suffering from a terminal illness, and he notes that some people—who he says speak on a philosophical rather than a clinical basis—condemn any effort to interfere with an individual's behavior against his wishes. He refers, of course, to Dr. Thomas Szasz and his co-believers in freedom of action circumscribed only by criminal proscriptions, not by mental health control of behavior.

Dr. Szasz represents the minority view and Dr. Murphy the majority; we tend to see—correctly or incorrectly—the would-be suicide as having mental illness and we treat him accordingly.

But other forms of self-destruction less sudden and less capable of being seen as the result of mental derangement are not interfered with in our society.

The law allows anyone who is not seen as incompetent to take extraordinary risks with his life—to jump or attempt to jump the Snake River Canyon, to compete in stock car races, to continue drinking after a diagnosis of cirrhosis of the liver, to smoke after a diagnosis of emphysema—and although such a course of action may be as truly suicidally motivated as an overdose or a self-inflicted gunshot wound, society elects not to interfere.

In previous generations, but not so very long ago, medicine differed from today's practice in important respects. First, doctors did as much harm as they did good, and well educated people—Thomas Jefferson among them—learned as much about medicine as they could to be free of the doctor's strange prescriptions and idiosyncratic kind of care. Avoiding medical treatment was not seen as necessarily pathological. Second, medical treatments caused extreme pain, and the choice of an individual to avoid medical help and to bear his illness or succumb to it without seeking aid was easily understandable. Third, medicine was much more pecuniarily oriented than it is today; there were no third-party payers, prepaid health plans, no Medicare and Medicaid, so a patient's decision not to seek the services of a doctor were seen as not different from a refusal to

[1] Murphy, G.: Suicide and the right to die, editorial, *Am. J. Psychiat.* *130*:4, April 1973, 472–473.

enter into any other kind of contract which required him to pay. Fourth, there was a doctrine of individual liberty and more recently—very recently—a doctrine of privacy, an area where the state should not enter. The idea that a patient has a right not to be treated has an honorable history; indeed even today a doctor who treats a patient without that patient's consent—short of an emergency situation and a patient unable to give consent to an important procedure—is guilty of a technical battery even where no actual damage has been done to the patient and there is no accusation of negligence. The well-established doctrine of informed consent as a prerequisite for treatment has been additionally emphasized since 1960 when courts began to insist on a broader disclosure to the patient and more patient participation in the decision-making process. The doctor is no longer the captain of a ship with authority over the mutinous crew member. Patients can participate in treatment decisions.

Certainly many people have elected not to take prescribed medical treatment. Christian Scientists and many others who avoided doctors functioned without medical care and exercised their right to refuse treatment without court intervention. They died, or lived, without any issue that brought their refusal to go along with doctors' recommendation into court.

Jehovah's Witnesses were another case. They wanted medical treatment, they did not avoid doctors and hospitals, but they did not want blood transfusions to be a part of that treatment. As hospital patients they represented an ambivalence that was hard to deal with—they wanted surgery but did not give consent to the transfusions without which the surgery might not be safe. Physicians were unwilling to proceed and risk allegations of negligence or malpractice; they sought court rulings on whether the patient had the right to refuse treatment. The issue could have been easily resolved if the patient had given up all claim to treatment—if he had just picked himself up or been picked up and had gone home—but courts did not seem to consider that option. Then Judge and now Chief Justice Warren Burger was one of the few judges who saw to the heart of this issue; in a famous dissent in a case in which a blood transfusion was ordered over the objection of the patient, he said that the right to be let alone included the right to refuse medical treatment even at great risk.

Although the Jehovah's Witnesses cases presented the courts with a strong ground for allowing the refusal of treatment, religious convictions guarded by the First Amendment, the courts refused to allow an absolute right to die. If a mother was pregnant—these cases arose before the abortion decision—the life of the fetus was considered so important and so involved with the mother's life that transfusions were ordered for the mother to save the life of the fetus. Similarly, if a mother or father had minor children, courts said they had the right to die but not to orphan small children. The protection of the fetus might not be seen as valid by today's courts since the abortion decision has given the mother the right to terminate the life of the fetus. Two other bases were found to limit the right to die, and both of these would still stand in any new case. If the patient was a minor it could not be presumed that his wish to die represented an informed opinion, and he would be forcibly transfused or treated so that when he reached his majority he would be in a position to make such life-and-death decisions on his own. Similarly, if a patient was unconscious or otherwise incompetent and the next-of-kin declined to consent for life-saving procedures, the court would not assume that if the patient were conscious and rational he would actually desire to die—even though he may have had a history of dedication to a religious sect that prohibits treatment—and the court would want to err on the side of life rather than to allow death.

It was in such a case that a New Jersey court said a few years ago, I think incorrectly, that, "There is no Constitutional Right to Die." A better expression of the law is to regard the Jehovah's Witnesses cases as clearly setting forth the right to die but somewhat limiting its application and basing it on the narrow ground of religious belief.

When life-saving techniques led doctors to the point where they were concerned with their abilities to keep patients alive for too long, they were not sure whether all their patients, or just their religious patients, had the right to refuse medical treatment. The problem was compounded by laws in some states—Florida is an example—which made doctors fear that if they did not force treatment on a patient and that patient died, they would be an accessory to that death.

The problem had been developing for at least fifty years, and particularly since 1926. That was the year that the intravenous drip technique of feeding was introduced for the maintenance of surgical patients. The new method also made it possible to maintain comatose medical patients more easily and more successfully than the earlier method of tube feeding. This new technique was a natural development from the improved methods of blood transfusion that were used in World War I which had application to surgical patients particularly. Eventually knowledge of electrolytes and the use of plasma during World War II would extend the applicability of intravenous transfusion and nourishment.

The World War II period also saw the introduction of sulfa drugs, and shortly afterwards, antibiotics. Doctors now had a combination of tools to treat debilitated patients and comatose and semicomatose patients; they could supply their nutritional needs and keep them from dying from the infections that had carried off so many seriously ill patients. Pneumonia had been called "the old man's friend" by generations of medical students and doctors; it ushered out those patients with progressive malignancies, with conditions associated with advanced age, and in comas. Now the old man's friend could be barred from the premises.

We have had many more methods of maintaining life—some of them the results of advances in electronic engineering. We have the cardiac pacemaker, electrical cardiac monitoring devices to alter out-of-synchronization heart rhythms. New knowledge of respiratory physiology and advances in oxygen therapy techniques allow patients to be supported through critical periods. Dialysis became available to compensate for kidney failure. Surgery, developing the radical new method of organ transplants, the kidney transplant, if not the heart transplant, allowed doctors to extend the lives of many patients. More recently the membrane lung, outside of the body, can increase the oxygenation of the blood and decrease the carbon dioxide level while lung tissue heals.

Before the introduction of these so-called extraordinary medical measures, doctors could relieve pain and they had some helpful remedies; they were particularly successful in intervening in life-threatening surgical conditions. Still, knowledge of medicine had not given the doctor much power to keep death at bay, and doctors were accustomed to fighting death hard but not fighting death long.

With new methods at their disposal, fending off death became a medical preoccupation. In the earlier period an occasional patient had been kept alive by diligent nursing and tube feeding. Comas of lengthy duration—one record coma of more than thirty years—had been maintained, the result of extremely diligent nursing care. Now methods were available—called "extraordinary" to distinguish them from traditional medical and nursing methods and which in many cases required the cooperation of many specialties and the facilities of large medical centers—by which doctors interested in setting a new record—for the maintenance of a patient in a state of coma longer than any previous case—could aim for inclusion in the *Guinness Book of World Records* with the unwitting assistance of the comatose patients. Some doctors seemed determined to try for this record. Dr. Robert Williams of the Department of Medicine, University of Washington, says that so-called passive euthanasia—the withdrawal of life support—is practiced with less frequency and with much greater delay than is desired by patients, their families, and others, and that although in the last eight years he has witnessed a rapid increase in the number of people with understanding and appreciation of passive euthanasia, "there remains a very strong opposition in the minds of some, especially among those who claim that such approaches constitute usurpation of the role of God." (Letter, *Commentary,* May 1974).

Many physicians, particularly those in modern hospitals with all the new medical hardware available—following the maxim that anything that is invented will find a use—ignored the direction of Arthur Hugh Clough, the Victorian poet, who told doctors more than a century ago:

> Thou shalt not kill; but need not strive
> Officiously to keep alive.

The Jehovah's Witnesses cases were efforts to come to a better legal delineation of the conditions which allowed and did not allow either an intelligent patient or someone acting in the interest of an incompetent patient to refuse potentially life-saving procedures. Although they told us that patients could make this decision for themselves if they were competent, the fact that this right was based on a religious ground raised the question of its application generally. These cases did not give us clear guidelines for incompetent patients who were terminally or irretrievably ill; they dealt with emergency situations where a blood transfusion might make the difference between life and death. They did not tell us whether a substituted consent could be the authority for the discontinuance of treatment in comatose or terminally ill patients. In contrast to the Jehovah's Witnesses cases where doctors urged treatment, some doctors, those not interested in the Guinness book, found themselves on the patient's side of the decision to discontinue treatment but not knowing if this placed them in legal peril. The Jehovah's Witnesses cases were misleading; they contained some hazy implication that if the rationale for allowing Jehovah's Witnesses to refuse treatment was their religious belief, those who did not have any such religious belief perhaps should be subject to coerced treatment. In the discussion that raged about whether there was a right to die, it was not pointed out that the Jehovah's Witnesses cases involved good prognosis patients. Most of the right to die situations, however, concern the very elderly, the fatally ill, the terminally ill.

Legal peril is not the only kind of peril; there is moral peril, too. But the traditional medical codes of ethics and the Nuremberg Code and the Helsinki Convention did not give clear guidance. The Hippocratic Oath states: "The regimen I adopt shall be for the benefit of my patients according to my ability and judgment, and not for their hurt or for any wrong. I will give no deadly drug to any, though it be asked of me, nor will I counsel such. . . ." It very clearly prohibits active euthanasia, but it does not directly deal with the termination of life-maintaining techniques, sometimes called passive euthanasia. The Declaration of Geneva, adopted by the General Assembly of the World Medical Association in 1948, emphasizes respect for life but goes no further: "I will practice my profession with conscience and dignity. The health of my patient will be my first consideration." and: "I will maintain the utmost respect for human life, from the time of conception; even under threat, I will not use my medical knowledge contrary to the laws of humanity."

The "Principles of Medical Ethics" of the American Medical Association which dates back to 1957 gives us even less of a guideline: "Ethical principles are basic and fundamental. Men of good conscience inherently know what is right or wrong, and what is to be done or to be avoided." A brief history of medical ethics accompanying the Principles does refer favorably to the Hippocratic Oath as "a living" and "workable" statement.

In the midst of uncertainty, doctors looked to the courts to confirm that they did not have to "strive officiously to keep alive." A Miami judge ruled that a seventy-two-year-old woman suffering from hemolytic anemia had the right to live or die with dignity and could refuse blood transfusions or surgery that would cause pain. The transfusions were only palliative and not remedial. The patient died the day after the judge's decision. Said the judge: "A person has the right not to suffer pain. A person has the right to live or die in dignity."

In Milwaukee a seventy-seven-year-old woman was allowed, after a bedside hearing in which the judge satisfied himself that she understood the effects of her decision, to refuse to consent to amputation of a leg that was gangrenous from a complication of diabetes. The hospital had insisted that she go through with the treatment. This judge said, "I believe we should leave [the patient] depart in God's own peace."

When the patient was incompetent and a relative believed it was best for him to die, courts were less supporting. A New York State Supreme Court Justice named the director of New York Hospital-Cornell Medical Center as the official guardian with authority to consent to the implantation of a new pacemaker battery for a seventy-nine-year-old patient after his wife had repeatedly refused to authorize the procedure. The patient was considered incompetent, unaware of his condition, incapable of making the decision. The wife had objected to the operation on the ground that he was "turning into a vegetable"; the court was influenced by testimony that the old gentleman, although deteriorated, was not in pain. The wife's reaction is given in a newspaper interview: "What has he got to live for: Nothing. He knows nothing, he has no memory whatsoever, he is turning into a vegetable. Isn't death better?"

When the courts felt that patients were incompetent, they were likely to presume that if they were competent they would want medical treatment. Two recent New Jersey cases ordered treatment—in one instance a hysterectomy, in the other a leg amputation. On the other hand, one recent Massachusetts case allowed a sixty-year-old state mental hospital patient diagnosed as schizophrenic to refuse a breast biopsy. While a patient in the hospital during a period when she was lucid, rational, and understood the possible consequences of her decision, she refused treatment. She continued to refuse treatment, but as time progressed this refusal was more and more based on delusional grounds. She said she feared the surgery because her aunt had died after similar surgery, which was not factual; she said that the surgery would interfere with her genital system and would prohibit a movie career. The court acknowledged this "right to die," which was presented to the court as part of the right of privacy, first recognized by the Supreme Court within recent years in a Connecticut case dealing with contraception and rapidly assuming prominence when it became the rationale for the Court's abortion stand in which the state was not allowed to impinge on an individual's private behavior. It further based the right to die upon the "right of a mature competent adult to refuse to accept medical recommendations that may prolong one's life and which, to a third person at least, appear to be in his best interests." It conceded that sometimes public policy requires treatment—

contagious illness would be an obvious example—and that the presence of minor children might lead to a different result, but it allowed the decision of this incompetent lady to prevail because it had reiterated her stand when competent.[2]

These decisions all come from courts of inferior jurisdiction; and decisions can be cited which go contrary to the Right to Die. Nevertheless, we can assume that we have a great deal of potential authority in the right to determine the length of our lives. If a right to die case ever got to the Supreme Court, we could cite Justice Burger's dissenting decision in a 1964 Jehovah's Witnesses case in which a blood tranfusion was ordered. The majority said a life hung in the balance and there was no time to research the patient's attitudes and to reflect on a judicial decision. Judge Skelly Wright said, "To refuse to act, only to find later that the law required action, was a risk I was unwilling to accept." But Burger's dissent quoted Justice Brandeis on the "right to be let alone—the most comprehensive of rights and the right most valued by civilized man," and it commented: "Nothing in this utterance suggests that Justice Brandeis thought an individual possessed these rights only as to *sensible* beliefs, *valid* thoughts, *reasonable* emotions, or *well-founded* sensations. I suggest that he intended to include a great many . . . ideas which do not conform, such as refusing medical treatment even at great risk." Since 1964 a long string of civil rights decisions in other areas would buttress Burger's minority view.[3]

The development of the kidney transplant operation and the fact that fresh kidneys gave a greater chance of a successful outcome have led to techniques of life support in patients who would otherwise have been allowed to die because they were no longer capable of the integrated functioning of heart, lungs, and brains and this condition was deemed irreversible. We thus have that medical anomaly, a so-called heart-beating cadaver, although the beating heart at one time would have militated against any determination of death. While theoretically not necessary in kidney transplants, a heart-beating cadaver so greatly enhances the likelihood of a successful transplant that, in fact, many programs use them exclusively.[4] Heart-beating cadavers are mandatory for transplantation of the heart and liver. The majority of medically acceptable cadaver donors are young patients who have sustained a head injury and are in an irreversible coma.

Because the wishes of the organ donor were not binding—they were subordinated to the wishes of the next of kin—it became necessary for statutes to

[2] Yetter, Alleged Incompetent, 41 Northamp. 67, 24 Fiduc. Rep. 1; see "Compulsory Medical Treatment and the Free Exercise of Religion," 42 Indiana L.J. 386.

[3] Application of President of Georgetown College, 331 F.2d 1010 (1964), 1016, cert. den. 377 U.S. 978 (1964).

[4] Lynn Banowsky, William Braun, and Magnus Magnusson, The medical and legal determination of death—its effect on cadaveric organ procurement, *The Journal of Legal Medicine*, November/December 1974, 44–48.

enable the potential donor to make the decision while he was competent to donate all or part of his body and have this wish be legally binding and supersede the wishes of the next of kin. The Uniform Anatomical Gift Act or a modified version of it is now law in all 50 states.

Defining when coma is irreversible and setting the time for the organ to be removed and for ordering the respirator to be turned off created legal, ethical, and moral complexities which began to be resolved when the Harvard School of Medicine Ad Hoc Committee's recommendations were made the basis of similar regulations in other medical centers. Although there has been some confusion in the public mind, these standards are to be applied not to the patient who we think deserves "death with dignity," the dying patient whose misery we do not want extended, but to patients who are in irreversible coma, who have had so much brain damage that we feel sure they will never return to consciousness. Because the patient's heart is beating he does not meet the traditional criteria of death, but since there is no discernible activity of the brain he does meet the criteria of brain death and it is assumed that brain death represents the death of the patient. If the patient is totally unaware of external stimuli, even intensely painful stimuli, there is no movement or breathing, over a period of at least one hour, and no elicitable reflexes, neurologists are willing to make a preliminary declaration of brain death. The tests must be repeated at a 24-hour interval before brain death can be finally declared. A flat electroencephalogram, particularly when repeated at an interval of 6, 12, or 24 hours—depending on the policy of the particular medical center—has great confirmatory value. Hypothermia and central nervous system depression caused by such agents as barbiturates must be ruled out before the criteria can be applied.

We have read in the papers of supposed cadavers who twitched or coughed at the point where their organ was about to be harvested—the word "harvested" has both a less personal and a more technicological feel about it than "cut out" or "excised"—and who were then saved from the knife. My neurosurgical friends tell me that in all these cases there had been a breach in the observation of strict criteria and that these patients did not thereafter recover but within the next few days went on to meet the brain death criteria.

There have been a few very interesting legal cases dealing with such patients—a California court held that the death of an auto victim was caused by the stopping of the respirator rather than by the injury which had caused brain death, but a later case in California and a Georgia case indicate that courts are willing to accept medical assurance that brain death is as useful a standard as heart and respiration death. The Georgia case, in which the doctors were held not to be the cause of the death of the patient who was already brain dead when they turned off the respirator, is now on appeal.

One of the problems with the brain death concept has been a misunderstanding on the part of courts, lawyers, journalists, and the general public; they have sometimes felt that brain death was a better standard of death than the conventional heart and respiration standard. Doctors have feared that if brain death is dignified by its use in a statute, such as the 1970 Kansas law which for the first time gave statutory recognition to brain death and subsequent laws passed in Maryland and Virginia, all patients might have to be observed over a period of time, possibly electroencephelogram determinations might have to be made on all patients, before death could be declared. Many doctors and medicine–law experts have advocated that brain death not be mentioned in statutes. The House of Delegates of the American Medical Association voted in 1972 to endorse the recommendation of its Judicial Council that at present a statutory definition of death is neither desirable nor necessary, that death be determined by the clinical judgment of the doctor using available and currently acceptable criteria. The use of generally accepted medical criteria would be relied on to protect the doctor declaring death on this basis. There would not then be the rigidity that is associated with statutory definitions. Doctors have recently also emphasized that EEG tracings are useful for confirmation, but they are not an integral part of the declaration of death determination and should not be considered to be mandated by statute or custom.

A new law in Georgia gets around some of the legal problems of statutory definitions; it states that use of the brain death standard does not supersede any previously accepted methods of determination: "The criteria for determining death authorized [in this statute] shall be cumulative to and shall not prohibit the use of other medically recognized criteria for determining death."

The subjects selected for prolongation of life because their organs are sought represent a much easier problem for us than the subjects who are unconscious or semiconscious who do not meet the criteria for brain death but who are still seen as incapable of recovering, and also the terminally ill patient. If a doctor sees the possibility of a meaningful recovery for this patient he is obligated to use all his skills to preserve the life of the patient. If the patient could see that he had no future he could elect not to have the doctor use further efforts. But "vegetative" patients are not in a position to make their intentions known, the doctors then have no firm guidelines. They turn to next-of-kin and ask them to decide, and although this is legally correct we are not always sure that the wishes of the next-of-kin coincide with those that the patient would express if he could.

This is the situation where we are concerned with quality of life, with which Joseph Fletcher has dealt; Fletcher wishes us to define "humanhood" and to say that when criteria of humanhood can not be found (such a minimal intelligence, self-awareness, self-control, and mentation) the patient is not human. The quality of the patient's life thus becomes the basis for stating whether life should be maintained.

Fletcher's ideas are a rationale for active euthanasia, the use of the physician to promote death, but up to this time the law has only allowed passive euthanasia, the withholding of medical helps, the discontinuance of life maintenance procedures. Sometimes the definition of passive euthanasia is stretched to cover situations where not only no medical help is given a patient but he is denied life support, intravenous nutrition, or even oral feeding. Babies in the neonatal period who have multiple defects are allowed to die over a period of time through starvation, and although this smacks more of active than of passive euthanasia it has not been legally challenged. Recently, however, when Dr. Kenneth Edelin used suffocation as part of an abortion or post-abortion technique—it has been argued that the suffocation occurred after the birth of the baby and therefore was not a part of the abortion—he was successfully prosecuted because some people at least felt he had gone past the passive and into the active area. But there are no important legal challenges to the right of the *adult and competent* patient to dictate his own treatment.

It is agreed then that the law gives a great deal of authority to discontinue life maintenance. The American Hospital Association, the American Medical Association, and some state medical societies have approved the concept that patients can withhold consent for life-saving procedures.[5]

How can that patient authority be exercised? As long as we are competent, there is no problem. Doctors should respect our wishes; if they do not we are free to dismiss them as our doctors. When we are no longer competent, the decision-making is left to next-of-kin. They usually hope to have the doctor make the decision, but although he may tell them what he thinks they should decide, they have the ultimate authority. Some doctors feel it is too guilt-provoking for families to have to decide to terminate life maintenance, but the law says it is the responsibility of patient or next-of-kin, not of the physician. The physician's input is his expression of opinion which is usually determinative. Next-of-kin may and often do have motives for wishing a death hastened, and if a doctor feels that this is against the best interests of the patient he can move to have a disinterested guardian appointed. This problem comes up with the newborn and infants. Recently in Maine parents refused to give consent to possibly life-saving surgery on a baby born with a malformation of his entire left side. The hospital physicians nevertheless obtained a court order and operated. The baby died. The President of the Maine Medical Association said the ruling could have "tremendous and far-reaching implications for any large hospital capable of sustaining life."[6] Some recent reports involve mentally deficient newborns who also have life-threatening congenital anomalies. Without the deficiency, parents invariably consent to the surgery. Should the presence of mental deficiency be a sufficient pretext for parents to refuse consent? When a baby starves to death over several weeks because the neonatal nursery orders are

[5] *Atlanta Journal and Constitution*, January 14, 1973, 3-B.
[6] *American Medical News*, March 11, 1974, 19.

that life support should not be maintained, the distinction between active and passive euthanasia becomes thin, and even those who oppose the assumption by the doctor of a new medical responsibility, the active hastening of death, might want to see a quick termination to this particular life.

We feel much better when we retain control over the manner of our death. This explains the popularity of the Living Will, prepared and distributed by the Euthanasia Educational Fund. But the document does not have legal effect. A will can be changed before the death of the testator; it only becomes final when the testator is dead. The Living Will concerns a living person; it represents a wish of the maker and not necessarily the present wish; it does not have the solemnity of a contract, deed, or testament. One suggestion recently made is that this be put in contract form, with a consideration on the part of both parties, but this would require that it no longer be a generally addressed "To whom it may concern" directive.[7] Another problem is that there is no guarantee that a Living Will's contents will be known to the hospital or doctor providing care. Who shall hold it? Who has the responsibility for making its contents known? The Living Will does have value in giving next-of-kin and physicians some concept of the attitude of the person toward life prolongation. One version reads in part:

> If the time comes when I can no longer take part in decisions for my own future, let this statement stand as the testament of my wishes. If there is no reasonable expectation of my recovery from physical or mental disability, I request to die and not be kept alive by artificial means or heroic measures. Death is as much a reality as birth, growth, maturity and old age–it is the one certainty. I do not fear death as much as I fear the indignity of deterioration, dependence and hopeless pain. I ask that drugs be mercifully administered to me for terminal suffering even if they hasten the moment of death.

I once suggested in an article that the only good way to ensure that our wishes be carried out, that we have the kind of final course that we would most want, is to match our hopes with the philosophical framework of our treating physicians.[8] If we seek out and trust ourselves to a doctor who shares our philosophy of life–whether it is to prolong life to the utmost or to hasten its end by liberal use of pain-killers–then the problem is resolved. An old lady in California sent me a rebuke; the tone implied, "Foolish boy." She said we are lucky to find any doctors at all, let alone a doctor we can get to know well enough to find out what his approach to philosophical questions may be.

Is the lady right? Should we be satisfied to have our dying attended by doctors who do not know our wishes? We have little alternative, but perhaps this is the greatest condemnation of modern medicine we can utter–the belief that medicine has become so depersonalized and the doctor–patient relationship so

[7] Smythe, J.: Antidysthanasia contracts: A proposal for legalizing death with dignity, 5 Pacific L.J., 738–763, 1974.

[8] Robitscher, J.: The right to die, *Hastings Center Report, 2*(4):11–14, September, 1972.

superficial that in a time of greatest need we are dependent on strangers and technicians. Now that the courts have given us some assurance that there is indeed a right to die, can we devise ways to implementing this right by making medical practice once again an integral part of a small society, not of a large impersonal society? Perhaps we cannot; my friend in California has insisted that I provide common-sense, not pie-in-the-sky, approaches to the problem of life prolongation. One approach, one that I do not recommend, is described in a recent issue of *Time* magazine—we have a new paramedical professional, the Death Companion who sits with "lonely, dying clients." After an eight-session course on the problems of bringing comfort to a dying stranger, the companions are available to sit for $7.50 an hour. The service is strictly commercial; its president has said, "We believe in free enterprise."[9]

We should be able to do better than that. If we cannot make medicine more personal and bring to doctors and to patients the satisfaction of relating to other humans at time of need, we can at least see this issue as demonstrating the wrong path that medicine has taken in its search for greater technological proficiency.

The differentiation of ordinary and extraordinary medical means has caused ethical and moral problems. At one time it had seemed to be a nice solution to the doctor's dilemma to say that he was obliged to provide the patient with all ordinary means of help—nutrition and routine nursing care—but that he was not obliged to use those extraordinary helps that have resulted from the medical advances that have been described. Monsignor Austin Vaughn has said, "It is permissible to stop the use of extraordinary means of prolonging life. But it is not permissible to intervene and bring about the death of an individual." The Lord Bishop of Exeter has said, "It is in general the Christian view that while there is a moral obligation to maintain life by all ordinary means, there is no obligation to use extraordinary means. Ordinary means are such actions as do not cause grave hardship to the patient and which offer a reasonable hope of success. Extraordinary means are means which involve a great expense, inconvenience, or hardship and which at the same time offer no reasonable expectation either of success or benefit." But one generation's extraordinary means is another's ordinary means, and in practice we define these in relativistic terms. It is ordinary to use dialysis for young people; it is extraordinary for old people. A Pennsylvania man had been in a coma for six years when in 1972 the doctor persuaded his father that the coma was irreversible and that an extraordinary means, of intravenous feeding, should be discontinued. The father gave his authorization, but he could not sleep that night, and the next morning asked that the feeding be resumed. What had been extraordinary had become ordinary.

The Lord Bishop of Exeter includes cost in his differentiating criteria of ordinary and extraordinary. Is cost a valid factor? Father Richard McCormick in

[9] *Time*, February 17, 1975, 68.

an article in the *Journal of the American Medical Association* has suggested that it is not the complexity of the technological advance or its cost that determines its extraordinary quality; he stresses the factor of hardship to the patient. Morally speaking, he says, ordinary means are those whose use does not entail grave hardship to the patient.[10] Dr. Irvine Page has said that if we wanted to keep alive all patients who could be kept alive, we could someday consume the entire national product in this pursuit. He opposes extraordinary means to prolong life unless some positive gain, such as the possibility of a resumption of a level of functional usefulness and some improvement of health, is a possibility.

No precise differentiation of ordinary and extraordinary is possible; the differentiation is to a large extent the responsibility of the individual doctor concerned with the individual patient.

Our decision not to prolong life is usually called passive euthanasia. Euthanasia connotes "mercy killing" to some, but it merely means a "good death." But possibly the use of the term euthanasia for the passive practice helps to break down the opposition to active euthanasia, so those who favor the passive and oppose the active sometimes suggest that we reserve the term euthanasia for induced death and use another term—such as "death with dignity"—for termination of prolongation procedures, for the concept of a good kind of death, a death with quality. Unless we make this distinction, they say, the opinion polls will continue to show increasing acceptance of active euthanasia on the part of both the general public and the physicians.

In 1950 only 36% of Americans said they approved of mercy killing for people with incurable diseases, according to the Gallup Poll. In 1973 this had risen to a majority, 53%, and of those thirty and under, 67% favored mercy killing.[11] One state legislature, Oregon, has had a bill proposed to allow mercy killing; it was later withdrawn. Six states have had passive euthanasia bills proposed to ensure that doctors would not incur any liability through their omitted acts. Most physicians do not favor such bills; they feel that passive euthanasia is now practiced and allowed and any new legal protection might cause notice and problems; in two states the measure have been voted down, and in four others they are bottled up in committees.[12]

Active euthanasia has many adherents in Scandinavia and England and it has won much legislative support there, although not majority support. Walter Alvarez, a physician-author identified with the Mayo Clinic, favors active euthanasia and regrets that in his medical career he was too inhibited and too conscious of the law and of medical tradition to put an end to the pain of some

[10] McCormick, R.: To save or let die: The dilemma of modern medicine, *JAMA 229:* 172–176, 1974.

[11] *New York Times,* August 2, 1973, p. 33.

[12] Euthanasia: No present future, *Medical World News,* April 23, 1973, 6.

patients.[13] Lawrence LeShan, a psychologist with the Ayer Foundation who has worked with dying patients, feels that physicians should not fall back on the Hippocratic Oath to justify a refusal to kill, and since he made this statement the Supreme Court has declared that some parts of the Hippocratic Oath are obsolete in terms of the needs of today. In 1973 the Court said that changing times relieved the physician from the necessity of conforming to that part of the Oath which prohibits abortion.[14]

Although active euthanasia continues to be forbidden and is criminal, doctors can come close to it in their efforts to relieve pain. The amount of narcotic needed to relieve severe pain depresses respiration, and this may be the factor that leads most directly to a patient's death. Doctors generally have no problem here as long as they prescribe drugs only in the range that good medical practice dictates; medical ethics has always put the alleviation of suffering on an equal footing with the prolongation of life as a medical goal. Pope Pius, no friend of euthanasia, has said that "the removal of pain and consciousness by means of drugs when medical reasons suggest it is permitted by religion and morality to both doctor and patient; even if the use of drugs will shorten life."

Active euthanasia in its several forms–voluntary, assisted, and involuntary–raises more problems. It has its articulate defenders; it can save pain; but it alters further the role of the doctor in promoting life–already altered by the new policy on abortion–and it prevents the patient from having the assurance that his doctor is his helper, not his executioner. The terrible pain that some patients undergo is the price that is exacted in order for all patients to feel trust in the doctor–patient relationship. The arguments against active euthanasia are equally compelling whether this active euthanasia is to be only at the request of the patient or if it can be imposed involuntarily by the action of a board, whether it involves the doctor's actually administering a lethal dose or only leaving it at the patient's bedside. The latter is what is sometimes called assisted euthanasia. It is legal in Switzerland, but it is seldom employed. Patients find it difficult to ask to be assisted in this way. Alice James, the sister of William and Henry, was dying from cancer when she wrote: "I am being ground slowly on the grim grindstone of physical pain, and on two nights I had almost asked for Katherine's lethal dose, but one steps hestitatingly along such unaccustomed ways, and endures from second to second."

Stewart Alsop, dying of leukemia, wrote that a patient suffering beyond existence should be given the opportunity to end his life, or, if he refused that option, to be allowed as much pain-killing drug, probably heroin, as he requires.[15] "If a human being must die, it is surely better that he die in the

[13] Alvarez, W.: Death with dignity, *The Humanist*, Sept.–Oct., 1971, 12–14.
[14] *Roe v. Wade*, 410 US 113, 35 L Ed 2d 147, 93 S Ct 705 (1973).
[15] Alsop, S.: The right to die with dignity, *Good Housekeeping*, August, 1974, p. 69.

illusion of painless pleasure—and heroin is very pleasurable—than in lonely agony."[16]

Most physicians would not support Alsop's first plan, but they would want to help the patient achieve as pain-free a death as he desires.

With a growing consensus that relief of pain and the termination of life-promoting devices are often proper, although killing another human is not, the concept of "death with dignity" has achieved great popularity. It is worth noting that its originator, Paul Ramsey, has written a piece called "The Indignity of 'Death with Dignity.'" In our effort to be mature, to show that we have learned to live with the idea of death, says Ramsey, we trivialize death by saying that death is simply a part of life, that death contributes to evolution.[17] I myself am guilty of such trivilization; I once gave a paper on the right to die and the obligation to die—our additional obligation to leave the earth's resources for use by others. Ramsey says rightly that such an approach is too denying of the grief, the shock, the pain, and the terror of death. There is dignity in our caring for the dying, he says, but death is not dignified and we cannot confer death with dignity on the dying; we can only provide them with some of the necessary but not sufficient conditions for achieving a good death. "If the dying die with a degree of nobility it will be mostly their doing in doing their own dying." We do not achieve a community with the dying by interposing between them and us the bloodless notion of "death with dignity"; we draw closer to them only if our concept includes the final indignity of death itself.[18]

The late Ernest Becker in his Pulitzer-prize-winning book, *The Denial of Death*, writes about the many ways we have to deny the finality, the tragedy, the fearsomeness of death. Becker and Ramsey emphasize the need to face our fears.

This is a timely warning. The movement to deal with death as merely another "fact of life" can make death so banal that we do not fight hard enough against it.

An editorial in *Ca—A Cancer Journal for Clinicians* reminds us that although adjustment and accommodation are acceptable approaches to the inevitable, it is "often difficult to know exactly what is inevitable. Here are but a few examples of what were once considered to be hopelessly irreversible: cardiac arrest, metastatic choriocarcinoma, 'thyroid storm,' acute leukemia in children, the impending rupture of an aortic aneurism and a heart damaged by rheumatic fever." The editorial concludes: "What once inevitably meant death, does no

[16] *Time*, June 10, 1974, 65.

[17] One recent addition to the rapidly growing literature on death and dying is called "Death as a Fact of Life." Henden, D.: *Death as a Fact of Life*, New York, W.W. Norton, 1973.

[18] Ramsey, P.: The indignity of "death with dignity," *Hastings Center Studies* *2*:47–62, 1974.

longer, and this fact should make physicians pause before too readily accommodating to dying and adjusting to the death of a patient."[19]

So problems still remain even though we have established a right to die. How should death be manipulated so as to allow medicine to remain within its traditional posture of promoting life? How can we live with the ambiguity and the responsibility that comes from a lack of definite guidelines on when to terminate life-saving procedures?

In gaining control over the prolongation of life, we have incurred the responsibility of determining at what point that death is most meaningful.

We will never find the perfect answer. We will not find a point we can define which separates lives terminated prematurely from those that are allowed to continue too long. We will not find the solution productive of only good and no evil. This is a lesson we have learned in other fields in recent years. We have begun to understand the dynamic nature of the forces that bear upon us and the difficulties of our choice. We see this same lesson in ecology, economics, diplomacy, and every other area in which we make choices. We cannot fight inflation without risking unemployment. We cannot employ the internal combustion engine without causing pollution.

These are the same kinds of decisions that medicine has to make. We have awakened from a dream of Social Darwinism, the concept that society is in an onward and upward evolution and that we will work our way to a point where what is unpleasant has been left behind and only the pleasant is in prospect.

Our new view recognizes forces and counterforces which interact with each other and which produce for us victory in defeat and defeat in victory; the view is a tragic view because it forces us to recognize that the solutions to our problems bring to us new and even more exquisite problems. If we are forced to face the fact that life has this tragic element—that dreams of conquering death and disease by replacing organs and prolonging life create new problems in the process of solving old ones—that also has its consoling aspects. We do have the choice to make use of them or not. We continue to have the age-old burden of man, the responsibility of making choices in a universe that has never been without its tragic component.

[19] Holleb, A.: A patient's right to die—the easy way out? *Ca—A Cancer Journal for Clinicians, 24*(4):256, 1974.

THE ALLOCATION OF EXOTIC MEDICAL LIFESAVING THERAPY

NICHOLAS RESCHER

I. The Problem

Technological progress has in recent years transformed the limits of the possible in medical therapy. However, the elevated state of sophistication of modern medical technology has brought the economists' classic problem of scarcity in its wake as an unfortunate side product. The enormously sophisticated and complex equipment and the highly trained teams of experts requisite for its utilization are scarce resources in relation to potential demand. The administrators of the great medical institutions that preside over these scarce resources thus come to be faced increasingly with the awesome choice: *Whose life to save?*

A (somewhat hypothetical) paradigm example of this problem may be sketched within the following set of definitive assumptions: We suppose that persons in some particular medically morbid condition are "mortally afflicted": It is virtually certain that they will die within a short time period (say ninety days). We assume that some very complex course of treatment (e.g., a heart transplant) represents a substantial probability of life prolongation for persons in this mortally afflicted condition. We assume that the facilities available in terms of human resources, mechanical instrumentalities, and requisite materials (e.g., hearts in the case of a heart transplant) make it possible to give a certain treatment–this "exotic (medical) lifesaving therapy," or ELT for short–to a certain, relatively small number of people. And finally we assume that a substantially greater pool of people in the mortally afflicted condition is at hand. The problem then may be formulated as follows: How is one to select within the pool of afflicted patients the ones to be given the ELT treatment in question; how to select those "whose lives are to be saved"? Faced with many candidates for an ELT process that can be made available to only a few, doctors and medical administrators confront the decision of who is to be given a chance at survival and who is, in effect, to be condemned to die.

This paper originally appeared in *Ethics: An International Journal of Social, Political and Legal Philosophy* (April 1969) and is reprinted with the permission of the author and The University of Chicago Press.

As has already been implied, the "heroic" variety of spare-part surgery can pretty well be assimilated to this paradigm. One can foresee the time when heart transplantation, for example, will have become pretty much a routine medical procedure, albeit on a very limited basis, since a cardiac surgeon with the technical competence to transplant hearts can operate at best a rather small number of times each week and the elaborate facilities for such operations will most probably exist on a modest scale. Moreover, in "spare-part" surgery there is always the problem of availability of the "spare parts" themselves. A report in one British newspaper gives the following picture: "Of the 150,000 who die of heart disease each year [in the U.K.], Mr. Donald Longmore, research surgeon at the National Heart Hospital [in London] estimates that 22,000 might be eligible for heart surgery. Another 30,000 would need heart and lung transplants. But there are probably only between 7,000 and 14,000 potential donors a year."[1] Envisaging this situation in which at the very most something like one in four heart-malfunction victims can be saved, we clearly confront a problem in ELT allocation.

A perhaps even more drastic case in point is afforded by long-term haemodialysis, an ongoing process by which a complex device—an "artificial kidney machine"—is used periodically in cases of chronic renal failure to substitute for a non-functional kidney in "cleaning" potential poisons from the blood. Only a few major institutions have chronic haemodialysis units, whose complex operation is an extremely expensive proposition. For the present and the foreseeable future the situation is that "the number of places available for chronic haemodialysis is hopelessly inadequate."[2]

The traditional medical ethos has insulated the physician against facing the very existence of this problem. When swearing the Hippocratic Oath, he commits

[1] Christine Doyle, "Spare-Part Heart Surgeons Worried by Their Success," *Observer*, May 12, 1968.

[2] J. D. N. Nabarro, "Selection of Patients for Haemodialysis," *British Medical Journal* (March 11, 1967), p. 623. Although several thousand patients die in the U.K. each year from renal failure—there are about thirty new cases per million of population—only 10 percent of these can for the foreseeable future be accommodated with chronic haemodialysis. Kidney transplantation—itself a very tricky procedure—cannot make a more than minor contribution here. As this article goes to press, I learn that patients can be maintained in home dialysis at an operating cost about half that of maintaining them in a hospital dialysis unit (roughly an $8,000 minimum). In the United States, around 7,000 patients with terminal uremia who could benefit from haemodialysis evolve yearly. As of mid-1968, some 1,000 of these can be accommodated in existing hospital units. By June 1967, a world-wide total of some 120 patients were in treatment by home dialysis. (Data from a forthcoming paper, "Home Dialysis," by C. M. Conty and H. V. Murdaugh. See also R. A. Baillod *et al.*, "Overnight Haemodialysis in the Home," *Proceedings of the European Dialysis and Transplant Association*, VI [1965], 99 ff.).

himself to work for the benefit of the sick in "whatsover house I enter."[3] In taking this stance, the physician substantially renounces the explicit choice of saving certain lives rather than others. Of course, doctors have always in fact had to face such choices on the battlefield or in times of disaster, but there the issue had to be resolved hurriedly, under pressure, and in circumstances in which the very nature of the case effectively precluded calm deliberation by the decision maker as well as criticism by others. In sharp contrast, however, cases of the type we have postulated in the present discussion arise predictably, and represent choices to be made deliberately and "in cold blood."

It is, to begin with, appropriate to remark that this problem is not fundamentally a medical problem. For when there are sufficiently many afflicted candidates for ELT then—so we may assume—there will also be more than enough for whom the purely medical grounds for ELT allocation are decisively strong in any individual case, and just about equally strong throughout the group. But in this circumstance a selection of some afflicted patients over and against others cannot *ex hypothesi* be made on the basis of purely medical considerations.

The selection problem as we have said, is in substantial measure not a medical one. It is a problem *for* medical men, which must somehow be solved by them, but that does not make it a medical issue—any more than the problem of hospital building is a medical issue. As a problem it belongs to the category of philosophical problems—specifically a problem of moral philosophy or ethics. Structurally, it bears a substantial kinship with those issues in this field that revolve about the notorious whom-to-save-on-the-lifeboat and whom-to-throw-to-the-wolves-pursuing-the-sled questions. But whereas questions of this just-indicated sort are artificial, hypothetical, and far-fetched, the ELT issue poses a *genuine* policy question for the responsible administrators in medical institutions, indeed a question that threatens to become commonplace in the foreseeable future.

Now what the medical administrator needs to have, and what the philosopher is presumably *ex officio* in a position to help in providing, is a body of *rational guidelines* for making choices in these literally life-or-death situations. This is an issue in which many interested parties have a substantial stake, including the responsible decision maker who wants to satisfy his conscience that he is acting in a reasonable way. Moreover, the family and associates of the man who is turned away—to say nothing of the man himself—have the right to an acceptable explanation. And indeed even the general public wants to know that what is being done is fitting and proper. All of these interested parties are

[3] For the Hippocratic Oath see *Hippocrates: Works* (Loeb ed.; London, 1959), I, p. 298.

entitled to insist that a reasonable code of operating principles provides a defensible rationale for making the life-and-death choices involved in ELT.

II. The Two Types of Criteria

Two distinguishable types of criteria are bound up in the issue of making ELT choices. We shall call these *Criteria of Inclusion* and *Criteria of Comparison,* respectively. The distinction at issue here requires some explanation. We can think of the selection as being made by a two-stage process: (1) the selection from among all possible candidates (by a suitable screening process) of a group to be taken under serious consideration as candidates for therapy, and then (2) the actual singling out, within this group, of the particular individuals to whom therapy is to be given. Thus the first process narrows down the range of comparative choice by eliminating *en bloc* whole categories of potential candidates. The second process calls for a more refined, case-by-case comparison of those candidates that remain. By means of the first set of criteria one forms a selection group; by means of the second set, an actual selection is made within this group.

Thus what we shall call a "selection system" for the choice of patients to receive therapy of the ELT type will consist of criteria of these two kinds. Such a system will be acceptable only when the reasonableness of its component criteria can be established.

III. Essential Features of an Acceptable ELT Selection System

To qualify as reasonable, an ELT selection must meet two important "regulative" requirements: it must be *simple* enough to be readily intelligible, and it must be *plausible,* that is, patently reasonable in a way that can be apprehended easily and without involving ramified subtleties. Those medical administrators responsible for ELT choices must follow a modus operandi that virtually all the people involved can readily understand to be acceptable (at a reasonable level of generality, at any rate). Appearances are critically important here. It is not enough that the choice be made in a *justifiable* way; it must be possible for people–*plain* people–to "see" (i.e., understand without elaborate teaching or indoctrination) that *it is justified,* insofar as any mode of procedure can be justified in cases of this sort.

One "constitutive" requirement is obviously an essential feature of a reasonable selection system: all of its component criteria–those of inclusion and those of comparison alike—must be reasonable in the sense of being *rationally defensible.* The ramifications of this requirement call for detailed consideration. But one of its aspects should be noted without further ado: it must be *fair*—it must treat relevantly like cases alike, leaving no room for "influence" or favoritism, etc.

IV. The Basic Screening Stage: Criteria of Inclusion (and Exclusion)

Three sorts of considerations are prominent among the plausible criteria of inclusion/exclusion at the basic screening stage: the constituency factor, the progress-of-science factor, and the prospect-of-success factor.

A. The Constituency Factor

It is a "fact of life" that ELT can be available only in the institutional setting of a hospital or medical institute or the like. Such institutions generally have normal clientele boundaries. A veterans' hospital will not concern itself primarily with treating non-veterans, a children's hospital cannot be expected to accommodate the "senior citizen," an army hospital can regard college professors as outside its sphere. Sometimes the boundaries are geographic—a state hospital may admit only residents of a certain state. (There are, of course, indefensible constituency principles—say race or religion, party membership, or ability to pay; and there are cases of borderline legitimacy, e.g., sex.[4]) A medical institution is justified in considering for ELT only persons within its own constituency, provided this constituency is constituted upon a defensible basis. Thus the haemodialysis selection committee in Seattle "agreed to consider only those applications who were residents of the state of Washington. . . . They justified this stand on the grounds that since the basic research . . . had been done at . . . a state-supported institution—the people whose taxes had paid for the research should be its first beneficiaries."[5]

While thus insisting that constituency considerations represent a valid and legitimate factor in ELT selection, I do feel there is much to be said for minimizing their role in life-or-death cases. Indeed a refusal to recognize them at all is a significant part of medical tradition, going back to the very oath of

[4] Another example of borderline legitimacy is posed by an endowment "with strings attached," e.g., "In accepting this legacy the hospital agrees to admit and provide all needed treatment for any direct descendant of myself, its founder."

[5] Shana Alexander, "They Decide Who Lives, Who Dies," *Life,* LIII (November 9, 1962), 102–25 (see p. 107).

Hippocrates. They represent a departure from the ideal arising with the institutionalization of medicine, moving it away from its original status as an art practiced by an individual practitioner.

B. The Progress-of-Science Factor

The needs of medical research can provide a second valid principle of inclusion. The research interests of the medical staff in relation to the specific nature of the cases at issue is a significant consideration. It may be important for the progress of medical science—and thus of potential benefit to many persons in the future—to determine how effective the ELT at issue is with diabetics or persons over sixty or with a negative RH factor. Considerations of this sort represent another type of legitimate factor in ELT selection.

A very definitely *borderline* case under this head would revolve around the question of a patient's willingness to pay, not in monetary terms, but in offering himself as an experimental subject, say by contracting to return at designated times for a series of tests substantially unrelated to his own health, but yielding data of importance to medical knowledge in general.

C. The Prospect-of-Success Factor

It may be that while the ELT at issue is not without *some* effectiveness in general, it has been established to be highly effective only with patients in certain specific categories (e.g., females under forty of a specific blood type). This difference in effectiveness—in the absolute or in the probability of success—is (we assume) so marked as to constitute virtually a difference in kind rather than in degree. In this case, it would be perfectly legitimate to adopt the general rule of making the ELT at issue available only or primarily to persons in this substantial-promise-of-success category. (It is on grounds of this sort that young children and persons over fifty are generally ruled out as candidates for haemodialysis.)

We have maintained that the three factors of constituency, progress of science, and prospect of success represent legitimate criteria of inclusion for ELT selection. But it remains to examine the considerations which legitimate them. The legitimating factors are in the final analysis practical or pragmatic in nature. From the practical angle it is advantageous—indeed to some extent necessary—that the arrangements governing medical institutions should embody certain constituency principles. It makes good pragmatic and utilitarian sense that progress-of-science considerations should be operative here. And, finally, the practical aspect is reinforced by a whole host of other considerations—including moral ones—in supporting the prospect-of-success criterion. The workings of each of these factors are of course conditioned by the ever-present element of limited availability. They are operative only in this context, that is, prospect of success is a legitimate consideration at all only because we are dealing with a situation of scarcity.

V. The Final Selection Stage: Criteria of Selection

Five sorts of elements must, as we see it, figure primarily among the plausible criteria of selection that are to be brought to bear in further screening the group constituted after application of the criteria of inclusion: the relative-likelihood-of-success factor, the life-expectancy factor, the family role factor, the potential-contributions factor, and the services-rendered factor. The first two represent the *biomedical* aspect, the second three the *social* aspect.

A. The Relative-Likelihood-of-Success Factor

It is clear that the relative likelihood of success is a legitimate and appropriate factor in making a selection within the group of qualified patients that are to receive ELT. This is obviously one of the considerations that must count very significantly in a reasonable selection procedure.

The present criterion is of course closely related to item *C* on the preceding section. There we were concerned with prospect-of-success considerations categorically and *en bloc.* Here at present they come into play in a particularized case-by-case comparison among individuals. If the therapy at issue is not a once-and-for-all proposition and requires ongoing treatment, cognate considerations must be brought in. Thus, for example, in the case of a chronic ELT procedure such as haemodialysis it would clearly make sense to give priority to patients with a potentially reversible condition (who would thus need treatment for only a fraction of their remaining lives).

B. The Life-Expectancy Factor

Even if the ELT is "successful" in the patient's case he may, considering his age and/or other aspects of his general medical condition, look forward to only a very short probable future life. This is obviously another factor that must be taken into account.

C. The Family Role Factor

A person's life is a thing of importance not only to himself but to others—friends, associates, neighbors, colleagues, etc. But his (or her) relationship to his immediate family is a thing of unique intimacy and significance. The nature of his relationship to his wife, children, and parents, and the issue of their financial and psychological dependence upon him, are obviously matters that deserve to be given weight in the ELT selection process. Other things being anything like equal, the mother of minor children must take priority over the middle-aged bachelor.

D. The Potential Future-Contributions Factor (Prospective Service)

In "choosing to save" one life rather than another, "the society," through the mediation of the particular medical institution in question–which should certainly look upon itself as a trustee for the social interest–is clearly warranted in considering the likely pattern of future *services to be rendered* by the patient (adequate recovery assumed), considering his age, talent, training, and past record of performance. In its allocations of ELT, society "invests" a scarce resource in one person as against another and is thus entitled to look to the probable prospective "return" on its investment.

It may well be that a thoroughly egalitarian society is reluctant to put someone's social contribution into the scale in situations of the sort at issue. One popular article states that "the most difficult standard would be the candidate's value to society," and goes on to quote someone who said: "You can't just pick a brilliant painter over a laborer. The average citizen would be quickly eliminated."[6] But what if it were not a brilliant painter but a brilliant surgeon or medical researcher that was at issue? One wonders if the author of the *obiter dictum* that one "can't just pick" would still feel equally sure of his ground. In any case, the fact that the standard is difficult to apply is certainly no reason for not attempting to apply it. The problem of ELT selection is inevitably burdened with difficult standards.

Some might feel that in assessing a patient's value to society one should ask not only who if permitted to continue living can make the greatest contribution to society in some creative or constructive way, but also who by dying would leave behind the greatest burden on society in assuming the discharge of their residual responsibilities.[7] Certainly the philosophical utilitarian would give equal weight to both these considerations. Just here is where I would part ways with orthodox utilitarianism. For–though this is not the place to do so–I should be prepared to argue that a civilized society has an obligation to promote the

[6] Lawrence Lader, "Who Has the Right to Live?" *Good Housekeeping* (January 1968), p. 144.

[7] This approach could thus be continued to embrace the previous factor, that of family role, the preceding item (*C*).

furtherance of positive achievements in cultural and related areas even if this means the assumption of certain added burdens.[8]

E. The Past Services-Rendered Factor (Retrospective Service)

A person's services to another person or group have always been taken to constitute a valid basis for a claim upon this person or group—of course a moral and not necessarily a legal claim. Society's obligation for the recognition and reward of services rendered—an obligation whose discharge is also very possibly conducive to self-interest in the long run—is thus another factor to be taken into account. This should be viewed as a morally necessary correlative of the previously considered factor of *prospective* service. It would be morally indefensible of society in effect to say: "Never mind about services you rendered yesterday—it is only the services to be rendered tomorrow that will count with us today." We live in very future-oriented times, constantly preoccupied in a distinctly utilitarian way with future satisfactions. And this disinclines us to give much recognition to past services. But parity considerations of the sort just adduced indicate that such recognition should be given *on grounds of equity*. No doubt a justification for giving weight to services rendered can also be attempted along utilitarian lines. ("The reward of past services rendered spurs people on to greater future efforts and is thus socially advantageous in the long-run future.") In saying that past services should be counted "on grounds of equity"—rather than "on grounds of utility"—I take the view that even if this utilitarian defense could somehow be shown to be fallacious, I should still be prepared to maintain the propriety of taking services rendered into account. The position does not rest on a utilitarian basis and so would not collapse with the removal of such a basis.[9]

As we have said, these five factors fall into three groups: the biomedical factors *A* and *B*, the familial factor *C*, and the social factors *D* and *E*. With items *A* and *B* the need for a detailed analysis of the medical considerations comes to the fore. The age of the patient, his medical history, his physical and psychological condition, his specific disease, etc., will all need to be taken into exact

[8] Moreover a doctrinaire utilitarian would presumably be willing to withdraw a continuing mode of ELT such as haemodialysis from a patient to make room for a more promising candidate who came to view at a later stage and who could not otherwise be accommodated. I should be unwilling to adopt this course, partly on grounds of utility (with a view to the demoralization of insecurity), partly on the non-utilitarian ground that a "moral commitment" has been made and must be honored.

[9] Of course the difficult question remains of the relative weight that should be given to prospective and retrospective service in cases where these factors conflict. There is good reason to treat them on a par.

account. These biomedical factors represent technical issues: they call for the physicians' expert judgment and the medical statisticians' hard data. And they are ethically uncontroversial factors–their legitimacy and appropriateness are evident from the very nature of the case.

Greater problems arise with the familial and social factors. They involve intangibles that are difficult to judge. How is one to develop subcriteria for weighing the relative social contributions of (say) an architect or a librarian or a mother of young children? And they involve highly problematic issues. (For example, should good moral character be rated a plus and bad a minus in judging services rendered?) And there is something strikingly unpleasant in grappling with issues of this sort for people brought up in times greatly inclined towards maxims of the type "Judge not!" and "Live and let live!" All the same, in the situation that concerns us here such distasteful problems must be faced, since a failure to choose to save some is tantamount to sentencing all. Unpleasant choices are intrinsic to the problem of ELT selection; they are of the very essence of the matter.[10]

But is reference to all these factors indeed inevitable? The justification for taking account of the medical factors is pretty obvious. But why should the social aspect of services rendered and to be rendered be taken into account at all? The answer is that they must be taken into account not from the *medical* but from the *ethical* point of view. Despite disagreement on many fundamental issues, moral philosophers of the present day are pretty well in consensus that the justification of human actions is to be sought largely and primarily–if not exclusively–in the principles of utility and of justice.[11] But utility requires reference of services to be rendered and justice calls for a recognition of services that have been rendered. Moral considerations would thus demand recognition of these two factors. (This, of course, still leaves open the question of whether the point of view provides a valid basis of action: Why base one's actions upon

[10] This in the symposium on "Selection of Patients for Haemodialysis," *British Medical Journal* (March 11, 1967), pp. 622–24. F. M. Parsons writes: "But other forms of selecting patients [distinct from first come, first served] are suspect in my view if they imply evaluation of man by man. What criteria could be used? Who could justify a claim that the life of a mayor would be more valuable than that of the humblest citizen of his borough? Whatever we may think as individuals none of us is indispensable." But having just set out this hard-line view he immediately backs away from it: "On the other hand, to assume that there was little to choose between Alexander Fleming and Adolf Hitler . . . would be nonsense, and we should be naive if we were to pretend that we could not be influenced by their achievements and characters if we had to choose between the two of them. Whether we like it or not we cannot escape the fact that this kind of selection for long-term haemodialysis will be required until very large sums of money become available for equipment and services [so that *everyone* who needs treatment can be accommodated]."

[11] The relative fundamentality of these principles is, however, a substantially disputed issue.

moral principles?—or, to put it bluntly—Why be moral? The present paper is, however, hardly the place to grapple with so fundamental an issue, which has been canvassed in the literature of philosophical ethics since Plato.)

VI. More Than Medical Issues Are Involved

An active controversy has of late sprung up in medical circles over the question of whether non-physician laymen should be given a role in ELT selection (in the specific context of chronic haemodialysis). One physician writes: "I think that the assessment of the candidates should be made by a senior doctor on the [dialysis] unit, but I am sure that it would be helpful to him—both in sharing responsibility and in avoiding personal pressure—if a small unnamed group of people [presumably including laymen] officially made the final decision. I visualize the doctor bringing the data to the group, explaining the points in relation to each case, and obtaining their approval of his order of priority.[12]

Essentially this procedure of a selection committee of laymen has for some years been in use in one of the most publicized chronic dialysis units, that of the Swedish Hospital of Seattle, Washington.[13] Many physicians are apparently reluctant to see the choice of allocation of medical therapy pass out of strictly medical hands. Thus in a recent symposium on the "Selection of Patients for Haemodialysis,"[14] Dr. Ralph Shakman writes: "Who is to implement the selection? In my opinion it must ultimately be the responsibility of the consultants in charge of the renal units . . . I can see no reason for delegating this responsibility to lay persons. Surely the latter would be better employed if they could be persuaded to devote their time and energy to raise more and more money for us to spend on our patients."[15] Other contributors to this symposium strike much the same note. Dr. F. M. Parsons writes: "In an attempt to overcome . . . difficulties in selection some have advocated introducing certain specified lay people into the discussions. Is it wise? I doubt whether a committee of this type can adjudicate as satisfactorily as two medical colleagues, particularly as successful therapy involves close cooperation between doctor and patient."[16] And Dr. M. A. Wilson writes in the same symposium: "The suggestion has been made that

[12] J. D. N. Nabarro, *op. cit.*, p. 622.

[13] See Shana Alexander, *op. cit.*

[14] *British Medical Journal* (March 11, 1967), pp. 622–24.

[15] *Ibid.*, p. 624. Another contributor writes in the same symposium, "The selection of the few [to receive haemodialysis] is proving very difficult—a true 'Doctor's Dilemma'—for almost everybody would agree that this must be a medical decision, preferably reached by consultation among colleagues" (Dr. F. M. Parsons, *ibid.*, p. 623).

[16] "The Selection of Patients for Haemodialysis," *op. cit.* (n. 10 above), p. 623.

lay panels should select individuals for dialysis from among a group who are medically suitable. Though this would relieve the doctor-in-charge of a heavy load of responsibility, it would place the burden on those who have no personal knowledge and have to base their judgments on medical or social reports. I do not believe this would result in better decisions for the group or improve the doctor–patient relationship in individual cases."[17]

But no amount of flag waving about the doctor's facing up to his responsibility—or prostrations before the idol of the doctor–patient relationship and reluctance to admit laymen into the sacred precincts of the conference chambers of medical consultations—can obscure the essential fact that ELT selection is not a wholly medical problem. When there are more than enough places in an ELT program to accommodate all who need it, then it will clearly be a medical question to decide who does have the need and which among these would successfully respond. But when an admitted gross insufficiency of places exists, when there are ten or fifty or one hundred highly eligible candidates for each place in the program, then it is unrealistic to take the view that purely medical criteria can furnish a sufficient basis for selection. The question of ELT selection becomes serious as a phenomenon of scale—because, as more candidates present themselves, strictly medical factors are increasingly less adequate as a selection criterion precisely because by numerical category-crowding there will be more and more cases whose "status is much the same" so far as purely medical considerations go.

The ELT selection problem clearly poses issues that transcend the medical sphere because—in the nature of the case—many residual issues remain to be dealt with once *all* of the medical questions have been faced. Because of this there is good reason why laymen as well as physicians should be involved in the selection process. Once the medical considerations have been brought to bear, fundamental social issues remain to be resolved. The instrumentalities of ELT have been created through the social investment of scarce resources, and the interests of the society deserve to play a role in their utilization. As representatives of their social interests, lay opinions should function to complement and supplement medical views once the proper arena of medical considerations is left behind.[18] Those physicians who have urged the presence of lay members on

[17] Dr. Wilson's article concludes with the perplexing suggestion—wildly beside the point given the structure of the situation at issue—that "the final decision will be made by the patient." But this contention is only marginally more ludicrous than Parson's contention that in selecting patients for haemodialysis "gainful employment in a well chosen occupation is necessary to achieve the best results" since "only the minority wish to live on charity" (*ibid.*).

[18] To say this is of course not to deny that such questions of applied medical ethics will invariably involve a host of medical considerations—it is only to insist that extramedical considerations will also invariably be at issue.

selection panels can, from this point of view, be recognized as having seen the issue in proper perspective.

One physician has argued against lay representation on selection panels for haemodialysis as follows: "If the doctor advises dialysis and the lay panel refuses, the patient will regard this as a death sentence passed by an anonymous court from which he has no right of appeal."[19] But this drawback is not specific to the use of a lay panel. Rather, it is a feature inherent in every *selection* procedure, regardless of whether the selection is done by the head doctor of the unit, by a panel of physicians, etc. No matter who does the selecting among patients recommended for dialysis, the feelings of the patient who has been rejected (and knows it) can be expected to be much the same, provided that he recognizes the actual nature of the choice (and is not deceived by the possibly convenient but ultimately poisonous fiction that because the selection was made by physicians it was made entirely on medical grounds).

In summary, then, the question of ELT selection would appear to be one that is in its very nature heavily laden with issues of medical research, practice, and administration. But it will not be a question that can be resolved on solely medical grounds. Strictly social issues of justice and utility will invariably arise in this area—questions going outside the medical area in whose resolution medical laymen can and should play a substantial role.

VII. The Inherent Imperfection (Non-Optimality) of any Selection System

Our discussion to this point of the design of a selection system for ELT has left a gap that is a very fundamental and serious omission. We have argued that five factors must be taken into substantial and explicit account:

A. Relative likelihood of success. Is the chance of the treatment's being "successful" to be rated as high, good, average, etc.?[20]

B. Expectancy of future life. Assuming the "success" of the treatment, how much longer does the patient stand a good chance (75 percent or better) of living—considering his age and general condition?

C. Family role. To what extent does the patient have responsibilities to others in his immediate family?

[19] M. A. Wilson, "Selection of Patients for Haemodialysis," *op. cit.*, p. 624.

[20] In the case of an ongoing treatment involving complex procedure and dietary and other mode-of-life restrictions—and chronic haemodialysis definitely falls into this category—the patient's psychological makeup, his willpower to "stick with it" in the face of substantial discouragements—will obviously also be a substantial factor here. The man who gives up, takes not his life alone, but (figuratively speaking) also that of the person he replaced in the treatment schedule.

D. Social contributions rendered. Are the patient's past services to his society outstanding, substantial, average, etc.?

E. Social contributions to be rendered. Considering his age, talents, training, and past record of performance, is there a substantial probability that the patient will–*adequate recovery being assumed*–render in the future services to his society that can be characterized as outstanding, substantial, average, etc.?

This list is clearly insufficient for the construction of a reasonable selection system, since that would require not only *that these factors be taken into account* (somehow or other), but–going beyond this–would specify *a specific set of procedures for taking account of them.* The specific procedures that would constitute such a system would have to take account of the interrelationship of these factors (e.g., *B* and *E*), and to set out exact guidelines as to the relevant weight that is to be given to each of them. This is something our discussion has not as yet considered.

In fact, I should want to maintain that there is no such thing here as a single rationally superior selection system. The position of affairs seems to me to be something like this: (1) It is necessary (for reasons already canvassed) to *have* a system, and to have a system that is rationally defensible, and (2) to be rationally defensible, this system must take the factors *A–E* into substantial and explicit account. But (3) the exact manner in which a rationally defensible system takes account of these factors cannot be fixed in any one specific way on the basis of general considerations. Any of the variety of ways that give *A–E* "their due" will be acceptable and viable. One cannot hope to find within this range of workable systems some one that is *optimal* in relation to the alternatives. There is no one system that does "the (uniquely) best"–only a variety of systems that do "as well as one can expect to do" in cases of this sort.

The situation is structurally very much akin to that of rules of partition of an estate among the relations of a decedent. It is important *that there be* such rules. And it is reasonable that spouse, children, parents, siblings, etc., be taken account of in these rules. But the question of the exact method of division–say that when the decedent has neither living spouse nor living children then his estate is to be divided, dividing 60 percent between parents, 40 percent between siblings versus dividing 90 percent between parents, 10 percent between siblings–cannot be settled on the basis of any general abstract considerations of reasonableness. Within broad limits, a *variety* of resolutions are all perfectly acceptable–so that no one procedure can justifiably be regarded as "the (uniquely) best" because it is superior to all others.[21]

[21] To say that acceptable solutions can range over broad limits is *not* to say that there are no limits at all. It is an obviously intriguing and fundamental problem to raise the question of the factors that set these limits. This complex issue cannot be dealt with adequately here. Suffice it to say that considerations regarding precedent and people's expectations, factors of social utility, and matters of fairness and sense of justice all come into play.

VIII. A Possible Basis for a Reasonable Selection System

Having said that there is no such thing as *the optimal* selection system for ELT, I want now to sketch out the broad features of what I would regard as *one acceptable* system.

The basis for the system would be a point rating. The scoring here at issue would give roughly equal weight to the medical considerations (A and B) in comparison with the extramedical considerations (C = family role, D = services rendered, and E = services to be rendered), also giving roughly equal weight to the three items involved here (C, D, and E). The result of such a scoring procedure would provide the essential *starting point* of our ELT selection mechanism. I deliberately say "starting point" because it seems to me that one should not follow the results of this scoring in an *automatic* way. I would propose that the actual selection should only be guided but not actually be dictated by this scoring procedure, along lines now to be explained.

IX. The Desirability of Introducing an Element of Chance

The detailed procedure I would propose—not of course as optimal (for reasons we have seen), but as eminently acceptable—would combine the scoring procedure just discussed with an element of chance. The resulting selection system would function as follows:

1. First the criteria of inclusion of Section IV above would be applied to constitute a *first phase selection group*—which (we shall suppose) is substantially larger than the number n of persons who can actually be accommodated with ELT.

2. Next the criteria of selection of Section V are brought to bear via a scoring procedure of the type described in Section VIII. On this basis a *second phase selection group* is constituted which is only *somewhat* larger—say by a third or a half—than the critical number n at issue.

3. If this second phase selection group is relatively homogeneous as regards rating by the scoring procedure—that is, if there are no really major disparities within this group (as would be likely if the initial group was significantly larger than n)—then the final selection is made by *random* selection of n persons from within this group.

This introduction of the element of chance—in what could be dramatized as a "lottery of life and death"—must be justified. The fact is that such a procedure would bring with it three substantial advantages.

First, as we have argued above (in Section VII), any acceptable selection system is inherently non-optimal. The introduction of the element of chance prevents the results that life-and-death choices are made by the automatic application of an admittedly imperfect selection method.

Second, a recourse to chance would doubtless make matters easier for the rejected patient and those who have a specific interest in him. It would surely be quite hard for them to accept his exclusion by relatively mechanical application of objective criteria in whose implementation subjective judgment is involved. But the circumstances of life have conditioned us to accept the workings of chance and to tolerate the element of luck (good or bad): human life is an inherently contingent process. Nobody, after all, has an absolute right to ELT–but most of us would feel that we have "every bit as much right" to it as anyone else in significantly similar circumstances. The introduction of the element of chance assures a like handling of like cases over the widest possible area that seems reasonable in the circumstances.

Third (and perhaps least), such a recourse to random selection does much to relieve the administrators of the selection system of the awesome burden of ultimate and absolute responsibility.

These three considerations would seem to build up a substantial case for introducing the element of chance into the mechanism of the system for ELT selection in a way limited and circumscribed by other weightier considerations, along some such lines as those set forth above.[22]

It should be recognized that this injection of *man-made* chance supplements the element of *natural* chance that is present inevitably and in any case (apart from the role of chance in singling out certain persons as victims for the affliction at issue). As F. M. Parsons has observed: "any vacancies [in an ELT program–specifically haemodialysis] will be filled immediately by the first suitable patients, even though their claims for therapy may subsequently prove less than those of other patients refused later."[23] Life is a chancy business and

[22] One writer has mooted the suggestion that: "Perhaps the right thing to do, difficult as it may be to accept, is to select [for haemodialysis] from among the medical and psychologically qualified patients on a strictly random basis" (S. Gorovitz, "Ethics and the Allocation on Medical Resources," *Medical Research Engineering,* V [1966], p. 7). Outright random selection would, however, seem indefensible because of its refusal to give weight to considerations which, under the circumstances, *deserve* to be given weight. The proposed procedure of superimposing a certain degree of randomness upon the rational-choice criteria would seem to combine the advantages of the two without importing the worst defects of either.

[23] "Selection of Patients for Haemodialysis," *op. cit.,* p. 623. The question of whether a patient for chronic treatment should ever be terminated from the program (say if he contracts cancer) poses a variety of difficult ethical problems with which we need not at present concern ourselves. But it does seem plausible to take the (somewhat antiutilitarian) view that a patient should not be terminated simply because a "better qualified" patient comes along later on. It would seem that a quasi-contractual relationship has been created through established expectations and reciprocal understandings, and that the situation is in this regard akin to that of the man who, having undertaken to sell his house to one buyer, cannot afterward unilaterally undo this arrangement to sell it to a higher bidder who "needs it worse" (thus maximizing the over-all utility).

even the most rational of human arrangements can cover this over to a very limited extent at best.

Bibliography[24]

S. Alexander. "They Decide Who Lives, Who Dies," *Life,* LIII (November 9, 1962), 102–25.

C. Doyle. "Spare-Part Heart Surgeons Worried by Their Success," *Observer* (London), May 12, 1968.

J. Fletcher. *Morals and Medicine.* London, 1955.

S. Gorovitz. "Ethics and the Allocation of Medical Resources," *Medical Research Engineering,* V (1966), 5–7.

L. Lader. "Who Has the Right To Live?" *Good Housekeeping* (January, 1968), pp. 85 and 144–50.

J. D. N. Nabarro, F. M. Parsons, R. Shakman, and M. A. Wilson. "Selection of Patients for Haemodialysis," *British Medical Journal* (March 11, 1967), pp. 622–24.

H. M. Schmeck, Jr. "Panel Holds Life-or-Death Vote in Allotting of Artificial Kidney," *New York Times,* May 6, 1962, pp. 1, 83.

G. E. W. Wolstenholme and M. O'Connor (eds.). *Ethics in Medical Progress.* London, 1966.

[24] I acknowledge with thanks the help of Miss Hazel Johnson, Reference Librarian at the University of Pittsburgh Library, in connection with the bibliography.

LIFE AND THE RIGHT TO LIFE

HENRY DAVID AIKEN

The aim of this paper is, in the first instance, one of clarification. That is, the aim of one whose philosophical vocation is directed not so much to questions of scientific fact, simply as such, as to questions concerning the manner in which the conceptual schemes are employed in science, and to the ways in which results achieved through their employment affect other activities, particularly those bearing upon our ordinary conduct as human beings. But the task of clarification subtends that of conduct: we want to go to our various enterprises with clearer heads so that we may make less confused and more rational decisions of policy.

Now in our thinking about the right to life and of the myriad problems which it raises, we frequently fail to distinguish two senses of the term *life* which, although undoubtedly related, have quite different meanings. Thus, for example, we often say of someone that he has led a full life, that his life was unfulfilled, or that he is full of life or lively. What do such locutions connote? For one thing life is not merely a matter of being alive in some purely biological or biophysical sense of the term. Something can be alive or be capable of life in the latter sense, yet not alive or capable of life in the sense implied in or by the expressions mentioned above. What we mean by a full life, I take it, is one full of significant experience and activity, and we apply it in the first instance only to human beings. An acorn might realize all its potentialities as a budding oak tree; yet we would not on that account think of it as having lived a full life. Just the contrary: the question of living or of having lived a full life doesn't arise in discussing acorns and oak trees. The reason, plainly, is that, as we normally conceive them, questions about the quality or character of their experience, except in mythic contexts, make no sense. Again, a person who is lively or full of life is one who is vivacious, active, eager for experience, and capable of participating in forms of activity which yield significant experience. Here too,

---

This paper originally appeared in *Ethical Issues in Human Genetics,* Bruce Hilton *et al.*, eds. (Plenum Press, New York: 1973). Reprinted with permission of the author and Plenum Publishing Company.

what we have in mind is the behavior of beings, conceived as conscious, purposeful, concerned with and engaged in various human activities. Life in this sense is a normative concept: it is something to be cherished, worthy of nurture and concern in its own right.

But there is another sense of the term "life," as employed in biology and the other so-called life sciences, which carries with it none of these connotations. In this sense, to be alive is simply to meet certain biophysical conditions. So understood, something is alive if its behavior, as a phenomenon, is explainable and predictable in accordance with certain covering scientific laws together with accurate descriptions of certain antecedent conditions. The variables required in giving an account of it may be more complex than those required in giving an account of non-living chemical changes. But that fact, if such it is, carries with it no normative implications whatever. In this, the biological sense, "life," like "zinc" or "entropy," is intrinsically a neutral descriptive term. Accordingly, no one has any intrinsic obligations with respect to it, save perhaps to observe it accurately and to explain its changes as best one can. No one, that is to say, has any obligations to it, to preserve it in being, to nurture it, to help it to realize its potentialities. In this sense, the right to life is entirely contingent. Its application depends entirely upon its bearing upon possibilities of life in the other, human and normative sense of the term.

Now a word must be said about the concept of a right. This concept involves, on the one side, the idea of a rightful claim, made in behalf of someone or another and, correlatively, that of responsible persons to whom alone such claims may be addressed and who are presumed to be held liable for their respect or fulfillment. Normally, rights are claimed only within contexts of developing social practices, and the same holds, whether the rights in question are legal or moral rights. A right does not exist merely because of someone's wish or aspiration. Commonly legal rights, in a society such as ours, are presumed to rest upon and to be justified in terms of more fundamental rights which we call "moral." And a legal right which has no independent moral justification is one which can be defended, if at all, only on the ground that its persistent violation would undermine the legal system to which, as a whole, we have some obligation to sustain. However, as in societies whose legal systems permit, say, the institution of slavery, it is plain that the legal rights of masters ought systematically to be reprobated and/or violated.

Why? Surely the reason is that such rights, or their observation, do violence to our notions of humanity and of the rights pertaining to a human being.

The adequacy of this answer turns, of course, upon our understanding of what it means to be a human being, a question to which I shall turn shortly. But before doing so, let me make a few additional remarks in defense of the preceding sketch of the notion of a right itself against certain objections to it which were made, explicitly or implicitly by certain other authors of this volume. According to them, a right amounts merely to an expectation and, hence, that X has a right if, and only if, he can expect society or a social group or another person to do something. This will not do. The society we live in creates many expectations which entail no corresponding right. Conversely, rights may be justly claimed, even when, alas, there is all too little reason to expect that the claims in question will generally be honored. The defeat of an expectation does not entail the defeat of a right or a claim regarding it; nor does the defeat of a right turn merely on the defeat of any accompanying expectations.

What is missing in the expectation theory of rights is precisely the notion of responsibility or obligation which, as I contend, that of a right essentially involves. When I make you a promise, for example, it does not suffice that you expect me to keep it. What is required is that you are entitled, *ceteris paribus,* to hold me liable for keeping it. And when I fail to keep it, you are entitled to demand an excuse or justification and, if none can be provided, then I must make amends in some appropriate way. Similarly, one may expect that people in a particular society will behave, or are likely to behave in certain ways toward living things without in the least being justified in inferring that they have, or consider themselves to have, any corresponding responsibilities or that the living things in question have any right or rights to live.

From a moral point of view, then, rights pertain primarily, although perhaps not exclusively, to human beings. Now it is true that we frequently use the term *man* as a virtual synonym for *human being.* Thus, when the 18th-century philosophers and publicists spoke of the "rights of man," it is evident that they were thinking of man as a human being, as a person entitled, as such, to realize certain potentialities and to exercise certain powers and, correspondingly, obliged, as a member of the human race, to respect (at the least of it) the rights of other human beings. But there is another use of the term *man* which carries with it no such implication. It does not follow from the fact that X meets the conditions required of anything which may be truly said to be a man, as the zoologist employs the term, that a claim is thereby established that X is a human being, entitled to the prerogatives and required to meet the obligations which our common humanity may entail.

Thus, we must distinguish between man conceived simply as a biological or biophysical phenomenon and man conceived as a human being. The former sense of the term *man* is no less neutral than that of the terms *rock* and *pine tree;* the latter is not. The Greeks, for example, were not unaware that barbarians were, like themselves, featherless bipeds; they simply did not regard barbarians as fully human. Similarly, racists may be excellent biologists who can recognize a black man in the neutral biological sense as a member of a certain species just as accurately as an anti-racist. The racist's difficulty is that he does not recognize the black man to be a full-fledged human being who ought therefore to be accorded all the rights which pertain to our common humanity. The question for the racists as well as for the humanists, in short, cannot be settled in purely empirical terms; for both it is a question of who or what must be or deserves to be treated as a human being. And in sum, man as human, is essentially and saliently a distinctive normative concept, whereas man, as the natural scientist employs the term, is not a normative concept at all.

This is not to deny, of course, that our conceptions of humanity—what is involved in the notion of being human and what is required of human beings in their dealings with one another—are capable of enlargement or diminution. Owing in part to our developing anthropological knowledge of the behavior of men in so-called primitive societies, to our grasp of the fact that primitive societies, like other societies, have histories, to our understanding that man's psychological and social endowments do not vary significantly with his color, that his capacities for human development are profoundly affected by his physical and social environments, and, more ominously, that scientific technological skills make it possible to reduce any individual to a condition little different from that of a plant, our notions of man's human possibilities and of their fragility are subject to a radical change.

Indeed, precisely because man, the organism, is so variously subject to manipulation, a good many pundits such as B. F. Skinner, now argue in effect that such concepts as human dignity and autonomy or agency should be consigned to the semantical wastebasket into which many others have already been thrown, including such notions as soul and spirit. In this connection it is also worth recalling that prescient technocrat, Jeremy Bentham, contended nearly two centuries ago that the same should also be the fate of such "fictitious entities" as moral, as distinct from legal, rights and responsibilities themselves. Were Bentham and Skinner to have their way, in fact, the whole problem of the right to live, in the moral and human sense, would disappear from discussions of public policies along with, whether they realize it or not, the entire system of practices comprehended under the heading of personal and/or human relations. In that case, however, the primary issues with which we are here concerned would simply disappear.

It is not for me to deny that such issues could disappear or that in certain quarters they may already be in the process of disappearing. No doubt, we might reach a situation in which the only essential relations in which we stand to others are relations of scientific observer to neutral object. However, even in that case, honest observers would still have obligations and rights in their scientific undertakings which make it difficult, if not impossible, to understand how they could conceive one another as mere input-output systems. Science itself is an essentially human activity which precludes the scientist from treating his fellow scientist merely as an object. For an object or phenomenon has no obligation to tell the truth, not to mention a right to tell it. If an Einstein regarded a Niels Bohr as nothing more than a rather subtle biophysical entity and vice versa, their putative "disagreements," from their own points of view, would degenerate by stages into minor collisions with respect to which questions of truth and validity, or honesty and perfidy, would be reduced to the same order as that of falling bodies which happen to get into one another's way. In fact, were an Einstein really to treat a Bohr as a system of inputs and outputs, he would eventually come so to regard himself. To that extent the self-corrective enterprise of scientific inquiry, with all the characteristic responsibilities and rights pertaining thereto, would lose its own integral meaning for the individuals involved.

Now the right to a human life—that is, to the life we call "human"—is of course fundamental, since it encompasses the whole life of the mind, intellectual, artistic, religious, moral and communal. It includes also the various forms of bodily gratification too often ignored by ascetic moralities which, mistakenly, assume that intellectual and other spiritual activities can go it on their own, regardless of our bodily needs and wants. As here understood, then, the right to life is the great encompassing right of which all other human rights are aspects. And even if the former is conceived, again mistakenly, as unconditional, the latter plainly are not. Their limitations are set by the other forms of being which belong, no less essentially, to an acceptable human existence.

The right to a human life, most certainly, is nullified unless there is also a right to life in the biological sense of the term. But this does not mean that the right to life in the latter sense is unconditional. Just the contrary. From a human point of view, the preservation of life, biologically, is at once conditional and contingent, like the right to fresh air. Other rights, such as the right to the liberties essential to a developed human existence, are also conditional, since their exercise affects other inherent dimensions of human beings. But they are plainly not merely contingent. For they are aspects of human life itself.

The significance of this point may be seen in the following way. In earlier times, men normally conceived themselves as belonging to a continuous cycle or (later) a progressive development of human existence. Their forbearers and successors were virtually as actual to them as their own contemporaries. In such circumstances the rights of the unborn, the dying, or even the dead were scarcely distinguishable from those of the living.

For better or worse, such assumptions can no longer be taken for granted. Nowadays, owing in part to our increasing mobility, physically and (at least, in Western countries) economically and socially, and hence owing to a new nomadism not contemplated in previous ages, men frequently have little or no sense of belonging to an enduring family or local community, let alone a nation or state. Our anonymous forefathers are homogenized, along with the traditions they represented, into a great semi-alien and pre-human past. And our successors, whose lives we can scarcely envisage, have become part of a largely problematic future which, if not tomorrow, then a few decades (or centuries) hence, may contain no human beings whatever.

The consequence is a conception of human existence in which, by stages, generalized potentialities are themselves reduced to mere possibilities and present actualities become overwhelmingly precious. Thus, whereas it appeared to our predecessors that the right of the fetus to continuing life was not less, or possibly even more, compelling than the right of its mother to hers, and the right of one's as yet unconceived successors was no less exigent than one's own, the presumptions of continuous human being and of the rights pertaining thereto are now radically foreshortened. For us, accordingly, the rights of man are the rights, in the first instance, of existing human beings. And it is their lives which, in the biological sense, are now objects of central moral concern. Indeed, it is a cruel joke to require that a fully living mother sacrifice her being for the sake of an unborn fetus whose continuing existence, not to say its promise of human life, remains problematic. Justice, in the modern world, does not command it; on the contrary, it is revolted at the prospect. Let it be shown that the unborn sons and daughters of woman will be capable of guaranteeing her rights to human being, and then we shall see what to make of their own. Let us make sure that the potentialities of the born stand a chance of realization before we commit ourselves to the putative rights of the unborn or unconceived.

Now I believe three interlocking and overlapping sets of rights are essential to our modern conception of human being. The first may be called rights to the forms of enlightenment essential to free agency. A free agent, however, is not simply one who is at liberty to do as he pleases, for a man may do as he likes and not be a fully free human being. A fully free human being is one who is able to distinguish between what he pleases and what his obligations commit him to and who realizes accordingly that those to whom he is obligated have rights to which his own pleasures or interests should, on occasion, be sacrificed. Accordingly, the rights necessary to the exercise of free agency include rights to participation in all of the human practices essential to the development and exercise of considered moral judgment and deliberation; rights to instruction in the main historical conceptions of human society, personal development, and of a good, or tolerable, life; rights to unimpeded analysis and criticism and hence to free inquiry and speculation; rights to the acquisition of all available knowledge concerning matters of fact indispensable to the formation of sound judgment; and, accordingly, rights to free speech, press, and use of all media whereby such knowledge is acquired and disseminated.

A second main group of rights are those pertaining to what J. S. Mill called "the inner life," for example, rights to artistic creation and appreciation, and rights to the formation of religious sensibilities, attitudes, and criticism, and to the foundation of associations for the cultivation of artistic and religious concerns; and, not least, rights to the formation of friendships and comradeships without which the life of the mind loses its substance.

But there is also a third group of rights, which I call "material rights," without whose enjoyment rights pertaining to free agency and the inner life become merely formal and claims concerning them bare pieces of rhetoric. Under this head are included rights pertaining to self-maintenance, to adequate food and shelter and, not least, fear from radical material deprivation or want. However, private philanthropy, as the whole history of the race demonstrates, never suffices to ensure such rights. Hence, we must include among the material rights of human beings, rights to the formation of political and social institutions designed to ensure human beings from material deprivation.

Minimal guarantees, however, are not enough, so long as there exist radical material inequalities among the members of society. Rich and poor can rarely meet as equal human beings, any more than great nations can meet with client states save on terms analogous to those of master and servant or slave. Accordingly, it is essential not only that individual men but also that political institutions, whether at the national or international level, acknowledge, in principle, that all human beings are entitled to roughly equivalent shares of the world's material goods and powers. And this, I take it, is an essential part of the indwelling meaning of the right to democracy or a democratic way of life.

Analytically, such rights of human beings are distinguishable, but in practice they cannot be sharply separated. They form natural *Gestalts* such that the diminution or abrogation of rights in one domain will normally result in corresponding diminutions and abrogations of rights elsewhere. To give absolute priority to material rights in a particular society aborts the lives of the human beings whose full stomachs are paid for at the expense of their inner lives. And the converse, generally speaking, is no less true.

By the same token, we can establish no platonic hierarchy of human rights, nor any principles which give invariant preference, in situations of conflict, to one right, or set of rights, in relation to another. In one society, at a particular time, the overwhelming obstacle to the common exercise of rights pertaining to enlightenment or to the inner life is widespread poverty and insecurity. In another, there may be widespread but mindless guarantees of material rights that leave the individuals involved spiritually and hence humanly inert.

Now, in barest outline, we have provided a context within which we may seriously begin to discuss the right to life for man in the biophysical sense which concerns this conference. As I have already suggested, this right is not only conditional but contingent. For even though biological life is a *sine qua non* for human life, it is not, as such, a part of human life. A man may live, or have promise of continuing to live, as an organism, yet in so doing lose, or else be devoid of, all promise of genuine human being. But the promise, or possibility, of human life depends not only on biological but also on social conditions of survival. Where such conditions are absent, the contingent right to biological life has a decreasing claim upon the consciences of enlightened men.

Already, even in Catholic countries, contraception is widely tolerated, or even encouraged, by enlightened persons on the ground that unrestricted conception itself aborts, or tends to abort, the human rights of the living. And the right to abortion of the fetus, although more problematic, is accepted on the grounds that the realization of its putative potentialities for human life preclude fulfillment of the rights of already living human beings. Actual human beings, and the rights pertaining thereto, come first. The alternative, which places no limitation whatever on the right of man to biological life, would be inevitably a counter-ideology of dispair which regards life as an abomination and its destruction a consummation devoutly to be wished: in short, nihilism. Such a counter-human ideology, pressed to its extreme, would lead to the position that either there are no human rights or else that the only human right is a right to immediate self-destruction or suicide.

I conclude first that claims regarding the right to biological survival are entirely contingent upon the ability of the individual in question to make, with the help of others, a human life for himself. This means that in circumstances where there exists no possibility of anything approaching a truly human life, the right to biological or physical survival loses its own *raison d'etre* and, hence, that the merciful termination of life, in the biophysical sense, is acceptable or perhaps even obligatory. Other things equal (which they rarely are), the rights of parents to accept the onus of caring for a radically defective child which has no capacity for enjoying a human life may be acknowledged. But then this is owing not to the child's right but to the human rights of its parents. And when the care of such a child seriously endangers the well-being of others, including in particular the members of its own family, parental rights (which are themselves, of course, always conditional) must give way to other, more exigent claims. The same holds true in the case of the lives of older individuals who, for one reason or another, can no longer function as human beings. But here, even more clearly than in the case of a radically defective fetus, it is the rights of others, including also members of society who have no familial relations to the individual in question, which must decide the issue.

Who shall be entitled to make decisions in such cases is of course a question of fundamental importance. But it is a different question, and until this fact is recognized, it cannot be rationally answered. With all respect, it seems to me that the state is usually far too coarse an institution to be entrusted with such decisions. At best, it may enact laws which make mercy killings legally permissible when conducted under proper auspices by just and competent persons, usually with the consent of other human beings whose own rights are directly involved.

Such an answer, I freely admit, is inadequate. This much, however, does seem certain—the consent of members of the individual's own family is never absolutely decisive. Indeed, in societies where overpopulation reduces most of its human inhabitants to a condition of perpetual misery, the consent of the family seems to me a matter of secondary importance.

In this connection, I would be less than candid if I did not at least raise the question of justifiable infanticide, even in circumstances where no genetic defect is discernible. Now the primary argument against such a practice has always been that healthy newborn infants have the capacity of developing into full-fledged human beings. And the right of the infant to life is based largely or entirely upon such presumptive potentialities. However, potentialities are not actualities, and where the rights of actual human beings are (for example) widely jeopardized by otherwise uncontrolled exponential population increases, then the above-mentioned argument loses much of its force. And in circumstances in which the potentialities of the infants in question promise only lives of endless misery and frustration for themselves, then the force of the arguments is further, if tragically, reduced. Human potentialities, in short, presuppose certain actual conditions of fulfillment. When such conditions cannot be met, the potentiality in question reverts to the status of a bare possibility, and bare possibilities afford no secure basis for a doctrine of human rights. But in candor I should also add that I have no clear notion of how a practice of justifiable infanticide might be properly implemented. And so I must leave the issue here unhappily unresolved. It is enough, at this stage, merely to raise it for your consideration.

It is time, however, to reverse the preceding line of argument. Granted that the right to life is conditional, many men are pointlessly denied their rights as human beings for indefensible reasons. And many are murdered daily for bad reasons or else for no reason at all.

Not only Nazi Germany but also every great nation-state in modern times has become involved in wholesale murder: that is, the extreme violation of the right to human life. And the more powerful the nation-state, the more is it in danger of becoming involved in wholesale murder.

What conclusion can we draw from this? For one thing, either that the concept of the sovereign nation-state is obsolete, or that nation-states and their putative legitimate rulers are subject to continual moral critique. The "national interest" is no longer acceptable, if it ever was, as a basic category of practical, not to say moral, reasoning. Nations, like individuals, are as they do, and deserve to be judged and treated accordingly. And their rulers, no less than their ordinary citizens, have, or should have, moral responsibilities not to their subjects or fellow-citizens only but to all humankind. There are such things as crimes against humanity. The essential trouble with the Nuremberg trials (for example) is not that they occurred but that they were administered exclusively by the victorious nations.

Nation-states, as such, have at best only a contingent and derivative moral authority. Accordingly, the contention that their subjects are always obliged to obey their *de facto* rulers is indefensible. The rights of considered dissent, disobedience, rebellion, and revolution are also indispensable, if derivative, human rights. Nor are they suspendable, as Justice Holmes emphasized, even in the presence of a "clear and present danger" to the nation-state itself. In short, neither citizens nor their rulers (including members of the judiciary and the military) are ever exempt from moral criticism and punishment when they engage in wholesale violations of human rights.

If the right to a human life is, as it should be, the object of our primary moral concern, then the major threat to it is not the conscienceless abortionist or the fetishist for whom every fetus is already a full-fledged human being, but the rulers of nation-states and their military and industrial auxiliaries for whom unlimited warfare, or the threat thereof, are acceptable conditions of national survival. National survival is in itself at most a contingent right which is limited by the human rights not only of the citizens of the nation in question but of all other human beings. The profound horror of modern war is not simply that it involves inescapably the large-scale killing of innocents, but that it destroys systematically the whole form of life we call human, and in so doing violates or degrades the humanity of the guilty as well as innocent.

In fairness, it should be asked whether the same thesis, consistently maintained, does not equally prohibit violent forms of resistance, rebellion, and in extremity, revolution. I shall not deny that revolution, like war, is frequently murderous. However, it rarely happens in our time that those who perform acts of revolutionary intent, not to mention those who are engaged in acts of resistance or rebellion, are in a position to destroy, in effect, an entire civilization and hence to negate all the conditions of meaningful human being. More saliently, their aim, when justified, is not the defense of a particular national interest or the interests of a particular social class, but rather the recognition and protection of human rights which have hitherto been ignored by the rulers of nation-states and their auxiliaries.

This in no way condones or excuses rebels and revolutionists who terrorize or degrade their oppressors, when they make them more beastly than they were before by acts of torture, by random assassinations, etc.—just the contrary. Fanon and his admirers to the contrary, there is no right to terrorize one's oppressors, if this means treating them simply as objects or things. For in doing so, the terrorist places himself automatically in a situation of terror which debases him no less than those he terrorizes.

I realize that the issues here are complex and that this brief excursion has done them scant justice. Thus, let me return to my main theme. We do not live in a paradise in which every featherless biped or proto-biped is or is automatically to be viewed and treated as human. The right to human life is part of the more inclusive right to a human life, a pattern of being which includes all the rights I have mentioned above. Abstracted from this context, I fail to see why the right to life among featherless bipeds is any more sacred than the right, if right there be, to life on the part of mosses, insects, or cancer cells.

As matters stand, we all move, unreasonably, between the fetishism of life and the fetishism of self-interest, whether it be the interest of the individual, the class, the nation, or the race, in both cases ignoring the arduous task of trying to be a human being.

DYING

ROBERT S. MORISON

The contemplation of death in the 20th century can tell us a good deal about what is right and what is wrong with modern medicine. At the beginning of the century death came to about 15 percent of all newborn babies in their first year and to perhaps another 15 percent before adolescence. Nowadays fewer than 2 percent die in their first year and the great majority will live to be over 70.

Most of the improvement can be explained by changes in the numbers of deaths from infectious disease, brought about in large part by a combination of better sanitation, routine immunization, and specific treatment with chemotherapeutic drugs or antibiotics. Also important, although difficult to quantify, are improvements in the individual's nonspecific resistance that are attributable to improved nutrition and general hygiene.

Infectious diseases typically attack younger people. Equally typically, although not uniformly, they either cause death in a relatively short time or disappear completely, leaving the individual much as he was before. That is what makes death from infection particularly poignant. The large number of people who used to die of infection died untimely deaths; they had not lived long enough to enjoy the normal human experiences of love, marriage, supporting a family, painting a picture, or discovering a scientific truth.

The medical profession and the individual physician clearly had every incentive to struggle endlessly against deaths of this kind. Every triumph over an untimely death was rewarded by the high probability of complete recovery and a long, happy and productive life. No wonder the profession developed an ethic that placed a preponderant emphasis on preserving life at all costs. No wonder also that it became preoccupied with the spectacular advances in science and technology that made such triumphs possible.

Nevertheless, it is still clear that we must all die sometime. As a matter of fact, the age at which the last member of a cohort, or age class, dies is now much the same as it was in biblical times. Whereas life expectancy at birth has improved by perhaps two and a half times, life expectancy at 70 has changed

---

This paper originally appeared in *Scientific American,* (Sept. 1973) and is reprinted with permission of the author and the publisher.

very little. (It is now approximately 12 years.) The difference is that in earlier times relatively few people managed to reach 70, whereas under present conditions nearly two-thirds of the population reach that age. As a result most people no longer die of some quick, acute illness but of the chronic deteriorations of old age. Only very recently has the general public or even the medical profession begun to realize that the attitudes and techniques developed in the battle against untimely death may not be entirely appropriate in helping the aged patient adapt to changed physiological and psychological circumstances.

The progress of technology puts in the physician's hands a constantly increasing number of things he can do for and to his aging patients. In the jargon of modern policy, the "options" have been greatly increased and the problem of therapy has become largely a problem of choice. Modern students of decision theory point out that all methods of choice making reduce ultimately to the making of value judgments. When a pediatrician encounters an otherwise normal child with a life-threatening sore throat, the value judgment is simple and immediate. The life is obviously worth saving at all costs and the only choice to be made is what antibiotic to use. At the other end of life, however, most patients present a varied mosaic of diminished and disordered function. For a man or woman over a certain age there is no such thing as complete recovery. Treatment directed at supporting one vital system may simply bring into greater prominence a more awkward or more painful disorder. Furthermore, many of the treatment options, unlike the treatment of acute infections that may threaten premature death, are far from simple and inexpensive. Instead they are often cumbersome, painful, and costly. The art, moreover, is constantly changing, so that it is hard to estimate the probable results of some of the most elaborate procedures.

For example, in managing the course of a patient with chronic cardiovascular disease the physician has available more or less conventional drugs that can strengthen and regularize the heartbeat, reduce the accumulation of excess fluid in the tissues, moderate the blood pressure, and relieve the pain of reduced circulation to the heart muscle. These results can be achieved rather simply, and such treatment has for years prolonged and enhanced the quality of the life of many people over 60. Now, however, the physician must also weigh the probable results of operations to install new heart valves, replace arteries, or substitute pacemakers, all at a substantial risk to the patient's life and with the certainty of considerable pain and disability. At the end of this line is the transplantation of an entire heart, with so many risks and costs that its benefits currently tempt even the most courageous only rarely.

If the disease process leaves the heart relatively untouched and concentrates on the central nervous system, other possibilities assert themselves. Medical technology can now substitute for all the life-supporting functions of the nervous system. Tubes into either the gastrointestinal tract or a vein take the place of eating, a similar tube into the bladder takes the place of normal elimination, and an artificial respirator takes the place of breathing. Various electronic devices can even keep the heart beating for weeks or months beyond its appointed time.

Thus it has come about that most therapeutic decisions for people above a certain age are not life-or-death decisions in any simple either-or sense. Inevitably the physician, the patient himself (if he is in a state to do so), or the patient's family must make more or less explicit judgments of the probable quality of the life that remains, as well as its probable length.

In this changed situation the severer critics of the medical profession go to considerable length to debunk the traditional mystery and wisdom of the physician as a decision maker. Among other things, they point out that there is nothing in the technological training of physicians that equips them to deal with questions of ethical, aesthetic, or human value. Some even favor legislation removing life-or-death decisions entirely from the hands of the physician and giving them to an ombudsman or to a committee of moral philosophers.

Even the closest friends of medicine must admit that the profession has brought much of the current criticism on itself by failing to maintain the balance between the technological and the humane that characterized the best physicians from the days of Hippocrates to roughly those of Sir William Osler (1849–1919). In a way physicians have been seduced, if not actually betrayed, by their very competence. They can do a great deal for their patients at a purely technological level and at the same time they face nothing but uncertainties when they confront the ineluctable ills of the spirit. Is it any wonder they rejoice in the one and neglect the other?

I have approached the problem of terminal illness, or the dying patient, in this somewhat circuitous fashion in order to show that the problems surrounding the deathbed are not quite so unprecedented or so unconnected with the rest of medicine as to require the development of entirely new attitudes or perhaps even a new profession of "thanatologists." Dying is continuous with living, and the questions that are now asked with such insistence at the bedside of the dying should also be at least in the back of the mind of those who attend the living in earlier stages. The physician must consider the quality of the life he is struggling to preserve and the probable effects of his therapy on this quality, whatever the age of the patient. He should also know how to help a patient of any age to accept circumstances that cannot be changed. The so-called terminally ill patient simply represents the limiting case. In this instance the value questions have become paramount.

For reasons that are not all easy to identify the past few years have seen an astonishing increase in public attention to death and dying. A recent bibliography listing the titles of both books and papers in scholarly journals on the subject is several pages long. It is a rare daily newspaper or popular periodical that has not published one such article or several. Approaches to the topic may be roughly separated into two classes: those that deal with making the patient's last days as physically and psychologically comfortable as possible and those that discuss the propriety of allowing or helping the patient to die at an appropriate time. Let us turn first to the care of the terminally ill.

Students of the process of dying have long emphasized the loneliness of the dying person. Not only is he destined to go where no one wants to follow but also the people around him prefer to pretend that the journey is really not going to take place. The practice of placing familiar articles and even animals, servants, and wives in the tomb or on the funeral pyre of the departed is testimony to man's desire to assuage the loneliness beyond. In "The Death of Ivan Ilych" Leo Tolstoy has given the classic description of the conspiracy of denial that so often surrounds the dying. The situation has certainly been made worse by the technological changes since Tolstoy's day. Then, at least, most people died at home, many of them surrounded by family and friends. Even if these attendants were primarily concerned with what would happen to themselves, as were those awaiting the death of the old Count Bezhukoi that Tolstoy also portrayed, they not only kept the patient from being physically alone but also made him quite conscious of being the center of attention.

Nowadays only a minority of people die at home, and the number is decreasing. Precise figures are hard to obtain, but the scattered studies that have been made agree that about half of all deaths occur in large general hospitals and a smaller but increasing number in nursing homes. Probably fewer than a third die at home, at work or in public places.

The past few years have seen a growing awareness that the big general hospital is not a very good place to die. Even though such hospitals have large staffs, most of these professionals are preoccupied with administrative matters and with the increasingly complicated technical aspects of keeping people alive. Surrounded by these busybodies, the dying patient is more often than not psychologically isolated. Recognizing that such an atmosphere is bad both for patients and for younger physicians in training, a few inspired physicians have developed special programs to instruct members of the hospital staff in the needs of the dying patient. Such efforts have been received with enthusiasm by the still quite limited numbers of physicians and students who have been exposed to them.

The usefulness of training hospital staff members to deal with the dying patient is a concept that is now spreading throughout the country. There is substantial hope that the next generation of physicians, nurses, and administrators will be much more understanding and helpful in meeting the special needs of the dying than their predecessors were. One of the leaders in this movement is an Illinois psychiatrist Elisabeth K. Ross. In an effort to inject something approaching methodological rigor into her understanding and teaching of the needs of the dying, she has distinguished five stages exhibited by dying patients: deniel, anger, bargaining (usually with God), depression, and acceptance. This effort toward intellectualization of the problem is admirable; at the same time it is probably true that what has been most influential in alleviating the loneliness of the dying is the warmth of Dr. Ross's sympathy and the intensity of her dedication to the effort.

A high proportion of the patients in nursing homes are destined to die there or to be removed only at the very last minute for intensive hospital care, yet few of these institutions appear to have given much thought to the special problems of the dying. The most striking exceptions are provided by what in England are called "hospices." The best-known of them is St. Christopher's, outside London. It was started by a physician who was also trained as a nurse. She appears to have combined the best of both professions in developing arrangements for taking care of seriously ill patients and providing a warm and understanding atmosphere for them. Because many of the patients at St. Christopher's suffer from malignant disease, special emphasis is put on the alleviation of pain. It has been found that success depends not only on providing the right drugs at the right time but also on developing an attitude of understanding, confidence and hope. Psychological support involves, among other things, deinstitutionalizing the atmosphere by encouraging members of the family, including children and grandchildren, both to "visit the patient" in a formal sense and to carry on such activities as may be usual for their age, so that the patient feels surrounded by ongoing normal life. Everyone who has observed the program or has been privileged to participate in it speaks appreciatively and even enthusiastically about its achievements.

Hospices also serve as centers for an active home-care program that is demonstrating the practicality of tending to many dying patients in the home, provided that the physical arrangements are satisfactory and that specialized help from outside is available for a few hours a day. Unfortunately most insurance plans, including the otherwise enlightened National Health Service in Britain, have tended to emphasize hospital care to the detriment of adequate support for proper care in the home.

Recent studies of home care for seriously ill patients suggest that in many cases it can be not only more satisfactory emotionally than hospital care but also considerably less costly. Much more information is needed before administrators of health plans can determine precisely how many and what kinds of personnel should be available for how long to deal with various kinds of home situation. Similarly, a few preliminary surveys of the technological ways of adapting the American home for the care of the chronically ill have been made. Here again there is enough information to suggest the importance of funds for special types of beds and wheelchairs, the installation of plumbing within easy range of the sickbed and so on. The data, however, are not yet precise enough to allow adequate planning or the calculation of insurance premiums adequate to cover such services.

In spite of the potential emotional and economic benefits of home care one cannot overlook the fact that many social and technological changes have made illness and death at home very different from what they were in the days of Ivan Ilych and Count Bezhukoi. It is not appropriate here to try to cast up an account of the costs and benefits of such factors as rising social and geographic mobility and the transition from the extended family to the nuclear family. When these costs are counted, however, it may be well to include mention of the increasing difficulty in finding a good place to grow old and die.

Let us next examine the question of when it becomes appropriate to die. No matter how considerate the physician, how supportive the institutional atmosphere, how affectionately concerned the family and friends or how well-adjusted the patient himself, there comes a time when all those involved must ask themselves just how much sense it makes to continue vigorously trying to postpone the inevitaorer Regrettably the literature addressed to the topics of death and dying often seems preoccupied more with dissecting the various ethical and legal niceties surrounding the moment and manner of death than with what one can do to make the last few months or years of life as rewarding as possible. No doubt this preoccupation is prompted by an apparent conflict between ancient taboos on the one hand and certain obvious commonsense considerations on the other.

Now, virtually no one who has thought about the matter at all, from John Doe to the Pope, feels that any absolute moral or legal obligation requires one to do everything one knows how to do in order to preserve the life of a severely deteriorated patient beyond hope of recovery. Indeed, in actual practice it now seems probable that only a small minority of patients have everything possible done for them right up to the moment of death. The difficulties, then, are not with the general principle but with how to arrange the details. First and foremost in presenting themselves are the theoretical and even metaphysical problems involved in the ways that men of goodwill attempt to justify actions that appear to violate the taboo against killing. Second, there are practical questions to be answered. How can the individual make sure that his wishes are carried out? Who is to make the decision if the patient is no longer competent to do so? What are the physician's responsibilities? What, for that matter, are his possible liabilities? How far should society go in attempting to protect the rights and regulate the behavior of the various parties?

No more than an outline of the theoretical problems can be presented here. Current discussion centers principally on three issues. First is the definition of death. Next is the difference, if any, between negative from practice thanasia and last is the definition of "extraordinary means."

The possibility of redefining death came into prominence a few years ago when a group of Boston physicians grew concerned about precisely when it was appropriate to remove a prospective donor's organs for transplantation. It rapidly became clear that "defining" death, for whatever purpose, is a complicated philosophical matter that admits of no easy resolution. What the members of the Boston group actually did was to devise an operational redefinition of the criteria to be applied in declaring that a person has died. The major difficulty they faced arose from the purely technical fact that it is now possible to maintain the function of the heart and the lungs by artificial means. The failure of these two vital functions can therefore no longer be regarded in all cases as the paramount criteria for pronouncing death, as was once set forth in all conventional medical-legal texts.

The Boston group instead recommended the use of a set of signs testifying to an essentially complete failure of a third set of functions: those of the nervous system. This proposal has been received with approval by a large number of physicians, theologians, and lawyers, and has now been included in the law of several states. The criteria as they stand are extremely rigid and involve the death of essentially all levels of the nervous system. They thus seem entirely adequate both to protect the patient against premature assaults in order to retrieve viable organs and at the same time to guard him against unduly zealous attempts to maintain elementary vital signs in the name of therapy.

There is another class of patients, however, in whom elementary vital signs have not failed, although the higher brain functions—thinking, communicating with others, and even consciousness itself—have departed. Such a patient constitutes a most distressing problem to families and physicians, and it has been suggested that a further revision of the criteria for a pronouncement of death might be used to justify the termination of active treatment in such cases. The idea is that what is really human and important about the individual resides in the upper levels of the nervous system, and that these attributes indeed die with the death of the forebrain.

The presumed merit of such a revision grows out of the way it avoids the basic ethical problem; obviously it makes no sense to go on treating what can rationally be defined as a corpse. The weight of opinion, however, seems to be against dealing with the question of cerebral incapacity in this oblique way. From several standpoints it seems preferable to face up to the fact that under these circumstances a patient may still be living in some sense, but that the obligation to treat the living is neither absolute nor inexorable.

No less a moral authority than the Pope appears to have lent his weight to this view and even to have spoken for most Christians when he announced a few years ago that the physician is not required to use "extraordinary means" to maintain the spark of life in a deteriorated patient with no evident possibility of recovery. Nevertheless, serious ambiguities still remain. At first the Vatican's phraseology appears to have been designed to allow the withdrawal of "heroic" and relatively novel procedures such as defibrillation, cardiac massage, and artificial respiration. More subtle analysts point out that there is no absolute scale of extraordinariness, and that what is or is not extraordinary can only be judged in relation to the condition of a given individual. Hence there is nothing extraordinary about using all possible means to keep alive a young mother who has suffered multiple fractures, severe hemorrhages, and temporary unconsciousness. Conversely, the term *extraordinary* may well be applied to relatively routine procedures such as intravenous feeding if the patient is elderly, has deteriorated, and has little hope of improvement. Thus the most active proponents of the doctrine of extraordinary means clearly interpret "extraordinary" to mean "inappropriate in the circumstances." Although many people who hold this position would disagree, it is not easy for an outsider to distinguish their interpretation from advocacy of what is sometimes called "negative euthanasia."

Negative euthanasia refers to withdrawal of treatment from a patient who as a result is likely to die somewhat earlier than he otherwise would. Many thoughtful and sensitive people who favor the principle dislike the term because it suggests that treatment is being withdrawn with an actual, if unspoken, intent to shorten the patient's life. These critics, who place a high value on the taboo against taking life, prefer to regard the withdrawal of active therapy as simply a matter of changing from a therapeutic regime that is inappropriate to one that is more appropriate under the circumstances. If death then supervenes, it is not regarded as the result of anything the physician has or has not done but simply as a consequence of the underlying illness. Thus the physician and those who have perhaps participated in his decision are protected from the fear that they are "playing God" or from similar feelings of guilt.

Many of those who favor negative euthanasia also recognize that appropriate care of the terminally ill may include "positive" procedures, such as giving morphine (which, among other effects, may advance the moment of death). Invoking what is known in Catholic circles as the doctrine, or law, of double effect, they regard such positive actions as permissible as long as the conscious intent is to achieve some licit purpose such as the relief of pain. This view in turn requires the drawing of an important moral distinction between "awareness of probable result" and "intent."

Other moralists, such as the blunter and more forthright proponents of what is called situational ethics, may dismiss such subtleties as irrelevant logic-chopping. In their view it is a mistake to extend the generalized taboos and abstract principles of the past to encompass the peculiarities of the 20th-century death scene. They prefer to focus attention on the scene itself and to do what seems best in terms of the probable results for all concerned. Perhaps not surprisingly, the situational ethicist, who derives much of his philosophical base from classical utilitarian, or consequential, ethics, sees relatively little difference between negative and positive euthanasia, that is, between allowing to die and causing to die.

For the sake of completeness let me inject my own opinion that although there may be only a trivial intellectual distinction between negative and positive euthanasia, it seems unwise and in any event useless at this time to enter into an elaborate defense of positive euthanasia. Although the principle has its enthusiastic advocates, their number is strictly limited. The overwhelming majority of physicians and certainly a substantial majority of laymen instinctively recoil from such active measures as prescribing a known poison or injecting a large bubble into a vein. There seems to be a point where simple human reactions supersede both legal sanction and rational analysis. As an example, very few New York physicians or nurses are anxious to exercise the right given them by the laws of the state to perform abortions as late as the 24th week of pregnancy. Furthermore, it appears that as a practical matter negative euthanasia, or the withdrawal of all active therapy except the provision of narcotics to subdue restlessness and pain, will in any case be followed in a reasonable length of time by the coming of what Osler termed near the turn of the century the "old man's friend": a peaceful death from bronchial pneumonia. Thus, in the terminology of the law courts, we need not reach the most difficult question.

In view of the prevailing theoretical uncertainties, it is not surprising that what is done to or for the dying patient varies widely from place to place and from physician to physician. It is impossible to be precise because very few scientific observations have been made. Perhaps the most careful study is one conducted by Diana Crane of the University of Pennsylvania, who asked a large number of physicians and surgeons what they would probably do in a series of precisely outlined clinical situations. From this and more anecdotal evidence it seems clear that very few do everything possible to prolong the lives of all their patients. At the other extreme, even fewer physicians would appear to employ active measures with the avowed intent of shortening life. In between there is an enormous range of decisions: to give or not to give a transfusion, to prescribe an antibiotic or a sulfa drug, to attach or disconnect a respirator or an artificial kidney, to install a cardiac pacemaker or to let the battery of one that is already installed run down. The overall impression gained both by the informed observer and by the sometimes despairing layman is that the median of all these activities lies rather far toward officiously keeping alive.

The reasons are obvious enough: the momentum of a professional tradition of preserving life at all costs, the reluctance of the physician and the layman to ignore ancient taboos or to impair the value of such positive concepts as the sanctity of life, and the ambiguities of the love-hate relationship between parents and children or husbands and wives. Finally, there is the continuing uncertainty about the legal position of a physician who might be charged with hastening the death of a patient by acts of either omission or commission.

Up to now, at least, the legal deterrent appears to have been something of a chimera. A conscientious search of the available literature in English has uncovered not one criminal action charging that a physician omitted treatment with the intent to shorten life. Indeed, there are surprisingly few actions that charge positive euthanasia. Even in the few actions that have been lodged, juries have shown a reluctance to convict when there is evidence that the action was undertaken in good faith to put the patient on the far side of suffering.

Approaching the problem from a somewhat different angle, although the definitive examples are few, there appears to be general agreement that the adult patient in full possession of his senses has every right to refuse treatment. It is somewhat less certain that such refusal is binding after a patient loses legal competence. Even less clear is the status of the expressed wish of a potential patient with respect to what he would want done in certain hypothetical future circumstances. Efforts to clarify the status of such communications, sometimes called "living wills," with physicians and relatives are being actively pursued.

Important though it may be to establish the rights and privileges of those foresighted enough to want to participate in the design of their own death, it must be admitted that such individuals now constitute only a trivial part of the population. The great majority prefer not to think about their own death in any way. Indeed, most people do not even leave a will directing what to do with their material possessions.

What, then, can be done for that large number of people likely to slip into an unanticipated position of indignity on a deathbed surrounded by busybodies with tubes and needles in their hands, ready to substitute a chemical or mechanical device for every item in the human inventory except those that make human life significant? In such instances, under ideal, or perhaps I should say idyllic, circumstances the attending physician would have known the patient and his family for a long time. Further, he would have sensed their conscious and unconscious wishes and needs, and drawing on his accumulated skills and wisdom, he would conduct the last illness so as to maximize the welfare of all concerned. Unfortunately under modern conditions few families have a regular physician of any kind and even fewer physicians possess the hypothetical virtues I have outlined.

However that may be, at least three approaches to the problem are being actively pursued at present. Foremost among them is the active discussion I have referred to. Not only the professional journals but also the monthly, weekly, and daily press are publishing numerous articles on death and dying. Radio and television programs have followed suit, and it must be a rare church discussion group that has not held at least one meeting devoted to death with dignity. At the very least such discussion must remove some of the reluctance to speak of death or even think about it. At best it must improve the possibility of communication and understanding between the patient and the physician. The resulting change in the climate of opinion cannot fail to make it easier to discard outmoded taboos in favor of the common sense of contemporary men.

Second, and equally important, are the formal and informal efforts to improve the education of physicians by redressing the imbalance between technical skill and human wisdom that has grown up during the present century. In addition to the kind of clinical concern for the dying exemplified by Dr. Ross, many medical schools are converting their courses in medical ethics from a guild-oriented preoccupation with fee splitting and other offenses against the in-group to a genuine concern for ethical values in the treatment of patients.

Third, there are the more formal attempts to clarify rights, responsibilities, and roles by means of legislation. Part of this effort is directed at establishing the obligation of a physician to follow the expressed wishes of his patient or, at the very least, at protecting the physician from liability if he does so. Other legislative proposals stem from a more or less explicit conviction that death is too important a matter to be left to physicians. Difficulty is often encountered, however, in finding a satisfactory alternative, and there is now much discussion of the relative merits of assigning ultimate authority to the next-of-kin, to an ombudsman, or to a committee of social scientists, philosophers, and theologians.

It is too early to predict how such suggestions will turn out, but there are some reasons for feeling that it may be well to go slow in formalizing what is bound to be a difficult situation and instead to redouble efforts toward developing the capacity of medical men and laymen to deal informally with the problem. In actual practice the conduct of a drawn-out terminal illness involves a series of small decisions, based on repeated evaluations of the physical and emotional condition of the patient and the attitudes, hopes, and fears of his family and friends. It is not easy to see how an outsider such as an ombudsman, much less a committee, can be very easily fitted into what is typically an unobtrusive incremental process. Concrete evidence on this point may be found with respect to the beginning of life in the attempts by opponents of contraception to inject either a sheriff or a bureaucrat, so to speak, into the bedroom. These attempts have not proved satisfactory and are more and more being denounced by the courts as invasions of privacy.

As long as progress is being made at the informal, grass-roots level, it may be just as well to refrain from drafting tidy legislative solutions to problems so profound. Whatever else may be said, it is obvious that changing attitudes toward death and dying provide an excellent paradigm of how changing technologies force on us the consideration of equally significant changes in value systems and social institutions.

A STATUTORY DEFINITION OF THE STANDARDS FOR DETERMINING HUMAN DEATH

AN APPRAISAL AND A PROPOSAL*

ALEXANDER M. CAPRON‡
AND LEON R. KASS §

In recent years, there has been much discussion of the need to refine and update the criteria for determining that a human being has died.[1] In light of medicine's

This paper originally appeared in *The University of Pennsylvania Law Review* (vol. 121), and is reprinted with the permission of the authors, *The University of Pennsylvania Law Review,* and Fred B. Rothman and Co.

*This Article grew out of discussions held by the Research Group (formerly Task Force) on Death and Dying of the Institute of Society, Ethics and the Life Sciences, a nonprofit organization engaged in interdisciplinary analysis of the issues generated by biomedical advances. The Research Group has been investigating various practical and philosophical problems in the "meaning" of death and the care of dying patients. Earlier drafts of the Article were discussed at meetings of the Research Group, and, although not the subject of formal approval by the group, the Article reflects the conclusions of the Research Group's members, who include Henry K. Beecher, M.D., Harvard University; Eric Cassell, M.D., Cornell University Medical College; Daniel Callahan, Ph.D., Institute of Society, Ethics and the Life Sciences; Renée C. Fox, Ph.D., University of Pennsylvania; Michael Horowitz, LL.B., New York, N.Y.; Hans Jonas, Ph.D., New School for Social Research; Irving Ladimer, S.J.D., American Arbitration Association; Marc Lappé, Ph.D., Institute of Society, Ethics and the Life Sciences; Robert Jay Lifton, M.D., Yale University; William F. May, Ph.D., Indiana Universtiy; Robert S. Morison, M.D., Cornell University; Paul Ramsey, Ph.D., Princeton University; Elisabeth Kübler-Ross, M.D., Chicago, Ill.; Alfred Sadler, M.D., Yale University; Blair Sadler, J.D., Yale University; Jane Schick, Ph.D., M.D., Cornell University Medical College; Robert Stevenson, Ph.D., Bionetics, Inc., Frederick, Md.; Robert Veatch, Ph.D., Institute of Society, Ethics and the Life Sciences. The work of the Research Group has been supported in part by a grant from the New York Foundation. The authors thank the members of the Research Group for their valuable suggestions and critical review of the manuscript and Sharmon Sollito, B.A., Institute of Society, Ethics and the Life Sciences, for her assistance in research.

‡Assistant Professor of Law, University of Pennsylvania. B.A. 1966, Swarthmore College; LL.B. 1969, Yale University. Member, District of Columbia Bar.

§Executive Secretary, Committee on the Life Sciences and Social Policy, National Research Council–National Academy of Sciences. B.S. 1958, University of Chicago; M.D. 1962, University of Chicago; Ph.D. 1967, Harvard University.

[1] *See, e.g.,* P. Ramsey, The Patient as Person 59-112 (1970); Louisell, *Transplantation:*

increasing ability to maintain certain signs of life artificially[2] and to make good use of organs from newly dead bodies, new criteria of death have been proposed by medical authorities.[3] Several states have enacted or are considering legislation to establish a statutory "definition of death,"[4] at the prompting of some members of the medical profession who apparently feel that existing, judicially-

---

Existing Legal Constraints, in Ethics in Medical Progress: With Special Reference to Transplantation 91-92 (G. Wolstenholme & M. O'Connor eds. 1966) [hereinafter cited as Medical Progress]; *Discussion* of Murray, *Organ Transplantation: The Practical Possibilities, in id.* 68 (comments of Dr. G. E. Schreiner), 71 (comments of Dr. M. F. A. Woodruff); Wasmuth & Stewart, *Medical and Legal Aspects of Human Organ Transplantation,* 14 Clev.-Mar. L. Rev. 442 (1965); Beecher, *Ethical Problems Created by the Hopelessly Unconscious Patient,* 278 New Eng. J. Med. 1425 (1968); Wasmuth, *The Concept of Death,* 30 Ohio St. L.J. 32 (1969); Note, *The Need for a Redefinition of "Death,"* 45 Chi.-Kent L. Rev. 202 (1969).

[2] A dramatic increase over the past twenty years in the use of extraordinary means of support such as artificial respirators for terminal patients is generally assumed by physicians, on the basis of observational and anecdotal evidence, but no quantitative studies of this phenomenon have been done. Telephone interview with Dr. Claude L'Enfant, Director, Division of Lung Diseases, National Heart and Lung Institute, Bethesda, Md., Oct. 16, 1972. For this reason it is not possible to offer any detailed estimate of the impact that the proposed statute, *see* text accompanying note 86 *infra,* would have on the resources allocated to patient care.

[3] *See, e.g.,* Ad Hoc Committee of the Harvard Medical School to Examine the Definition of Brain Death, *A Definition of Irreversible Coma,* 205 J.A.M.A. 337 (1968) [hereinafter cited as *Irreversible Coma*]; *Discussion* of Murray, *Organ Transplantation: The Practical Possibilities,* in Medical Progress, *supra* note 1, at 69-74 (remarks of Drs. G. P. J. Alexandre, R. Y. Calne, J. Hamburger, J. E. Murray, J. P. Revillard & G. E. Schreiner); *When Is a Patient Dead?,* 204 J.A.M.A. 1000 (1968) (editorial); *Updating the Definition of Death,* Med. World News, Apr. 28, 1967, at 47.

In an earlier article, our Research Group appraised the proposed new medical criteria for the determination of death and discussed some of the sources of public concern. Task Force on Death and Dying, Institute of Society, Ethics and the Life Sciences, *Refinements in Criteria for the Determination of Death: An Appraisal,* 221 J.A.M.A. 48 (1972) [hereinafter cited as *Refinements in Criteria*]. In discussing the procedures used to establish the new criteria, the article concluded:

> Clearly, these matters of decisionmaking and the role of law need further and widespread discussion. The acceptability of any new concept or criteria of death will depend at least as much on the acceptability of the procedure by which they are adopted as on their actual content.

Id. 53.

[4] Kan. Stat. Ann. § 77-202 (Supp. 1971); Maryland Sessions Laws ch. 693 (1972). Bills are presently pending in Florida, H. 551, 2d Fla. Legis. (n.s.) (1971), Illinois, H. 1586, 77th Gen. Assemb., 1st Sess. (1971), and Wisconsin, S. 550, Biennial Sess. (1971).

Section IV of this Article argues that terms such as "defining death" and "definition of death" are extremely ambiguous, and that the ambiguity is an important cause of misunderstanding and confusion regarding the propriety of legislation in this area. Though it would be desirable not to use such terms, they are too well established in professional

framed standards might expose physicians, particularly transplant surgeons, to civil or criminal liability.[5] Although the leading statute in this area[6] appears to create more problems than it resolves,[7] some legislation may be needed for the protection of the public as well as the medical profession, and, in any event, many more states will probably be enacting such statutes in the near future.[8]

I. Background

Courts and physicians can no longer assume that determining whether and when a person has died is always a relatively simple matter. The development and use of sophisticated machinery to maintain artificially both respiration and circulation has introduced difficulties in making this determination in some instances. In such cases, the use of a cardiac pacemaker or a mechanical respirator renders doubtful the significance of the traditional "vital signs" of pulse, heartbeat, and respiratory movements as indicators of continuing life. Similarly, the ability of an organ recipient to go on living after his own heart has

---

and public discourse on these matters to be eliminated. For convenience, and often deliberately to emphasize the problem of ambiguity, we occasionally use these terms in quotation marks. For an explanation of four different levels of specificity which may be intended when the term "definition of death" is employed, see notes 57-60 *infra* & accompanying text.

[5] *See, e.g.,* Taylor, *A Statutory Definition of Death in Kansas,* 215 J.A.M.A. 296 (1971) (letter to the editor), in which the principal draftsman of the Kansas statute states that the law was believed necessary to protect transplant surgeons against the risk of "a criminal charge, for the existence of a resuscitated heart in another body should be excellent evidence that the donor was not dead [under the "definition" of death then existing in Kansas] until the operator excised the heart." *Cf.* Kapoor, *Death & Problems of Transplant,* 38 Man. B. News 167, 177 (1971); Baker, *Liability and the Heart Transplant,* 6 Houston L. Rev. 85, 97-101 (1968). The specter of civil liability was raised in *Tucker v. Lower,* a recent action brought by the brother of a heart donor against the transplantation team at the Medical College of Virginia. *See* notes 42-50 *infra* & accompanying text.

[6] Kan. Stat. Ann. § 77-202 (Supp. 1971); see notes 74-88, 98-101 *infra* & accompanying text for a discussion of this statute.

[7] *See* notes 74-85 *infra* & accompanying text.

[8] In addition to the state medical societies, *see* Taylor, *supra* note 5, others have advocated a statutory definition of death. "Medical researchers and M.D.'s involved in transplants must break with their traditional reluctance to seek statutory changes in the definition of death or find themselves floundering in a morass of court suits in coming years." 15 Drug Research Rep., June 7, 1972, at 1. Moreover, once a statute is enacted on a new subject in one state, there seem to be direct and indirect pressures for it to be viewed as a model for adoption in other states. *Cf. id.* 5, which notes that the Virginia statute allowing a medical examiner to provide a decedent's organs for transplantation is "the first of its kind in the nation" and terms it "a model for other states to emulate," when it is in fact poorly conceived and nearly incomprehensible.

been removed and replaced by another's has further undermined the status of the beating heart as one of the most reliable—if not *the* most reliable—signs that a person is still alive. In addition, the need of transplant surgeons to obtain organs in good condition from cadavers has stimulated the search for tests that would permit the death of the organism as a whole to be declared before the constituent organs have suffered extensive deterioration. Consequently, new criteria for judging a person dead have been proposed and are gaining acceptance among physicians. The most prominent are those formulated in 1968 by the Harvard Medical School's Ad Hoc Committee to Examine the Definition of Brain Death, chaired by Dr. Henry K. Beecher.[9]

The Harvard Committee described in considerable detail three criteria of "irreversible coma": (1) "unreceptivity and unresponsivity" to "externally applied stimuli and inner need"; (2) absence of spontaneous muscular movements or spontaneous respiration; and (3) no elicitable reflexes.[10] In addition, a flat (isoelectric) electroencephalogram was held to be "of great confirmatory value" for the clinical diagnosis.[11] Although generally referred to as criteria for "cere-

[9] *See Irreversible Coma, supra* note 3. In addition to Dr. Beecher, the committee consisted of nine other physicians, an historian, a lawyer, and a theologian, all Harvard University faculty members. *Id.*

[10] *Id.* 337/38-

[10] *Id.* 337-38.

[11] The Harvard committee spelled out its central conclusions as follows:

> An organ, brain or other, that no longer functions and has no possibility of functioning again is for all practical purposes dead. Our first problem is to determine the characteristics of a *permanently* nonfunctioning brain.
>
> A patient in this state appears to be in deep coma. The condition can be satisfactorily diagnosed by points 1, 2, and 3 to follow. The electroencephalogram (point 4) provides confirmatory data, and when available it should be utilized. In situations where for one reason or another electroencephalographic monitoring is not available, the absence of cerebral function has to be determined by purely clinical signs, to be described, or by absence of circulation as judged by stand-still of blood in the retinal vessels, or by absence of cardiac activity.
>
> 1. *Unreceptivity and Unresponsitivity.*—There is a total unawareness to externally applied stimuli and inner need and complete unresponsiveness—our definition of irreversible coma. Even the most intensely painful stimuli evoke no vocal or other response, not even a groan, withdrawal of a limb, or quickening of respiration.
>
> 2. *No Movements or Breathing.*—Observations covering a period of at least one hour by physicians is [*sic*] adequate to satisfy the criteria of no spontaneous muscular movements or spontaneous respiration or response to stimuli such as pain, touch, sound, or light. After the patient is on a mechanical respirator, the total absence of spontaneous breathing may be established by turning off the respirator for three minutes and observing whether there is any effort on the part of the subject to breathe spontaneously. (The respirator may be turned off for this time provided that at the start of the trial period the patient's carbon dioxide tension is within the normal range, and provided also that the patient has been breathing room air for at least 20 minutes prior to the trial.)

bral death" or "brain death," these criteria assess not only higher brain functions but brainstem and spinal cord activity and spontaneous respiration as well. The accumulating scientific evidence indicates that patients who meet the Harvard criteria will not recover and on autopsy will be found to have brains which are obviously destroyed,[12] and supports the conclusion that these criteria may be

---

3. *No reflexes.*–Irreversible coma with abolition of central nervous system activity is evidenced in part by the absence of elicitable reflexes. The pupil will be fixed and dilated and will not respond to a direct source of bright light. Since the establishment of a fixed, dilated pupil is clear-cut in clinical practice, there should be no uncertainty as to its presence. Ocular movement (to head turning and to irrigation of the ears with ice water) and blinking are absent. There is no evidence of postural activity (decerebrate or other). Swallowing, yawning, vocalization are in abeyance. Corneal and pharyngeal reflexes are absent.

As a rule the stretch of tendon reflexes cannot be elicited; i.e., tapping the tendons of the biceps, triceps, and pronator muscles, quadriceps and gastrocnemius muscles with the reflex hammer elicits no contraction of the respective muscles. Plantar or noxious stimulation gives no response.

4. *Flat Electroencephalogram* –Of great confirmatory value is the flat or isoelectric EEG. We must assume that the electrodes have been properly applied, that the apparatus is functioning normally, and that the personnel in charge is competent. We consider it prudent to have one channel of the apparatus used for an electrocardiogram. This channel will monitor the ECG so that, if it appears in the electroencephalographic leads because of high resistance, it can be readily identified. It also establishes the presence of the active heart in the absence of the EEG. We recommend that another channel be used for a noncephalic lead. This will pick up space-borne or vibration-borne artifacts and identify them. The simplest form of such a monitoring noncephalic electrode has two leads over the dorsum of the hand, preferably the right hand, so the ECG will be minimal or absent. Since one of the requirements of this state is that there be no muscle activity, these two dorsal hand electrodes will not be bothered by muscle artifact. The apparatus should be run at standard gains 10μv/mm, 50μv/5 mm. Also it should be isoelectric at double this standard gain which is 5μv/mm or 25μv/5 mm. At least ten full minutes of recording are desirable, but twice that would be better.

It is also suggested that the gains at some point be opened to their full amplitude for a brief period (5 to 100 seconds) to see what is going on. Usually in an intensive care unit artifacts will dominate the picture, but these are readily identifiable. There shall be no electroencephalographic response to noise or to pinch.

All of the above tests shall be repeated at least 24 hours later with no change.

The validity of such data as indications of irreversible cerebral damage depends on the exclusion of two conditions: hypothermia (temperature below 90 F. [32.2 C.]) or central nervous system depressants, such as barbiturates.

Irreversible Coma, supra note 3, at 337-38.

[12] In the largest single study of patients with flat E.E.G.'s of twenty-four hours' duration, which involved 2639 comatose patients without anesthetic doses of c.n.s. depressants, not one recovered. Silverman, Masland, Saunders & Schwab, *Irreversible Coma Associated with Electrocerebral Silence,* 20 Neurology 525 (1970). In an unreported study on 128 individuals who fulfilled the Harvard clinical criteria, postmortem examinations showed their brains to be destroyed. Unpublished results of E. Richardson, reported in *Refinements in Criteria, supra* note 3, at 50-51.

useful for determining that death has occurred. The Harvard Committee's views were apparently well received in the medical community.[13] Not all physicians have been enthusiastic, however. Professor David Rutstein of the Harvard Medical School, for example, expressed concern over "this major ethical change [which] has occurred right before our eyes . . . with little public discussion of its significance."[14]

Not surprisingly, disquiet over the change in medical attitude and practice arose in lay as well as medical circles.[15] The prospect of physicians agreeing amongst themselves to change the rules by which life is measured in order to salvage a larger number of transplantable organs met with something short of universal approval.[16] Especially with increasing disenchantment over heart transplantation (the procedure in which the traditional criteria for determining death posed the most difficulties), some doubt arose whether it was wise to adopt measures which encouraged a medical "advance" that seemed to have gotten ahead of its own basic technology. Furthermore, many people—doctors included—found themselves with nagging if often unarticulated doubts about how to proceed in the situation, far more common than transplantation, in which a long-comatose patient shows every prospect of "living" indefinitely with arti-

[13] One member of the committee has observed that "since the publication of the report, the clinical recommendations have been accepted and followed on a worldwide basis in a most gratifying fashion." Curran, *Legal and Medical Death—Kansas Takes the First Step,* 284 New Eng. J. Med. 260 (1971). Dr. Beecher recently noted that legal doubts have prevented uniform acceptance of the Harvard Committee's views. "Almost every (doctor) on the East Coast has accepted irreversible brain damage as the criterion for death, whereas most West Coast physicians do not for fear of suits." Ross, *Death with Dignity,* The Washington Post, Aug. 9, 1972, at B-15, col. 1 (quoting Dr. Beecher); for a fuller account of the testimony, see *Hearings on Death with Dignity Before the Senate Special Comm. on Aging,* 92d Cong., 2d Sess. (1972).

[14] Rutstein, *The Ethical Design of Human Experiments,* 98 Daedalus 523, 526 (1969). Leaders of the Netherlands Red Cross Society's Organ Transplantation Committee argue that only "total absence of the brain's functional capacity" and not "irreversible coma" indicates that death has occurred and state the Dutch position that the Harvard criteria "are grounds for stopping treatment and letting the patient die," but not for declaring death. Rot & van Till, *Neocortical Death after Cardiac Arrest,* 2 Lancet 1099-100 (1971) (letter to the editor).

[15] *See, e.g.,* Arnold, Zimmerman & Martin, *Public Attitudes and the Diagnosis of Death,* 206 J.A.M.A. 1949 (1968) [hereinafter cited as Arnold]; Biörck, *When is Death?,* 1968 Wis. L. Rev. 484, 490-91; N.Y. Times, Sept. 9, 1968, at 23, col. 1 (quoting Drs. F. C. Spencer & J. Hardy); *The Heart: Miracle in Cape Town,* Newsweek, Dec. 18, 1967, at 86-87.

[16] [C]ertain actions by transplant surgeons in establishing time of death on death certificates and hospital records have shaken public confidence. Coroners have denounced them in the press for signing a death certificate in one county when the beating heart was removed a day later in a far-off city. The public wonders what the "item" was that was transplanted across the state line and later registered as a person in the operating room record. Corday, *Life-Death in Human Transplantation,* 55 A.B.A.J. 629, 632 (1969).

ficial means of support.[17] As a result of this growing public and professional concern, elected officials,[18] with the encouragement of the medical community,[19] have urged public discussion and action to dispel the apprehension created by the new medical knowledge and to clarify and reformulate the law. Some commentators, however, have argued that public bodies and laymen in general have no role to play in this process of change.[20] Issue is therefore joined on at least two points: (1) ought the public to be involved in "defining" death? and (2) if so, how ought it to be involved–specifically, ought governmental action, in the form of legislation, be taken?[21]

II. *Public Involvement or Professional Prerogative?*

In considering the possible need for and the desirability of public involvement, the central question appears to be to what extent, if at all, the "defining" of death is a medical matter, properly left to physicians because it lies within their particular sphere of competence. The belief that the matter of "defining death" is wholly medical is frequently expressed, and not only by physicians.[22]

[17] [M]any people are now maintained in a sort of twilight state by the use of machines which do the work of their lungs or their heart while they are completely unconscious. . . . Many of these people will never resume an independent existence away from the machines. . . . One has to decide therefore when to switch off the machines, and this question arises quite independently of considerations about transplants. *Discussion* of Murray, *Organ Transplantation: The Practical Possibilities,* in Medical Progress, *supra* note 1, at 71 (comments of Dr. M. F. A. Woodruff).

[18] *See, e.g.,* Mondale, *Health Science and Society,* 117 Cong. Rec. S3708 (daily ed. Mar. 24, 1971); H. Res. 2830, 2d Fla. Legis. (n.s.) (1971) (resolution introduced by Rep. Sackett, M.D., to create a commission to study death).

[19] *See, e.g.,* Taylor, *supra* note 5, at 296; Arnold, *supra* note 15, at 1954; Corday, *Definition of Death: A Double Standard,* Hospital Tribune, May, 4, 1970, at 8; Halley & Harvey, *On an Interdisciplinary Solution to the Legal-Medical Definitional Dilemma in Death,* 2 Indiana Legal F. 219, 227 (1969).

[20] *See, e.g.,* Kennedy, *The Kansas Statute on Death: An Appraisal,* 285 New Eng. J. Med. 946 (1971); *What and When is Death?,* 204 J.A.M.A. 539, 540 (1968) (editorial).

[21] To some extent this formulation of the problem is, of course, unrealistic, since "public action" (*i.e.,* action by an official public body), in the form of a court decision, can come about at the instance of a private litigant regardless of any policy reasons in favor of public inaction. Although it may therefore be impossible to avoid creating "law" on the subject, there might still be no significant public involvement if the courts restricted themselves merely to endorsing conclusions reached by private groups, such as those representing physicians. Kennedy's support for judicial involvement in "defining" death seems to operate on that premise. *See* Kennedy, *supra* note 20, at 947. *See also* note 36 *infra* & accompanying text.

[22] *See, e.g.,* Kennedy, *supra* note 20, at 947; Berman, *The Legal Problems of Organ Transplantation,* 13 Vill. L. Rev. 751, 754 (1968); Sanders & Dukeminier, *Medical*

Indeed, when a question concerning the moment at which a person died has arisen in litigation, common law courts have generally regarded this as "a question of fact" for determination at trial on the basis (partially but not exclusively) of expert medical testimony.[23] Yet the standards which are applied in arriving at a conclusion, although based on medical knowledge, are established by the courts "as a matter of law."[24]

Advance and Legal Lag: Hemodialysis and Kidney Transplantation, 15 U.C.L.A.L. Rev. 357, 409 (1968) [hereinafter cited as Sanders]; National Conference of Commissioners on Uniform State Laws, Handbook and Proceedings of the Annual Conference 192 (1968); *cf.* Sadler, Sadler & Stason, *The Uniform Anatomical Gift Act: A Model for Reform*, 206 J.A.M.A. 2501 (1968). The ad hoc Harvard Committee, composed largely of physicians, came to the same conclusion. *See Irreversible Coma, supra* note 3, at 339.

[23] *See* Thomas v. Anderson, 96 Cal. App. 2d 371, 215 P.2d 478 (1950). In that appeal, the court was called upon to decide whether the trial judge had erred in holding inapplicable to the case a provision of the California Probate Code based on the Uniform Simultaneous Death Act which provided for the equal distribution of the property of two joint tenants "[w]here there is no sufficient evidence that [they] have died otherwise than simultaneously. . . ." The court cited the definition in *Black's Law Dictionary* that "death is the cessation of life; the ceasing to exist; defined by physicians as a total stoppage of the circulation of the blood, and a cessation of the animal and vital functions consequent thereon, such as respiration, pulsation, etc.," and went on to observe that "death occurs precisely when life ceases and does not occur until the heart stops beating and respiration ends. Death is not a continuing event and is an event that takes place at a precise time." *Id.* at 375, 215 P.2d at 482. It concluded that the "question of fact" as to which of the two deceased men died first had been correctly determined by the trial court in light of "sufficient evidence" given by nonmedical witnesses concerning the appearance of the men on the evening in question.

[24] Smith v. Smith, 229 Ark. 579, 587, 317 S.W.2d 275, 279 (1958) (quoting 41 Am. Jur. *Husbands and Wives* § 244 (1938)). The Smiths, a childless couple who by will had each left his or her estate to the other, were involved in an automobile accident. Mr. Smith apparently died immediately, but when assistance arrived Mrs. Smith was unconscious, and she remained so in the hospital for seventeen days. Thereafter, Mr. Smith's administrator petitioned for the construction of the wills, alleging:

> That as a matter of modern medical science, your petitioner . . . will offer the Court competent proof that the [Smiths] lost their power to will at the same instant, and that their demise as earthly human beings occurred at the same time in said automobile accident, neither of them ever regaining any consciousness whatsoever.

Id. at 582, 317 S.W.2d at 277. The Supreme Court of Arkansas upheld the trial court's dismissal of the petition as a matter of law on the ground that "it would be too much of a strain on credulity for us to believe any evidence offered to the effect that Mrs. Smith was dead, scientifically or otherwise, unless the conditions set out in the [*Black's Law Dictionary* (4th ed.)] definition existed." *Id.* at 586-87, 317 S.W.2d at 279. The court took "judicial notice that one breathing, though unconscious, is not dead," *id.* at 589, 317 S.W.2d at 281, and concluded that Mrs. Smith's death was therefore not simultaneous with her husband's.

Cf. In re Estate of Schmidt, 261 Cal. App. 2d 262, 67 Cal. Rptr. 847 (1968). *Schmidt*, like *Thomas* and *Smith*, involved an inheritorship issue under the Uniform Simultaneous

Thus while it is true that the application of particular criteria or tests to determine the death of an individual may call for the expertise of a physician, there are other aspects of formulating a "definition" of death that are not particularly within medical competence. To be sure, in practice, so long as the standards being employed are stable and congruent with community opinion about the phenomenon of death, most people are content to leave the matter in medical hands.[25] But the underlying extra-medical aspects of the "definition" become visible, as they have recently, when medicine departs (or appears to depart) from the common or traditional understanding of the concept of death. The formulation of a concept of death is neither simply a technical matter nor one susceptible of empirical verification. The idea of death is at least partly a philosophical question, related to such ideas as "organism," "human," and "living." Physicians *qua* physicians are not expert on these philosophical questions, nor are they expert on the question of which physiological functions decisively identify a "living, human organism." They, like other scientists, can suggest which "vital signs" have what significance for which human functions. They may, for example, show that a person in an irreversible coma exhibits "total unawareness to externally applied stimuli and inner need and complete unresponsiveness,"[26] and they may predict that when tests for this condition yield the same results over a twenty-four-hour period there is only a very minute chance that the coma will ever be "reversed."[27] Yet the judgment that "total unawareness . . . and complete unresponsiveness" are the salient characteristics of death, or that a certain level of risk of error is acceptable, requires more than technical expertise and goes beyond medical authority, properly understood.

The proposed departure from the traditional standards for determining death not only calls attention to the extra-medical issues involved, but is itself a source of public confusion and concern. The confusion can perhaps be traced to the fact that the traditional signs of life (the beating heart and the expanding chest) are manifestly accessible to the senses of the layman, whereas some of the new criteria require sophisticated intervention to elicit latent signs of life such as

---

Death Act. The court of appeals found that there was sufficient eyewitness testimony by laymen to support the trial court's conclusion that Mrs. Schmidt survived her husband by some minutes, and it found no fault in the use of the *Black's Law Dictionary* "definition of death" despite the argument that it "is an anachronism in view of the recent medical developments relating to heart transplants," since there was no evidence that the deceased were resuscitable. *Id.* at 273, 67 Cal. Rptr. at 854 (dictum).

[25] *See* Arnold, *supra* note 15, at 1950, in which the public's "nearly complete acceptance" of professional practice in this century until cardiac transplantation began is contrasted with the great concern manifested in the 19th century and earlier, before embalming became routine, largely because of the fear of premature burial.

[26] *Irreversible Coma, supra* note 3, at 337.

[27] *See* note 12 *supra.*

brain reflexes. Furthermore, the new criteria may disturb the layman by suggesting that these visible and palpable traditional signs, still useful in most cases, may be deceiving him in cases where supportive machinery is being used. The anxiety may also be attributable to the apparent intention behind the "new definition," which is, at least in part, to facilitate other developments such as the transplantation of cadaver organs. Such confusion and anxiety about the standards for determining death can have far-reaching and distressing consequences for the patient's family, for the physician, for other patients, and for the community at large.[28] If the uncertainties surrounding the question of determining death are to be laid to rest, a clear and acceptable standard is needed. And if the formulation and adoption of this standard are not to be abdicated to the medical fraternity under an expanded view of its competence and authority, then the public and its representatives ought to be involved.[29] Even if the medical profession takes the lead—as indeed it has—in promoting new criteria of death, members of the public should at least have the opportunity to review, and either to affirm or reject the standards by which they are to be pronounced dead.

III. What Manner of Public Involvement?

There are a number of potential means for involving the public in this process of formulation and review, none of them perfect. The least ambitious or comprehensive is simply to encourage discussion of the issues by the lay press, civic groups, and the community at large. This public consideration might be directed or supported through the efforts of national organizations such as the American Medical Association, the National Institutes of Health, or the National

[28] *See* Sanders, *supra* note 22, at 407-09; 3 M. Houts & I. H. Haut, Courtroom Medicine § § 1.02(3) (a)-(g) (1971). As long as the legal standard is ambiguous, the possibility exists that the processes of criminal, as well as civil, justice will be impeded. *See, e.g.*, D. Meyers, The Human Body and the Law 116-18 (1970) (discussing an unreported British case, *Regina v. Potter,* in which a manslaughter defendant was convicted of common assault upon proof that surgeons had removed a kidney from the decedent while he was being maintained on a respirator and before he had been found to be "dead"); *Trial to Test M.D.'s Role in Death of Heart Donor,* A.M.A. News, Nov. 11, 1968, at 2 (man charged with manslaughter raised as defense surgeons' removal of victim's heart when he was kept alive by artificial means).

[29] Matte, *Law, Morals, and Medicine: A Method of Approach to Current Problems,* 13 J. For. Sci. 318, 331-32 (1968). *See also* note 19 *supra.*

> A theoretical risk of illegal conduct exists in the present state of the law. The law is apparently waiting for a social and theological consensus on this point [of "defining" death] . . . . The theologians, the philosophers and the physicians will have to formulate the judgment of propriety here before it is crystallized into a definite statutory rule.

Discussion of Louisell, *Transplantation: Existing Legal Constraints,* in Medical Progress, *supra* note 1, at 99 (comments of Prof. D. W. Louisell).

Academy of Sciences.[30] A resolution calling for the establishment of an ad hoc body to evaluate public attitudes toward the changes wrought by biomedical advances has been sponsored by Senator Mondale since 1967 and was adopted by the Senate in December 1971.[31] Mondale's proposed National Advisory Commission on Health Science and Society, under the direction of a board of fifteen members of the general public and professionals from "medicine, law, theology, biological science, physical science, social science, philosophy, humanities, health administration, government, and public affairs," would conduct "seminars and public hearings" as part of its two-year study.[32] As important as it is to ventilate the issues, studies and public discussions alone may not be adequate to the task. They cannot by themselves dispel the ambiguities which will continue to trouble decisionmakers and the public in determining whether an artificially-maintained, comatose "patient" is still alive.

A second alternative, reliance upon the judicial system, goes beyond ascertaining popular attitudes and could provide an authoritative opinion that might offer some guidance for decisionmakers. Reliance on judge-made law would, however, neither actively involve the public in the decisionmaking process nor lead to a prompt, clear, and general "definition." The courts, of course, cannot speak in the abstract prospectively, but must await litigation, which can involve considerable delay and expense, to the detriment of both the parties and society. A need to rely on the courts reflects an uncertainty in the law which is unfortunate in an area where private decisionmakers (physicians) must act quickly and irrevocably. An ambiguous legal standard endangers the rights—and in some cases the lives—of the participants. In such circumstances, a person's choice of one course over another may depend more on his willingness to test his views in court than on the relative merits of the courses of action.[33]

[30] For example, early in the debate over heart replacement the Board on Medicine of the National Academy issued a "Statement on Cardiac Transplantation," but addressed itself primarily to the need for caution in the spread of the operation to medical centers which were not suited to carrying it out scientifically. 18 News Report of the National Academy of Sciences 1 (Mar. 1968).

[31] S. J. Res. 75, 92d Cong., 1st Sess. (1971), in 117 Cong. Rec. S20,089-93 (daily ed. Dec. 2, 1971). *See also* note 18 *supra*. The joint resolution is now in the House Committee on Interstate Commerce.

[32] S. J. Res. 75, 92d Cong., 1st Sess. (1971), in 117 Cong. Rec. S20,090 (daily ed. Dec. 2, 1971).

[33] For example, suppose that transplant surgeons were willing to employ a neurological definition of death, although most other physicians continued to use the "traditional" definition because of the unsettled nature of the law. If (*ex hypothesis*) those surgeons were less averse to the risks of testing their position in litigation, because of their temperament, training, values and commitments, or desire for success, their "courage" could lead to patients being declared dead prematurely according to the traditional standard.

Once called upon to "redefine" death–for example, in a suit brought by a patient's relatives or, perhaps, by a revived "corpse" against the physician declaring death–the judiciary may be as well qualified to perform the task as any governmental body. If the issue could be resolved solely by a process of reasoning and of taking "judicial notice" of widely known and uncontroverted facts, a court could handle it without difficulty. If, on the other hand, technical expertise is required problems may arise. Courts operate within a limited compass–the facts and contentions of a particular case–and with limited expertise; they have neither the staff nor the authority to investigate or to conduct hearings in order to explore such issues as public opinion or the scientific merits of competing "definitions."[34] Consequently, a judge's decision may be merely a rubberstamping of the opinions expressed by the medical experts who appear before him.[35] Indeed, those who believe that the "definition of death" should be left in the hands of physicians favor the judicial route over the legislative on the assumption that, in the event of a law suit, the courts will approve "the consensus view of the medical profession"[36] in favor of the new standards. Leaving the task of articulating a new set of standards to the courts may prove unsatisfactory, however, if one believes, as suggested previously, that the formulation of such standards, as opposed to their application in particular cases, goes beyond the authority of the medical profession.[37]

Uncertainties in the law are, to be sure, inevitable at times and are often tolerated if they do not involve matters of general applicability or great moment. Yet the question of whether and when a person is dead plainly seems the sort of issue that cannot escape the need for legal clarity on these grounds. Therefore, it is not surprising that although they would be pleased simply to have the courts

[34] *See, e.g.,* Repouille v. United States, 165 F.2d 152, 153 (2d Cir. 1947) (L. Hand, J.), 154 (Frank, J., dissenting).

[35] Because of the adversary nature of the judicial process, testimony is usually restricted to the "two sides" of an issue and may not fairly represent the spectrum of opinion held by authorities in the field.

[36] Kennedy, *supra* note 20, at 947. Kennedy's reliance on a medical "consensus" has a number of weaknesses, which he himself seems to acknowledge: (1) there may be "a wide range of opinions" held by doctors, so that "there need not necessarily be only one view" on a subject which is supported by the medical community, in part because (2) the "usual ways" for these matters to be "discussed and debated" are not very clear or rigorous since (3) the "American medical profession is not all that well regulated" unlike its British counterpart and (4) is not organized to give "official approval" to a single position or (5) to give force to its decision, meaning (6) that "the task will be assumed by some other body, most probably the legislature." *Id.*

[37] *Cf.* Blocker v. United States, 288 F.2d 853, 860 (D.C. Cir. 1961) (en banc) (Burger, J., concurring in the result) (criticizing psychiatrists' attempt to alter legal definition of "mental disease").

endorse their views, members of the medical profession are doubtful that the judicial mode of lawmaking offers them adequate protection in this area.[38] There is currently no way to be certain that a doctor would not be liable, criminally or civilly, if he ceased treatment of a person found to be dead according to the Harvard Committee's criteria but not according to the "complete cessation of all vital functions" test presently employed by the courts. Although such "definitions" were adopted in cases involving inheritors' rights and survivorship[39] rather than a doctor's liability for exercising his judgment about when a person has died, physicians have with good reason felt that this affords them little assurance that the courts would not rely upon those cases as precedent.[40] On the contrary, it is reasonable to expect that the courts would seek precedent in these circumstances. Adherence to past decisions is valued because it increases the likelihood that an individual will be treated fairly and impartially; it also removes the need to relitigate every issue in every case. Most importantly, courts are not inclined to depart from existing rules because to do so may upset the societal assumption that one may take actions, and rely upon the actions of others, without fear that the ground rules will be changed retroactively.[41]

Considerations of precedent as well as other problems with relying on the judicial formulation of a new definition were made apparent in *Tucker v. Lower*,[42] the first case to present the question of the "definition of death" in the context of organ transplantation. Above all, this case demonstrates the uncertainty that is inherent in the process of litigation, which "was touch and go

[38] *See* note 19 *supra.*

[39] *See* notes 23-24 *supra; cf.* Gray v. Sawyer, 247 S.W.2d 496 (Ky. 1952).

[40] *See* Taylor, *supra* note 5, at 296. *But cf.* Kennedy, *supra* note 20, at 947.

[41] "[R]ules of law on which men rely in their business dealings should not be changed in the middle of the game . . ." Woods v. Lancet, 303 N.Y. 349, 354, 102 N.E.2d 691, 695 (1951). It must be admitted, however, that such principles usually find their most forceful articulation when the court is about to proceed on the counter-principle that when necessary the common law will change with the times to achieve justice. (In *Woods*, for example, the New York Court of Appeals overruled its prior decision in Drobner v. Peters, 232 N.Y. 220, 133 N.E. 567 (1921), in order to permit a child to sue for prenatal injuries.) Although in this country, at least, strict adherence to precedent has been less true on the civil side than on the criminal (where the courts hold closer to the doctrine of *nullum crimen sine lege* than do English courts), it is probably fair to state that judges are more likely to depart from precedent in order to *create* a new cause of action than they are to reject an existing standard and thereby destroy a cause; to adjust the "definition of death" to the perhaps changing views of the medical profession would be to derogate the rights of those litigants injured by declarations of death which departed from previously accepted standards.

[42] Tucker v. Lower, No. 2831 (Richmond, Va., L. & Eq. Ct., May 23, 1972).

for the medical profession"[43] as well as the defendants. *Tucker* involved a $100,000 damage action against Drs. David Hume and Richard Lower and other defendant doctors on the Medical College of Virginia transplant team, brought by William E. Tucker, whose brother's heart was removed on May 25, 1968, in the world's seventeenth human heart transplant. The plaintiff claimed that the heart was taken without approval of the next of kin and that the operation was commenced before his brother had died. On the latter point, William Tucker offered evidence that his brother was admitted to the hospital with severe head injuries sustained in a fall and that after a neurological operation he was placed on a respirator. At the time he was taken to the operating room to have his organs removed "he maintained vital signs of life, that is, . . normal body temperature, normal pulse, normal blood pressure and normal rate of respiration."[44] Based on the neurologist's finding that the brother was dead from a neurological standpoint, the respirator was turned off and he was pronounced dead. The defendants moved to strike the plaintiff's evidence and for summary judgment in their favor, but the trial judge denied the motions.

> The function of This Court is to determine the state of the law on this or any other subject according to legal precedent and principle. The courts which have had occasion to rule upon the nature of death and its timing have all decided that death occurs at a precise time, and that it is defined as the cessation of life; the ceasing to exist; a total stoppage of the circulation of the blood, and a cessation of the animal and vital functions consequent thereto such as respiration and pulsation.[45]

The court adhered to "the legal concept of death" and rejected "the invitation offered by the defendants to employ a medical concept of neurological death in establishing a rule of law."[46] The court ruled that the jury would be allowed to assess damages if it concluded "that the decedent's life was terminated at a time earlier than it would ordinarily have ended had all reasonable medical efforts been continued to prolong his life."[47]

When he sent the case to the jurors, however, the judge permitted them to consider all possible causes of death, including injury to the brain as well as cessation of breathing or heartbeat, and a verdict was returned for the defendants. Unfortunately, the discrepancy between the initial ruling and the subse-

[43] 15 Drug Research Rep., June 7, 1972, at 1.

[44] Tucker v. Lower, No. 2831, at 4 (Richmond, Va., L. & Eq. Ct., May 23, 1972).

[45] *Id.* at 8 (citations omitted).

[46] *Id.*

> While it is recognized that none of the cases cited above involved transplants, to employ a different standard in this field would create chaos in other fields of the law and certainly it cannot be successfully argued that there should be one concept of death which applies to one type of litigation while an entirely different standard applies in other areas.

Id. at 8-9.

[47] *Id.* at 11.

quent instructions to the jury did little to resolve the legal uncertainty. The plaintiff has announced that he plans to appeal to the Supreme Court of Virginia,[48] and the creation of a clear and binding rule will depend on the action of that court.[49]

In declining the defendants' suggestion that he adopt a standard based on neurological signs, the judge stated that application for "such a radical change" in the law should be made "not to the courts but to the legislature wherein the basic concepts of our society relating to the preservation and extension of life could be examined and, if necessary, reevaluated."[50] A statutory "definition" of death would have notable advantages as an alternative to a judicial promulgation. Basically, the legislative process permits the public to play a more active role in decisionmaking and allows a wider range of information to enter into the framing the criteria for determining death. Moreover, by providing prospective guidance, statutory standards could dispel public and professional doubt, and could provide needed reassurance for physicians and patients' families, thereby reducing both the fear and the likelihood of litigation for malpractice (or even for homicide).

The legislative alternative also has a number of drawbacks, however. Foremost among these is the danger that a statute "defining" death may be badly drafted. It may be either too general or too specific, or it may be so poorly worded that it will leave physicians or laymen unsure of its intent. There is also the danger that the statutory language might seem to preclude future refinements that expanding medical knowledge would introduce into the tests and procedures for determining death. The problem of bad draftsmanship is compounded by the fact that a statute once enacted may be difficult to revise or repeal, leaving to the slow and uncertain process of litigation the clarification of its intent and meaning.[51] By contrast, although judges usually espouse the doctrine of stare decisis, flexibility over time is a hallmark of the common law. An additional practical problem is the possibility that the statutes enacted may reflect primarily the interests of powerful lobbying groups—for example, state medical societies or transplant surgeons. This possibility—similar to the danger of judicial "rubberstamping" of medical experts' opinions—may be avoided by legislatures' holding open and well-publicized hearings at which sociologists,

[48] N.Y. Times, May 27, 1972, at 15, col. 5; *id.*, June 4, 1972, § 4, at 7, col. 1.

[49] As one medical journal, which favors legislative formulation of a "definition," said of the decision of the Richmond court: "It applies only to cases coming before that court and can be reversed on appeal or overriden by contrary decisions handed down in higher courts." 15 Drug Research Rep., June 7, 1972, at 1.

[50] Tucker v. Lower, No. 2831, at 10 (Richmond, Va., L. & Eq. Ct., May 23, 1972).

[51] The general durability of statutes has the backhanded advantage, however, of emphasizing for the public as well as for legislators the importance of a thorough thrashing out of the issues in hearings and legislative debates.

lawyers, theologians, and representatives of various viewpoints are also called upon to testify.

Professor Ian Kennedy has suggested the further danger that a statutory "definition," rather than protecting the public may leave it vulnerable to physicians who through "liberal interpretation and clever argument" might take actions "just within the letter if not the spirit of the law."[52] Kennedy would rely instead on the medical profession's generalized "consensus view"[53] of the proper "definition of death." It is, however, far from clear why physicians who would violate a statute are unlikely to depart from such an informal "consensus," which may or may not eventually be sanctioned by the courts. Legislation will not remove the need for reasoned interpretation–first by physicians and perhaps then by judges–but it can restrict the compass within which they make their choices to one which has been found acceptable by the public.

Finally, the legislative route may reduce the likelihood that conflicting "definitions" of death will be employed in different jurisdictions in this country. Theoretically, uniformity is also possible in judicial opinions, but it occurs infrequently. If the formulation and reception of the Uniform Anatomical Gift Act provide any precedent, the Commissioners on Uniform State Laws appear to be well situated to provide leadership in achieving an intelligent response to changes in medical procedure.[54]

In sum, then, official action, as opposed to mere discussion of the issues, is needed if the conflict between current medical practice and present law is to be eliminated. A reformulation of the standards for determining death should thus be undertaken by either courts or legislatures. There are strengths and weaknesses in both law-creating mechanisms, but on balance we believe that if legislators approach the issues with a critical and inquiring attitude, a statutory "definition" of death may be the best way to resolve the conflicting needs for definiteness and flexibility, for public involvement and scientific accuracy.[55]

[52] Kennedy, *supra* note 20, at 947.

[53] *Id.*

[54] Completed in July 1968 by the Commissioners on Uniform State Laws and approved by the American Bar Association in August of that year, the Uniform Anatomical Gift Act was adopted with only minor changes in 40 jurisdictions including the District of Columbia in 1969; by the end of 1971, the Act had been adopted in the remaining 11 states. For a detailed discussion of the national acceptance of the Act see Sadler, Sadler & Stason, *Transplantation and the Law: Progress Toward Uniformity*, 282 New Eng. J. Med. 717 (1970). *See also* Brickman, *Medico-Legal Problems with the Question of Death*, 5 Calif. W.L. Rev. 110, 122 (1968) (urging Commissioners to draft uniform act on "the procedures for determining death").

[55] This is, of course, not to say that a judge faced with a case to decide should hold back from engaging in the sort of analysis, or reaching the conclusions about a proper "definition," presented here. As Professor Clarence Morris once observed, the age-old argument that a legislature has a "superior opportunity" to frame general rules should not

Moreover, since pressures for a legislative response to the problem appear to be mounting,[56] careful examination of the proper scope and content of such a statute seems to be called for.

IV. What Can and Should Be Legislated?

Arguments both for and against the desirability of legislation "defining" death often fail to distinguish among the several different subjects that might be touched on by such legislation. As a result, a mistaken impression may exist that a single statutory model is, and must be, the object of debate. An appreciation of the multiple meanings of a "definition of death" may help to refine the deliberations.

Death, in the sense the term is of interest here, can be defined purely formally as the transition, however abrupt or gradual, between the state of being alive and the state of being dead.[57] There are at least four levels of "definitions" that would give substance to this formal notion; in principle, each could be the subject of legislation: (1) the basic concept or idea; (2) general physiological standards; (3) operational criteria; and (4) specific tests or procedures.[58]

---

foreclose judicial reform of the law where the legislature has failed to act. A judge has, after all, "no reliable way of knowing" that legislative action will ever be forthcoming, and if he acts in a way the legislature finds erroneous, his mistake can be set right by statute. Morris, *Liability for Pain and Suffering,* 59 Colum. L. Rev. 476, 482 (1959).

[56] *See* note 8 *supra.* It would certainly be preferable for state legislatures and the Uniform Act Commissioners to begin work on laws now, rather than risking the enactment of "emergency legislation hastily contrived in response to public pressure and emotional reaction to [a] particular medical calamity." Matte, *supra* note 29, at 332; *cf.* Woodside, *Organ Transplantation: The Doctor's Dilemma and the Lawyer's Responsibility* 31 Ohio St. L.J. 66, 96 (1970).

[57] For a debate on the underlying issues see Morison, *Death: Process or Event?,* 173 Science 694 (1970); Kass, *Death as an Event: A Commentary on Robert Morison,* 173 Science 698 (1971).

[58] To our knowledge, this delineation of four levels has not been made elsewhere in the existing literature on this subject. Therefore, the terms "concept," "standard," "criteria," and "tests and procedures" as used here bear no necessary connection to the ways in which others may use these same terms, and in fact we recognize that in some areas of discourse, the term "standards" is more, rather than less, operational and concrete than "criteria"–just the reverse of our ordering. Our terminology was selected so that the category we call "criteria" would correspond to the level of specificity at which the Ad Hoc Harvard Committee framed its proposals, which it called and which are widely referred to as the "new *criteria*" for determining death. We have attempted to be consistent in our use of these terms throughout this Article. Nevertheless, our major purpose here is not to achieve public acceptance of our terms, but to promote awareness of the four different levels of a "definition" of death to which the terms refer.

The *basic concept* of death is fundamentally a philosophical matter. Examples of possible "definitions" of death at this level include "permanent cessation of the integrated functioning of the organism as a whole," "departure of the animating or vital principle," or "irreversible loss of personhood." These abstract definitions offer little concrete help in the practical task of determining whether a person has died but they may very well influence how one goes about devising standards and criteria.

In setting forth the *general physiological standard(s)* for recognizing death, the definition moves to a level which is more medico-technical, but not wholly so. Philosophical issues persist in the choice to define death in terms of organ systems, physiological functions, or recognizable human activities, capacities, and conditions. Examples of possible general standards include "irreversible cessation of spontaneous respiratory and/or circulatory functions," "irreversible loss of spontaneous brain functions," "irreversible loss of the ability to respond or communicate," or some combination of these.

Operational criteria further define what is meant by the general physiological standards. The absence of cardiac contraction and lack of movement of the blood are examples of traditional criteria for "cessation of spontaneous circulatory functions," whereas deep coma, the absence of reflexes, and the lack of spontaneous muscular movements and spontaneous respiration are among criteria proposed for "cessation of spontaneous brain functions" by the Harvard Committee.[59]

Fourth, there are the *specific tests and procedures* to see if the criteria are fulfilled. Pulse, heart beat, blood pressure, electrocardiogram, and examination of blood flow in the retinal vessels are among the specific tests of cardiac contraction and movement of the blood. Reaction to painful stimuli, appearance of the pupils and their responsiveness to light, and observation of movement and breathing over a specified time period are among specific tests of the "brain function" criteria enumerated above.

There appears to be general agreement that legislation should not seek to "define death" at either the most general or the most specific levels (the first and fourth). In the case of the former, differences of opinion would seem hard to resolve, and agreement, if it were possible, would provide little guidance for practice.[60] In the case of the latter, the specific tests and procedures must be kept open to charges in medical knowledge and technology. Thus, arguments concerning the advisability and desirability of a statutory definition of death are usually confined to the two levels we have called "standards" and "criteria," yet often without any apparent awareness of the distinction between them. The need for flexibility in the face of medical advance would appear to be a persuasive argument for not legislating any specific operational criteria. More-

[59] *See* notes 3, 10 *supra*.

[60] *Cf.* Robertson, *Criteria of Death*, 175 Science 581 (1972) (letter to the editor).

over, these are almost exclusively technical matters, best left to the judgment of physicians. Thus, the kind of "definition" suitable for legislation would be a definition of the general physiological standard or standards. Such a definition, while not immutable, could be expected to be useful for a long period of time and would therefore not require frequent amendment.

There are other matters that could be comprehended in legislation "defining" death. The statute could specify who (and how many) shall make the determination. In the absence of a compelling reason to change past practices, this may continue to be set at "a physician,"[61] usually the doctor attending a dying patient or the one who happens to be at the scene of an accident. Moreover, the law ought probably to specify the "time of death." The statute may seek to fix the precise time when death may be said to have occurred, or it may merely seek to define a time that is clearly after "the precise moment," that is, a time when it is possible to say "the patient is dead," rather than "the patient has just now died." If the medical procedures used in determining that death has occurred call for verification of the findings after a fixed period of time (for example, the Harvard Committee's recommendation that the tests be repeated after twenty-four hours), the statute could in principle assign the "moment of death" to either the time when the criteria were first met or the time of verification. The former has been the practice with the traditional criteria for determining death.[62]

Finally, legislation could speak to what follows upon the determination. The statute could be permissive or prescriptive in determining various possible subsequent events, including especially the pronouncement and recording of the death, and the use of the body for burial or other purposes.[63] It is our view that these matters are best handled outside of a statute which has as its purpose to "define death."[64]

[61] *Cf.* Uniform Anatomical Gift Act § 7(b).

[62] *See* note 99 *infra* & accompanying text.

[63] If . . . sound procedures for stating death are agreed to and carried out, then theologians and moralists and every other thoughtful person should agree with the physicians who hold that it is *then* permissible to maintain circulation of blood and supply of oxygen in the corpse of a donor to preserve an *organ* until it can be used in transplantation. Whether one gives the body over for decent burial, performs an autopsy, gives the cadaver for use in medical education, or uses it as a "vital organ bank" are all alike procedures governed by decent respect for the bodies of deceased men and specific regulations that ensure this. The ventilation and circulation of organs for transplant raises no question not already raised by these standard procedures. None are life-and-death matters.
P. Ramsey, The Patient as Person 72 (1970).

[64] Nevertheless, a statutory "definition" of death would most appropriately be codified with the provisions on the procedures to be followed to certify death, undertake post-mortem examinations, and so forth. For the reasons given below, the statute "defining" death ought not to be appended to the Uniform Anatomical Gift Act or other "special purpose" laws, however. *See* notes 65, 79-80 *infra* & accompanying text.

V. Principles Governing the Formulation of a Statute

In addition to carefully selecting the proper degree of specificity for legislation, there are a number of other principles we believe should guide the drafting of a statute "defining" death. First, the phenomenon of interest to physicians, legislators, and laymen alike is human death. Therefore, the statute should concern the death of a human being, not the death of his cells, tissues or organs, and not the "death" or cessation of his role as a fully functioning member of his family or community. This point merits considerable emphasis. There may be a proper place for a statutory standard for deciding when to turn off a respirator which is ventilating a patient still clearly alive, or, for that matter, to cease giving any other form of therapy.[65] But it is crucial to distinguish this question of "when to allow to die?" from the question with which we are here concerned, namely, "when to declare dead?" Since very different issues and purposes are involved in these questions, confusing the one with the other clouds the analysis of both. The problem of determining when a person is dead is difficult enough without its being tied to the problem of whether physicians, or anyone else, may hasten the death of a terminally-ill patient, with or without his consent or that of his relatives, in order to minimize his suffering or to conserve scarce medical resources.[66] Although the same set of

[65] *See* Potter, *The Paradoxical Preservation of a Principle,* 13 Vill. L. Rev. 784, 791 (1968):

> What type of questions are entailed in the debate concerning when a comatose patient should be declared dead? Medical questions and answers are only one element of the decisionmaking process. Medical skill may be used to establish that a patient has now entered and is likely to remain in a certain condition. But medical personnel along with the other members of the community must then ask: "What are we to do with patients in this condition?" The answer to that question does not flow directly from any medical knowledge. It is a question of social policy which must be decided by the entire community. Implementation of the communal policy may be left in the hands of physicians, but they act as agents of the communal conscience.

See generally Note, *Death with Dignity: A Recommendation for Statutory Change,* 22 U. Fla. L. Rev. 368 (1970); Fletcher, *Legal Aspects of the Decision Not to Prolong Life,* 203 J.A.M.A. 65 (1968); Sharpe & Hargest, *Lifesaving Treatment for Unwilling Patients,* 36 Fordham L. Rev. 695 (1968); Note, *The Dying Patient: A Qualified Right to Refuse Medical Treatment,* 7 J. Fam. L. 644 (1967); Elkinton, *The Dying Patient, The Doctor and the Law,* 13 Vill. L. Rev. 740 (1968); Biörck, *supra* note 15, at 488-90.

[66] The ease with which the two questions can become confused is demonstrated by the following "general definition of human death" proposed in Halley & Harvey, *Medical vs. Legal Definitions of Death,* 204 J.A.M.A. 423, 425 (1968);

> Death is irreversible cessation of *all* of the following: (1) total cerebral function, (2) spontaneous function of the respiratory system, and (3) spontaneous function of the circulatory system.
>
> Special circumstances may, however, justify the pronouncement of death when consultation consistent with established professional standards have been obtained and

social and medical conditions may give rise to both problems, they must be kept separate if they are to be clearly understood.

Distinguishing the question "is he dead?" from the question "should he be allowed to die?" also assists in preserving continuity with tradition, a second important principle. By restricting itself to the "is he dead?" issue, a revised "definition" permits practices to move incrementally, not by replacing traditional cardiopulmonary standards for the determination of death but rather by supplementing them. These standards are, after all, still adequate in the majority of cases, and are the ones that both physicians and the public are in the habit of employing and relying on. The supplementary standards are needed primarily for those cases in which artificial means of support of comatose patients render the traditional standards unreliable.

Third, this incremental approach is useful for the additional and perhaps most central reason that any new means for judging death should be seen as just that and nothing more—a change in method dictated by advances in medical practice, but not an alteration of the meaning of "life" and "death." By indicating that the various standards for measuring death relate to a single phenomenon legislation can serve to reduce a primary source of public uneasiness on this subject.[67] Once it has been established that certain consequences—for example, burial, autopsy, transfer of property to the heirs, and so forth—follow from a determination of death, definite problems would arise if there were a number of "definitions" according to which some people could be said to be "more dead" than others.

There are, of course, many instances in which the law has established differing definitions of a term, each framed to serve a particular purpose. One wonders, however, whether it does not appear somewhat foolish for the law to offer a number of arbitrary definitions of a natural phenomenon such as death. Nevertheless, legislators might seek to identify a series of points during the process of dying, each of which might be labelled "death" for certain purposes. Yet so far as we know, no arguments have been presented for special purpose standards except in the area of organ transplantation. Such a separate "defini-

---

when valid consent to withhold or stop resuscitative measures has been given by the appropriate relative or legal guardian.

The authors seem to have realized the mistake in making the state of being dead (rather than the acceptance of imminent death) depend on the "consent" of a relative or guardian, and this aspect of the "definition of death" is absent from their subsequent writings. *See, e.g.*, Halley & Harvey, *Law-Medicine Comment: The Definitional Dilemma of Death*, 37 J. Kan. B. Ass'n 179, 185 (1968); *cf.* D. Meyers, *supra* note 28, at 135-36 (criticizing Halley and Harvey's second definition for its internal inconsistency).

67 *See* notes 15, 16 *supra*. The way in which cardiopulmonary and brain functions relate to each other and to the phenomenon of death is explored in note 89 *infra*.

tion of death," aimed at increasing the supply of viable organs, would permit physicians to declare a patient dead before his condition met the generally applicable standards for determining death if his organs are of potential use in transplantation. The adoption of a special standard risks abuse and confusion, however. The status of prospective organ donor is an arbitrary one to which a person can be assigned by relatives[68] or physicians and is unrelated to anything about the extent to which his body's functioning has deteriorated. A special "definition" of death for transplantation purposes would thus need to be surrounded by a set of procedural safeguards that would govern not only the method by which a person is to be declared dead but also those by which he is to be classified as an organ donor.[69] Even more troublesome is the confusion over the meaning of death that would probably be engendered by multiple "definitions."[70] Consequently, it would be highly desirable if a statute on death could avoid the problems with a special "definition." Should the statute happen to facilitate organ transplantation, either by making more organs available or by making prospective donors and transplant surgeons more secure in knowing what the law would permit, so much the better.[71]

If, however, more organs are needed for transplantation than can be legally obtained, the question whether the benefits conferred by transplantation justify the risks associated with a broader "definition" of death should be addressed directly[72] rather than by attempting to subsume it under the question "what is

[68] Uniform Anatomical Gift Act § 2(c). For example, if a special standard were adopted for determining death in potential organ donors, relatives of a dying patient with limited financial means might feel substantial pressure to give permission for his organs to be removed in order to bring to a speedier end the care given the patient.

[69] The Uniform Anatomical Gift Act, which establishes procedures for the donation of organs by an individual or his relatives, appears to operate on the premise that "death" will be determined by standards which are generally accepted and applied in the ordinary course of events; it does not undertake to "define" death. *But cf.* note 100 *infra.*

[70] For instance, suppose that Mr. Smith, a dying patient in University Hospital, is found to be immunologically well matched with Mr. Jones, a University Hospital patient awaiting a heart transplant. Under the special transplantation "definition" Smith is then declared dead, but just as the surgeons are about to remove Smith's heart, Jones suddenly dies. The doctors then decide that Smith is no longer needed as an organ donor. His condition does not meet the standards for declaring death in nondonors. Is Smith "dead" or "alive"?

[71] This would be the case if the generally applicable standards for determining death permit organs to be removed at a time when they are still useable for transplantation purposes. The "definition" suggested by the Article meets this objective, we believe.

[72] Much of the public's fear of premature excision arises from the failure to distinguish the general practitioner's and the transplant surgeon's meaning of the term 'death'. It would be desirable to distinguish the two formally, and use different terms.

Hillman & Aldridge, *Towards a Legal Definition of Death,* 116 Sol. J. 323, 324 (1972) [hereinafter cited as Hillman]. These British medical-legal commentators suggest that

death?" Such a direct confrontation with the issue could lead to a discussion about the standards and procedures under which organs might be taken from persons near death, or even those still quite alive, at their own option[73] or that of relatives, physicians, or representatives of the state. The major advantage of keeping the issues separate is not, of course, that this will facilitate transplantation, but that it will remove a present source of concern: it is unsettling to contemplate that as you lie slowly dying physicians are free to use a more "lenient" standard to declare you dead if they want to remove your organs for transplantation into other patients.

Fourth, the standards for determining death ought not only to relate to a single phenomenon but should also be applied uniformly to all persons. A person's wealth or his "social utility" as an organ donor should not affect the way in which the moment of his death is determined.

Finally, while there is a need for uniformity of application at any one time, the fact that changes in medical technology brought about the present need for "redefinition" argues that the new formulation should be flexible. As suggested in the previous section, such flexibility is most easily accomplished if the new "definition" confines itself to the general standards by which death is to be determined and leaves to the continuing exercise of judgment by physicians the establishment and application of appropriate criteria and specific tests for determining that the standards have been met.

"irreversible brain damage," which would include patients with no higher brain activity but continued spontaneous respiration, be recognized as a ground for removal of organs prior to ordinary death. They contemplate that certain "essential safeguards" be incorporated into a statute on "irreversible brain damage" to avoid abuse of this category. *Id.* 325.

Prior to the first heart transplant in France, a special "definition" was enacted to remove any uncertainty about the permissibility of removing a beating heart from a "dead" donor. In April 1968 the government decreed a "definition of clinical death" for use with organ donors, based on a flat electroencephalogram of ten minutes duration which was taken to show that an artificially maintained patient lacks "function in the higher nervous centers." D. Meyers, *supra* note 28, at 113. Meyers seems to question this approach; he believes that the public must be shown

> not just that the brain has been irreparably damaged, but also that the extent of this damage is absolutely inconsistent with continued maintenance of independent life in the individual. If electro-enphalograph testing can in fact show this, then it is a valuable definitional tool in ascertaining clinical death; but the medical profession as yet appears somewhat divided on its reliability. In such circumstances, the public cannot be expected to accept the evidence of an electro-encephalographic reading as part of a legislative definition of death.

Id. 135.

[73] *See, e.g.,* Blachly, *Can Organ Transplantation Provide an Altruistic-Expiatory Alternative to Suicide?,* 1 Life-Threatening Behavior 6 (1971); Scribner, *Ethical Problems of Using*

VI. The Kansas Statute

The first attempt at a legislative resolution of the problems discussed here was made in 1970 when the State of Kansas adopted "An Act relating to and defining death."[74] The Kansas statute has received a good deal of attention; similar legislation was enacted in the spring of 1972 in Maryland and is presently under consideration in a number of other jurisdictions.[75] The Kansas legislation, which was drafted in response to developments in organ transplantation and medical support of dying patients, provides "alternative definitions of death,"[76] set forth in two paragraphs. Under the first, a person is considered "medically and legally dead" if a physician determines "there is the absence of spontaneous respiratory and cardiac function and . . . attempts at resuscitation are considered hopeless."[77] In the second "definition," death turns on the absence of spon-

---

Artificial Organs to Sustain Human Life, 10 Trans. Am. Soc. Artif. Internal Organs 209, 211 (1964) (advocating legal guidelines to permit voluntary euthanasia for purpose of donating organs for transplantation).

[74] Law of Mar. 17, 1970, ch. 378, [1970] Kan. Laws 994 (codified at Kan. Stat. Ann. § 77-202 (Supp. 1971)). It provides in full:

> A person will be considered medically and legally dead if, in the opinion of a physician, based on ordinary standards of medical practice, there is the absence of spontaneous respiratory and cardiac function and, because of the disease or condition which caused, directly or indirectly, these functions to cease, or because of the passage of time since these functions ceased, attempts at resuscitation are considered hopeless; and, in this event, death will have occurred at the time these functions ceased; or
>
> A person will be considered medically and legally dead if, in the opinion of a physician, based on ordinary standards of medical practice, there is the absence of spontaneous brain function; and if based on ordinary standards of medical practice, during reasonable attempts to either maintain or restore spontaneous circulatory or respiratory function in the absence of aforesaid brain function, it appears that further attempts at resuscitation or supportive maintenance will not succeed, death will have occurred at the time when these conditions first coincide. Death is to be pronounced before artificial means of supporting respiratory and circulatory function are terminated and before any vital organ is removed for purposes of transplantation.
>
> These alternative definitions of death are to be utilized for all purposes in this state, including the trials of civil and criminal cases, any laws to the contrary notwithstanding.

[75] *See* note 4 *supra.* In the Maryland law, which is nearly identical to its Kansas progenitor, the phrase "in the opinion of a physician" was deleted from the first paragraph, and the phrase "and because of a known disease or condition" was added to the second paragraph following "ordinary standards of medical practice." Maryland Sessions Laws ch. 693 (1972). Interestingly, Kansas and Maryland were also among the first states to adopt the Uniform Anatomical Gift Act in 1968, even prior to its official revision and approval by the National Conference of Commissioners on Uniform State Laws.

[76] Note 74 *supra.*

[77] *Id.* In using the term "hopeless," the Kansas legislature apparently intended to indicate that the "absence of spontaneous respiratory and cardiac function" must be irreversible

taneous brain function if during "reasonable attempts" either to "maintain or restore spontaneous circulatory or respiratory function," it appears that "further attempts at resuscitation or supportive maintenance will not succeed."[78] The purpose of the latter "definition" is made clear by the final sentence of the second paragraph:

> Death is to be pronounced before artificial means of supporting respiratory and circulatory function are terminated and *before any vital organ is removed for the purpose of transplantation.*[79]

The primary fault with this legislation is that it appears to be based on, or at least gives voice to, the misconception that there are two separate phenomena of death. This dichotomy is particularly unfortunate because it seems to have been inspired by a desire to establish a special definition for organ transplantation, a definition which physicians would not, however, have to apply, in the draftsman's words, "to prove the irrelevant deaths of most persons."[80] Although there is nothing in the Act itself to indicate that physicians will be less concerned with safeguarding the health of potential organ donors, the purposes for which the Act was passed are not hard to decipher, and they do little to inspire the average patient with confidence that his welfare (including his not being prematurely declared dead) is of as great concern to medicine and the State of Kansas as is the facilitation of organ transplantation.[81] As Professor Kennedy cogently observes, "public disquiet [over transplantation] is in no way allayed by the existence in legislative form of what appear to be alternative definitions of death."[82] One hopes that the form the statute takes does not reflect a conclusion on the part of the Kansas legislature that death occurs at two distinct points during the process of dying.[83] Yet this inference can be derived from the Act, leaving open the prospect "that X at a certain stage in the

---

before death is pronounced. In addition to being rather roundabout, this formulation is also confusing in that it might be taken to address the "when to allow to die?" question as well as the "is he dead?" question. *See* note 85 *infra* & accompanying text.

[78] Note 74 *supra.*

[79] *Id.* (emphasis added).

[80] Taylor, *supra* note 5, at 296.

[81] *Cf.* Kass, *A Caveat on Transplants,* The Washington Post, Jan. 14, 1968, § B, at 1, col. 1; *Discussion* of Murray, *Organ Transplantation: The Practical Possibilities,* in Medical Progress, *supra* note 1, at 67 (comments of Dr. T. E. Starzl): "[T]he new risk is introduced [by the use of cadaver organs] that the terminal care of such potential donors may be adversely influenced by the events which are expected to follow after death, which might conceivably remove whatever small chance there might have been for survival."

[82] Kennedy, *supra* note 20, at 947.

[83] General use of the term "resuscitation" might suggest the existence of a common notion that a person can die once, be revived (given life again), and then die again at a later

process of dying can be pronounced dead, whereas Y, having arrived at the same point, is not said to be dead."[84]

The Kansas statute appears also to have attempted more than the "definition" of death, or rather, to have tried to resolve related questions by erroneously treating them as matters of "definition." One supporter of the statute praises it, we think mistakenly, for this reason: "Intentionally, the statute extends to these questions: When can a physician avoid attempting resuscitation? When can he terminate resuscitative efforts? When can he discontinue artificial maintenance?"[85] To be sure, "when the patient is dead" is one obvious answer to these questions, but by no means the only one. As indicated above, we believe that the question "when is the patient dead?" needs to be distinguished and treated separately from the questions "when may the doctor turn off the respirator?" or "when may a patient—dying yet still alive—be allowed to die?"

VII. A Statutory Proposal

As an alternative to the Kansas statute we propose the following:

> A person will be considered dead if in the announced opinion of a physician, based on ordinary standards of medical practice, he has experienced an irreversible cessation of spontaneous respiratory and circulatory functions. In the event that artificial means of support preclude a determination that these functions have ceased, a person will be considered dead if in the announced opinion of a physician, based on ordinary standards of medical practice, he has experienced an irreversible cessation of spontaneous brain functions. Death will have occurred at the time when the relevant functions ceased.

This proposed statute provides a "definition" of death confined to the level of *general physiological standards,* and it has been drafted in accord with the five

---

time—in other words, that death can occur at two or more distinct points in time. But resuscitation only restores life "from *apparent* death or unconsciousness." Webster's Third New International Dictionary 1937 (1966) (emphasis added). The proposed statute, text accompanying note 88 *infra,* takes account of the possibility of resuscitation by providing that death occurs only when there has been an *irreversible* cessation of the relevant vital bodily functions. *Cf.* 3 M. Houts & I. H. Haut, Courtroom Medicine § 1.01 (3) (d) (1971):

> The ability to resuscitate patients after apparent death, coupled with observations that in many cases the restoration was not to a state of consciousness, understanding and intellectual functioning, but merely to a decerebrate, vegetative existence, and with advances in neurology that have brought greater, though far from complete, understanding of the functions of the nervous system, has drawn attention to the role of the nervous system in maintaining life.

[84] Kennedy, *supra* note 20, at 948.

[85] Mills, *The Kansas Death Statute: Bold and Innovative,* 285 New Eng. J. Med. 968 (1971).

principles set forth above in section V. First, the proposal speaks in terms of the *death of a person.* The determination that a person has died is to be based on an evaluation of certain vital bodily functions, the permanent absence of which indicates that he is no longer a living human being. By concentrating on the death of a human being as a whole, the statute rightly disregards the fact that some cells or organs may continue to "live" after this point,[86] just as others may have ceased functioning long before the determination of death. This statute would leave for resolution by other means the question of when the absence or deterioration of certain capacities, such as the ability to communicate, or functions, such as the cerebral, indicates that a person may or should be allowed to die without further medical intervention.

Second, the proposed legislation is predicated upon the single phenomenon of death. Moreover, it applies uniformly to all persons,[87] by specifying the circumstances under which each of the standards is to be used rather than leaving this to the unguided discretion of physicians. Unlike the Kansas law, the model statute does not leave to arbitrary decision a choice between two apparently equal yet different "alternative definitions of death."[88] Rather, its second standard is applicable only when "artificial means of support preclude" use of the first. It does not establish a separate kind of death, called "brain death." In other words, the proposed law would provide two standards gauged by different functions, for measuring different manifestations of the same phenomenon. If cardiac and pulmonary functions have ceased, brain functions cannot continue; if there is no brain activity and respiration has to be maintained artificially, the same state (*i.e.*, death) exists.[89] Some people might prefer a single standard, one based either on cardiopulmonary or brain functions. This

[86] *Cf.* F. Moore, Transplant 27-36 (1972).

[87] Differences in the exact mode of diagnosing death will naturally occur as a result of differing circumstances under which the physician's examination is made. Thus, the techniques employed with an automobile accident victim lying on the roadside at night may be less sophisticated than those used with a patient who has been receiving treatment in a well-equipped hospital.

[88] Kan. Stat. Ann. § 77-202 (Supp. 1971).

[89] [L]ife is supported by the smooth and integrated function of three principal systems: circulatory, respiratory and nervous. . . . So long as the integrated function of these three systems continues, the individual lives. If any one of them ceases to function, failure of the other two will shortly follow, and the organism dies. In any case it is *anoxia,* or deprivation of oxygen, that is the ultimate cause of death of cells: in central nervous system failure, because the impulses which maintain respiration cease; in cardiac failure, because oxygenated blood is not moved to the cells; and in respiratory failure, because the blood, although circulating, is not releasing carbon dioxide nor replenishing oxygen in the lungs. Although other organs, such as the liver and kidneys, perform functions essential to life, their failure does not *per se* result in immediate death; it results, rather, in the

would have the advantage of removing the last trace of the "two deaths" image, which any reference to alternative standards may still leave. Respiratory and circulatory indicators, once the only touchstone, are no longer adequate in some situations. It would be possible, however, to adopt the alternative, namely that death is *always* to be established by assessing spontaneous brain functions. Reliance only on brain activity, however, would represent a sharp and unnecessary break with tradition. Departing from continuity with tradition is not only theoretically unfortunate in that it violates another principle of good legislation suggested previously, but also practically very difficult, since most physicians customarily employ cardiopulmonary tests for death and would be slow to change, especially when the old tests are easier to perform,[90] more accessible

---

eventual failure of one of the three systems described, and is thus only in indirect cause of death.

3 M. Houts & I. H. Haut, Courtroom Medicine § 1.01(2)(a) (1971).

It has long been known that, even when a patient loses consciousness and becomes areflexive, he may recover if heartbeat and breathing continue, but if they do not there is no hope of recovery. Thus, death came to be equated with the absence of these two "vital signs," although what was being detected was really the permanent cessation of the integrated functioning of the circulatory, respiratory, and nervous systems. In recent years, the traditional concept of death has been departed from, or at least severely strained, in the case of persons who were dead according to the rationale underlying the traditional standards in that they had experienced a period of anoxia long enough to destroy their brain functions, but in whom respiration and circulation were artifically re-created. By recognizng that such artificial means of support may preclude reliance on the traditional standards of circulation and respiration, the statute proposed here merely permits the logic behind the long-existing understanding (*i.e.*, integrated trisystemic functioning) to be served; it does not create any "new" type of death. Practically, of course, it accomplishes this end by articulating the "new" standard of "irreversible cessation of spontaneous brain functions," as another means of measuring the existing understanding. Dr. Jean Hamburger has observed, "After the guillotine has cut off a criminal's head, it is possible now to keep the heart and lungs going for days. Do you think that such a person is dead or alive?" *Discussion* of Louisell, *Transplantation: Existing Legal Constraints,* in Medical Progress, *supra* note 1, at 100. The purpose of the "new" standard is to make it clear that the answer to Hamburger's question is unequivocably that the person is dead. *Cf.* Gray v. Sawyer, 247 S.W.2d 496 (Ky. 1952) (newly discovered evidence that blood was gushing from decedent's decapitated body is significant proof that she was still alive following an accident); Biörck, *supra* note 15, at 485; Note, *supra* note 1, at 206.

[90] The clinical signs of irreversible loss of brain functions are probably not a great deal more difficult to elicit than the traditional signs of death are to detect, although the former are less accessible since they require active intervention to be educed and are not susceptible of mere observation. Aside from the taking of an electroencephalogram, the tests involved (such as tickling the cornea, irrigating the ear with ice water, and tapping the tendons with a reflex hammer) are fairly simple, but unlike the customary tests (such as listening for heartbeat with a stethoscope, seeing if a mirror held by the nose and mouth is clouded by breathing, and measuring pulse), they require equipment which a physician may be less likely to have at hand.

and acceptable to the lay public, and perfectly adequate for determining death in most instances.

Finally, by adopting standards for death in terms of the cessation of certain vital bodily functions but not in terms of the specific criteria or tests by which these functions are to be measured, the statute does not prevent physicians from adapting their procedures to changes in medical technology.[91]

A basic substantive issue remains: what are the merits of the proposed standards? For ordinary situations, the appropriateness of the traditional standard, "an irreversible cessation of spontaneous respiratory and circulatory functions,"[92] does not require elaboration. Indeed, examination by a physician may be more a formal than a real requirement in determining that most people have died. In addition to any obvious injuries, elementary signs of death such as absence of heartbeat and breathing, cold skin, fixed pupils, and so forth, are usually sufficient to indicate even to a layman that the accident victim, the elderly person who passes away quietly in the night, or the patient stricken with a sudden infarct has died.[93] The difficulties arise when modern medicine

[91] For example, it remains to be determined whether an electroencephalographic reading is necessary for an accurate diagnosis, as many now hold, or whether it should be regarded as having only "confirmatory value," as urged by the Harvard Committee. *See* note 11 *supra.*

[92] This language, taken from the proposed statute, is intended as a succinct summary of the standard now employed in ordinary circumstances. Of course, the requirement that the cessation of these functions be *irreversible* cannot be emphasized too strongly. A physician may be needed to make this determination in some cases—and to apply the means necessary to reverse a temporary cessation caused by a heart attack or the like. But laymen are also aware of the significance of the requirement as is indicated by the common practice of giving "first aid," in the form of artificial respiration, to restore breathing in victims of mishaps, particularly drowning, electric shock, and poisoning.

Two British commentators suggest that legislation "defining" death also prescribe the resuscitative efforts required to be made before death may be declared. Hillman, *supra* note 72, at 325. We believe it is enough to demand "irreversibility," as a consequence of which whatever attempts at resuscitation are established by current standards of good medical practice would be compelled.

[93] The statute provides that the determination of death depends on "the announced opinion of a physician." This raises two distinct sorts of questions. First, which physician's opinion is decisive? As previously observed, text accompanying note 64 *supra,* under "ordinary standards of medical practice" the physician declaring death would be the patient's own attending physician; this is particularly true of a patient who is receiving cardiopulmonary support in a hospital. Since, however, circumstances such as an automobile accident may arise in which death will have to be determined by a physician who had not previously attended the decedent, it was thought best to cast the language in terms of "a physician."

Second, questions may arise concerning the determination of death by nonphysicians. In an emergency, laymen may sometimes have to decide whether death has occurred, and to act on that determination, as in deciding whether to attempt to rescue someone who may or may not have already died. The proposed statute does nothing to change that

intervenes to sustain a patient's respiration and circulation. As we noted in discussing the Harvard Committee's conclusions, the indicators of brain damage appear reliable, in that studies have shown that patients who fit the Harvard criteria have suffered such extensive damage that they do not recover.[94] Of course, the task of the neurosurgeon or physician is simplified in the common case where an accident victim has suffered such gross, apparent injuries to the head that it is not necessary to apply the Harvard criteria in order to establish cessation of brain functioning.

The statutory standard, "irreversible cessation of spontaneous brain functions," is intended to encompass both higher brain activities and those of the brainstem. There must, of course, also be no spontaneous respiration; the second standard is applied only when breathing is being artificially maintained. The major emphasis placed on brain functioning, although generally consistent with the common view of what makes man distinctive as a living creature, brings to the fore a basic issue: What aspects of brain function should be decisive? The question has been reframed by some clinicians in light of their experience with patients who have undergone what they term "neocortical death" (that is, complete destruction of higher brain capacity, demonstrated by a flat E.E.G.). "Once neocortical death has been unequivocally established and the possibility of any recovery of consciousness and intellectual activity [is] thereby excluded, . . . although [the] patient breathes spontaneously, is he or she alive?"[95] While patients with irreversible brain damage from cardiac arrest seldom survive more than a few days, cases have recently been reported of survival for up to two and one-quarter years.[96] Nevertheless, though existence in this state falls far short of a full human life, the very fact of spontaneous respiration, as well as

---

practice or to alter any liability that might result under such circumstances, but merely specifies that an official determination must rest on "the opinion of a physician." This is consistent with existing state laws on the procedures by which death is "certified." These provisions, as well as ordinary medical practices, make it unnecessary to spell out in the model statute the exact manner in which the physician's opinion should be recorded or certified in the medical files or official documents.

[94] *See* note 12 *supra* & accompanying text.

[95] Brierley, Adams, Graham & Simpsom, *Neocortical Death After Cardiac Arrest,* 2 Lancet 560, 565 (1971) [hereinafter cited as Brierley]. In addition to a flat (isoelectric) electroencephalogram, a "neuropathological examination of a biopsy specimen . . . from the posterior half of a cerebral hemisphere" provides further confirmation. *Id.* The editors of a leading medical journal question "whether a state of cortical death can be diagnosed clinically." Editorial, *Death of a Human Being,* 2 Lancet 590 (1971). *Cf.* note 14 *supra.*

[96] Brierley and his colleagues report two cases of their own in which the patients each survived in a comatose condition for five months after suffering cardiac arrest before dying of pulmonary complications. They also mention two unreported cases of a Doctor Lewis, in one of which the patient survived for 2¼ years. Brierley, *supra* note 95, at 565.

coordinated movements and reflex activities at the brainstem and spinal cord levels, would exclude these patients from the scope of the statutory standards.[97] The condition of "neocortical death" may well be a proper justification for interrupting all forms of treatment and allowing these patients to die, but this moral and legal problem cannot and should not be settled by "defining" these people "dead."

The legislation suggested here departs from the Kansas statute in its basic approach to the problem of "defining" death: the proposed statute does not set about to establish a special category of "brain death" to be used by transplanters. Further, there are a number of particular points of difference between them. For example, the proposed statute does not speak of persons being "medically and legally dead," thus avoiding redundancy and, more importantly, the mistaken implication that the "medical" and "legal" definitions could differ.[98] Also, the proposed legislation does not include the provision that "death is to be pronounced before" the machine is turned off or any organs removed. Such a *modus operandi,* which was incorporated by Kansas from the Harvard Committee's report, may be advisable for physicians on public relations grounds, but it has no place in a statute "defining" death. The proposed statute already provides that "Death will have occurred at the time when the relevant functions ceased."[99] If supportive aids, or organs, are withdrawn after this time,

[97] The exclusion of patients without neocortical function from the category of death may appear somewhat arbitrary in light of our disinclination to engage in a philosophical discussion of the basic concepts of human "life" and "death." *See* text accompanying notes 57-60 *supra.* Were the "definition" contained in the proposed statute a departure from what has traditionally been meant by "death," such a conceptual discussion would clearly be in order. But, as this Article has tried to demonstrate, our intention has been more modest: to provide a clear restatement of the traditional understanding in terms which are useful in light of modern medical capabilities and practices. *See* note 89 *supra.*

A philosophical examination of the essential attributes of being "human" might lead one to conclude that persons who, for example, lack the mental capacity to communicate in any meaningful way, should be regarded as "not human" or "dead." It would nevertheless probably be necessary and prudent to treat the determination of that kind of "death" under special procedures until such time as medicine is able routinely to diagnose the extent and irreversibility of the loss of the "central human capacities" (however defined) with the same degree of assurance now possible in determining that death has occurred. Consequently, even at the conceptual level, we are inclined to think that it is best to distinguish the question "is he dead?" from such questions as "should he be allowed to die?" and "should his death be actively promoted?"

[98] The use of the word "legally" (as in "a person will be considered legally dead") in a law defining death is redundant. Besides, if there were a distinction between a "medical" and a "legal" standard of death, a statute could only legislate the legal standard. Consequently, the adjectives "medical" and "legal" are unnecessary as well as potentially misleading. *Cf.* Halley & Harvey, *Medical vs. Legal Definition of Death,* 204 J.A.M.A. 423 (1968).

[99] It is necessary to state a standard for judging *when* death occurred for disputes, typically concerning inheritance or rights of survivorship, in which the exact time of death is a

such acts cannot be implicated as having caused death. The manner in which, or exact time at which, the physician should articulate his finding is a matter best left to the exigencies of the situation, to local medical customs or hospital rules, or to statutes on the procedures for certifying death or on transplantation if the latter is the procedure which raises the greatest concern of medical impropriety. The real safeguard against doctors killing patients is not to be found in a statute "defining" death. Rather, it inheres in physicians' ethical and religious beliefs, which are also embodied in the fundamental professional ethic of *primum non nocere* and are reinforced by homicide and "wrongful death" laws and the rules governing medical negligence applicable in license revocation proceedings or in private actions for damages.

The proposed statute shares with the Kansas legislation two features of which Professor Kennedy is critical. First, it does not require that two physicians participate in determining death, as recommended by most groups which set forth suggestions about transplantation. The reasons for the absence of such a provision should be obvious. Since the statute deals with death in general and not with death in relation to transplantation, there is no reason for it to establish a general rule which is required only in that unusual situation. If particular dangers lurk in the transplantation setting, they should be dealt with in legislation on that subject, such as the Uniform Anatomical Gift Act.[100] If all current means of determining "irreversible cessation of spontaneous brain functions" are inherently so questionable that they should be double-checked by a second (or

---

decisive factor. The proposed statute, in accordance with existing practice, *see* text accompanying note 62 *supra,* fixes the time of death as the point at which the person actually dies, not the point at which the diagnosis is confirmed. This approach conforms to the commonsense understanding that both a man who dies in a coal mine and cannot be found for 24 hours and one who dies in a hospital where the practice is to require confirmation of the diagnosis by repeating the tests after 24 hours have been dead for a day before their deaths can be pronounced with certainty. The statutory phrase "relevant functions" refers to whichever functions are being measured: cardiopulmonary functions in the usual case, or brain functions where the others are obscured by the artificial means being employed.

[100] In fact, § 7(b) of the Uniform Anatomical Gift Act calls only for one physician: "The time of death [of a donor] shall be determined by a physician who attends the donor at his death, or, if none, the physician who certifies the death."

In *Tucker v. Lower* (*see* notes 42-50 *supra* & accompanying text) the defendants argued that this provision amounted to a "definition" of death (death is when a physician says you're dead), although Virginia had not adopted the Act until 1970, two years after the transplantation of the plaintiff's brother's heart. The court rejected this argument since "neither the decedent nor anyone acting on his behalf had made a gift of any part of his body" and the Act was therefore inapplicable. The reasons for rejecting the defendant's suggestion seem to us to go deeper; they have been presented throughout this Article and are summarized in the concluding section.

third, fourth, etc.) physician to be trustworthy, or if a certain means of measuring brain function requires as a technical matter the cooperation of two, or twenty, physicians, then the participation of the requisite number of experts would be part of the "ordinary standards of medical practice" that circumscribe the proper, non-negligent use of such procedures. It would be unfortunate, however, to introduce such a requirement into legislation which sets forth the general standards for determining who is dead, especially when it is done in such a way as to differentiate between one standard and another.

Kennedy's second objection, that a death statute ought to provide "for the separation and insulation of the physician (or physicians) attending the patient donor and certifying death, from the recipient of any organ that may be salvaged from the cadaver," is likewise unnecessary.[101] As was noted previously, language that relates only to transplantation has no place in a statute on the determination of death.

VIII. Conclusion

Changes in medical knowledge and procedures have created an apparent need for a clear and acceptable revision of the standards for determining that a person has died. Some commentators have argued that the formulation of such standards should be left to physicians. The reasons for rejecting this argument seem compelling: the "definition of death" is not merely a matter for technical expertise, the uncertainty of the present law is unhealthy for society and physicians alike, there is a great potential for mischief and harm through the possibility of conflict between the standards applied by some physicians and those assumed to be applicable by the community at large and its legal system, and patients and their relatives are made uneasy by physicians apparently being free to shift around the meaning of death without any societal guidance. Accordingly, we conclude the public has a legitimate role to play in the formulation and adoption of such standards. This article has proposed a model statute which bases a determination of death primarily on the traditional standard of final respiratory and circulatory cessation; where the artificial maintenance of these functions precludes the use of such a standard, the statute authorizes that death be determined on the basis of irreversible cessation of spontaneous brain functions. We believe the legislation proposed would dispel public confusion and concern and protect physicians and patients, while avoiding the creation of "two types of death," for which the statute on this subject

[101] Kennedy, *supra* note 20, at 949. Again, § 7(b) of the Uniform Anatomical Gift Act covers this point adequately: "The physician [who declares death] shall not participate in the procedures for removing or transplanting a part."

first adopted in Kansas has been justly criticized. The proposal is offered not as the ultimate solution to the problem, but as a catalyst for what we hope will be a robust and well-informed public debate over a new "definition." Finally, the proposed statute leaves for future resolution the even more difficult problems concerning the conditions and procedures under which a decision may be reached to cease treating a terminal patient who does not meet the standards set forth in the statutory "definition of death."

DEATH

PROCESS OR EVENT?

BARBARA MACKINNON

One of the central issues in recent bioethical literature is concerned with determining the time of death, or "so to speak" *redefining* death in the light of problems introduced by modern medical methods of preserving life and controlling death.[1] There now exist many fewer cases in which a person breathes his last and is suddenly dead. Apart from accidental or sudden death, today most people die in hospitals surrounded and aided by vast amounts of modern medical technology. Through this technology man is able to control in many cases the manner of and rate at which death takes place. The three great life-sustaining body systems–respiratory, circulatory, and nervous–when left to themselves fail almost simultaneously because of their physiological interdependence. Yet now

---

This paper was originally read at the Pacific Meeting of the American Philosophical Association, March 1975.

[1] Some recent discussions of this issue include: Robert M. Veatch, "Brain Death: Welcome Definition–or Dangerous Judgment?" *Hastings Center Report,* Vol. 2, No. 5 (November, 1972), 10–13; Alexander M. Capron, "Determining Death: Do We Need a Statute?" *Hastings Center Report,* Vol. 3, No. 1 (February, 1973), 6–7; A. M. Capron and L. R. Kass, "A Statutory Definition of the Standards for Determining Human Death," *University of Pennsylvania Law Review,* Vol. 121 (1972), 87–118; Jos. Fletcher, "New Definitions of Death," *Prism* (January, 1974), 13ff; Eric Cassell, "Being and Becoming Dead," in *Death in the American Experience,* ed. by Arien Mack (New York: Shocken Books, 1973); Daniel C. Maguire, "Death, Legal and Illegal," *Atlantic Monthly,* Vol. 233, No. 2 (February, 1974), 72–85; and "The Freedom to Die," *Commonweal,* Vol. 96 (August, 1972), 423–27; "Dealing with Death," *Medical World News* (May, 1971), 30–36 (A brief but comprehensive review of the burgeoning field of thanatology, the leaders in the field, and its concerns); Jonas B. Robitscher, "The Right to Die," *Hastings Center Report,* Vol. 2, No. 4 (September, 1972), 11–14; D. Maguire et al. "A Plea for Beneficent Euthanasia," *The Humanist,* Vol. 34, No. 4 (July, 1974), 4–11; P. Ramsay and L. R. Kass, "Death with Dignity," *Hastings Center Report,* Vol. 4, No. 3 (July, 1974), 16–18; Herman Feifl, MD, ed., *Meanings of Death* (New York: McGraw-Hill, 1959); Donald R. Cutler, ed. *Updating Life and Death* (Boston: Beacon Press, 1969); R. W. Wertz, *Readings on Ethical and Social Issues in Biomedicine* (Englewood Cliffs, New Jersey: 1973); E. Kübler-Ross, *On Death and Dying* (New York: Macmillan, 1969); and The Harvard Medical School's Ad Hoc Committee to Examine the Definition of Brain Death, chaired by Dr. Henry K. Beecher, *Journal of the American Medical Association,* Vol. 205 (1968), 337ff.

heart and lungs can be kept functioning artificially even in cases where all central nervous system activity has ceased. This situation raises peculiar and important practical problems for medicine and law. Consider the quandary of the person who must turn off a respirator; is he simply acknowledging that the patient on it is no longer living, or has he helped to bring on his death?[2] Or consider this case: A man shot in the head has resulting irreparable brain damage, his heart being kept beating and his breathing supported artificially until his heart can be removed for transplant. In the ensuing court trial the defense claims that the defendant who shot the victim did not kill him, for his heart was still beating. Rather the doctor who removed his heart for transplant killed him.[3] Or consider the case of the man whose heart stopped beating for twenty-three minutes, and who is now claiming to have died and returned from the dead.[4]

While discussion of these problems has clarified some of the issues involved, it has rarely shown much awareness of, or ability to deal with, the underlying philosophical problems. Thus there remain some residual confusions which prevent the present discussion from being the enlightened debate that it should be. One partial exception to this is a debate between Robert S. Morison and Leon R. Kass at the 1971 meeting of the American Association for the Advancement of Science.[5] Their debate centered on the question of whether death is an event which occurs all at once, or a process occurring over a more or less extended period of time. This debate is one which I believe presents a fruitful focus for philosophical analysis. Let me first summarize the position of the two participants in this debate.

According to Morison, the traditional or literary view of death as an event or "the thief which comes in the night" came about as a result of committing

[2] Cf. the case of Bruce Tucker and the Medical College of Virginia. He suffered a massive brain injury and his vital signs were maintained artificially until his heart and kidneys were removed for transplant. The case and ensuing legal problems are related in Veatch, *op. cit.*

[3] This case occurred recently in Oakland, California. The jury in the case found the defendant guilty following the judge's instruction that a person is dead when the brain ceases to function. At this time California did not have a "brain-death" statute, as had Kansas since 1970 and as is pending in other state legislatures. The case can be followed in the May, 1974, issues of the San Francisco *Examiner* and *Chronicle*.

[4] Victor Solow of Marinac, New York, recently claimed this feat as reported by a San Francisco newscast in June, 1974.

[5] The symposium, "Problems in the Meaning of Death," took place at the AAAS meeting in Chicago, December 29, 1970, and is published in *Science*, Vol. 173 (August 20, 1971), pp. 694–702. Both of these men are associated with the Institute of Society, Ethics and the Life Sciences, Hastings on the Hudson, New York, 10706. The *Hastings Center Report* is a bimonthly publication of this institute. Dr. Kass is also executive secretary of the Committee on the Life Sciences and Social Policy of the National Research Council, National Academy of Sciences, Washington, D.C., and Professor Morison is also professor of science and society at Cornell University.

something like what Whitehead called the "fallacy of misplaced concreteness," reifying or personifying the qualities of living or dead things.[6] While this ploy may provide many an interesting literary metaphor, to regard death in this way, Morison believes, is to set up an artificial discontinuity in what is actually a continuous process. Human life consists rather of the continuous processes of growth and decay of a complex and highly differentiated integrated organism whose final failure does not occur all at once. Some parts begin failing rather early in life, and certain cells continue to live after the time when most would agree that the person is dead.[7] Moreover, the practical implications of continuing to regard death as an event, Morison believes, are beginning to be more crucial. Doing so simply does not let us face many modern medical problems as we should. If we could accept death as a continuous process, we would see that "as the complexity and richness of the interactions of a human individual wax and wane, his value can be seen to change in relation to other values."[8] Just as at the beginning of life the potentiality for life increases, so at the end it decreases. As life increases it becomes more valuable and as it decreases it becomes less valuable, less worth living. It is the quality of life, in particular the complexity and richness, that is valued, not life absolutely. Morison states: "The value of human life varies with time and circumstance."[9] If such a position is accepted, he continues, those concerned with certain decisions about death will no longer feel obliged to continue to preserve a life which is barely present at great cost to the patient and to others. They will no longer be required to devote scarce medical resources to those who are terminally ill or dying at the expense of those who still have much life potential. The organs of the dying person, for example, may be more valuable to another than to the person himself. Judgments about such things should depend, he suggests, on a weighing of benefits against costs. At times the life and life expectancies of an individual may be outweighed by the costs that continuing to preserve his life would bring both to himself and to those immediately and more distantly affected by this effort. To regard life as of absolute value and something to be preserved at all costs diverts us from making good decisions in these matters.

In opposition to this view Kass, the second participant in the debate, argues that we should not and need not abandon the traditional concept of death as event, nor the attempt to find adequate criteria for determining when this event

[6] Morison, *op. cit.*, p. 694. Whitehead's discussion of the fallacy of misplaced concreteness may be found in *Science and the Modern World* (New York: Macmillan, 1967), pp. 51–55.

[7] Cf. Eliot Slater, "Death: The Biological Aspect," in *Euthanasia and the Right to Die*, ed. by A.B. Downing (Los Angeles: Nash Pub. Co., 1969), pp. 49–60, and U.S. VonEuler, "Physiological Aspects of Aging and Death," in *The End of Life*, A Nobel Conference, ed. by John D. Poslansky (Amsterdam: North Holland Publ. Co., 1973), pp. 23–35.

[8] Morison, *op. cit.*, p. 696.

[9] *Ibid.*, p. 697.

has occurred.[10] Confusion results from not recognizing that it is the organism as a whole that dies, not its parts separately.[11] When the central organization has disintegrated, the organism is dead. For example, he points to the distinction which biologists make between "somatic death," or death of the whole, and "necrosis," or death of the parts. Death is thus of "the organism as an integrated functional unit."[12] Kass would also have us retain the traditional view of the incommensurable value of human life. He opposes the notion that the patient's life may no longer be worth living or preserving in terms of the results of some cost-benefit analysis. However, one may still recognize the person's right to a good death. In fact, Kass holds that his position gives a better basis for giving primary consideration to the welfare of the dying patient.[13]

What I would like to do here is consider three problem areas in which philosophical clarification is required to make continued discussion of the issue more enlightened and complete. Thus let us consider in turn: (1) what death would have to be to be an *event;* (2) whether the difficulty of locating an event called death is due to an *indeterminacy* in nature or in our knowledge; (3) whether any practical considerations do or should bear on the *meaning* of the concept or term "death."

Can Death be an Event?

Descriptively speaking, for death to be an event it would have to be something like a boundary line separating the before-condition of a living organism from the after-condition of a dead organism or corpse. This would imply that the changes that occur during life are changes within a living organism and that the change at death is something quite different. Such a common-sense view may provide support for Plato's comment that life, when it is present in a body, cannot stand the approach of death and flees.[14] While no intention is made here to support the theory of forms that underlies such comments, there does seem to be some agreement with our ordinary reluctance to speak of a person as being partly dead or partly alive. One is either living or dead—even if we may have difficulty deciding in some cases just which is the case.

While common sense may support the view that death is an event, in the sense just spoken of, it is *not* so clear that it implies a view of death as an instantaneous happening. However, this is the view of the nature of an event

[10] Kass, *op. cit.*, p. 698.
[11] *Ibid.*, p. 699.
[12] Kass, *op. cit.*, p. 699. Kass refers for this distinction to H.C. Hopps, *Principles of Pathology* (New York: Appleton-Century-Crofts, 1959), p. 78.
[13] Kass, *op. cit.*, p. 701.
[14] Plato, *Phaedo.*

implied in Morison's position. He notes that like other stages in life, death is not a "discontinuous sharply identifiable instantaneous configuration(s)."[15]

However, when we look to recent physical and philosophical interpretations of time and change, we find that this view of the nature of an event is challenged. Both kinetic theory and nuclear physics speak of events, but view them as taking some minimal interval or amount of time; there is a minimal time with which both molecular and nuclear collisions (events) are associated. No physical event is instantaneous in the sense of taking place in an instant which has no extension. Death, then, as a physical event would also take some minimal amount of time.

The question of the nature of an event raises a whole range of problems associated with the meaning of time and its relation to change. Contemporary process philosophers, for example, use the word *event* to portray not a theory of discrete instants but of continuity and process. For such theories, there are no durationless instants. In fact, these philosophers question the analogy between time and a geometrical line, duration being thought to be quite different from mathematical continuity which they view as simply discontinuity infinitely repeated. For Whitehead, an event is basically a happening with a spatio-temporal spread associated with the life span of an actual entity which takes some time to reach a peak of development.[16] According to such views, for death to be an event it need not occur instantaneously. Death could remain an event which marks a difference between a living and dead organism, and yet take at least some minimal amount of time or have a certain duration. All I would do here, however, is make the point that the debate over whether death is a process or event is very much dependent for its resolution upon a clarification of basic views of time and change. Our purpose here is not to settle that issue but to point out such areas needing further analysis.

Death and a Problem of Indeterminacy

Another issue in this debate which needs further clarification concerns the question of whether the difficulty of locating an event called death is due to an indeterminacy in nature or in our knowledge. According to Morison, the tinker-

[15] Morison, *op. cit.*, p. 694.

[16] Whitehead, *Science and the Modern World,* and *Process and Reality.* A fine summary and a comment in its own right is the article by Milic Capek in change in *The Encyclopedia of Philosophy,* ed. Paul Edwards (New York: Macmillan, 1967), Vol. 2, pp. 75–79. One problem that remains with this type of theory of change and time, as Capek points out, is the difficulty of "synthesizing conceptually the continuity of becoming and the individuality of events." This is a problem which we also have here if we intend to speak of death as an event in the process sense of the term.

ing of man with his own death "has made it obvious that death is not really a very easily identifiable event or configuration."[17] From this he draws the conclusion that the new ability to control death simply reveals what has been the case all along, viz., that death is not an event but a process. There is a similar indeterminacy at the beginning of life, he believes. Human life begins "inconspicuously, unconsciously, and at an unknown time."[18] However, he argues from the inconspicuousness of the early beginning of a person's life and our inability to know actually when any life has begun to the conclusion that the beginning is actually no definite event at all. In Morison's argument, there is obviously a confusion between an indeterminacy in nature itself (that both at the beginning of a person's life and at its end there is no one event of demarcation) and an indeterminacy in our knowledge (that we cannot find the point of beginning or ending, that it does not jump out at us, that it is not easily knowable).

Kass, on the other hand, seems to recognize this confusion, and argues that this tinkering of man with his own machinery has simply made it more difficult for man to know when the so-called moment of death has arrived; it obscures what is happening. If the indeterminacy were in nature, then there really would be no natural distinction between living and non-living or dead organism, and to speak of death would be merely a convention of speech (something he believes Morison is actually implying).[19] To try to define death in such a situation would then be useless, for there would be no definite "moment" of death. If, on the other hand, the indeterminacy is in our knowledge, then we may continue to try our best to decide what the time of death is, and what happenings mark its occurrence. New definitions in terms of "brain-death" may then be the direction in which to proceed. This again raises formidable philosophical issues, such as the question of whether there is any other nature for us than nature as we know it. A whole range of problems connected with the subject–object dilemma surface here, and again we can only point to this problem area as one which bears on the issue at hand. However, it would seem that common sense has the right lead in its conviction that there is a difference between saying that we do not *know* whether or not a person is dead and that in fact he may not *be* either dead or alive.

The Practical Meaning of "Death"

A third philosophical issue underlying this debate for which, I believe, philosophical analysis could be helpful is concerned with the nature of the

[17] Morison, *op. cit.*, p. 695.
[18] *Ibid.*, p. 696.
[19] Kass, *op. cit.*, p. 699.

meaning of the term or concept, "death," and the bearing of practical considerations on this meaning. If we take the meaning of a term as in some way connected with its use and question the use of this particular term, we shall find again that it is ordinarily used to designate something that happens to a person which differentiates his being a living and his being a dead person. It marks a difference which actually seems to occur regardless of our present difficulty in knowing when it has occurred. In his attempt to discredit the notion of death as an event, Morison implies that there is no such natural difference and that "death" is but a convention of speech used for certain practical purposes.[20] Surely there are practical purposes which use of the term "death" has. There are burial customs which need a death to have occurred in order to proceed, insurance benefits can be paid only at death, and the caring for a person in an attempt to save or lengthen his life needs to have a time when it can be discontinued. If, as Morison suggests, death is a process which begins with aging and does not end until the last cell is dead, and if these just-mentioned practical occurrences can take place only if the end of the process is reached, then it would seem that either our customs or our ordinary use of the term "death" would have to be changed.

Of course there is just such a practical issue at hand which may be a partial cause at least of a change in the meaning of the term. In order for the medical practice of transplants of certain organs to continue, it is imperative that "death" be defined in such a way that it takes place earlier or on a different basis than formerly thought, e.g., no longer as heart-death but brain-death. The other alternative, viz., namely to change people's beliefs that one should not take an organ from a person until he is dead, would be a more difficult task to accomplish. For Morison, however, this is not the issue to be pursued. Rather what we want to know, he believes, is when life is not worth preserving, or at what time before that last cell is dead can we morally and responsibly cease trying to preserve a person's life. We need not call that point death—or we may do so, recognizing that it is with this practical purpose in mind that we do so.

According to Kass, on the other hand, there is a major difference between the question of when a person has died and when life is no longer worth living. The first is a factual question for medical science, while the second is a question of value or utility, a socio-moral question. The one is a matter of the true and the other a matter of the useful or good.[21] Morison is concerned basically with the latter, according to Kass, with the determination of the value of life in terms of utility through a cost-benefit analysis. Kass's comments seem to be basically valid and social needs such as the handling of life insurance benefits, and the cost of terminal hospital care or the need for body parts as transplants should not

[20] *Ibid.*, p. 699.
[21] *Ibid.*

govern our redefinition of death.[22] However, there is some question as to whether the factual question is a question which medical science alone can answer.

However, there is another reason for redefining a term in the light of practical considerations, viz., because these practical elements constitute the meaning of terms. Consider the theories of meaning of pragmatism and operationalism According to Peirce, the meaning of a term *is* its practical or sensible effects, i.e., how one believes the object will react to expected handling.[23] The meaning of death would then consist of such sensible aspects of the occurrence that we expect, as well as such ways that we ourselves habitually react to the event. Operationalism raises the question of whether or not we can define or understand the meaning of death apart from the operations by which we measure it. Can we distinguish the concept of death from the criteria for determining that one has died? Can we say what death is apart from the cessation of heart beat, respiration, or brain waves, for example?[24] Is death simply these effects or the operations by which we measure them, or is it something else which we know by these means? Again we run into philosophical issues which need to be considered more thoroughly in any complete treatment of the issue at hand.

Thus we seem in each of these three areas of consideration to have delved into philosophical matters crucial to any adequate treatment of the initial question about the nature of death, and yet we have had to leave the matter there without final resolution. However, our analysis was intended only to do just that, viz., to point out some philosophical issues underlying the debate, to discuss them briefly, and thereby to set a direction for further discussion of this problem.

Nevertheless we may have been able to arrive at *some* tentative conclusions. The alternatives of regarding death as an event that occurs more or less instantaneously (or as a definite occurrence) and regarding it as a more vaguely defined

[22] Paul Ramsey believes this to be "an ethical consideration of considerable importance." He states, "If in the practical order we need to separate between the physician who is responsible for the care of a prospective donor and the physician who is responsible for a prospective recipient, do we not need in the intellectual order to keep the question of the definition of death equally discrete from the use of organs in transplantation?" "On Updating Death," *op. cit.*, p. 51.

[23] Cf. P.W. Bridgman, *The Logic of Modern Physics* (New York: Macmillan, 1927), and C.S. Peirce, "How to Make Our Ideas Clear," in *Classic American Philosophers*, Max H. Fisch ed. (New York: Appleton-Century-Crofts, 1951), and elsewhere.

[24] A very fine discussion of this distinction and of the Harvard Medical School Report on determining the time of death can be found in Ramsey, *op. cit.*, pp. 35ff. He notes, for example, that although brain-death may become the basis for the meaning of death, the criteria for determining that it has occurred may continue to involve respiration and circulation. For example, one may use spontaneous heart function as a way of knowing whether there is any discernable central nervous system activity.

process occurring over an extended period of time are not the only ones. Death could be an *event* which is a definite happening causing a dramatic change and yet take some time. To pinpoint one moment or instant of its occurrence may thus be the wrong course to take. Furthermore, the beliefs of common sense that death is an actual occurrence in nature should be respected, and thus the attempt to find out more about its character and the time of its occurrence is not mistaken. While aging, and even dying, may be processes, death could still be thought of as an event. Finally, while modern medical technology has raised new problems for the practice of medicine concerning the terminally ill, as well as for the laws that concern death, to redefine death as a way out of these problems or for whatever practical purposes is less than honest. Nevertheless, there are ways in which practical matters may enter right into the meaning of death, such as that described by operational and pragmatic theories of meaning.

Finally, while the distinction between death as process and death as event has proven an interesting basis on which to try to build a case concerning the value of human life, it is the opinion of this author that this is not the best place to locate such an argument. Rather than rely upon the notion of death as an event to ground such a value, this writer would suggest a more Kantian interpretation of a person as a uniquely autonomous creature with respect for him flowing from his ability to direct his life from within, i.e., his autonomy. Such a basis would need to be supplemented by a theory of the historical and organic existence of a person and a treatment of problems concerned with a conflict of personal rights. This would differ in important ways from utilitarian theories such as that of Morison for which human life may be highly valued but only insofar as it promotes the welfare of the greatest number of people, or the social or common good, or the greatest utility in some sense of that term. However, while this is where the argument should be located, we can do no more here than state this view without the further amplification and argumentation needed to support it.

Such a view as I suggest would include giving primary consideration to the terminally ill patient in decisions about terminal care for him. However, it would also respect the patient's right to die as well as to live in a way that he himself chose, and thus voluntary euthanasia would be permitted.[25] This view would be

[25] It is interesting to note that the AMA has also sanctioned "Dignified Death" at their 1973 clinical meeting, adopting the guideline that "the cessation of the employment of extraordinary means to prolong the life of the body when there is irrefutable evidence that biological death is imminent is the decision of the patient and his immediate family." However, they remain opposed to direct euthansia, supposedly voluntary or non-voluntary. Moreover they also concluded that "a statutory definition of death is neither desirable nor necessary—death shall be determined by the clinical judgment of the physician, using the necessary available and currently accepted criteria." One wonders about the basis for "currently accepted" criteria, and why these should not be made a matter of law.

based primarily on the premise that the purpose of law is simply to protect persons from being harmed by others and not to regulate private morality.[26] Legal regulations here would merely protect persons from being exploited or being forced to act against their will. Nor is this to say that any such choice would be morally praiseworthy, for the legal and moral spheres are separable even if in some case overlapping. Those cases in which no choice has been or can be made by the patient, and falling thus under the category of possible non-voluntary euthanasia or mercy killing, would involve greater problems. In some of these cases we may be justified in letting a person die by withholding so-called extraordinary means of survival from him. The traditional distinctions between ordinary and extraordinary and between active and passive measures still seem to offer some aid. At the same time human life may retain its near-absolute value, granted another base than can be found in the distinction of death as event and death as process.[27]

[26] This view is in basic agreement with section 13 of the Wolfenden Committee Report of the British Academy, March, 1959, printed in the *Proceedings of the British Academy*, Vol. 45, and reprinted in *The Enforcement of Morals* by Lord P. Devlin (London: Oxford University Press, 1965). It states that the function of criminal law is "to preserve public order and decency, to protect the citizen from what is offensive or injurious and to provide sufficient safeguards against exploitation and corruption of others . . ." This view has a near counterpart in the 1955 Model Penal Code of the American Law Institute. A fine discussion of some of the problems of this view can be found in "The Enforcement of Morals" by Ernest Nagel, in *Moral Problems in Contemporary Society*, ed. by Paul Kurtz (Buffalo, New York: Prometheus Books, 1969), pp. 137–160. Cf. also Anthony Flew, "The Principle of Euthanasia," in Downing, *op. cit.*, pp. 30–48. Flew emphasizes the need to protect the patient's choice in voluntary euthanasia and states that voluntary must mean the "strong, constant, and unequivocally expressed wish of the affected candidates themselves" (p. 30). In the same volume one can find a suggested legal document designed to protect such rights (pp. 197–206).

[27] The question of the absolute value of life bears on other issues than that which we have discussed in this article. However, there is much validity in the comments by David Maguire that "life is *the* good thing and the precondition of all good things. Any decision to end it in any context, for self or for another, must be slow, deliberate, and reverential." Cited from "Death, Legal and Illegal," *op. cit.*, p. 85.

FURTHER READINGS

For lists of further sources see: Danner K. Clouser, *Abortion and Euthanasia: An Annotated Bibliography,* with Arthur Zucker (Society for Health and Human Values, Philadelphia, 1974); Euthanasia Educational Fund, "Euthanasia–An Annotated Biblography," (Euthanasia Educational Fund, 250 W. 57th St., New York, N.Y. 10019, May, 1970); Richard A. Kalish, "Death and Dying: A Briefly Annotated Bibliography," in *The Dying Patient* eds., Orville G. Brim, Jr. et al. (New York: The Russell Sage Foundation, 1970), pp. 323–80.

Books

A. Alvarez, *The Savage God: A Study of Suicide* (New York: Random House, 1972); Ernest Becker, *The Denial of Death* (New York: Free Press, 1973); Ann Cartwright et al., *Life Before Death* (London: Routledge and Kegan Paul, 1973); Jacques Choron, *Death and Western Thought* (New York: Collier Books, 1973), and also by this author, *Suicide* (New York: Charles Scribners and Sons, 1972); Diana Crane, *The Social Aspects of the Prolongation of Life,* Social Science Frontiers (New York: The Russell Sage Foundation, 1969); Donald Cutler, ed., *Updating Life and Death* (Boston: Beacon Press, 1969); J. D. Douglas, *The Social Meaning of Suicide* (Princeton: Princeton University Press, 1967); A. B. Downing, *Euthanasia and the Right to Die* (New York: Humanities Press, 1970); E. Durkheim, *Suicide* (Glencoe: Free Press, 1951); N. L. Farberow, and E. S. Shneidman, eds., *The Cry for Help* (New York: McGraw-Hill, 1961); Herman Feifel, *The Meaning of Death* (New York: McGraw Hill, 1965); Joseph Fletcher, *Morals and Medicine* (Boston: Beacon Press, 1954); Milton McC. Gatch, *Death: Meaning and Mortality in Christian Thought and Contemporary Culture* (New York: Seabury Press, 1969); Ivan K. Goldberg et al. eds., *Psychopharmacologic Agents for the Terminally Ill and Bereaved* (New York: Columbia University Press, 1973); Geoffrey Gorer, *Death, Grief and Mourning* (New York: Doubleday, 1965); Jonathan Gould and Lord Craigmyle, eds., *Your Death Warrant? The Implications of Euthanasia* (New York: Arlington House, 1973); David Hendin, *Death as a Fact of Life* (New York: Norton, 1973); Elisabeth Kübler-Ross, *On Death and Dying* (New York: Macmillan, 1969); Austin H. Kutscher and Michael R. Goldberg, *Caring for the Dying Patient and His Family: A Model for Medical Education–Medical Center Conferences* (New York: Health Sciences Publishing Corp., 1973); Arien Mack, ed., *Death in the American Experience* (New York: Schocken, 1973); Robert E. Neale, *The Art of Dying* (New York: Harper & Row, 1973); Leonard Pearson, ed., *Death and*

Dying: Current Issues in the Treatment of the Dying Person (Cleveland: Case Western Reserve Univ. Press, 1969); Paul Ramsey, *The Patient as Person* (New Haven: Yale University Press, 1970); H. L. P. Resnick, ed., *Suicidal Behaviors* (Boston: Little, Brown, 1968); Edwin S. Shneidman, *Deaths of Man* (New York: Quadrangle Books, 1973); Edwin S. Shneidman, ed., *Essays in Self-Destruction* (New York: Science House, 1967), and also by this editor, *On the Nature of Suicide* (San Francisco: Jossey-Bass, 1969); Edwin S. Shneidman and Norman L. Farberow, *Clues to Suicide* (New York: McGraw-Hill, 1959), and this author with Normal L. Farberow and Robert E. Litman, *The Psychology of Suicide* (New York: Science House, 1970); A. Solzhenitsyn, *Cancer Ward* (New York: Bantam Books, 1969); Sophocles, *Antigone;* David Sudnow, *Passing On: The Social Organization of Dying* (Englewood Cliffs, N.J.: Prentice-Hall, 1960); Leo Tolstoy, *Death of Ivan Illyich* (New York: Signet Books, New American Library, 1960); Arnold Toynbee et al., *Man's Concern with Death* (New York: McGraw-Hill, 1968); Robert M. Veatch, *Death, Dying and the Biological Revolution* (forthcoming); Arthur Winter, ed., *The Moment of Death: A Symposium* (Springfield, Ill.: Charles C. Thomas, 1965); Alfred Worcester, *The Care of the Aging, The Dying and The Dead,* second edition (Springfield, Ill.: Charles C Thomas, 1961).

Articles

John Agate, "Care of the Dying in Geriatric Departments," *Lancet* (Feb. 17, 1973), pp. 364–66; J. F. Alderete et al., "Irreversible Coma: A Clinical, Electroencephalographic, and Neuropathological Study," *Transactions of the American Neurological Association 93* (1968), pp. 16–20; J. D. Arnold et al., "Public Attitudes and the Diagnosis of Death," *Journal of the American Medical Association 206* (Nov. 25, 1968), pp. 1949–54; F. J. Ayd, Jr., "The Hopeless Case: Medical and Moral Considerations," *Journal of the American Medical Association 181* (1962), pp. 1099–1102; B. Bard and J. Fletcher, "The Right To Die," *Atlantic* (April, 1968), pp. 59–64; David Barton, "Death and Dying: A Psychiatrist's Perspective," *Soundings 55* (Winter, 1972), pp. 459–71; William H. Baughman, "Euthanasia: Criminal Tort, Constitutional and Legislative Questions," *Notre Dame Lawyer 48* (1973), pp. 1202–60; D. P. Becker et al., "An Evaluation of the Definition of Cerebral Death," *Neurology 20* (1970), pp. 459–62; Henry K. Beecher, "After the 'Definition of Irreversible Coma'," *New England Journal of Medicine 281* (Nov. 6, 1970–71), other articles by this author are: "Definitions of Life and Death for Medical Science and Practice," *Annals of the New York Academy of Sciences 169* (Jan. 21, 1970), pp. 471–74, and "Ethical Problems Created by the Hopelessly Unconscious Patient," *New*

England Journal of Medicine 278 (1968), pp. 1425–30, and "Procedures for the Appropriate Management of Patients Who May Have Supportive Measures Withdrawn," *Journal of the American Medical Association 209* (1969), p. 405; Sissela Bok et al., "The Dilemmas of Euthanasia," *BioScience 23* (August, 1973), pp. 461–78; G. Biorck, "On the Definitions of Death," *World Medical Journal* 14 (Sep./Oct., 1967), pp. 137–39; P. Braunstein et al., "A Simple Bedside Evaluation for Cerebral Blood Flow in the Study of Cerebral Death: A Prospective Study on 34 Deeply Comatose Patients," *Journal of Roentgenology, Radium Therapy and Nuclear Medicine 118* (August, 1973), pp, 757–67; J. B. Brierley et al., "Neocortical Death after Cardiac Arrest," *Lancet* (Sep. 11, 1971), pp. 560–65; Howard W. Brill, "Death with Dignity: A Recommendation for Statutory Change," *University of Florida Law Review 12* (Winter, 1970), pp. 368–83; Norman Brown et al., "The Preservation of Life," *Journal of the American Medical Association 221* (Jan. 5, 1970), pp. 76–82; Norman L. Cantor, "A Patient's Decision to Decline Life-Saving Medical Treatment: Bodily Integrity versus the Preservation of Life," *Rutgers Law Review 26* (Winter, 1972), pp. 228–64; Alexander M. Capron, "Determining Death: Do We Need a Statute?" *Hastings Center Report 3* (Feb., 1973), pp. 6–7; Eric J. Cassell, "Death and the Physician," *Commentary* (June, 1969), pp. 73–79, and also "Permission to Die," *BioScience 23* (August, 1973), pp. 475–78; V. U. Collins, "Limits of Medical Responsibility in Prolonging Life: Guides to Decisions," *Journal of the American Medical Association 206* (1968), pp. 389–92; W. J. Curran, "Legal and Medical Death: Kansas Takes the First Step," *New England Journal of Medicine 284* (1971), pp. 260–61; Raymond S. Duff and A. G. M. Campbell, "Moral and Ethical Dilemmas in the Special-Care Nursery," *New England Journal of Medicine 289* (Oct. 25, 1973), pp. 890–94; Herman Feifel et al., "Physicians Consider Death," *Proceedings American Psychological Association Convention* (1967), pp. 201–02; George Fletcher, "Legal Aspects of the Decision Not to Prolong Life," *Journal of the American Medical Association 203* (Jan. 1, 1968), pp. 65–68, and also "Prolonging Life," *Washington Law Review 42* (1967), pp. 999–1016; John Fletcher, "Attitudes Towards Defective Newborns," *Hastings Center Studies 2* (Jan., 1974), pp. 21–32; John Freeman and Robert E. Cooke, "Is There a Right to Die–Quickly?" *Journal of Pediatrics 80* (Spring, 1972), pp. 904–08; Emil J. Freireich, "The Best Medical Care for the 'Hopeless' Patient," *Medical Opinion* (Feb., 1972), pp. 51–55; Gerald J. Gruman, "An Historical Introduction to Ideas About Voluntary Euthanasia: With a Bibliographic Survey and Guides for Interdisciplinary Studies," *Omega 4* (Summer, 1973), pp. 87–138; Edward J. Gurney, "Is There a Right to Die? A Study of the Law of Euthanasia," *Cumberland Sanford Law Review 3* (Summer, 1972), pp. 235–61; James M. Gustafson, "Mongolism, Parental Desires, and the Right to Life," *Perspectives in Biology and Medicine 16* (Summer, 1973), pp. 529–57; M. M. Halley and W. F. Harvey, "Medical vs. Legal Definitions of

Death," *Journal of the American Medical Association 204* (March 6, 1968), pp. 423–25, and also by these authors, "Law–Medicine Comment: The Definitional Dilemma of Death," *Journal of the Bar Association of the State of Kansas 37* (Fall, 1968), p. 179, and by these same authors with others, "Definition of Death," *New England Journal of Medicine 279* (1968), p. 834; Sheila Hancock et al., "Care of the Dying," *British Medical Journal* (Jan. 6, 1973), pp. 29–41; Harvard Medical School, Ad Hoc Committee of the Harvard Medical School to Examine the Definition of Brain Death, "A Definition of Irreversible Coma," *Journal of the American Medical Association 205* (1968), pp. 337–40; Leonard J. Hertzberg, "Cancer and the Dying Patient," *American Journal of Psychiatry 128* (Jan., 1972), pp. 806–10; Dallas M. High, "Death: Its Conceptual Elusiveness," *Soundings 55* (Winter, 1972), pp. 438–58; David C. Humphrey, "Dissection and Discrimination: The Social Origins of Cadavers in America. 1760–1915," *Bulletin of the New York Academy of Medicine 49* (Sep., 1973), pp. 819–27; Ivan Illich, "The Political Uses of Natural Death," *Hastings Center Studies 2* (Jan., 1974), pp. 3–20; Franz J. Ingelfinger, "Bedside Ethics for the Hopeless Case," *New England Journal of Medicine 289* (Oct. 25, 1973), pp. 914–15; Institute of Society, Ethics and the Life Sciences, Task Force on Death and Dying, "Refinements on Criteria for the Determination of Death," *Journal of the American Medical Association 221* (July 3, 1972), pp. 48–53; B. Kahana and E. Kahana, "Attitudes of Young Men and Women Toward Awareness of Death," *Omega 3* (Feb., 1972), pp. 37–44; Gerald Kelly, "The Duty of Using Artificial Means of Preserving Life," *Theological Studies 11* (June, 1950), pp. 203–20, and also "The Duty to Preserve Life," *Theological Studies 12* (Dec., 1951), pp. 550–56; Ian McColl Kennedy, "The Kansas Statute on Death–An Appraisal," *New England Journal of Medicine 285* (1971), pp. 946–50; J. Kimura et al., "The Isoelectric Electroencephalogram: Significance in Establishing Death in Patients Maintained on Mechanical Respirators," *Archives of Internal Medicine 121* (1968), pp. 511–17; Ronald Koenig, "Dying vs. Well-Being," *Omega 4* (Fall, 1973), pp. 181–94; J. Korein et al., "On the Diagnosis of Cerebral Death: A Prospective Study," *Electroencephalography and Clinical Neurophysiology 27* (1969), p. 700; Daniel Maguire, "The Freedom to Die," *Commonweal 96* (August 11, 1972), pp. 423–27; David Lester, "Attitudes Toward Death Today and Thirty-five Years Ago," *Omega 2* (August, 1971), p. 168; Helge H. Mansson, "Justifying the Final Solution," *Omega 3* (May, 1972), pp. 79–87; William May, "Attitudes Toward the Newly Dead," *Hastings Center Studies 1* (No. 1, 1973), pp. 3–13; Don Harper Mills, "The Kansas Death Statute–Bold and Innovative," *The New England Journal of Medicine 285* (Nov. 21, 1971), pp. 968–69; Robert S. Morison, "Dying," *Scientific American 229* (Sep., 1973), pp. 55–62; F. P. McKegney and P. Lange, "The Decision to No Longer Live on Chronic Hemodialysis," *American Journal of Psychiatry 128* (Sep., 1971), pp. 267–74; F. D. Moore, "Medical Responsibility for the Prolongation of Life," *Journal of the American Medical Association 206* (1968), pp. 384–86; Lucy Griscom Morgan, "On Drinking the Hemlock," *Hastings Center Report 1* (Dec., 1971), pp. 4–5; Robert E. Neale, "Between the Nipple and the Everlasting Arms," *Archives of the Foundation of Thanatology 3* (Spring, 1971), pp. 21–30; Russell Noyes and Terry A. Travis, "The Care of Terminally Ill Patients," *Archives of Internal Medicine 132* (Oct., 1973), pp. 607–11; Talcott Parsons, Renée C. Fox, and Victor M. Lisz, "The 'Gift of Life' and Its Reciprocation," *Social Research 39* (1972), pp. 367–415; David Peretz et al., "Survey of Physicians' Attitudes Toward Death and Bereavement: Comparison of Psychiatrists and Non-Psychiatrists," *Journal of Thanatology 1* (March, 1971), p. 91; Pius XII, "The Pope Speaks, Prolongation of Life," *Osservatore Romano 4* (1957), pp. 393–98; Ralph B. Potter, "The Paradoxical Preservation of a Principle," *Villanova Law Review 13* (1968), pp. 784–92; P. W. Pretzel, "Philosophical and Ethical Considerations of Suicide Prevention," *Bulletin of Suicidology* (July, 1968), pp. 30–38; *Proceedings of the Royal Society of Medicine*, report of a meeting, "Euthanasia," *63* (July, 1970), pp. 659–70; Warren T. Reich and Harmon Smith, commentators, "Case Studies in Bioethics: On the Birth of a Severely Handicapped Infant," *Hastings Center Report* (Sep., 1973), pp. 10–12; Jonas Robitscher, "The Right to Die," *Hastings Center Report 2* (Sep., 1972), pp. 11–14; Don A. Rockwell and William O'Brien, "Physicians' Knowledge and Attitudes About Suicide," *Journal of The American Medical Association 225* (Sep. 10, 1973), pp. 1347–49; S. D. Rosoff et al., "The EEG in Establishing Brain Death: A 10-Year Report with Criteria and Legal Safeguards in the 50 States," *Electroencephalography and Clinical Neurophysiology 24* (1968), pp. 283–84; Walter W. Sackett, "Death with Dignity," *Medical Opinion and Review 5* (June, 1969), pp. 25–31; Anthony Shaw, "Dilemmas of 'Informed Consent' in Children," *New England Journal of Medicine 289* (Oct. 25, 1973), pp. 885–90; D. Silverman et al., "Irreversible Coma Associated with Electrocerebral Silence," *Neurology 20* (1970), pp. 525–33; Helen Silving, "Euthanasia: A Study in Comparative Criminal Law," *University of Pennsylvania Law Review 103* (1954), pp. 350–89; G. Smith and E. D. Smith, "Selection for Treatment on Spina Bifida Cystica," *British Medical Journal* (Oct. 27, 1973), pp. 189–204; Anselm Strauss, and Barney Glaser, and Jeanne Quint, "The Nonaccountability of Terminal Care," *Hospitals 38* (Jan., 1964), pp. 73–87; Thomas S. Szasz, "The Ethics of Suicide," *The Antioch Review 31* (Spring, 1971), pp. 7–17; Judith Thompson, "Killing, Letting Die, and the Trolley Problem," *Monist* (Spring, 1975), and also by this author, "Rights and Deaths," *Philosophy and Public Affairs* Vol. 2, No. 2 (Winter, 1973); Uniform Anatomical Gift Act: Final Draft, July 30, 1968, prepared by National Conference of Commissioners on Uniform State Laws, in *The Moment of Death: A Symposium* ed. A. Winter (Springfield, Ill.: Charles C Thomas, 1969); Robert M. Veatch, "Brain Death: Welcome Definition–or Dangerous Judgment?" *Hastings Center Report 2* (Nov., 1972), pp. 10–13, and also by this author: "Death and Dying," *U.S. Catholic* (April, 1972), pp. 6–13, and "Is Death Moral: Technology's Impact on Death," in *Morality and Social Conflicts* ed. Lucia M. Palmer (forthcoming); Carl E. Wasmuth, Jr., "The Concept of Death," *Ohio Law Journal 30* (1969), pp. 32–60; Thomas A. Wassmer, "Between Life and Death: Ethical and Moral Issues Involved in Recent Medical Advances," *Villanova Law Review 13* (1968), pp. 759–83; David White, "Death Control," *New Society 22* (Nov. 30, 1972), pp. 502–05; Laurens P. White, ed., "Care of Patients with Fatal Illness," *Annals of the New York Academy of Sciences 164* (Dec., 1969), pp. 635–896; Robert H. Williams, "Our Role in the Generation, Modification and Termination of Life," *Journal of the American Medical Association 209* (August 11, 1969), pp. 914–17.

INDEX

MIX
Papier aus verantwortungsvollen Quellen
Paper from responsible sources
FSC® C105338

If you have any concerns about our products,
you can contact us on
ProductSafety@springernature.com

In case Publisher is established outside the EU,
the EU authorized representative is:
Springer Nature Customer Service Center GmbH
Europaplatz 3, 69115 Heidelberg, Germany

Printed by Libri Plureos GmbH
in Hamburg, Germany